Intuitive Analog
Circuit Design

Intuitive Analog Circuit Design

Marc T. Thompson, Ph.D.

ELSEVIER

AMSTERDAM • BOSTON • HEIDELBERG • LONDON
NEW YORK • OXFORD • PARIS • SAN DIEGO
SAN FRANCISCO • SINGAPORE • SYDNEY • TOKYO

Newnes is an imprint of Elsevier

Newnes

Newnes is an imprint of Elsevier
The Boulevard, Langford Lane, Kidlington, Oxford OX5 1GB, UK
225 Wyman Street, Waltham, MA 02451, USA

First edition 2006

Notice

British Library Cataloguing in Publication Data
A catalogue record for this book is available from the British Library

Library of Congress Cataloging-in-Publication Data
A catalogue record for this book is available from the Library of Congress

ISBN: 978-0-12-405866-8

For information on all Newnes publications visit our website at store.elsevier.com

Printed and bound in the US
14 15 16 17 18   10 9 8 7 6 5 4 3 2 1

Dedication

In memoriam

In memory of Noah Michael Thompson (2006–2008)
…We hardly knew ye.
   Love from Dad, Mom and Sophie

and

Merrill Dean Thompson (1939–2013)
Thanks, Dad, for inspiring a love of engineering in me from an early age.

Contents

CHAPTER 12 **Basic Operational Amplifier Topologies**
and a Case Study **465**

Preface to the Second Edition

Changes in the second edition

The author and the editors received many comments from readers regarding the content of the first edition of this book. Based, partly, on these concerns, the following additions were made to the second edition:

Chapter 2: a section on the "logarithmic decrement", a very useful technique, used primarily by mechanical engineers to estimate the damping ratio of a pole pair, from measurements of the transient response, was added.

Chapter 5: an example of tuned transistor amplifier was added.

Chapter 7: the section on the emitter follower was extensively expanded to discuss the dreaded high-frequency emitter follower oscillation. Lab experiments were shown where a 2N3904 emitter follower unintentionally oscillates at 100 MHz. Many other examples also were added to this chapter.

Chapter 8: sections illustrating the bad effects of parasitic inductance on current mirror speed were introduced.

Chapter 9: the chapter was significantly expanded with the description of JFETS and JFET amplifiers. Several more MOS amplifiers (including the shunt peaked MOS amplifier) were added.

Chapter 10: lab experiments illustrating charge control concepts were added.

Chapter 11: the chapter on feedback systems was extensively written with new concepts and numerous new examples. Lab experiments showing the effects of capacitive loading on op-amps were added.

Chapter 14: sections were added on active filters, and on passive implementation of delay lines.

Chapter 16: a completely new chapter on electrical noise was added.

Chapter 17: some transmission line experiments were added. A section on the use and abuses of SPICE simulation was added.

Software used by the author

Throughout, circuit simulation examples were redone in LTSPICE, and the LTSPICE.cir files are provided to the reader. LTSPICE is copyrighted by Linear Technology Corporation. In Chapter 14, some active filters were designed using Texas Instruments' FilterPro software, version 3.10. This software is copyrighted by Texas Instruments, Inc. Other simulations were done using MATLAB. MATLAB is a registered trademark of The Mathworks, Inc.

Thanks

I hope, in this preface, to give thanks to those who have directly and indirectly inspired me to learn analog stuff over the years.

Thanks go to my undergraduate and graduate professors at MIT who taught me the basics, and the not so basics: Prof Jim Roberge, Prof Harry Lee, Prof Dick Thornton, and Prof Kim Vandiver.

Thanks also to the hands-on teaching assistants: Leo Casey, Tom Lee, and Dave Trumper.

And to those of us who suffered and TA'd together: Tracy Clark, Kent Lundberg, and Dave Perreault

And to Dr Jeff Roblee (Precitech, Keene, NH) who has long tutored me on mechanical and thermal stuff.

Thanks to my editor on another Elsevier text, the late Bob Pease, wacky analog guru.

Also, thanks to the excellent Elsevier editorial team, who did outstanding work on short deadlines despite having the Atlantic Ocean between us.

Extra-special thanks go to my 20 or so students in my 2012 grad class, Worcester Polytechnic Institute's occasionally-taught ECE529, Special Topics, "Analog Circuits and Intuitive Design Methods". The students serve as willing and capable editors for this second edition. They did an excellent job finding my typos and white lies. Only time will tell if they found all of my errors and omissions.

Marc Thompson

Harvard, Massachusetts,

July 2013.

From a Next Generation Analog Designer (?)

Schematics courtesy of Sophie Madeleine Thompson, May 26, 2013.
Harvard, Massachusetts.

Introduction and Motivation

1

"The world, as we all know, is analog"
—Seen outside teaching assistant's office in the 5th floor lab, MIT building 38, mid-1980s

IN THIS CHAPTER

▶ This chapter serves as an introduction to the philosophy and topical coverage of this book. A very brief history of transistor development, invention of the analog integrated circuit (IC) and operational amplifier advances are given.

The need for analog designers

There is an inexorable trend in recent years to "go digital"—in other words, to do more and more signal processing in the digital domain due to a purported design flexibility. However, the world is an analog place and the use of analog processing allows electronic circuits to interact with the physical world. Not discounting the importance of digital signal processing (DSP) and other digital techniques, there are many analog building blocks such as operational amplifiers, transistor amplifiers, comparators, analog-to-digital (A/D) and digital-to-analog (D/A) converters, phase-locked loops, and voltage references (to name just a few) that are still used and will be used far into the future. Therefore, there is a continuing need for course development and education covering basic and advanced principles of analog circuit design.

One reason why analog electronic circuit design is so interesting is that it encompasses so many different disciplines. Here is a partial "shopping list", in no particular order, of disciplines encompassed by the broad field of analog circuit design:

- *Analog filters*: discrete or ladder filters, active filters, switched capacitor filters, and crystal filters.
- *Audio amplifiers*: power op-amps and output (speaker driver) stages.
- *Oscillators*: including LC, crystal, relaxation, and feedback oscillators; phase-locked loops; and video demodulation. We may also include in this category

Intuitive Analog Circuit Design. http://dx.doi.org/10.1016/B978-0-12-405866-8.00001-2

unintentional oscillators such as emitter followers that may oscillate at high frequencies.

- *Device fabrication and device physics*: metal oxide–semiconductor field-effect transistors (MOSFETs), bipolar transistors, diodes, insulated gate bipolar transistors (IGBTs), silicon-controlled rectifiers (SCRs), metal oxide–semiconductor (MOS)-controlled thyristors, etc.
- *IC fabrication*: operational amplifiers, comparators, voltage references, phase-locked loops (PLLs), etc.
- *Analog-to-digital (A/D) interface*: A/D and D/A, voltage references.
- *Radio-frequency (RF) circuits*: RF amplifiers, filters, mixers, and transmission lines; cable television (TV).
- *Controls*: control system design and compensation, servomechanisms, and speed controls.
- *Power electronics*: this field requires knowledge of MOSFET drivers, control system design, personal computer (PC) board layout, and thermal and magnetic issues; motor drivers; as well as device fabrication of transistors, MOSFETs, IGBTs, and SCRs.
- *Medical electronics*: instrumentation (electrocardiogram and nuclear magnetic resonance), defibrillators, and implanted medical devices.
- *Simulation*: SPICE and other circuit simulators.
- *PC board layout*: this requires knowledge of inductance and capacitive effects, grounding, shielding, and PC board design rules.
- Use *of circuit analogies*: using mechanical, magnetic, thermal, or acoustic "circuits" to model the behavior of systems.

Since we live in a world where more and more digital processing is taking place, analog designers must also become comfortable with digital-processing concepts so that we can all work together. In the digital world, some subsystem designs are based on analog counterparts. When designing a digital filter, one often first designs an analog prototype and then through an A/D transformation the filter is converted to the digital domain. For example, a bilinear transformation may be used where a filter designed in the s-domain (analog, using inductors, capacitors, and/or active elements) is transformed to a filter in the z-domain (digital, with gain elements and delays).

This technique stems in part from the fact that designers are in general more comfortable working in the analog domain when it comes to filtering. It is very easy to design a second-order analog Butterworth filter (you can find the design in any number of textbooks or analog filter cookbooks) but the implementation in the digital domain requires additional steps or other simulation tools.

Also, at sufficiently high frequencies, a digital transmission line or a high-speed signal trace on a PC board must be treated as a distributed analog system with traveling waves of voltage and current. Increasing density of digital ICs and faster switching speeds are adding to the challenges of good PC board design due to extra power requirements and other issues such as ground bounce.

The bottom line is that it behooves even digital designers to know something about analog design.

Some early history of technological advances in analog integrated circuits

The era of semiconductor devices can arguably be traced back as far as Dr Julius Lilienfeld, who has several U.S. patents giving various MOS structures (Figure 1.1). In three patents, Dr Lilienfeld gave structures of the MOSFET, metal-semiconductor field-effect transistor (MESFET), and other MOS devices.

We entered the bipolar transistor semiconductor era over 50 years ago with early work in solid-state physics and the invention of the bipolar transistor, and significant technological advances in analog circuit design and device fabrication are still being made. In 1947−1948, Bardeen, Brattain, and Shockley demonstrated the first bipolar transistors (Figure 1.2).[1]

The first ICs were produced around 1959 by teams at Fairchild Semiconductor and Texas Instruments (TI) (Figure 1.3). TI claims invention of the IC, with J.S. Kilby's U.S. patent "Miniaturized Electronic Circuits" #3,138,743, filed February 6, 1959. Workers at Fairchild filed for a patent on the first planar IC (arguably more easily

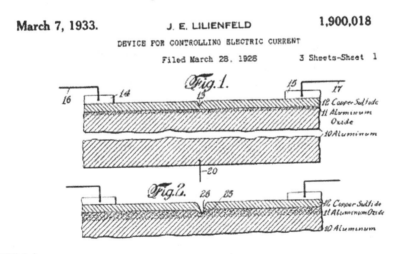

FIGURE 1.1

Excerpt from Lilienfeld's U.S. patent 1,900,018[2] (1933).

[1]See U.S. Patent #2,569,347, "Circuit Element Utilizing Semiconductive Materials", issued September 25, 1951, to William Shockley. Bardeen, Brattain, and Shockley shared the 1956 Nobel prize in physics for their discoveries related to the transistor. An excellent description of semiconductor transistor physics is given in Shockley's Nobel lecture "Transistor Technology Evokes New Physics", dated December 11, 1956.

[2]Lilienfeld had three patents in succession covering basic MOS transistor structures.

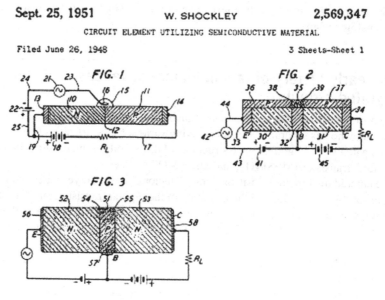

Sept. 25, 1951   W. SHOCKLEY   2,569,347

CIRCUIT ELEMENT UTILIZING SEMICONDUCTIVE MATERIAL.

Filed June 26, 1948   3 Sheets-Sheet 1

FIGURE 1.2

Excerpt from Shockley's U.S. patent 2,569,347 (1951).

manufactured than the TI invention) shortly after; see R.N. Noyce, "Semiconductor Device-and-Lead Structure", U.S. patent # 2,981, 877, filed July 30, 1959.[3]

These ICs had minimum feature sizes of around 125 μm. Since then, device geometries have gotten smaller and smaller with the invention and rapid improvements in the IC. Moore's law, named for Fairchild and Intel founder Gordon Moore, predicts that the density of transistor packaging in ICs doubles approximately every 18 months, a trend that has been proved to be remarkably accurate over the past 30 years.

At the time of this writing,[4] IC manufacturers are using 22-nm manufacturing processes, and smaller transistor sizes are anticipated. Smaller size allows the packaging of more and more complicated structures in a given die area. Researchers[5] are also actively working on three-dimensional IC structures in an attempt to pack more and more functionality into a given die volume.

After the invention of the IC around 1958–1959 by workers at TI and Fairchild, the first IC-operational amplifiers were introduced in the early to mid 1960s. The first commercially successful op-amps were the Fairchild μA709 (1965) and the National

---

[3]Full text and images of patents are available from the U.S. Patent office, http://www.uspto.gov.
[4]Second edition, Fall, 2012; note that in 1983, a typical minimum linewidth in ICs was 1.5 μm (1500 nm). Manufacturers are aiming for smaller gate lengths. See Gordon Moore's paper "The Role of Fairchild in Silicon Technology in the Early Days of 'Silicon Valley'" where the history of Fairchild IC development is recounted.
[5]See, e.g. Matrix Semiconductor, Inc. and *A Vertical Leap for Microchips* by Thomas H. Lee.

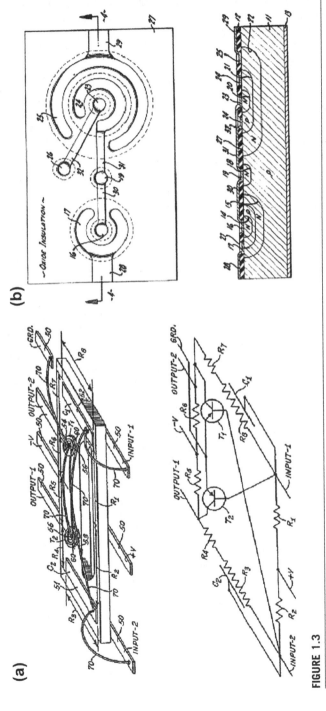

FIGURE 1.3

Diagrams from competing IC patents[5] from Texas Instruments (a) and Fairchild (b).

LM101 (1967), designed by the legendary analog wizard Bob Widlar.[6] These devices had a voltage offset of a few millivolts and a unity gain bandwidth of around 1 MHz and required external components for frequency compensation. Soon after (1968), the ubiquitous Fairchild μA741, the industry's first internally compensated op-amp, was introduced and became a bestseller. In the 741 op-amp, a 30-pF compensating capacitor was integrated onto the chip using metal oxide technology. It was a "plug and play" op-amp solution as opposed to the LM101 because this compensating capacitor was added internal to the IC.[7] The corresponding price reductions and specification improvements of the monolithic IC op-amps as compared to the earlier discrete designs (put forth, for instance by Philbrick)[8] made these IC op-amps instant successes.

Since that time, op-amps have been designed and introduced with significantly better voltage offset and bandwidth specifications, as well as improvements in other specifications such as input current, slew rate, common mode range, and the like. Field-effect transistor input op-amps became available in the 1970s with lower input current than their bipolar counterparts. Novel topologies such as the current-feedback op-amp have been introduced with success for high-speed applications.[9] Typical high-speed op-amps today have bandwidths of hundreds of megahertz.[10] Power op-amps[11] exist that can drive speakers or other heavy resistive or inductive loads

---

[6]The earlier μA702 op-amp was designed by Widlar and introduced in 1963 by Fairchild but never achieved much commercial success. Widlar went back to the drawing board and came up with the 709 around 1965; it was the first op-amp to cost less than $10. After a salary dispute with Fairchild, Widlar moved to National Semiconductor where he designed the LM101 and later improved the design resulting in the LM101A (1968). Details and history of the LM101 and 709 are given in Widlar's paper "Design Techniques for Monolithic Operational Amplifiers" with citation given at the end of this chapter.

[7]The 741 does not need an external compensation capacitor as did previous op-amps such as the LM101 and the 709. The "plug and play" ease of use of the 741 apparently offsets the fact that under most applications with closed-loop gains greater than 1 the device is overcompensated. More details on op-amp topologies are given in a later chapter in this book. Details and history of the 709, LM101, and 741 op-amps are also given in Walt Jung's *IC Op-Amp Cookbook*, 3rd edition, pp. 75—98. There are excellent discussions of op-amp external compensation given in James K. Roberge's classical text on operational amplifiers.

[8]For instance, the Philbrick K2-W op-amp, made with discrete components (vacuum tubes!), and sold from 1951 to 1971. It had a small signal bandwidth of around 300 kHz and an open-loop gain of 10,000 or so. The units were priced at around $22. See the article by Bob Pease, "What's all this K2-W Stuff, Anyway?" Philbrick also made the P2, a low-input current discrete operational amplifier built with a handful of transistors and other discrete components and priced at around $200. See "The Story of the P2—The First Successful Solid-State Operational Amplifier with Picoampere Input Currents" by Bob Pease, found in *Analog Circuit Design Art Science and Personalities*, edited by Jim Williams.

[9]The current-feedback op-amp does not have constant gain-bandwidth product limitation as does the standard voltage feedback op-amp.

[10]See, e.g. National's LM6165 with a gain-bandwidth product (GBP) of 725 MHz, the Linear Technology LT1818 with GBP = 400 MHz, or the Analog Devices AD8001 with GBP = 600 MHz.

[11]One example is the National LM12, also designed by Bob Widlar. An excellent IEEE paper discussing the design of the LM12 is "A Monolithic Power Op Amp" with citation at the end of this chapter. The LM12 has recently been obsoleted.

with several amperes of load current. Low-power op-amps with submilliwatt standby power dissipation are now commonplace. Rail-to-rail op-amps are now available.

These advances have opened new applications and product markets for devices based on analog signal processing and DSP. Currently, cellular telephone, cable TV, and wireless internet technologies are driving the business in RF analog circuit design and miniature handheld power electronics. Low-power devices enable the design of battery-powered devices with long battery life.

Digital vs. analog implementation: designer's choice

In many instances, functions that might be implemented in the digital domain would be difficult, costly, and power hungry to implement as compared to a relatively simple analog counterpart. Next we will consider a number of examples of functions that can be implemented in the analog domain with a few components (and no clock circuits...).

First, consider the design of a logarithmic amplifier. A log amp could be built with a DSP with lookup tables and the like, or in the analog world, we can exploit the well-known logarithmic/exponential voltage−current relationship between base−emitter voltage and collector current of a bipolar transistor operated in the forward-active region, as given by:

$$I_C \approx I_S e^{\frac{qV_{BE}}{kT}}$$
$$V_{BE} \approx \frac{kT}{q} \ln\left(\frac{I_C}{I_S}\right) \tag{1.1}$$

This relationship holds over many orders of magnitude of transistor collector current. Therefore, one can use a transistor PN junction to implement a low-cost logarithmic amplifier (Figure 1.4). The input−output transfer function of this circuit, assuming an ideal transistor and op-amp, is:

$$v_o = -\frac{kT}{q} \ln\left(\frac{v_i}{RI_s}\right) \tag{1.2}$$

This circuit provides an output voltage that is proportional to the natural logarithm of the input voltage. An implementation in the digital domain would be considerably more involved.

The same principles can be applied to create analog multipliers, dividers, and square root circuits, as Barrie Gilbert[12] showed some time ago. Let us look at the circuit in Figure 1.5. The dotted line follows a loop of base−emitter voltages; by using

---

[12]See Barrie Gilbert's original references, including: "A Precise Four-Quadrant Multiplier with Subnanosecond Response", *IEEE Journal of Solid-State Circuits*, December 1968, pp. 365−373. The "translinear principle" shows us how to calculate.

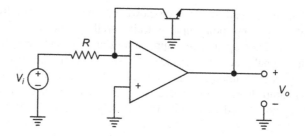

FIGURE 1.4

Simple logarithmic amplifier.

the "translinear principle" (discussed later in this book) we can find the relationship between the various transistor–collector currents given that there is a loop of V_{BES}:

$$I_{C1}I_{C2} = I_{C3}I_{C4} \qquad (1.3)$$

This means we can express the output current I_0 as the square root of the product of the two inputs:

$$I_0 = \sqrt{I_1 I_2} \qquad (1.4)$$

Now, consider the design of a fifth-order elliptic low-pass filter[13] with a cutoff frequency of 5 MHz. This is a typical specification for a video low-pass filter used for antialiasing, where a typical analog bandwidth of the signal is around 5 MHz and the A/D sampling rate in the system may be 13.5 MHz. You can

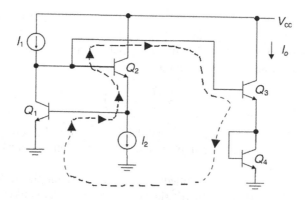

FIGURE 1.5

A translinear circuit, with dotted line showing the "Gilbert loop". The output current I_0 is equal to the square root of the product of the two inputs I_1 and I_2.

---

[13]Yes, you really can design analog filters with inductors.

implement this sharp-cutoff filter[14] with just a handful of discrete components (Figure 1.6). Note that in this filter the source and termination resistances are each 75 Ω, corresponding to the characteristic impedance of a typical video BNC cable. Again, an implementation in the digital domain would be significantly more complicated, especially if a high-frequency cutoff is required.

My digital designer friends would argue that the preceding functions could easily be provided by a DSP. They are perhaps right. But it is hard to argue against the fact

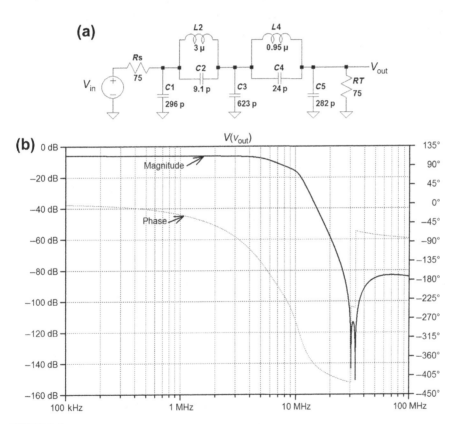

FIGURE 1.6

Fifth-order elliptic ladder filter with 5 MHz, −3 dB cutoff frequency. (a) Circuit. (b) Frequency response. Note that the DC response of the filter is −6 dB due to the 75-Ω source-termination resistive divider. (For color version of this figure, the reader is referred to the online version of this book.)

---

[14]Elliptic filters are commonly used in analog video filtering for antialiasing where a very sharp-cutoff transition band is required. Elliptic and other ladder filter designs are in tabulated form in Anatol Zverev's *Handbook of Filter Synthesis*. And, yes, you can build practical analog filters using inductors. This particular filter has an attenuation of ½ (or −6 dB) due to the 75-Ω resistive divider.

that it is cost effective and elegant to build a log amp with a very wide dynamic range with a transistor and a few other components. And you cannot generate timing waveforms with a circuit much simpler than a 555 timer, designed in the early 1970s and still used more than 40 years later.

So, why do we become analog designers?

One possible answer to this hypothetical question is to note that in any given analog design problem there is not one absolute, unique, and correct answer, or "perfect" design. As a matter of fact, if you think that you have arrived at the unique, perfect solution in the analog domain you are undoubtedly mistaken. In the analog design space, there are infinite possibilities of how to implement a given function. The challenge, and eventually (hopefully!) the reward, to the analog designer is to meet these requirements in a given design space meeting cost, size, and/or performance constraints.

Note on nomenclature in this text

In this text, there is a nomenclature that is used with regard to signals. In general, a transistor terminal voltage in an operating amplifier has a DC operating point and a small signal variation about that operating point. The nomenclature used in the case of a transistor base–emitter voltage is:

$$v_{BE} = V_{BE} + v_{be} \qquad (1.5)$$

where v_{BE} (small "v" and capital "BE") is the total variable, V_{BE} (capital "V" and capital "BE") is the DC operating point, and v_{be} (small "v" and small "be") is the small-signal variation.

Note on coverage in this book

It is impossible to cover all aspects of analog design in a single textbook. Rather, in this text, I have attempted to provide a potpourri of important techniques, tricks, and analysis tools that I have found useful in designing real-world analog circuits. Where necessary, mathematical derivations for theoretical techniques are given. In other areas, intuitive techniques and analogies are used to "map" solutions from one design domain to the analog design domain, hopefully bypassing extensive mathematical derivations.

Other references, including textbooks and scholarly journals, have been provided at the end of each chapter for the reader to explore the topics in more depth. The author has provided explanatory notes including opinions with some of the citations.

It is assumed that the reader has a familiarity with Laplace transforms, pole-zero plots, Bode plots, the concept of system step response, and a basic understanding of differential equations. Chapter 2 of this book reviews these signal processing basics. These fundamentals are needed to foster understanding of the more advanced topics in later chapters.

Further reading

[1]   Analog Devices. *Nonlinear circuits handbook*. Analog Devices; 1976. Good coverage of logarithmic amplifiers and other nonlinear analog circuits.

[2]   Bardeen J, Brattain WH. The transistor, a semiconductor triode. *Phys Rev* 1948;**74**(2):230−1. Reprinted in *Proc IEEE* 1998;**86**(1):29−30.

[3]   Bondyopadhyay P. In the beginning. *Proc IEEE* 1998;**86**(1):63−77.

[4]   Bondyopadhyay P. W = Shockley, the transistor pioneer—portrait of an inventive genius. *Proc IEEE* 1998;**86**(1):191−217.

[5]   Brinkman W, Haggan D, Troutman W. A history of the invention of the transistor and where it will lead us. *IEEE J Solid-State Circuits* 1997;**32**(12):1858−65.

[6]   Brinkman W. The transistor: 50 glorious years and where we are going. In: 1997 IEEE international solid-state circuits conference. Feb 6−8, 1997. pp. 22−6.

[7]   Fullagar D. A new high performance monolithic operational amplifier. *Fairchild Appl Brief* 1968. Description of the 741 op-amp by its designer.

[8]   Jung W. *IC op-amp cookbook*. 3rd ed. SAMS; 1995.

[9]   Kilby J. The integrated circuit's early history. *Proc IEEE* 2000;**88**(1):109−11.

[10] Lee TH. A vertical leap for microchips. *Sci Am* January 13, 2002. This note describes Matrix Technologies' efforts and motivation for producing 3D integrated circuits. Tom Lee was a protégé of Jim Roberge at MIT.

[11] Manglesdorf C. The changing face of analog IC design. *IEEE Trans Fundam* 2002;**E85-A**(2):282−5.

[12] Manglesdorf C. The future role of the analog designer. ISSCC 93, session WE3, pp. 78−9.

[13] Melliar-Smith CM, Borrus MG, Haggan D, Lowrey T, Sangiovanni-Vincentelli A, Troutman W. The transistor: an invention becomes a big business. *Proc IEEE* 1998;**86**(1):86−110.

[14] Moore G. The role of Fairchild in silicon technology in the early days of "Silicon Valley". *Proc IEEE* 1998;**86**(1):53−62.

[15] Moore G. Cramming more components onto integrated circuits. *Electronics* 1965:114−7. The original paper where "Moore's Law" was first published.

[16] National Semiconductor. Log converters. National Application Note AN-30.

[17] National Semiconductor. Linear applications handbook.

[18] Pearson GL, Brattain WH. History of semiconductor research. *Proc IRE* 1955;**43**(12):1794−806.

[19] Pease B. What's all this K2-W stuff, anyhow? *Electron Des* 2003.

[20] Pease B. "The story of the P2—the first successful solid-state operational amplifier with picoampere input currents. In: Williams J, editor. *Analog circuit design: art, science and personalities*. Butterworth-Heinemann; 1991. p. 67−78. A history and description of the K2-W operational amplifier, a discrete op-amp sold in the 50s and 60s by Philbrick, and built with vacuum tubes, and the P2, a discrete-transistor Philbrick operational amplifier.

[21] Pease R. *Troubleshooting analog circuits*. Butterworth-Heinemann; 1991. We'll miss you, Bob.

[22] Perry TS. For the record: Kilby and the IC. *IEEE Spectrum* 1988;**25**(13):40–1. Description of the history of Kilby's invention.

[23] Roberge JK. *Operational amplifiers, theory and practice*. John Wiley; 1975. A classic text on op-amps, now out of print but available in lots of libraries.

[24] Sah CT. Evolution of the MOS transistor—from conception to VLSI. *Proc IEEE* 1988;**76**(10):1280–326.

[25] Schaller R. Moore's law: past, present and future. *IEEE Spectrum* 1997;**34**(6):52–9.

[26] Shockley W. Transistor technology evokes new physics. 1956 Physics Nobel Prize Lecture December 11, 1956. An amazingly interesting and clear description of transistors and transistor physics.

[27] Shockley W. Transistor electronics—imperfections, unipolar and analog transistors. *Proc IEEE* 1997;**85**(12):2055–80.

[28] Shockley W. Theory of p-n junctions in semiconductors and p-n junction transistors. *Bell Syst Tech J* 1949;**28**(7):436–89.

[29] Shockley W. Electrons, holes and traps. *Proc IRE* 1958;**46**(6):973–90.

[30] Siebert WMC. *Circuits, signals and systems*. McGraw-Hill; 1986.

[31] Small J. General-purpose electronic analog computing: 1945–1965. *IEEE Ann Hist Comput* 1993;**15**(2):8–18.

[32] Soloman J. A tribute to Bob Widlar. *IEEE J Solid-State Circuits* 1991;**26**(8):1087–9. Tribute and anecdotes about Bob Widlar.

[33] Sporck C, Molay RL. *Spinoff: a personal history of the industry that changed the world*. Saranac Lake Publishing; 2001.

[34] Sugii T, Watanabe K, Sugatani S. Transistor design for 90-nm generation and beyond. *Fujitsu Sci Technol J* 2003;**39**(1):9–22.

[35] United States Patent Office. Website: http://www.uspto.gov.

[36] Warner R. Microelectronics: its unusual origin and personality. *IEEE Trans Electron Devices* 2001;**48**(11):2457–67.

[37] Widlar RJ. Design techniques for monolithic operational amplifiers. *IEEE J Solid-State Circuits* 1969;**SC-4**(4):184–91. Describes in detail the designs of the 709 and LM101A op-amps.

[38] Widlar RJ. Monolithic op amp—the universal linear component. National semiconductor linear applications handbook. Application Note no. AN-4.

[39] Widlar RJ, Yamatake M. A monolithic power op-amp. *IEEE J Solid-State Circuits* 1988;**23**(2). A detailed description of the design and operation of the LM12, an integrated circuit power op-amp designed in the 1980s. Applications of the LM12 are given in National application note AN-446, "A 150W IC Op Amp Simplifies Design of Power Circuits," also written by Widlar and Yamatake.

[40] Williams AB, Taylor FJ. *Electronic filter design handbook LC, active and digital filters*. 2nd ed. McGraw-Hill; 1988.

[41] Williams J, editor. *The art and science of analog circuit design*. Butterworth-Heinemann; 1998.

[42] Williams J. *Analog circuit design: art, science and personalities*. Butterworth-Heinemann; 1991. These books edited by Jim Williams present a potpourri of design tips, tricks, and personal histories by some of the most noteworthy analog designers in the world.

[43]　Zverev A. *Handbook of filter synthesis*. John Wiley; 1967. An excellent overview of analog ladder filter design with special emphasis on Butterworth, Bessel, Chebyshev and elliptic implementations. Lowpass to highpass and lowpass to bandpass transformations are also discussed in detail.

U.S. Patents[15]

[1]　Bardeen J, Brattain WH. Three-electrode circuit element utilizing semiconductive materials. U.S. Patent # 2,524,035, filed June 7, 1948; issued October 3, 1950.

[2]　Kilby JS. Miniaturized electronic circuits. U.S. Patent # 3,138,743, filed February 6, 1959; issued June 23, 1964.

[3]　Lilienfeld J. Method and apparatus for controlling electric currents. U.S. Patent # 1,745,175, filed October 8, 1926; issued January 28, 1930; Amplifier for electric currents. U.S. Patent # 1,877,140, filed December 8, 1928; issued September 13, 1932; Device for controlling electric current. U.S. Patent # 1,900,018, filed March 28, 1928; issued March 7, 1933.

[4]　Noyce RN. Semiconductor device-and-lead structure. U.S. Patent # 2,981,877, filed July 30, 1959; issued April 25, 1961.

[5]　Shockley W. Circuit element utilizing semiconductive materials. U.S. Patent # 2,569,347, filed June 17, 1948; issued October 3, 1950; Semiconductor amplifier. U.S. Patent # 2,502,488, filed September 24, 1948; issued April 5, 1950. Reprinted in *Proc IEEE* 1998;**86**(1):34−6.

---

[15]All patents listed are available on the U.S. Patent and Trademark website: www.uspto.gov or from Google Patents.

Review of Signal Processing Basics

2

"Make everything as simple as possible, but not simpler."

—Albert Einstein

IN THIS CHAPTER

▶ In this chapter, the basics of signal processing and analysis are covered. The important tools offered such as transfer functions in the Laplace domain, the concept of poles and zeros, first- and second-order systems, and their step and impulse responses are reviewed, as these concepts are needed by the reader in later chapters.

Review of Laplace transforms, transfer functions, and pole-zero plots

The transfer function of a system relates the input and output. The transfer function and pole-zero plot of any linear time-invariant (LTI) system can be found by replacing all electronic components with their impedance expressed in the Laplace domain and solving the resultant circuit. For instance, the transformation from the circuit domain to the Laplace (or s) domain is made by making the following substitutions of circuit elements:

Circuit Domain	Laplace (s) Domain
Resistance R	R
Inductance L	Ls
Capacitance C	$\frac{1}{Cs}$

The resultant transformed circuit also expresses a differential equation; the differential equation can be found by making the substitution:

$$s \Rightarrow \frac{\mathrm{d}}{\mathrm{d}t} \qquad (2.1)$$

Intuitive Analog Circuit Design. http://dx.doi.org/10.1016/B978-0-12-405866-8.00002-4

This transfer function of any lumped LTI system always works out to have polynomials in the Laplace[1] variable s. For instance, a typical transfer function with multiple poles and zeros has the form:

$$H(s) = \frac{a_n s^n + a_{n-1} s^{n-1} + \cdots + a_1 s + 1}{b_m s^m + b_{m-1} s^{m-1} + \cdots + b_1 s + 1} \qquad (2.2)$$

The values of s where the denominator of $H(s)$ becomes zero are the *poles* of $H(s)$, and the values of s where the numerator of $H(s)$ becomes zero are the *zeros* of $H(s)$. In this case, there are n zeros and m poles in the transfer function.

The values of s where poles and zeros occur can be either real or imaginary. Real-axis poles result in step responses expressed as simple exponentials (of form $e^{-t/\tau}$ where τ is the time constant) without overshoot and/or ringing. Complex poles always exist in pairs and can result in overshoot and ringing in the transient response of the circuit if damping is sufficiently low.

As an example of the use of the Laplace technique, consider the simple circuit of Figure 2.1(a), with two resistors and a capacitor. We can transform this circuit to the Laplace domain by recognizing that the impedance of a capacitor becomes $1/(Cs)$, resulting in the circuit shown in Figure 2.1(b).

Using Figure 2.1(b) and the resistive divider relationship, the input/output transfer function $H_1(s)$ is:

$$H_1(s) = \frac{v_o(s)}{v_i(s)} = \frac{R_2 + \frac{1}{Cs}}{R_1 + R_2 + \frac{1}{Cs}} = \frac{R_2 Cs + 1}{(R_1 + R_2)Cs + 1} \qquad (2.3)$$

Does this transfer function make sense? At very low frequencies approaching zero (which we can evaluate by making the substitution $s \to 0$), the transfer function is approximately:

$$H_1(s)|_{s \to 0} \approx 1 \qquad (2.4)$$

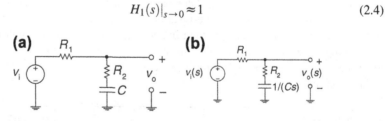

FIGURE 2.1

Simple *RC* circuit for finding system function and poles and zeros. (a) Original circuit. (b) Circuit transformed to the Laplace domain. Note that the capacitor has been transformed into the Laplace domain resulting in an impedance of $1/Cs$.

---

[1]A "lumped" circuit has any number of resistors, capacitors, inductors, or dependent sources operating at frequencies low enough where wave phenomena and transmission line effects can be ignored. Remember that you can form this polynomial by taking your original circuit and replacing each inductor with a component with impedance Ls and replacing each capacitor with a Laplace component with impedance $1/(Cs)$, and solving normally using nodal analysis or other techniques.

This makes sense, since the capacitor becomes an open circuit at zero frequency and the entire input signal is passed through to the output. At very high frequencies (or when $s \to \infty$), the capacitor shorts out and the transfer function is approximately:

$$H_1(s)|_{s \to \infty} \approx \frac{R_2}{(R_1 + R_2)} \tag{2.5}$$

This also makes sense, since at very high frequencies this circuit looks like a simple voltage divider with R_1 and R_2. The pole and zero of $H_1(s)$ are found by:

$$s_{\text{pole}} = -\frac{1}{(R_1 + R_2)C}$$

$$s_{\text{zero}} = -\frac{1}{R_2 C} \tag{2.6}$$

The pole-zero plot for this transfer function is shown in Figure 2.2.

This same procedure can be used for more complicated circuits with any number of resistors, capacitors, inductors, and dependent sources.

First-order system response

Common first-order systems are the voltage-driven RC circuit (Figure 2.3(a)) and the current-driven RC circuit (Figure 2.3(b)). Topologically, these circuits result in output responses that are the same.

The step responses for the circuits are:

$$\text{Voltage} - \text{driven } RC$$

$$v_o(t) = V\left(1 - e^{-t/\tau}\right)$$

$$i_r(t) = \frac{V}{R} e^{-t/\tau}$$

$$\tau = RC \tag{2.7}$$

$$\text{Current} - \text{driven } RC$$

$$v_o(t) = IR\left(1 - e^{-t/\tau}\right)$$

$$i_r(t) = I e^{-t/\tau}$$

$$\tau = RC$$

The risetime of the voltage or current is defined as the time it takes the response to rise from 10% of the final value to 90% of the final value. For the first-order system, this risetime is given by:

$$\tau_R = 2.2\tau \tag{2.8}$$

This definition of 10–90% risetime is shown in the step response of a generic lumped first-order system in Figure 2.4.

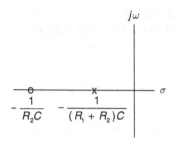

FIGURE 2.2

Pole-zero plot for simple RC circuit above with a low-frequency pole and a higher frequency zero.

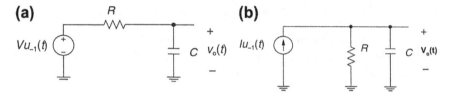

FIGURE 2.3

First-order RC circuits.[2] (a) Voltage-driven by input voltage source. (b) Current-driven by input current source.

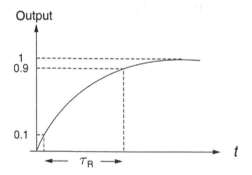

FIGURE 2.4

Unity step response of first-order system showing 10–90% risetime.

The bandwidth is defined as the frequency at which the magnitude of the output response to an AC input signal drops to 70.7% (or −3 dB) of the DC value. For the first-order system, the transfer function can be expressed as:

$$H(s) = \frac{1}{\tau s + 1}$$

$$\angle H(s) = -\tan^{-1}(\omega\tau)$$

(2.9)

[2]Note the notation that $u_{-1}(t)$ is the unit step.

We find that the -3 dB bandwidth is (in both radians per second and hertz):

$$\omega_h = \frac{1}{\tau}$$

(2.10)

$$f_h = \frac{\omega_h}{2\pi}$$

where ω_h is in radians per second and f_h is in Hertz. From these relationships, we can derive the relationship between bandwidth (in Hz) and risetime, which is exact for a first-order system:

$$\tau_R f_h = 0.35$$

(2.11)

In Figure 2.5, we see the unit step response and the frequency response for a first-order system with cutoff frequency 1 rad/s.

The result of Eqn (2.11) is a useful one to remember. From this we can see that the risetime of a first-order system with bandwidth 1 MHz is 350 ns.[3]

First-order system approximations at low and high frequency

The transfer function of a first-order system can be expressed as:

$$H(s) = \frac{1}{\tau s + 1}$$

(2.12)

where τ is the "time constant" of this first-order system. The magnitude $/H(j\omega)/$, angle $\angle H(j\omega)$ and group delay $G(j\omega)$[4] of this transfer function are found by:

$$H(j\omega) = \frac{1}{j\omega\tau + 1}$$

$$|H(j\omega)| = \frac{1}{\sqrt{(\omega\tau)^2 + 1}}$$

(2.13)

$$\angle H(j\omega) = -\tan^{-1}(\omega\tau)$$

$$G(j\omega) = \frac{-d\angle H(j\omega)}{d\omega} = \frac{\tau}{1 + (\omega\tau)^2}$$

---

[3]For instance, in an order-of-magnitude measurement if you make a scope measurement of risetime of 1 μs, you know the order-of-magnitude bandwidth is 1 MHz (actually 350 kHz).

[4]Group delay is a measure of how much time delay the frequency components in a signal undergo. Mathematically, the group delay of a system is the negative derivative of the phase with respect to omega. To find group delay for the first-order system, we make use of the identity of the derivative of the arctangent: $\frac{d}{dx}(\tan^{-1}u) = \frac{1}{1+u^2}\frac{du}{dx}$

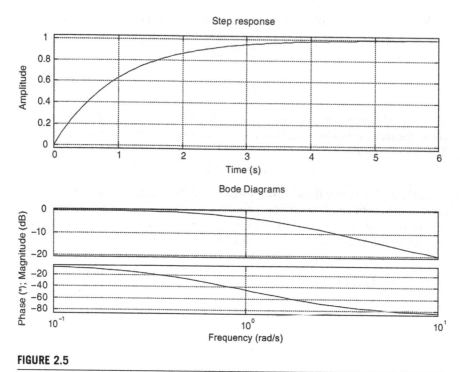

FIGURE 2.5

First-order system step and frequency responses for a system with cutoff frequency 1 rad/s.

Now, let us note how this single pole behaves at low and high frequencies (where low and high are defined in relation to the pole frequency). For high frequencies, where $\omega\tau \gg 1$, we can approximate the magnitude $|H(j\omega)|$, phase $\angle H(j\omega)$, and group delay $G(j\omega)$ of the transfer functions as:[5]

$$|H(j\omega)|_{\omega\tau\gg1} \approx \frac{1}{\omega\tau}$$

$$\angle H(j\omega)_{\omega\tau\gg1} \approx -\frac{\pi}{2} + \frac{1}{\omega\tau} \qquad (2.14)$$

$$G(j\omega)_{\omega\tau\gg1} \approx \frac{1}{\omega^2\tau}$$

For high frequencies, the magnitude of the transfer function decreases proportionally with frequency, corresponding to a -20 dB/decade roll-off. The angle approaches $-\pi/2$ radians, or $-90°$. The group delay decreases with frequency; higher frequency components are delayed less than lower frequency components.

---

[5]We make use of several trigonometric identities here. For instance, the power series expansion for arctangent is $\tan^{-1}(x) = \pi/2 - 1/x + 1/(3x^3) - ...$ for $x > 1$.

For low frequencies, where $\omega\tau \ll 1$, we can approximate the transfer functions as:[6]

$$|H(j\omega)|_{\omega\tau\ll1} \approx 1 - \frac{1}{2}(\omega\tau)^2 \approx 1$$

$$\angle H(j\omega)_{\omega\tau\ll1} \approx -\omega\tau \tag{2.15}$$

$$G(j\omega)_{\omega\tau\ll1} \approx \tau$$

For low frequencies, the magnitude of the transfer function is approximately unity; however, there is some finite phase shift due to the pole. For instance, at a frequency 10 times lower than the pole frequency, the magnitude is 0.995, while there is -0.1 radian (or $-5.7°$) phase shift. Also, at frequencies much lower than the pole frequency, the negative phase shift is approximately linear with the frequency. This means that a high-frequency pole behaves approximately as a time delay for low frequencies, as shown by the group delay calculation. The negative phase shift can have consequences in, for instance, feedback loops where the extra phase shift can cause oscillations.

The group delay for a first-order low-pass filter with cutoff frequency $\omega_c = 1$ rad/s is shown in Figure 2.6. Note that at low frequencies (low compared to the cutoff

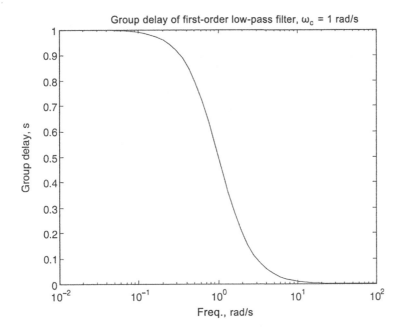

FIGURE 2.6

Group delay of first-order low-pass filter with cutoff frequency $\omega_c = 1$ rad/s. Note that at very low frequencies ($\omega \ll \omega_c$), the group delay is approximately 1 s.

[6]Here we make use of the power series expansion: $\frac{1}{\sqrt{1+x}} \approx 1 - \frac{x}{2}$ for $x \ll 1$.

frequency), the group delay is approximately flat. Therefore, for frequencies in this region, the low-pass filter behaves approximately as a constant time delay. For example, a sine wave at 10^{-1} rad/s (0.0159 Hz) would be delayed going through this filter by 1 s. A sine wave at 10 rad/s (1.59 Hz) would be delayed by roughly 0.01 s.

First-order system step response for short times ($t \ll RC$)

Let us examine a first-order step response for time short compared to the time constant of the system. We know that the step response for the first-order system of Figure 2.7 is:

$$v_0(t) = V\left(1 - e^{\frac{-t}{RC}}\right) \tag{2.16}$$

where V is the amplitude of the input step. Let us assume that the input steps at time $t = 0$; let us examine the behavior of this step response for times shorter than RC. We can use the series expansion for e^x, which is:

$$e^x = 1 + x + \frac{x^2}{2!} + \frac{x^3}{3!} + \cdots \tag{2.17}$$

Therefore, for $t \ll RC$ we can approximate the step response as:

$$v_0(t) = V\left(1 - \left(1 - \frac{t}{RC} - \left(\frac{1}{2!}\right)\left(\frac{t}{RC}\right)^2 - \cdots\right)\right) \approx V\left(\frac{t}{RC}\right) \text{ if } t \ll RC \tag{2.18}$$

Therefore, the step response "looks" linear in the initial time period $t \ll RC$ after $t = 0$. The capacitor current is roughly constant for $t \ll RC$:

$$i(t) = \frac{V - v_0(t)}{R} \approx \frac{V}{R}\left(1 - \frac{t}{RC}\right) \text{ if } t \ll RC \tag{2.19}$$

We can use this result to help us analyze the operation of a full-wave rectifier (Figure 2.8) using circuit simulation[7]. The input sine wave is a 60 Hz, 120 VRMS

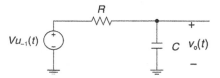

FIGURE 2.7

First-order RC circuit.

---

[7]This plot was done using PowerSim's PSIM power electronics simulator. Other circuit simulation plots throughout this book were done using Linear Technology's LTSPICE.

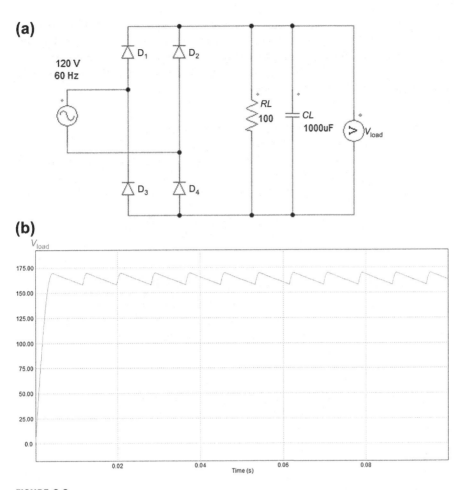

FIGURE 2.8

A 60-Hz full-wave rectifier. (a) Circuit.[8] (b) PSIM simulation for output voltage
($V_{load+} - V_{load-}$). (For color version of this figure, the reader is referred to the online version
of this book.)

waveform,[8] which has peaks at ±170 V. During the positive peaks of the sine wave,
D_1 and D_4 turn ON and charge up the 1000-μF load capacitor. During negative peaks
of the input sine wave, D_2 and D_3 are on. Therefore, the fundamental component of
the ripple of the output load voltage occurs at 120 Hz.

---

[8]This terminology means volts root-mean-square. The RMS value of a sine wave with peaks at
±170 V is approximately 120 V RMS. Alternate terminology would be 120 VAC, which means an
AC sine wave with value 120 VRMS.

Note that in this circuit we have an exponential decay with RC time constant $(100\ \Omega) \times (1000\ \mu F) = 100$ ms. Since this time constant is much longer than the period of the fundamental ripple frequency ($1/120$ Hz $= 8.3$ ms), the decay of the exponential "looks" approximately linear, as shown in the PSIM plot. We can very roughly calculate the amplitude of the ripple voltage on the capacitor using the following reasoning:

- The output voltage (ignoring ripple) is approximately 170 V.
- The current in the load resistor is approximately 170 V/100 $\Omega = 1.7$ A.
- The capacitor is discharging for (very roughly) 8.3 ms when all diodes are OFF and load current is supplied from the hold-up capacitor.
- Using $I = Cdv/dt$, we find that the ripple voltage on the capacitor is:

$$\Delta v_o \approx \frac{I_L T_{\text{discharge}}}{C} \approx \frac{(1.7)(8.3 \times 10^{-3})}{1000 \times 10^{-6}} \approx 14\ \text{V} \tag{2.20}$$

First-order system with extra high-frequency pole

How does a system behave if it has a dominant low-frequency real-axis pole and a second real-axis pole at a much higher frequency? Specifically, let us first consider a first-order system with a pole at -1 rad/s and a transfer function:

$$H(s) = \frac{1}{(s+1)} \tag{2.21}$$

Let us next investigate how the system behaves if it has a higher order pole added at -10 rad/s, resulting in a modified transfer function $H'(s)$:

$$H'(s) = \frac{1}{(s+1)(0.1s+1)} \tag{2.22}$$

The pole-zero plot is shown in Figure 2.9 for a system $H'(s)$ with a dominant pole at -1 rad/s and a second higher frequency pole at -10 rad/s. We expect the step

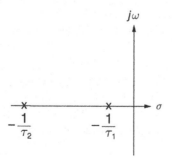

FIGURE 2.9

First-order system pole-zero plot for $H'(s)$ with a dominant pole at frequency $-1/\tau_1 = -1$ rad/s and a higher frequency pole at $-1/\tau_2 = -10$ rad/s.

response and frequency response of this system to be dominated by the pole at −1 rad/s. For instance, we expect a 10–90% risetime of ∼2.2 s, and a −3 dB bandwidth of 1 rad/s.

The step response (Figure 2.10) shows that the risetime is indeed dominated by the low-frequency pole. However, we note a subtlety: the modified system shows that in the time range 0−∼3 s the output of the modified system lags the output of the original system. This is consistent with our previous analysis that high-frequency poles behave as time delays for low frequencies compared to the pole frequency. In the case of a high-frequency pole at −10 rad/s, we expect the output of $H'(s)$ to lag that of $H(s)$ by approximately 0.1 s by Eqn (2.15).

A comparison of the frequency response plots of a first-order system with $H'(s)$ is shown in Figure 2.11. We note that, at low frequencies, the frequency response is dominated by the low-frequency pole, as expected. We also note that the magnitude responses match very well, while there is some extra phase shift in the phase response due to the high-frequency pole.

Second-order systems

A second-order system is one where there are two poles. For second-order systems consisting of resistors and capacitors (without any inductors or dependent sources),

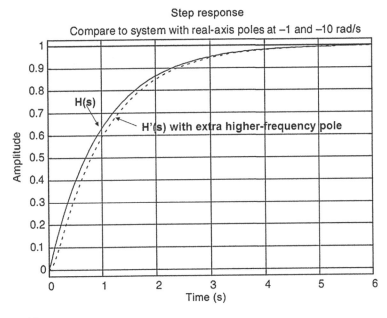

FIGURE 2.10

First-order system with added high-frequency pole (step response). Note that the step response of the system with the added pole $H'(s)$ lags the step response of the original system $H(s)$.

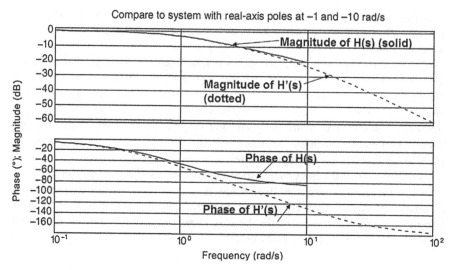

FIGURE 2.11

First-order system transfer function with added high-frequency pole (frequency response). Note that at frequencies below that of the low-frequency pole, the two magnitude transfer functions match well.

the poles lie on the real axis. For this special case, there is no possibility of overshoot or ringing in the step response.

Other second-order systems can be built with coupled, lumped energy-storage mechanisms or with dependent sources, which may result in overshoot and ringing in the transient response, provided that losses (i.e. "damping") are not too high. Some examples of systems that may be potentially underdamped are:

- Mass and spring
- *LC* circuit
- Rotational inertia and torsional spring
- *RC* circuits with op-amp feedback
- Magnetic suspensions.

Second-order mass–spring system (ideally undamped, damping ratio $\zeta = 0$)

A mass–spring system is shown in Figure 2.12. For simplicity, we will ignore any damping by assuming that the spring is ideal and that there is no friction due to wind resistance. When this system vibrates, energy is alternately stored in the kinetic energy of the mass and in the stretch of the spring. If we consider y to be the displacement of the spring from free-hanging equilibrium, the force that the spring exerts on the mass is:

$$f_y = -ky \tag{2.23}$$

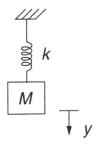

FIGURE 2.12

Undamped mass–spring system. The mass M (in kilograms) is coupled to a spring with spring constant k (in Newton/meter). The resultant oscillation frequency is $\sqrt{k/M}$ (rad/s).

Newton's law applied to the moving mass is:

$$f_y = -ky = M\frac{d^2y}{dt^2} \tag{2.24}$$

This results in the differential equation of motion for this mass spring of:

$$M\frac{d^2y}{dt^2} + ky = 0 \tag{2.25}$$

Let us guess a sinusoidal solution (since the second derivative of a sinusoid is proportional to the negative of the same sinusoid):[9]

$$y(t) = Y_0 \sin(\omega t) \tag{2.26}$$

Putting this guess into the differential equation and canceling out the sinusoid results in:

$$M\left(-\omega^2 Y_0 \sin(\omega t)\right) + k(Y_0 \sin(\omega t)) = 0 \tag{2.27}$$

The solution of this equation is:

$$-M\omega^2 + k = 0 \Rightarrow \omega = \sqrt{\frac{k}{M}} \tag{2.28}$$

Therefore, this system will vibrate with a frequency (in radians/second) $\omega = \sqrt{k/M}$, which we recognize as the oscillation frequency of a single-degree-of-freedom mass–spring system.

In a later chapter, we discuss in some detail coupled mass–spring systems and compare them to coupled LC circuits.

---

[9]We can put our intuition to work here. We know that if we excite a mass–spring system (by stretching the spring and then letting go of the mass) that the mass vibrates sinusoidally with a constant frequency. Ignoring losses in the spring and air friction, this system will vibrate sinusoidally forever.

A second-order *RLC* electrical system with damping, $\zeta < 1$

A second-order series *RLC* circuit is shown in Figure 2.13(a). This circuit has some damping due to the resistor. Transforming to the Laplace domain results in the circuit of Figure 2.13(b). If the resistance *R* is small (and we will define what "small" means later on in this chapter), this system is "underdamped". The transfer function can be expressed in standard form, as a function of natural frequency and damping ratio, as:

$$H(s) = \frac{v_o(s)}{v_i(s)} = \frac{\frac{1}{Cs}}{R + Ls + \frac{1}{Cs}} = \frac{1}{LCs^2 + RCs + 1} = \frac{1}{\frac{s^2}{\omega_n^2} + \frac{2\zeta s}{\omega_n} + 1}$$

$$= \frac{\omega_n^2}{s^2 + 2\zeta \omega_n s + \omega_n^2} \tag{2.29}$$

This equation has been put into second-order system standard form where ω_n is the "natural frequency" of the circuit and ζ is the "damping ratio". In the series *RLC* circuit, the natural frequency and damping ratio are:

$$\omega_n = \frac{1}{\sqrt{LC}}$$

$$\zeta = \frac{\omega_n RC}{2} = \frac{1}{2} \frac{R}{\sqrt{\frac{L}{C}}} = \frac{1}{2} \left(\frac{R}{Z_o} \right) \tag{2.30}$$

The natural frequency is the oscillation frequency if there is no damping and is an indication of the relative speed of response of the system. The damping ratio tells you how oscillatory (or not) the step response is and how peaky (or not) the frequency response is. Note that the damping ratio depends on the value of the series resistor *R* compared to the "characteristic impedance" Z_o. In resonant *LC* circuits, this term Z_o crops up again and again.

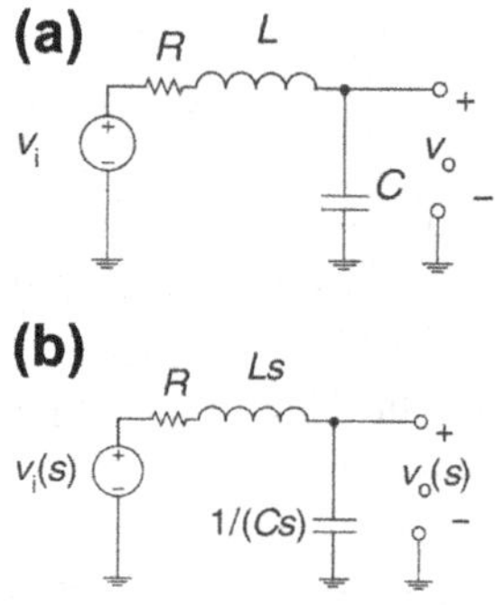

FIGURE 2.13

Second-order series *RLC* system. (a) Circuit. (b) Circuit transformed to the Laplace domain.

The magnitude and phase of the frequency response are given by:

$$H(j\omega) = \frac{1}{\frac{2j\zeta\omega}{\omega_n} + \left(1 - \frac{\omega^2}{\omega_n^2}\right)}$$

$$|H(j\omega)| = \frac{1}{\sqrt{\left(\frac{2\zeta\omega}{\omega_n}\right)^2 + \left(1 - \frac{\omega^2}{\omega_n^2}\right)^2}} \qquad (2.31)$$

$$\angle H(j\omega) = -\tan^{-1}\frac{\left(\frac{2\zeta\omega}{\omega_n}\right)}{\left(1 - \frac{\omega^2}{\omega_n^2}\right)} = -\tan^{-1}\left(\frac{2\zeta\omega\omega_n}{\omega_n^2 - \omega^2}\right)$$

A plot of this transfer function magnitude and phase for varying values of damping ratio ζ is shown in Figure 2.14. Note that as the damping ratio decreases, the peak of the frequency response[10] (which occurs at a frequency near ω_n) increases.

Frequency response for natural frequency = 1 and various damping ratios

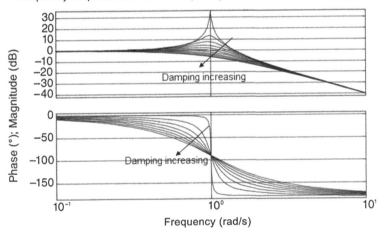

FIGURE 2.14

Frequency response for second-order systems, for damping ratios $\zeta = 0.01, 0.11, 0.21 \ldots$ 1.01; natural frequency $\omega_n = 1$. Note that for low damping there is significant peaking in the frequency response near 1 rad/s.

---

[10]We can show that the frequency at which the transfer function has its peak is $\omega_p = \omega_n\sqrt{1 - 2\zeta^2}$ for $\zeta < 0.707$. This frequency is close to the natural frequency ω_n if the damping ratio is small. The magnitude of the transfer function at this peak frequency is $M_p = \frac{1}{2\zeta\sqrt{1-\zeta^2}}$ for $\zeta < 0.707$.

For series resistance R larger than $2Z_0$, the damping ratio is larger than 1. In this "overdamped" case, the two poles are on the negative real axis. The poles of $H(s)$ in the overdamped case are found by:

$$s_{poles} = -\omega_n\left[-\zeta \pm \sqrt{\zeta^2 - 1}\right] \tag{2.32}$$

Note that as the damping ratio ζ varies, the pole locations vary as well. For zero damping (i.e. $\zeta = 0$), the poles lie on the $j\omega$ axis at $\pm j\omega_n$, resulting in a very oscillatory transient response at the natural frequency.

For critical damping (with $\zeta = 1$), both poles are on the real axis at the same location (at location $s = -\omega_n$). The overdamped case occurs for $\zeta > 1$; in this case, there is one low-frequency real pole at less than $-\omega_n$ rad/s and a higher frequency pole at higher than $-\omega_n$ rad/s. Pole locations for the underdamped, critically damped, and overdamped cases are shown in Figure 2.15. We also see how the roots move as damping is increased. The path that the roots take as damping increases is called the "root locus".

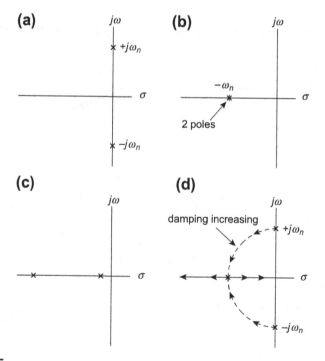

FIGURE 2.15

Pole-zero plot for series RLC circuit. (a) Underdamped circuit with $R \ll 2Z_0$. (b) Critically damped with $R = 2Z_0$. (c) Overdamped with $R \gg 2Z_0$. (d) Locus of roots as damping increases.

Now, what happens if we excite this system exactly at the natural frequency,[11] or at $\omega = \omega_n$? The frequency response is:

$$|H(j\omega)|_{\omega=\omega_n} = \frac{1}{2\zeta}$$

$$\angle H(j\omega)_{\omega=\omega_n} = -\frac{\pi}{2}$$

(2.33)

For low damping ratio ζ, the frequency response shows a very peaky response at the natural frequency.

In Figure 2.16, we see the relationship between the pole locations, the natural frequency ω_n, and the damping ratio ζ. The natural frequency is the length of the vector from the origin to either of the complex poles. The imaginary part of the pole is the "damped natural frequency" ω_d; this is the frequency of oscillation when the poles are excited. The real part of the poles $(-\zeta\omega_n)$ sets the rate at which the oscillation envelope decays.

Quality factor "*Q*"

Another commonly used metric for evaluating second-order systems is *quality factor* or "*Q*". Quality factor is defined as:

$$\omega \frac{E_{\text{stored}}}{P_{\text{diss}}}$$

(2.34)

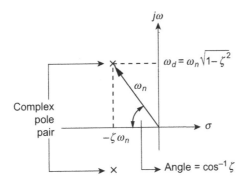

FIGURE 2.16

Relationship between natural frequency ω_n and damping ratio ζ for a pole pair in the s-plane. The natural frequency is the length of the vector from the origin to one of the poles. ω_d is the damped natural frequency.

[11]Sometimes called exciting the system "at resonance" or at the frequency where the inductive and capacitive reactances cancel each other.

where E_{stored} is the peak stored energy in the system and P_{diss} is the average power dissipation. First, let us look at a series resonant network (Figure 2.17(a)). The calculation for quality factor at resonance is:

$$\omega = \frac{1}{\sqrt{LC}}$$

$$E_{stored} = \frac{1}{2}LI_{pk}^2$$

$$P_{diss} = \frac{1}{2}I_{pk}^2 R \tag{2.35}$$

$$Q = \left(\frac{1}{\sqrt{LC}}\right)\left(\frac{\frac{1}{2}LI_{pk}^2}{\frac{1}{2}I_{pk}^2 R}\right) = \frac{\sqrt{\frac{L}{C}}}{R} = \frac{Z_0}{R}$$

In the series resonant circuit, as resistance R increases, quality factor decreases. This is because a large resistor provides more damping.

Next, let us consider a parallel RLC network (Figure 2.17(b)). The calculation for Q at resonance is:

$$\omega = \frac{1}{\sqrt{LC}}$$

$$E_{stored} = \frac{1}{2}CV_{pk}^2$$

$$P_{diss} = \frac{1}{2}\frac{V_{pk}^2}{R} \tag{2.36}$$

$$Q = \left(\frac{1}{\sqrt{LC}}\right)\left(\frac{\frac{1}{2}CV_{pk}^2}{\frac{1}{2}\frac{V_{pk}^2}{R}}\right) = \frac{R}{\sqrt{\frac{L}{C}}} = \frac{R}{Z_0}$$

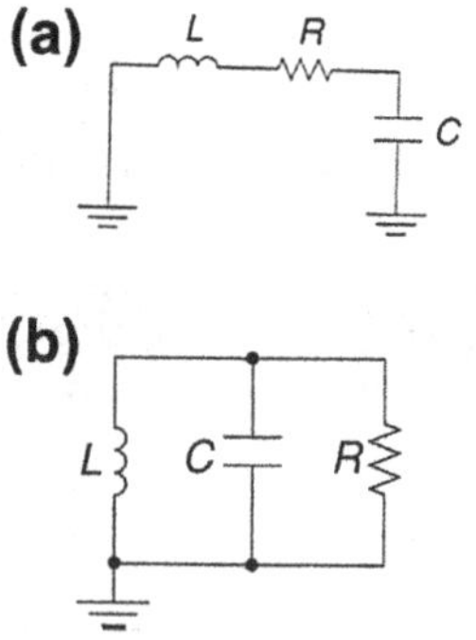

FIGURE 2.17

Resonant networks. (a) Series resonant RLC network. (b) Parallel resonant.

In the parallel resonant circuit, a small resistor provides more damping.
Next, we see that quality factor is related to damping ratio by:

$$Q = \frac{1}{2\zeta} \tag{2.37}$$

Let us rewrite the previous equation for the transfer function of a series resonant circuit, this time using quality factor:

$$H(s) = \frac{1}{\frac{s^2}{\omega_n^2} + \frac{2\zeta s}{\omega_n} + 1} = \frac{1}{\frac{s^2}{\omega_n^2} + \frac{s}{\omega_n Q} + 1} \tag{2.38}$$

The magnitude of this transfer function is:

$$|H(s)| = \frac{1}{\sqrt{\left(1 - \frac{\omega^2}{\omega_n^2}\right)^2 + \left(\frac{\omega}{\omega_n Q}\right)^2}} \tag{2.39}$$

At the natural frequency ($\omega = \omega_n$), the magnitude of the transfer function is:

$$|H(s)|_{\omega=\omega_n} = Q \tag{2.40}$$

Transient response of second-order system with damping ratio $\zeta < 1$

Next, we will look at the transient response of a second-order system with damping. A second-order LCR system is shown in Figure 2.18. The input is a step $v_i(t)$ at time $t = 0$. Let us assume, at first, that the resistance is small enough so that the system is underdamped (i.e. that it has complex poles, or equivalently $\zeta < 1$).

The step response (for $\zeta < 1$) can be expressed as:[12]

$$v_o(t) = 1 - \frac{e^{-\zeta\omega_n t}}{\sqrt{1 - \zeta^2}} \sin(\omega_d t + \theta)$$

$$\omega_d = \omega_n \sqrt{1 - \zeta^2} \tag{2.41}$$

$$\theta = \tan^{-1}\left(\frac{\sqrt{1 - \zeta^2}}{\zeta}\right)$$

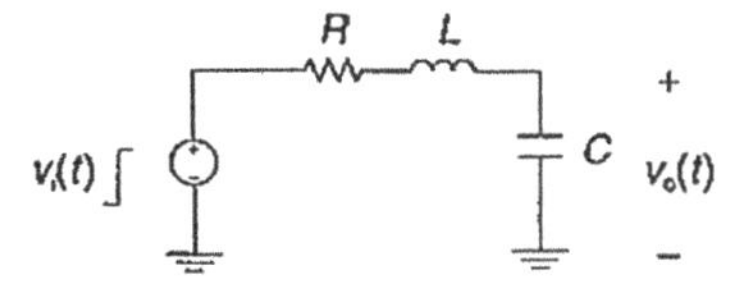

FIGURE 2.18

Second-order system with voltage step input.

---

[12]For more detail, see e.g. James Roberge, *Operational Amplifiers: Theory and Practice*, John Wiley, 1975.

where ω_d is the "damped natural frequency", which is the frequency of the oscillations. We note that the envelope of the sinusoidal oscillation has an exponential decay with time constant $1/(\zeta\omega_n)$. The magnitude of the peak of the step response is:

$$P_0 = 1 + e^{\frac{-\pi\zeta}{\sqrt{1-\zeta^2}}} \tag{2.42}$$

For instance, for a damping ratio of 0.2, the peak overshoot for a second-order system is approximately 1.6. The ring frequency is the imaginary part of the complex poles, and the real part of the complex poles controls the decay time of the ring envelope.

Remembering the calculations for natural frequency and damping ratio:

$$\omega_n = \frac{1}{\sqrt{LC}}$$

$$\zeta = \frac{R}{2\sqrt{\dfrac{L}{C}}} \tag{2.43}$$

we see that for series resistance less than twice the characteristic impedance $\sqrt{L/C}$, the system is underdamped. The step response for a second-order system with $\omega_n = 1$ and damping ratio varying from 0.01 to 1.01 is shown in Figure 2.19. The step response eventually settles out to a final value of 1 V with a decay time depending on the damping.

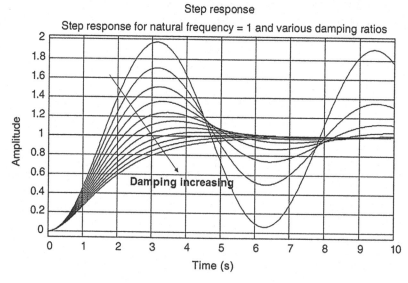

Step response

FIGURE 2.19

Step response for second-order systems, for damping ratios 0.01, 0.11, 0.21 ... 1.01.

Example 2.1: Bandwidth and risetime of a second-order system

Previously in this chapter, we showed that the product of bandwidth and risetime $\tau_R f_h = 0.35$ for a first-order system. How do we figure out the equivalent relationship for a second-order system?

A solution: first, we recognize that a second-order system is characterized by both natural frequency ω_n and damping ratio ζ. We could resort to numerical computation, but in this example we will solve this approximately by using circuit simulation and a graphical approach.

In Figure 2.20(a), we see the second-order system, in this case with a natural frequency $\omega_n = 1$ rad/s and damping ratio that varies as the resistor varies. The resultant step response and frequency response is shown in Figure 2.20(b).

In Figure 2.21, we see the resultant product $\tau_R f_h$ plotted as a function of damping ratio for the second-order system. What we note from this graph is that the product $\tau_R f_h$ is still approximately 0.35, but this approximation becomes less accurate at low damping ratios.

Free vibration of damped, second-order system

How does a damped second-order system vibrate if we set an initial condition? For instance, if we jolt a second-order system with a short pulse (i.e. an "impulse") how does the system vibrate?

Shown in Figure 2.22 is the impulse response of our second-order system, this time excited by a unit impulse (a unit impulse is a short pulse of voltage at the input, with an amplitude of 1).

Logarithmic decrement[13]

Quite often you would like to know the damping ratio and/or Q of a second-order system, given the step response or impulse response. One interesting technique in analysis that is commonly used in the world of mechanical engineering is the logarithmic decrement. The logarithmic decrement gives us a simple way to estimate the damping ratio of an underdamped resonant system by measuring the step or impulse response.

Recall that a second-order system can be characterized by natural frequency ω_n and damping ratio damp. The transfer function of a second-order system is:

$$H(s) = \frac{1}{\frac{s^2}{\omega_n^2} + \frac{2\zeta}{\omega_n}s + 1} \tag{2.44}$$

---

[13]Thanks to Prof. J. Kim Vandiver from MIT who introduced this method to me when I was a Mechanical Vibrations student in graduate school. The technique of logarithmic decrements is primarily used by mechanical engineers, but still has use for those of us in the E.E. world.

(a)

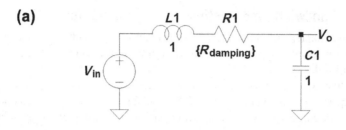

(b)

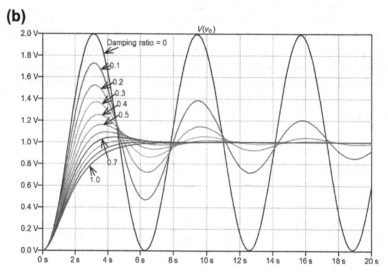

(c)

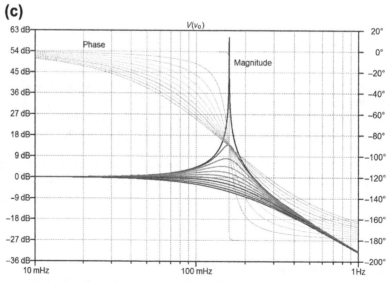

FIGURE 2.20

Second-order system with natural frequency $\omega_n = 1$ radian/second, and damping ratio varied from 0.05 to 1.0. (a) LTSPICE circuit. (b) Step response. (c) Frequency response. (For color version of this figure, the reader is referred to the online version of this book.)

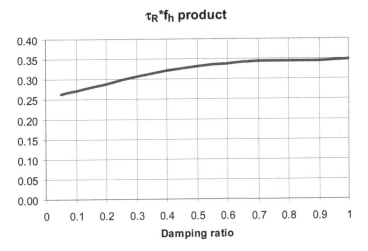

FIGURE 2.21

Plotted product $\tau_R f_h$ for a second-order system as a function of damping ratio ζ. (For color version of this figure, the reader is referred to the online version of this book.)

The impulse response of this system can be expressed as:

$$v_o(t) = Ae^{-\zeta\omega_n t}\cos(\omega_d t + \phi) \tag{2.45}$$

where A is a constant and ϕ is a phase shift. We can focus on the exponential part of this expression to figure out how the oscillation decays as a function of the damping ratio. In Figure 2.23 we see the impulse response of an RLC system, with the amplitudes of the various oscillations marked $x_1, x_2, \ldots, x_n$. Let us ask ourselves how the amplitude of successive oscillations decreases as time increases:

$$\frac{x_n}{x_{n+1}} = \frac{Ae^{-\zeta\omega_n t_n}}{Ae^{-\zeta\omega_n (t_n+T_D)}} = e^{\zeta\omega_n T_D} \tag{2.46}$$

where T_D is the period of oscillation. We note that the period of oscillation is related to the natural frequency as:

$$T_D = \frac{2\pi}{\omega_d} = \frac{2\pi}{\omega_n\sqrt{1-\zeta^2}} \tag{2.47}$$

Therefore, the ratio by which the amplitudes decrease from cycle to cycle (the "logarithmic decrement" δ) is:

$$\delta = \ln\left(\frac{x_n}{x_{n+1}}\right) = \zeta\omega_n T_D = \frac{2\pi\zeta}{\sqrt{1-\zeta^2}} \tag{2.48}$$

For small damping ratios ($\zeta \ll 1$), we can estimate this logarithmic decrement as:

$$\delta \approx 2\pi\zeta \quad \text{for } \zeta \ll 1 \tag{2.49}$$

(a)   **Second-order system with varying damping ratio**
In this circuit, damping ratio = $R_{damp}/2$
Natural frequency = 1 rad/s

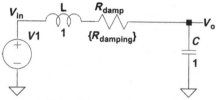

.tran 0 25 0 0.0001
file: **Second-order system impulse response varying damping.asc**
.step param $R_{damping}$ list 1e-6 0.2 0.4 0.6 0.8 1 1.2 1.4 1.6 1.8 2

(b)

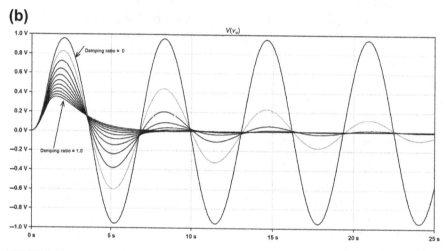

FIGURE 2.22

Impulse response of second-order system with varying damping ratio ζ. (a) Circuit. (b) Unit impulse response. (For color version of this figure, the reader is referred to the online version of this book.)

Sometimes it is more convenient to count multiple cycles in the logarithmic decrement estimation, especially if the damping ratio is very small and the amount of amplitude change from cycle to cycle is small. In this case, we see the ratio of the first amplitude to the $(k+1)$th amplitude is:

$$\frac{x_1}{x_{k+1}} = \frac{Ae^{-\zeta\omega_n t_n}}{Ae^{-\zeta\omega_n(t_n+kT_D)}} = e^{\zeta\omega_n(kT_D)} \tag{2.50}$$

The resultant logarithmic decrement is:

$$\delta = \left(\frac{1}{k}\right)\left(\frac{2\pi\zeta}{\sqrt{1-\zeta^2}}\right) \tag{2.51}$$

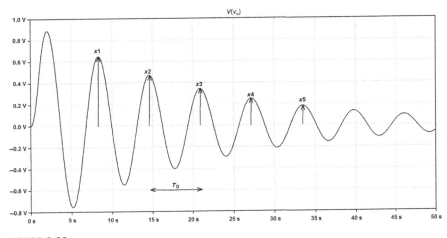

FIGURE 2.23

Waveform illustrating the use of the logarithmic decrement.

Example 2.2: Using the logarithmic decrement

We will illustrate the method by use of an example. Let us find the damping ratio of the waveform of Figure 2.23 (which happens to have a damping ratio of 0.05). The first and fifth amplitudes are:

$$x_1 = 0.65$$
$$x_5 = 0.18$$

(2.52)

The resultant logarithmic decrement is:

$$\delta = \left(\frac{1}{4}\right)\ln\left(\frac{x_1}{x_{4+1}}\right) = \left(\frac{1}{4}\right)\ln\left(\frac{0.65}{0.18}\right) = 0.32$$

(2.53)

The estimated damping ratio is:

$$\zeta \approx \frac{\delta}{2\pi} \approx 0.051$$

(2.54)

Higher order systems

Most systems in nature have more than two poles, but in many cases, there is a dominant pole or a dominant pole pair. So, for quick calculations, most systems we encounter in nature may be approximated as either first or second order.

Second-order electrical system with extra high-frequency poles

How does a second-order system behave if there are one or more extra poles added? Well, the response depends on how close to the resonant poles the extra added pole(s) are. As we have seen previously, a high-frequency pole has a magnitude of approximately 1.0 with a negative phase shift that decreases approximately proportionally with the frequency (corresponding to a time delay).

Shown in Figure 2.24 is the step response for a second-order system with natural frequency $\omega_n = 1$ radian/second and damping ratio $\zeta = 0.5$. Superimposed on this plot is a second plot showing the effects of an additional pole at -10 rad/s. Note that the two plots match up fairly well, with the exception that the plot for the system with the extra pole is delayed (by approximately 0.1 s) in the risetime area. This corresponds to the finite delay caused by the high-frequency pole.

In Figure 2.25, we see the effects of a lower frequency pole added to the same system. This time, the extra pole is at -2 rad/s. We note that the effect on the original system is more pronounced this time. This is expected, since the added pole is not that much higher than the natural frequency of the original system.

Second-order system with widely spaced real-axis poles

The pole plot of a second-order system with widely spaced negative real-axis poles is shown in Figure 2.26.

The transfer function for this system is:

$$H(s) = \frac{1}{(\tau_1 s + 1)(\tau_2 s + 1)} \tag{2.55}$$

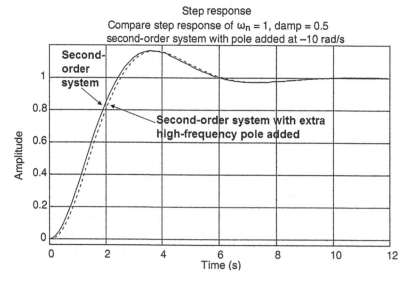

FIGURE 2.24

Step responses for second-order systems with natural frequency $\omega_n = 1$ rad/s and damping ratio $\zeta = 0.5$, with and without extra pole at -10 rad/s.

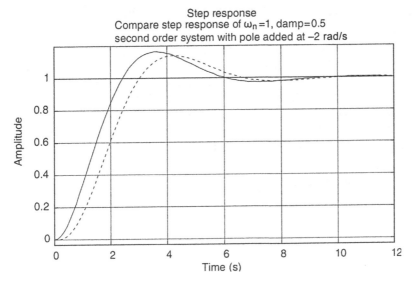

FIGURE 2.25

Step response for second-order systems with $\omega_n = 1$ rad/s and damping ratio $\zeta = 0.5$, with extra pole at -2 rad/s. Solid line: original second-order system. Dashed line: second-order system with extra higher frequency pole added.

where τ_1 is the time constant associated with the dominant, low-frequency pole and τ_2 is the time constant of the faster (nondominant) pole. Multiplying out the denominator results in:

$$H(s) = \frac{1}{\tau_1 \tau_2 s^2 + (\tau_1 + \tau_2)s + 1} \tag{2.56}$$

In this case, the second-order system behaves as two cascaded first-order systems.

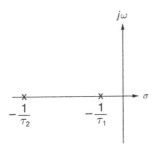

FIGURE 2.26

Second-order system with two widely spaced real-axis poles.

Finding approximate pole locations from the transfer function denominator

We will now go off on a little bit of a tangent, which will help us in further chapters when we will be approximating the bandwidth of circuits. Let us assume that we have a system and that we know that this system has three poles that are on the real axis and are widely spaced, as shown in Figure 2.27. We have a low-frequency pole at frequency $-1/\tau_1$, a high-frequency pole at $-1/\tau_3$, and an intermediate frequency pole at $-1/\tau_2$, or, said another way, the pole time constants are related as $\tau_1 \gg \tau_2 \gg \tau_3$. The transfer function of this system is:

$$H(s) = \frac{1}{(\tau_1 s + 1)(\tau_2 s + 1)(\tau_3 s + 1)}$$

$$= \frac{1}{\tau_1 \tau_2 \tau_3 s^3 + (\tau_1 \tau_2 + \tau_1 \tau_3 + \tau_2 \tau_3)s^2 + (\tau_1 + \tau_2 + \tau_3)s + 1} \tag{2.57}$$

This transfer function is of the form:

$$H(s) = \frac{1}{a_3 s^3 + a_2 s^2 + a_1 s + 1}$$

$$a_3 = \tau_1 \tau_2 \tau_3$$

$$a_2 = \tau_1 \tau_2 + \tau_1 \tau_3 + \tau_2 \tau_3 \tag{2.58}$$

$$a_1 = \tau_1 + \tau_2 + \tau_3$$

Let us call the pole locations p_{low}, p_{medium}, and p_{high}, for "low", "medium", and "high" frequency. We find that the approximate pole locations are:

$$p_{\text{low}} \approx -\frac{1}{a_1} \approx -\frac{1}{\tau_1 + \tau_2 + \tau_3} \approx -\frac{1}{\tau_1}$$

$$p_{\text{high}} \approx -\frac{a_2}{a_3} \approx -\frac{\tau_1 \tau_2 + \tau_1 \tau_3 + \tau_2 \tau_3}{\tau_1 \tau_2 \tau_3} \approx -\frac{\tau_1 \tau_2}{\tau_1 \tau_2 \tau_3} \approx -\frac{1}{\tau_3} \tag{2.59}$$

$$p_{\text{medium}} \approx \frac{a_1}{a_2} \approx -\frac{\tau_1 + \tau_2 + \tau_3}{\tau_1 \tau_2 + \tau_1 \tau_3 + \tau_2 \tau_3} \approx -\frac{\tau_1}{\tau_1 \tau_2} \approx -\frac{1}{\tau_2}$$

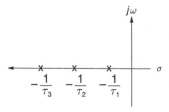

FIGURE 2.27

Pole plot of three widely spaced poles on the negative real axis.

In fact, we can extend this methodology to denominator polynomials of any degree where the pole locations are all on the negative real axis and are widely spaced. In this case, the pole locations are:

$$p_k \approx -\frac{a_{k-1}}{a_k}$$

$$a_0 = 1$$

(2.60)

As an example, consider the fifth-order polynomial

$$H(s) = \frac{1}{10^{-10}s^5 + 1.111 \times 10^{-6}s^4 + 1.122 \times 10^{-3}s^3 + 1.122 \times 10^{-1}s^2 + 1.111s + 1}$$

(2.61)

Using our approximation formula, we find the approximate pole locations:

$$p_5 \approx -\frac{1.111 \times 10^{-6}}{10^{-10}} \approx -11,110 \text{ rad/s}$$

$$p_4 \approx -\frac{1.122 \times 10^{-3}}{1.111 \times 10^{-6}} \approx -1010 \text{ rad/s}$$

$$p_3 \approx -\frac{1.122 \times 10^{-1}}{1.122 \times 10^{-3}} \approx -100 \text{ rad/s}$$

(2.62)

$$p_2 \approx -\frac{1.111}{1.122 \times 10^{-1}} \approx -9.9 \text{ rad/s}$$

$$p_1 \approx -\frac{1}{1.111} \approx -0.9 \text{ rad/s}$$

In this example, the actual pole locations[15] are at $-1, -10, -100, -1000$ and $-10,000$ rad/s. We find that the methodology gives us a rough approximation of the pole locations if we are correct in our initial assumption that the poles are on the real axis and are widely spaced.

Review of resonant electrical circuits

Consider a parallel *LC* circuit as shown in Figure 2.28. This is a standard resonant circuit that will oscillate at frequency ω_r as the energy sloshes back and forth between electric storage in the capacitor and magnetic storage in the inductor. Let us assume that the initial condition in the circuit is that the inductor current is zero and the capacitor voltage is some voltage v_c. The current in the inductor is given by the differential equation:

$$\frac{di_L}{dt} = \frac{v_c}{L}$$

(2.63)

[15]If you do not believe me, multiply it out for yourself or solve it using MATLAB.

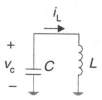

FIGURE 2.28

Undamped parallel LC resonant circuit.

Similarly, the voltage on the capacitor can be found by:

$$\frac{dv_c}{dt} = \frac{-i_L}{C} \tag{2.64}$$

We can solve for the voltage on the capacitor $v_c(t)$ by differentiating the capacitor equation and substituting for di/dt, resulting in:

$$\frac{d^2 v_c}{dt^2} = -\frac{1}{C}\frac{di_L}{dt} = -\frac{v_c}{LC} \tag{2.65}$$

Now, we can find the resonant frequency by guessing that the voltage $v_c(t)$ is sinusoidal with $v_c(t) = V_0 \sin \omega t$. Putting this into the equation for capacitor voltage results in:

$$-\omega^2 \sin(\omega t) = -\frac{1}{LC} \sin(\omega t) \tag{2.66}$$

This means that the resonant frequency is the standard (as expected) resonance:

$$\omega_r^2 = \frac{1}{LC} \tag{2.67}$$

Use of energy methods to analyze undamped resonant circuits

Using energy methods we can easily solve many problems in electrical and mechanical engineering. Energy is stored in capacitors in an electric field, in inductors in a magnetic field, and in moving objects, springs, and thermal masses. The stored energies in these various storage modes are found in Table 2.1.

By using energy methods, we can find the ratio of maximum capacitor voltage to maximum inductor current in the resonant circuit of Figure 2.28 by using conservation of energy. Assuming that the capacitor is initially charged to V_0 volts and remembering that capacitor stored energy $E_c = \frac{1}{2}CV^2$ and inductor stored energy is $E_L = \frac{1}{2}LI^2$, we can write the following:

$$\frac{1}{2}CV_0^2 = \frac{1}{2}LI_0^2 \tag{2.68}$$

Table 2.1 List of Symbols

Storage Mode	Relationship	Comments
Capacitor/electric field storage	$E_{elec} = \frac{1}{2}CV^2$	C = capacitance. V = voltage.
Inductor/magnetic field storage	$E_{mag} = \frac{1}{2}LI^2 = \int \frac{B^2}{2\mu_0}dV$	L = inductance; I = current; B = magnetic flux density (Tesla).
Kinetic energy	$E_k = \frac{1}{2}Mv^2$	M = mass, v = velocity.
Rotary energy	$E_r = \frac{1}{2}I\omega^2$	I = mass moment of inertia (kg m^2). ω is rotary speed in radians/second.
Spring	$E_{spring} = \frac{1}{2}kx^2$	k = spring constant (N/m). x = spring displacement.
Potential energy	$\Delta E_p = Mg\Delta h$	Δh = height change.
Thermal energy	$\Delta E_T = C_{TH}\Delta T$	C_{TH} = thermal capacitance (J/K). ΔT is change in temperature (K).

What does this mean about the maximum magnitude of the inductor current? Well, we can solve for the ratio of V_0/I_0 resulting in:

$$\frac{V_0}{I_0} = \sqrt{\frac{L}{C}} \equiv Z_0 \tag{2.69}$$

The term "Z_0" is defined as the characteristic impedance of a resonant circuit. Let's assume that we have an inductor–capacitor circuit with $C = 1\,\mu\text{F}$ and $L = 1\,\mu\text{H}$. This means that the resonant frequency is 10^6 rad/s (or 166.7 kHz) and that the characteristic impedance is $1\,\Omega$. Shown in Figure 2.29 is a simulation of this circuit with initial condition of 1 V capacitor voltage and zero inductor current. As expected, the resonant frequency is 166 kHz and the maximum inductor current is 1 A. Note that the capacitor voltage and inductor current are 90° out of phase. This is characteristic of an undamped resonant circuit operating at resonance.

We could also use energy methods to analyze the oscillation frequency of the simple mass–spring system evaluated before. The maximum potential energy stored in a spring is:

$$E_p = \frac{1}{2}k(Y_0)^2 \tag{2.70}$$

The maximum kinetic energy stored in the moving mass is:

$$E_k = \frac{1}{2}Mv^2 = \frac{1}{2}M(\omega Y_0)^2 \tag{2.71}$$

If we equate maximum kinetic and potential energy, the result is:

$$\frac{1}{2}k(Y_0)^2 = \frac{1}{2}M(\omega Y_0)^2 \rightarrow \omega = \sqrt{\frac{k}{M}} \tag{2.72}$$

(a)

Parallel undamped LC with initial conditions 1 V on capacitor and zero inductor current

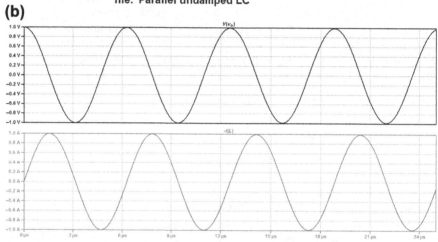

.tran 0 25 us 0 1 n
file: Parallel undamped LC

(b)

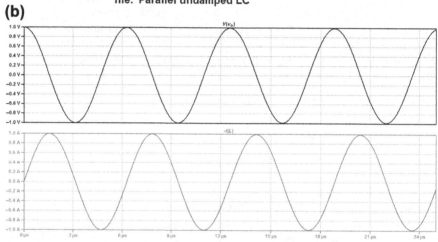

FIGURE 2.29

LTSPICE output for parallel undamped LC circuit (a) with $L = 1$ µH and $C = 1$ µF and initial conditions with the capacitor charged to 1 V and zero inductor current. (b) The resultant oscillation frequency is ~ 167 kHz. (For color version of this figure, the reader is referred to the online version of this book.)

Risetime for cascaded systems

For multiple systems in cascade (Figure 2.30), we can estimate out the overall system risetime if we know the risetime of the individual blocks. The risetimes

FIGURE 2.30

Cascaded systems.

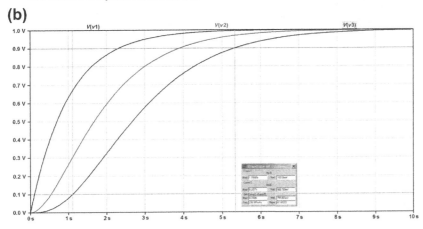

(a)

.tran 0 10 0 0.0001
file: Risetime of systems in cascade.asc

(b)

FIGURE 2.31

Three identical cascaded systems. (a) LTSPICE circuit. (b) Risetimes of three node voltages V_1, V_2, and V_3. Voltage V_3 has a measured 10–90% risetime of 4.27 s. (For color version of this figure, the reader is referred to the online version of this book.)

do not simply add; for instance, for N systems wired in series, each with its own risetime τ_{R1}, τ_{R2}, ...,τ_{RN}, the overall risetime of the cascade τ_R is:[16]

$$\tau_R \approx \sqrt{\tau_{R1}^2 + \tau_{R2}^2 + \tau_{R3}^2 + \cdots \tau_{RN}^2} \qquad (2.73)$$

Using the risetime addition rule, we find that a cascade of N identical stages has an estimated risetime equal to $\sqrt{N}$ times the risetime of an individual stage.

Example 2.3: Risetime of three systems in cascade

In Figure 2.31(a), we have three identical RC circuits, buffered in-between with ideal buffers. The risetime of each first-order system is 2.2 s (corresponding to

---

[16]A detailed mathematical treatment for risetimes of cascaded systems can be found in William Siebert's *Circuits, Signals and Systems*, Chapter 16, and also in Tom Lee's *The Design of CMOS Radio Frequency Integrated Circuits*, Chapter 7. Note that the risetime addition rule only holds true if the step responses of the individual systems are well behaved, which is discussed in more detail in Tom Lee's book. Furthermore, the risetime addition rule is approximate.

an RC time constant of 1 s). In Figure 2.31(b), we see how the risetime of the voltage of each successive stage increases. Using our rule for estimating the rise-time of cascaded systems (which is approximate) we calculate the risetime of voltage V_3, we calculate ($\sqrt{3} \times 2.2$), or a risetime of 3.8 s. LTSPICE shows that the risetime is actually 4.23 s.

Chapter 2 problems
Problem 2.1

A single-pole low-pass filter (Figure 2.32) has $R = 1$ kΩ and $C = 1000$ pF. Find the input–output transfer function $H(s)$ and calculate the 10–90% risetime and −3 dB bandwidth of this circuit.

Problem 2.2

Find and plot the angle and group delay of the single-pole filter of Problem 2.1.

Problem 2.3

Simulate the circuit of Problem 2.1 and verify your results in Problems 2.1 and 2.2.

Problem 2.4

Design a second-order RLC circuit with an input–output transfer function exhibiting a natural frequency $\omega_n = 10^6$ rad/s and a damping ratio $\zeta = 0.1$. What is the Q of this circuit?

Problem 2.5

For the circuit of Problem 2.4, plot the magnitude and angle of the transfer function. Correlate the simulated peak in the magnitude response with circuit Q.

Problem 2.6

A mass–spring system has mass $M = 10$ kg and spring constant $k = 10$ N/m. What is the oscillation frequency of this system?

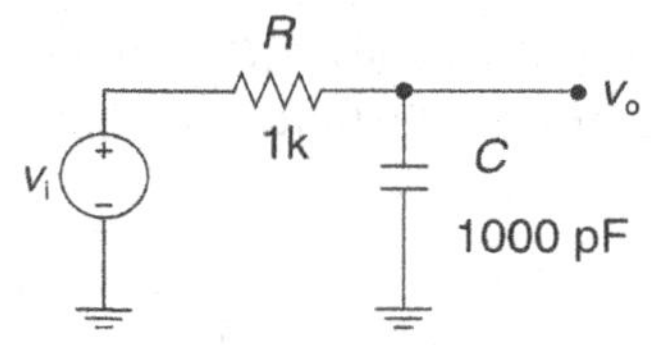

FIGURE 2.32

Single-pole filter for Problem 2.1.

Problem 2.7

An undamped parallel resonant circuit has $L = 10\ \mu H$ and $C = 100\ \mu F$. At time $t = 0$, the capacitor is charged to 1 V and the inductor current is zero. Find and plot capacitor voltage and inductor current for $t > 0$. What is the peak inductor current?

Problem 2.8

In the RC circuit of Figure 2.33, the circuit is driven by a 100-V input step $V_{in}(t) = 100u_{-1}(t)$.

a. Sketch and label resistor current $i(t)$ and capacitor voltage $v_c(t)$.
b. How much energy is dissipated in the resistor during the capacitor charging process?
c. Increase the resistor to 10 kΩ and repeat the experiment. After you apply the step and the capacitor is fully charged, how much energy is dissipated in the resistor?
   (*Hint: You can do (2) and (3) the easy way or the hard way. The easy way considers the energy stored and/or charge stored in the capacitor.*)
   The circuit with $R = 10{,}000$ is now driven by a 100-V square wave at 10 Hz (for $V_{in}(t)$, $V_{max} = 100$ V, $V_{min} = 0$ V).
d. Sketch and label resistor current $i(t)$ and capacitor voltage $v_c(t)$.
e. Approximately how much power is dissipated in the resistor?
f. Increase the frequency to 1 MHz. After many RC time constants, approximately how much power is dissipated in the resistor? (*Hint: What frequency components are in a 50% duty cycle square wave?*)
   Now, the circuit with $R = 10$ kΩ is driven with an AC voltage source $v_{in}(s)$.
g. Draw the Bode plot of v_o/v_{in}.
h. What is the -3 dB bandwidth, in Hz of v_o/v_{in}?
i. What is the 10–90% risetime?
j. Derive an expression relating the bandwidth and risetime.

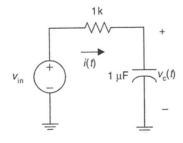

FIGURE 2.33

Circuit for Problem 2.8.

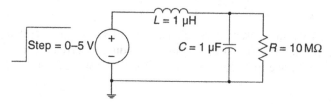

FIGURE 2.34

Circuit for Problem 2.9.

Problem 2.9

For the *LC* circuit in Figure 2.34,

a. Calculate the resonant frequency.
b. Sketch the step response of the capacitor voltage, labeling values and axes.

Problem 2.10

a. Find the 10–90% risetime of the low-pass filter of Figure 2.35(a).
b. The low-pass filter circuit from part (a) is cascaded with an identical circuit, with a buffer in-between to ensure that one stage does not load down the other. Find the 10–90% risetime of the overall cascade.
c. Verify these results using LTSPICE.

Problem 2.11

You have a series *RC* circuit configured as a one-pole low-pass filter and are attempting to ascertain the capacitance value of the circuit by measuring the step response.

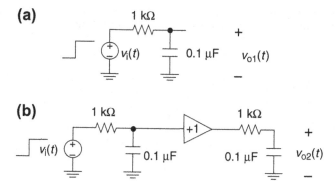

FIGURE 2.35

(a) Single-pole system for Problem 2.10. (b) Two cascaded first-order systems with unity gain buffer in-between.

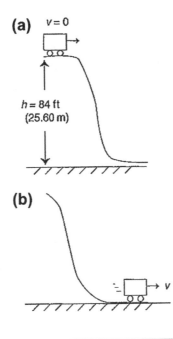

(a) $v = 0$

$h = 84$ ft
(25.60 m)

(b)

→ v

FIGURE 2.36

Energy method problem. (a) Roller coaster on top of 84-ft-high hill. (b) Roller coaster at the bottom of hill.

You know that the resistance value is 100 Ω. All you have available is an oscilloscope with a bandwidth of 100 MHz and a step generator with a very fast risetime. Using the oscilloscope, you measure a 10–90% risetime of 5 ns. What is the capacitance value? (Assume that the scope does not present any loading to the circuit.)

Problem 2.12

A roller coaster is initially at rest at the top of an 84-ft (25.6-m) tall hill (Figure 2.36(a)). The coaster is then gently pushed over the edge and travels to the bottom of the hill as shown in Figure 2.36(b).

a. What is the speed of the coaster (in meters/second) at the bottom of the hill, assuming that mechanical friction and air friction are negligible?
b. What is the speed of the coaster at the bottom of the hill, if the speed at the top of the hill is 10 m/s?

Further reading

[1] Andrews J. *Low-pass risetime filters for time domain applications*. Picosecond Pulse Labs, Application note #AN-7a; 1999.
[2] Beranek LL. *Acoustics*. Acoustical Society of America; 1954.

[3] *CRC standard mathematical tables.* 28th ed. CRC Press; 1987.

[4] Guillemin E. *Introductory circuit theory.* John Wiley; 1953.

[5] Johnson H. Risetime of lossy transmission lines. *EDN* October 2, 2003:32.

[6] Lee TH. The design of CMOS radio-frequency integrated circuits.

[7] Rao SS. *Mechanical vibrations.* 3rd ed. Addison-Wesley; 1995.

[8] Roberge J. *Operational amplifiers: theory and practice.* John Wiley; 1975.

[9] Senturia SD, Wedlock BD. *Electronic circuits and applications.* reprinted by Krieger; 1993.

[10] Siebert WMcC. *Circuits, signals and systems.* McGraw-Hill; 1986.

Review of Diode Physics and the Ideal (and Later, Nonideal) Diode

3

"Seek simplicity and distrust it."

—**Alfred North Whitehead**

IN THIS CHAPTER

▶ The basics of bipolar devices are covered, including basic semiconductor physics;[1] the concepts of electron and hole flow in semiconductors; the differences between drift and diffusion flow, generation, and recombination; and the effects of semiconductor doping on carrier concentrations. We finish with a discussion of the ideal diode and illustrate how a diode not only can conduct forward current but also can block reverse voltage. Detailed mathematical derivations are avoided wherever possible. However, enough mathematical detail is given so that the reader can discern the important scaling laws and functional dependencies of the ideal diode. At the end of the chapter, we will discuss some of the factors that result in nonideal behavior in diodes. We will conclude with a discussion of load lines, a useful method for solving for the operating point of circuits with nonlinear devices. The load-line technique will be useful in later chapters in analyzing transistors.

[1]We will not go into the quantum mechanics of semiconductors, which provides the rigorous analyses. The simpler models developed in this chapter hopefully will give insight into the basics of semiconductor operation. Excellent reviews of semiconductor physics are given in Shockley's and Bardeen's 1956 Nobel prize lectures, with reference given at the end of this chapter.

Current flow in insulators, good conductors, and semiconductors

In nature, from the point of view of the ease of producing current flow in a material, there are three broad classes of materials: insulators, conductors, and semiconductors. Semiconductors and metals can support significant current flow but the charge movement mechanisms are different in the two types of materials. A measure of how "good" an electrical conductor is can be quantified by the material property electrical resistivity and/or its inverse, electrical

Intuitive Analog Circuit Design. http://dx.doi.org/10.1016/B978-0-12-405866-8.00003-6

conductivity. Electrical resistivity[2] is a measure of how well a given material conducts current. If there are lots of free charged carriers available, a material is deemed a good conductor.

Insulators are materials such as quartz, rubber, plastics, and certain ceramics that do not support current flow very well. In other words, the electrical resistivity of an insulating material is very high. The resistance to current flow is very high in insulators because there are very few free charges available to contribute to current flow.

Good conductors support current flow easily. In other words, to get substantial current flow in a good conductor you do not have to supply very much driving voltage. Some metals, such as copper, aluminum, gold, and silver, are very good conductors because they have a sea of free electrons, each with a negative charge $-q$ associated with them,[3] available to support current flow. Typical metals have electrical conductivity many orders of magnitude higher than those of semiconductors or insulators. Most metals are good conductors; however, none of them are perfect conductors,[4] at least at room temperature.

Let us do a simple calculation of the resistance of a copper rod (Figure 3.1). The electrical resistance of the rod[5] is:

$$R = \frac{l}{\sigma A} \tag{3.1}$$

where l is the length, A is the cross-sectional area, and σ is the electrical conductivity (which has units of $\Omega^{-1}\mathrm{m}^{-1}$). Let us connect a battery to the metal rod as shown in Figure 3.1. The current flow is left to right in the metal rod (corresponding to electron flow from right to left[6]). For a metal rod of length 0.1 m and cross-sectional area 10^{-4} m^2, the resistance at room temperature[7] is approximately 16.9 μΩ.

In terms of electrical resistivity, somewhere in-between conductors and insulators are semiconductors. These materials are moderately good at supporting current flow. Typical semiconductors are silicon (Si), germanium (Ge), and gallium arsenide

---

[2]If you look in technical references (such as the *Handbook of Chemistry and Physics*), you can find electrical conductivity listed for materials. Electrical resistivity has units of ohms-meters. Electrical conductivity is the inverse of electrical resistivity.

[3]The charge on an electron is $-q$, where $q \approx 1.6 \times 10^{-19}$ C.

[4]There are compounds, such as niobium–titanium, which become superconducting at cryogenic temperatures (e.g. liquid helium temperature, at 4.2 K). Scientists are searching for superconductors (conductors whose electrical resistivity goes to zero if the temperature is lowered below a critical temperature). As of yet, no room-temperature superconductors have been discovered. The electrical conductivity of copper at room temperature is approximately $5.9 \times 10^7/\Omega$ m, while that of aluminum is approximately $3.5 \times 10^7/\Omega$ m. Note that these values of electrical conductivity vary significantly with the specific alloy and that electrical conductivity of metals increases as operating temperature decreases.

[5]This calculation ignores any resistance due to contacts at the ends of the rod.

[6]Note that the charge on the electron is negative. Hence, when electrons flow from right to left, the net current flow is from left to right. The directions of current flow are shown by convention.

[7]Electrical resistivity is the inverse of electrical conductivity. The electrical resistivity of copper at room temperature is approximately 1.7×10^{-8} Ω m. The temperature coefficient of resistivity of copper is approximately $+0.4\%/°C$.

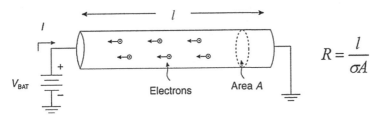

$$R = \frac{l}{\sigma A}$$

FIGURE 3.1

Metal rod with current flow due to applied battery voltage. In the metal, electric current is carried by the flow of electrons. Electrons flowing from right to left carry a negative charge, and hence the net current flow is from left to right.

(GaAs). An important difference between semiconductors and metallic conductors is the method by which charged carriers convey current. In semiconductors, two types of charged carriers contribute to current flow. They are electrons (negatively charged particles with charge $-q$) and holes (positively charged particles with charge $+q$). An *intrinsic* semiconductor is a semiconductor material that has not been doped with impurities. In the next section, we will discuss electrons and holes in semiconductors.

Electrons and holes

Shown in an idealized two-dimensional way in Figure 3.2 is a silicon lattice.[8] Each silicon atom has four outer shell ("valence band") electrons and four shared outer electrons with the nearest atomic neighbors. Hence there are a total of eight "covalent" bonds per atom. This idealized model of the silicon lattice and the resultant concepts of electron and hole motion gloss over many subtleties associated with quantum mechanical effects. However, this idealized model has been proved to be useful, and surprisingly accurate, in explaining current flow in diodes and transistors.

---

[8]The structures of silicon and germanium lattices are actually diamond-type crystal lattices with atoms having covalent bonds to four nearest neighbors. This picture is an idealized, flattened-out depiction of the actual lattice. This method of explaining approximately carrier flow in semiconductors was put forth by physicist William Shockley in the early days of semiconductors (about 1950). The laws of physics that explain electronic bonding are quantum mechanical in nature, and hence Schrodinger's wave equation must be solved. Shockley in describing carrier motion in transistors says, "The laws of physics which explain the behavior of the electron-pair bond are *quantum* laws and employ *wave-mechanics* to describe the motion of the electrons." See W. Shockley *Transistor Electronics: Imperfections, Unipolar and Analog Transistors*. Hence the models shown in Figure 3.2 and Figure 3.3 are a bit cartoonish in nature; however, they capture the essence of the physics involved in carrier flow in semiconductors.

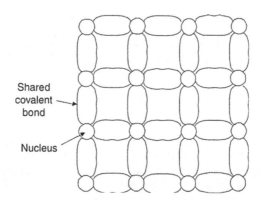

FIGURE 3.2

Structure of idealized 2D silicon lattice with all covalent bonds filled. Note that the actual 3D lattice has a "diamond cubic" shape.

In the ideal lattice, all the covalent bonds are filled and the associated electrons are tightly bound to their neighboring nuclei. Therefore, no "free" charges are available to contribute to current flow.[9]

Now, in a real-world semiconductor, some fraction of these covalent bonds breaks due to the random thermal energy of the silicon lattice (Figure 3.3). If a voltage is applied to the sample, these free electrons can flow and support a net current. The resultant space where the electron was before the bond was broken is called a *hole*. A hole carries a net positive charge of $+q$, and holes can *also* move about in the silicon lattice. In other words, the hole behaves like a charge carrier much like the electron, except that the hole carries a positive charge. The total current flow in a semiconductor is hence a combination of the current due to free electrons and the current due to free holes, or:

$$J = J_e + J_h \tag{3.2}$$

where J is total current density (expressed as amps per square meter[10]) and J_e and J_h are the electron and hole currents, respectively.

In a perfect silicon lattice without any added impurities, which we will call an *intrinsic semiconductor*, the number of free electrons and the number of free holes are equal, because free electrons and holes are created in pairs. We will call the number of electrons per unit volume in an intrinsic semiconductor n and the number of holes per unit volume p. From this simple reasoning,

---

[9]This is how a semiconductor behaves at very low temperatures. At cryogenic temperatures, for example, few outer shell bonds are broken (since the lattice is so cold and there is not enough thermal energy to shake electrons free).

[10]It is simple to find the total current in a semiconductor if you can calculate the current density J. Just multiply current density J (with units of amperes per square meter) by the total cross-sectional area A through which the current flows: $I = JA$.

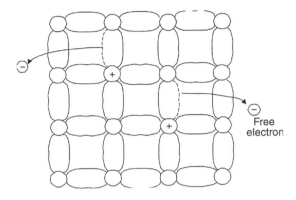

FIGURE 3.3

Silicon lattice with some broken covalent bonds. Each broken bond allows a negatively charged electron and a positively charged hole to move.

we find that the number of holes equals the number of electrons in an intrinsic semiconductor, or:

$$n = p \equiv n_i \tag{3.3}$$

We have now defined the *intrinsic carrier concentration* n_i, which is the number of bonds that are broken in the intrinsic semiconductor.

Furthermore, we also find that, as the temperature rises, the number of broken bonds increases since there is more thermal agitation. Therefore, in the intrinsic semiconductor, the intrinsic carrier concentration (we will call it $n_i(T)$) is a function of temperature,[11] or:

$$np = n_i^2(T) \tag{3.4}$$

In silicon, a typical value for intrinsic carrier concentration n_i at room temperature is $\sim 1.5 \times 10^{10}/cm^3$. This intrinsic carrier concentration is a strong function of temperature and increases with temperature.

Let us do a simple thought experiment and connect a battery to a piece of semiconductor material, in this case, silicon (Figure 3.4). The intrinsic silicon has some fraction[12] of its total outer shell bonds broken and hence there is concentration n_i of free electrons and free holes in the sample. These free charges can move and

---

[11]This result is called the "mass action" law and has a counterpart in chemistry. For instance, in an acidic solution, the concentrations of hydrogen (H^+) and hydroxyl (OH^-) also follow the mass action law.

[12]Let us take a look at the fraction of total bonds that are broken. Silicon contains about 5×10^{22} atoms/cc of volume. At room temperature, there are about $n_i \approx 1.5 \times 10^{10}$ free electrons and free holes per cubic centimeter. So, a *very* small fraction (approximately one part in 3.33×10^{12}) of total electron bonds are broken. However, this small fraction has a drastic effect on current flow in semiconductors.

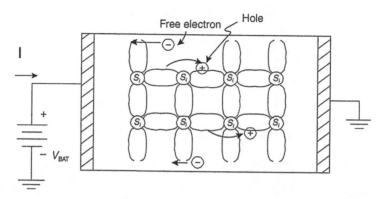

FIGURE 3.4

Battery connected to a piece of semiconductor (silicon, Si) material. The semiconductor has current flow both due to negatively charged electrons and positively charged holes. In the picture shown, electrons move from right to left and holes move from left to right, due to the force on the charges from the applied electric field from the battery.

contribute to conduction current. The electric field set up by the battery causes the free electrons to flow to the left and causes holes to flow to the right.[13] This total current adds up to the battery terminal current I as shown.

Drift, diffusion, recombination, and generation

When a charge carrier is free in a semiconductor, several things can happen to it. It can *drift* due to a force applied to it from an electric field. It can *diffuse* if there is a density gradient. If an electron meets a hole, the electron and hole can *recombine* wherein the electron fills the hole. While this is happening, new carriers are being *generated* due to thermal agitation or external excitation.[14] All four mechanisms happen simultaneously in a semiconductor. The details of the balance of drift, diffusion, recombination, and generation tell us the net current in a piece of semiconductor.

Drift

Shown in Figure 3.5(a) is an electron in an applied electric field **E**. The electron has an electrostatic force acting on it. In this picture, the force acts from right to left if the

---

[13]We can think of a hole as a void that gets filled when a covalent bond jumps from one location to another. In this simplified way, we can think of a "hole" as a current-carrying particle.

[14]For instance, if you shine light on a piece of semiconductor or heat it up, you will generate electron/hole pairs.

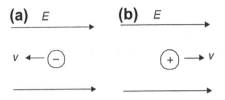

FIGURE 3.5

Charged carriers in an electric field. (a) Electron drift, in a direction opposite to that of the electric field. (b) Hole drift, in the same direction as the electric field.

electric field vector **E** points from left to right. The electron has an average velocity given by:

$$v = -\mu_n \mathbf{E} \tag{3.5}$$

where μ_n is a material constant[15] called the *mobility* of the electron.[16] The velocity has a minus sign because the drift motion of the electron is opposite of the direction of the applied electric field. Figure 3.5(a) shows a hole in an applied electric field **E**. The hole has a force acting on it in the same direction as the applied field. The velocity of the hole is given by:

$$v = \mu_p \mathbf{E} \tag{3.6}$$

where μ_p is the mobility of the hole. Note that the mobilities of electrons and holes are different. Typically, the mobility of electrons is two to three times higher than that of holes in semiconductors.[17]

Now, let us consider the case where we have a lot of holes in the presence of an electric field. The drift current density due to holes is:

$$J_{h,drift} = q\mu_p p \mathbf{E} \tag{3.7}$$

where p is the concentration of holes (holes per unit volume) and **E** is the electric field. Similarly, drift current density due to electrons is:

$$J_{e,drift} = q\mu_n n \mathbf{E} \tag{3.8}$$

---

[15]Note that this "constant" is not really constant at all. It varies somewhat with temperature, doping level, and level of the applied electric field. You might ask why the electron has an average velocity v and why it does not keep accelerating to higher and higher velocity. A perfectly periodic lattice at absolute zero would not scatter a free electron. However, above absolute zero, the lattice vibrates and the moving electron bangs into the lattice, slowing it down. The moving electrons can also be scattered by impurities in the lattice. The average time between collisions is denoted τ_c. The electron reaches an average drift velocity proportional to the electric field. The mobility of a free particle is given by $\mu = q\tau_c/m_{eff}$ where q is electronic charge and m_{eff} is the effective mass of the moving particle.

[16]Mobility has units of square centimeters per volt-second.

[17]In silicon, at modest doping levels and electric fields, typical numbers for mobilities are $\mu_n \approx 1360$ cm^2/V s and $\mu_p \approx 500$ cm^2/V s. See, e.g. R. W. Pierret, *Modular Series on Solid State Devices*, vol. 1, "*Semiconductor Fundamentals*", p. 58.

Diffusion

Fick's law of diffusion describes how particles under random thermal motion tend to spread[18] from a region of higher concentration to a region of lower concentration. This principle is illustrated by opening a perfume bottle in the corner of a closed room. If you wait long enough, the perfume odor will permeate the room because the perfume molecules have diffused from one side of the room to the other, from a region of high concentration to a region of low concentration. Mathematically, three-dimensional diffusion is characterized by Fick's diffusion law, which states that the diffusion flux is proportional to the concentration gradient, as:

$$F = -D\nabla C \qquad (3.9)$$

where C is the concentration of the diffusing particles, F is the diffusion flux (particles per square meter per second), and D is the diffusion constant, which has units of cm^2 per second. For a one-dimensional problem, Fick's law reduces to:

$$F = -D\frac{dC}{dx} \qquad (3.10)$$

Therefore, charged particles tend to flow down a concentration gradient. This diffusion process also occurs in PN junctions whenever there are gradients of free charged carriers.

We can work out the form (but not the detail) of Fick's law by considering a thought experiment. Consider a region of space where there is a changing concentration of free charges, in this case, holes (Figure 3.6). The holes are undergoing random thermal motion. For instance, at $x = -x_0$, on an average, one half of the holes are traveling to the left and one half are traveling to the right. The same is

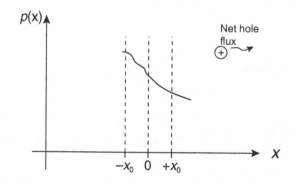

FIGURE 3.6

Hole concentration gradient, resulting in a net hole flux from left to right.

---

[18]A good, intuitive explanation of diffusion is given in R. W. Pierret, *Modular Series on Solid State Devices*, vol. 1, "*Semiconductor Fundamentals*", pp. 67–71 and also in R. Feynman et al., *The Feynman Lectures on Physics*, vol. 1, Chapter 43.

true at $x = +x_0$. In order to find the net current at $x = 0$, we recognize that the current at $x = 0$ is the sum of the current from the left plus the current from the right, or:

$$J(x = 0) = k[p(x = -x_0) - p(x = +x_0)] = -k(2x_0)\left[\frac{dp}{dx}\right] \qquad (3.11)$$

where k is some constant that makes the units work out. Note that the net current at $x = 0$ is proportional to the difference in the concentrations at $x = -x_0$ and $x = +x_0$. Through a mathematical manipulation[19] above we see that the current is also proportional to dp/dx or the gradient of the concentration. We now recognize the familiar form for diffusion current for holes:

$$J_{\text{h,diff}} = -qD_p\frac{dp}{dx} \qquad (3.12)$$

where q is the electronic charge and D_p is the diffusion constant[20] for holes. Using a similar derivation, we can find the electron diffusion current:

$$J_{\text{e,diff}} = qD_n\frac{dn}{dx} \qquad (3.13)$$

Let us do a diffusion thought experiment, illustrated in Figure 3.7. At time $t = 0$, a high concentration of particles (in this case, electrons) exists at $x = 0$. These particles can be created by illuminating a piece of semiconductor, or by other mechanisms. The particles are in random thermal motion; some diffuse to the left and to the right.

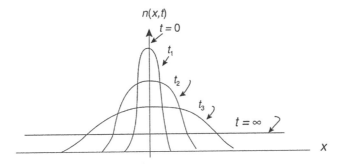

FIGURE 3.7

Illustration of a diffusion thought experiment. At time $t = 0$ there is a very high concentration of particles at $x = 0$. The particle concentration spreads out and varies with time as shown, with $t_3 > t_2 > t_1$, etc.

[19]This hand-waving derivation glosses over details of collisions and mean free path, etc., which is well discussed in Feynman's lecture 43; see reference at the end of this chapter.

[20]Approximate values for diffusion constants in silicon at room temperature are $D_n \approx 35$ cm²/s and $D_p \approx 12.5$ cm²/s. Interestingly enough, the diffusion and drift coefficients satisfy the "Einstein relation", which relates the values of diffusion and mobility coefficients as: $\frac{D_p}{\mu_p} = \frac{D_n}{\mu_n} = \frac{kT}{q}$. We see cropping up again and again in semiconductor physics the "thermal voltage" kT/q, which is approximately 26 mV at room temperature.

The concentration of particles $n(x, t)$ at various times is shown. At $t = t_1$, the maximum concentration at $x = 0$ has dropped, and the particles have spread to the left and the right. Further smearing of the particle concentration occurs at t_2 and t_3. As time reaches infinity, the concentration is the same everywhere, and diffusion ends. A closed-form solution for this diffusion problem exists;[21] the electron concentration everywhere is:

$$n(x, t) = \frac{A}{\sqrt{4\pi D_n t}} e^{\frac{x^2}{4D_n t}} + n_0 \qquad (3.14)$$

where A is a constant, D_n is the diffusion constant for electrons, and n_0 is the equilibrium concentration of electrons.

In a famous experiment devised in 1949 called the Shockley–Haynes experiment[22] (Figure 3.8(a)), an area in a piece of semiconductor material was illuminated while an electric field E_x was applied. The electric field is proportional to

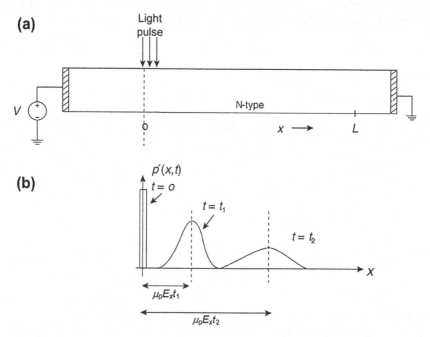

FIGURE 3.8

(a) Cartoon illustration of the Haynes–Shockley experiment. (b) The concentration of extra holes ($p'(x, t)$) moves to the right with speed $\mu_p E_x$ where μ_p is the mobility of the holes.

---

[21] See, e.g. R. B. Adler et al., *Introduction to Semiconductor Physics*, SEEC Volume 1, p. 174, 1964.
[22] This was the first direct observation of minority carrier drift and diffusion dynamics in semiconductors. Described in great detail in Shockley's Nobel lecture (1956); J. R. Haynes and W. Shockley, *Physical Review*, vol. 75, p. 691 (1949); and J. R. Haynes and W. Shockley, "The Mobility and Life of Injected Holes and Electrons in Germanium", *Physical Review*, vol. 81, p. 835, 1951. It is also described in SEEC Volume 1, referenced earlier.

the voltage applied to the left side of the sample, and the electric field points from left to right.

At time $t = 0$, a light source is turned on and creates a high concentration P_0 of holes at $x = 0$. The light is turned off and the concentration of holes diffuses due to the concentration gradient and drifts to the right due to the applied electric field. A detector is used to measure the hole concentration at $x = L$. The solution for the hole concentration is given by:

$$p(x, t) = \frac{P_0}{2\sqrt{\pi D_p t}} e^{-\left[\frac{(x - \mu_p E_x t)^2}{4 D_p t} + \frac{t}{\tau_p}\right]} + p_{no} \tag{3.15}$$

The evolution of the concentration of holes in the sample is shown in Figure 3.8(b). Using this result we can find the drift coefficient for holes μ_p by viewing how fast the pulse center moves from left to right. The drift coefficient is:

$$\mu_p = \frac{v_x}{E_x} \tag{3.16}$$

where v_x is the velocity of the center of the pulse from left to right. The diffusion coefficient D_p can be found by viewing how fast the pulse spreads.

Generation and recombination

What happens when a free electron traveling through the lattice happens to meet a hole? The electron will tend to fill the hole, and the electron and hole effectively annihilate each other. This process is known as *recombination*. Similarly, the process by which carrier pairs are created is called *generation*. Carrier pairs can be created by heating up the lattice or by illuminating the semiconductor with light. It must be understood that in equilibrium in a semiconductor, the processes of generation and recombination are always going on and are in balance.

Comment on total current in semiconductors

In a semiconductor, the total current flowing is a combination of the currents due to drift and the currents due to diffusion. What makes this a complicated process is that in a given transistor, there are drift and diffusion components due to both holes and electrons. So, there are four components of total current that we need to consider. We will look at how this works later on, but for now let us consider the effects of semiconductor doping.

Effects of semiconductor doping

When a semiconductor material is doped, very small quantities of selected impurities are selectively added, which significantly alters the balance of holes and electrons that exists in an intrinsic semiconductor.

Donor-doped material

Typical dopants for N-type semiconductors are periodic table (Figure 3.9) column V elements such as phosphorus and arsenic. These column V elements have five outer shell electrons. When a phosphorus atom displaces a silicon atom in the lattice, there is one extra outer shell electron available; this extra electron is fairly free to move about the lattice and acts as a mobile particle with a net negative charge of $-q$.

Doping silicon with phosphorus or arsenic (Figure 3.10(a)) increases the free-electron population above the intrinsic electron concentration n_i. Another effect is to lower the equilibrium concentration of holes; for high doping levels, there are lots of free electrons available to recombine with holes and hence the population of holes goes down. The net effect of all of this is an N-type material at equilibrium with significant doping[23] ($N_D \gg n_i$). The equilibrium carrier concentrations are:[24]

$$n_{no} \approx N_D$$

$$p_{no} \approx \frac{n_i^2}{N_D} \tag{3.17}$$

III	IV	V	VI
5	6	7	8
B	C	N	O

	13	14	15	16
IIB	Al	Si	P	S

30	31	32	33	34
Zn	Ga	Ge	As	Se

48	49	50	51	52
Cd	In	Sn	Sb	Te

FIGURE 3.9

Section of the periodic table, showing column III and column V elements, which are commonly used as dopants in semiconductors. "Acceptor" dopants are from column III (boron, aluminum, and gallium) and "donor" dopants are from column V (phosphorous and arsenic).

[23] This assumes that the doping is not so high that the semiconductor becomes *degenerate*. For a discussion of degenerate semiconductor statistics, see, e.g. Robert F. Pierret, *Modular Series on Solid State Devices*, volume 1, "*Semiconductor Fundamentals*", published by Addison-Wesley, Reading Mass., 1983.

[24] In this terminology, n_{no} is the equilibrium concentration of free electrons in an N-type material, and p_{no} is the equilibrium concentration of holes in the N-type material. This same nomenclature is used in Paul Gray and Campbell Searle, *Electronic Principles Physics, Models and Circuits*, published by John Wiley, 1967.

(a)

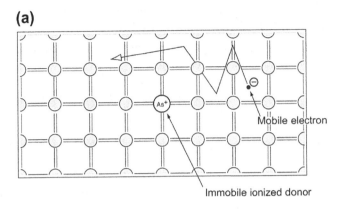

Immobile ionized donor

(b)

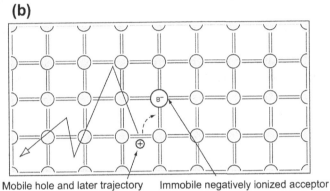

Mobile hole and later trajectory   Immobile negatively ionized acceptor

FIGURE 3.10

Donor- and acceptor-doped semiconductors. (a) Donor-doped semiconductor, showing free electrons that can contribute to current flow. (b) Acceptor-doped semiconductor, with free hole.

Acceptor-doped material

Boron is in column III of the periodic table of elements and hence has three outer shell electrons and is a typical dopant for P-type semiconductors. Boron is therefore an "acceptor" donor. When a boron atom displaces a silicon atom in the lattice (Figure 3.10(b)), there is one less outer shell electron available to generate covalent bonds. The result is that there is a quantum "void" called a "hole", which acts as a mobile particle with a net positive charge. For P-type semiconductors, the equilibrium hole and electron concentrations are:

$$p_{po} \approx N_A$$

$$n_{po} \approx \frac{n_i^2}{N_A}$$

$$(3.18)$$

PN junction under thermal equilibrium

We can now put together these concepts of carrier flow, creation, and annihilation—that is, drift, diffusion, generation, and recombination—to show how a diode operates. First we will consider the case of "thermal equilibrium"; the diode has no connections and no current flow and is sitting on a laboratory bench. Furthermore, there is no external excitation on the diode, such as a light source.

Let us consider an idealized PN diode structure. A one-dimensional model of a bipolar junction diode, composed of adjacent P- and N-regions with conducting contacts on either end is shown in Figure 3.11(a), along with the diode circuit model. The anode of the diode is the connection to the P-region, and the cathode is the connection to the N-region. We will discuss the "depletion region" as shown in the diagram near the PN junction later on. Let us consider an "abrupt" junction[25] where the doping changes abruptly at $x=0$ as shown in Figure 3.11(b). On the P-side of the junction is a concentration of N_A acceptors and on the N-side of the junction is a concentration of N_D donors.

Initially, let us consider the behavior of this diode without any contacts at either end, and hence no applied voltage. Let us do a thought experiment and

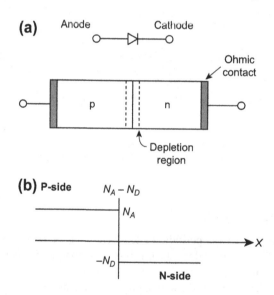

FIGURE 3.11

Ideal PN junction geometry. (a) Diode, showing anode side (P-side) and cathode side (N-side). (b) Doping levels on either side of the abrupt junction, with N_A acceptors on the P-side and N_D donors on the N-side. Near the junction is a "depletion region".

[25]This is an approximation to a junction where the doping levels change very abruptly from P-type to N-type at $x=0$.

consider what happens if we bring P-material in close proximity to the N-material. When the P-material is brought in proximity to the N-material, the high concentration of holes in the P-material tends to diffuse into the N-material and the high concentration of electrons in the N-material tends to diffuse into the P-material.

If diffusion was the only carrier motion process, eventually the concentrations of electrons and holes would be the same throughout the entire diode. However, when holes diffuse from the P-side to the N-side, fixed negative charges in the silicon lattice are uncovered. When electrons diffuse from the N-side to the P-side, fixed positive charges in the lattice are uncovered. Hence, near the junction there is a net negative charge on the P-side. By the same reasoning, there is a net positive charge near the junction on the N-side. The fixed negative charge on the P-side tends to oppose further diffusion of electrons from the N-side to the P-side. Likewise, the fixed positive charges on the N-side tend to oppose diffusion of holes from the P-side to the N-side. As we will see later on, we can quantify this effect by noting that a barrier electric field is set up, which opposes diffusion. The equilibrium concentration of electrons and holes throughout the diode is shown graphically in Figure 3.12.

This net charge near the junction creates an electric field. This field acts as a barrier to further current flow across the junction. The charge profile, electric field, and potential for this junction at equilibrium are shown in Figure 3.13.[26] In equilibrium, there is a delicate balance between drift and diffusion in the diode. There are four current components (drift and diffusion for electrons and holes), but in the end, there is no diode current.

This drift–diffusion process results in a region near the junction that is nearly devoid of free carriers. This so-called depletion region extends from $-x_p$ to $+x_n$ as shown in the diagrams.

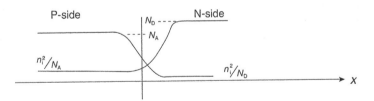

FIGURE 3.12

Carrier concentrations in a PN junction diode without any applied voltage bias. On the N-side, the concentration of electrons is approximately N_D far away from the barrier and the concentration of holes is n_i^2/N_D. On the P-side, the concentration of holes far away from the barrier is N_A and the concentration of electrons far from the barrier is n_i^2/N_A.

---

[26]Note that these pictures are drawn for an abrupt junction, where it is assumed that doping on the N-side is N_D and doping on the P-side is N_A.

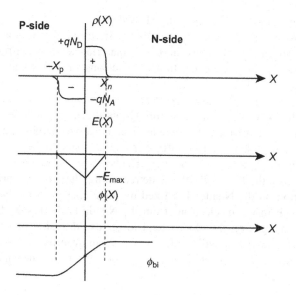

FIGURE 3.13

Ideal PN junction in thermal equilibrium, in a one-dimensional approximation. These plots show the depletion region charge density $\rho(x)$, electric field $E(x)$, and junction potential $\phi(x)$.

We can find the electric field $E(x)$ by using the one-dimensional form of Gauss's law, which is:

$$\frac{dE(x)}{dx} = \frac{-\rho(x)}{\varepsilon_{Si}} \qquad (3.19)$$

where $\rho(x)$ is the electric charge density and ε_{Si} is the dielectric permittivity of the silicon material. Since the charge density $\rho(x)$ is piecewise constant on the P- and N-sides, the electric field is piecewise linear on the P- and N-sides, as shown. The maximum electric field is found at the junction at $x = 0$ and is given by:

$$|E_{max}| = \frac{qN_D x_p}{\varepsilon_{Si}} = \frac{qN_A x_n}{\varepsilon_{Si}} \qquad (3.20)$$

We can find the junction potential similarly by integrating the electric field, since

$$\frac{d\phi(x)}{dx} = -E(x) \qquad (3.21)$$

At zero bias, there is a voltage ϕ_{bi}, called the *built-in potential*, across the depletion region of the junction. This voltage is a result of an electric field that opposes the diffusion of mobile electrons and holes across the junction. The built-in potential is given by:

$$\phi_{bi} = \frac{kT}{q} \ln\left(\frac{N_A N_D}{n_i^2}\right) \qquad (3.22)$$

We will see that the term kT/q comes up over and over in semiconductors, where k is the Boltzmann's constant, T is the absolute temperature in Kelvin, and q is the electronic charge. This "thermal voltage" is approximately 26 mV at room temperature.

PN junction under applied forward bias

When we forward bias a diode, we apply a positive voltage between the anode (P-side contact) and the cathode (N-side contact) of the diode. Under forward bias, the delicate balance between drift and diffusion is altered significantly and net current flows in the diode. The application of a forward bias voltage (i.e. a positive potential applied to the P-side relative to the N-side) reduces the amplitude of the barrier electric field at the junction. When this barrier is reduced, it is easier to inject holes from the P-side to the N-side and electrons from the N-side to the P-side. This causes a net forward current to flow.

Shown in Figure 3.14 are the junction charge, electric field, and voltage under forward ($V > 0$), reverse ($V < 0$), and no bias ($V = 0$) conditions. When there is no bias—i.e. no voltage applied to the diode—the voltage across the diode depletion region is just the built-in potential ϕ_{bi}. There is no net diode current because all components of drift and diffusion current in the diode exactly cancel each other. The electric field barrier at the junction is as shown.

Under reverse bias, the applied voltage adds to the junction built-in potential ϕ_{bi}. This in turn means that, under reverse bias, the electric field barrier is higher, and the depletion region is wider. Under higher reverse bias, the depletion region width increases, and the maximum electric field increases as well. If you keep increasing the reverse bias, the electric field increases until junction breakdown is reached. Junction breakdown in silicon occurs at an electric field of the order of 3×10^5 V/cm.

Under forward bias, the electric field barrier is decreased by the applied voltage. This reduction in electric field barrier allows holes to be injected from the P-side to the N-side and electrons to be injected from the N-side to the P-side. The injected carriers then diffuse away from the junction, setting up a net diode current—called the diode *forward* current. As we will see later on, there is an exponential relationship between diode voltage and current when the diode is forward biased.

Shown in Figure 3.15 are the concentrations of minority carriers[27] on the N- and P-sides of the semiconductor, with the diode under forward bias. Under forward

---

[27]Let us discuss why the carriers shown are "excess minority carriers". Consider the concentration of n' electrons on the P-side of the diode. The n' electrons are minority carriers since electrons are in the minority on the P-side of the diode. This, of course, assumes that the number of carriers injected is small compared to the majority carrier concentration N_A, or $n' << N_A$. This condition is termed in the literature as *low-level injection*. We call the plotted carriers n' "excess" because these carriers are in excess of the equilibrium concentration n on the P-side. Remember that the equilibrium concentrations of carriers on the P-side are $p \approx N_A$ and $n \approx n_i^2/N_A$.

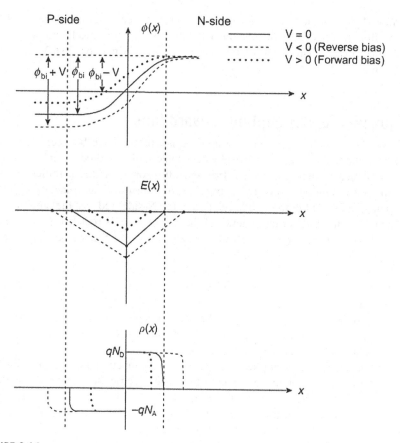

FIGURE 3.14

Depletion region under zero, forward, and reverse bias showing depletion region charge density $\rho(x)$, electric field $E(x)$, and junction potential $\phi(\chi)$. Under reverse bias, the width of the depletion region grows, and the maximum electric field grows as well. Under forward bias, the converse is true.

bias, the electric field barrier due to the depletion region is lowered, electrons are injected from the N-side to the P-side, and holes are injected from the P-side to the N-side. The injected electrons on the P-side diffuse to the left. The injected holes on the N-side diffuse to the right. This diffusion results in a net diode current from left to right. Detailed considerations[28] show us that there is an exponential

---

[28]See, e.g. Pierret's *Modular Series on Semiconductor Devices, Volume II, The PN Junction Diode*, pp. 48–50. The factor kT/q (with k = Boltzmann's constant, T = absolute temperature in Kelvin, and q = electronic charge) comes up again and again in semiconductors. The value of kT/q is approximately 26 mV at room temperature (300 K).

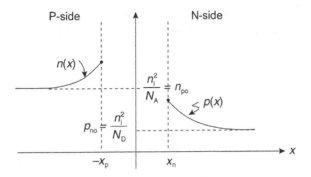

FIGURE 3.15

Minority carrier concentrations in a PN junction under forward bias. n_{po} is the concentration of electrons far from the junction on the P-side. p_{no} is the concentration of holes far from the junction on the N-side.

relationship between diode voltage and the excess minority carrier concentrations at the edge of the depletion regions, or:

$$n'\left(-x_p\right) = n(x) - n_{po} = A\left(e^{\frac{qV_D}{kT}} - 1\right)$$

$$p'(x_n) = p(x) - p_{no} = B\left(e^{\frac{qV_D}{kT}} - 1\right)$$

(3.23)

where A and B are constants having to do with doping levels and the intrinsic carrier concentration,[29] n_{po} is the electron concentration on the P-side far from the junction, and p_{no} is the hole concentration on the N-side far from the junction.

We are now in a position to piece together arguments for the exponential voltage–current relationship in a forward-biased diode. The keys to realizing this are as follows:

- Holes are injected from the P-side to the N-side, and electrons are injected from the N-side to the P-side.
- These injected holes result in a concentration gradient on either side of the depletion region.
- The minority carrier concentrations $n(x)$ and $p(x)$ at the edge of the depletion region follow an exponential relationship with the diode voltage V_D.
- This concentration gradient results in a diffusion current of excess minority carriers. In essence, the injected carriers diffuse away from the junction. As the

[29] Again, detailed considerations show that $A = n_i^2/N_A$ and $B = n_i^2/N_D$, where N_A is the acceptor concentration on the P-side and N_D is the donor concentration on the N-side.

carriers diffuse away, they recombine with majority carriers. The shape of the curve follows an exponential decay away from the junction as:[30]

$$p'(x) = p(x = x_n)e^{\frac{-x'}{L_p}}$$

• The minority carrier currents are dominated by diffusion. In other words, the minority diffusion current is larger than the minority drift current. This is due to the fact that the minority concentration is small (shown before due to the law of mass action). However, the minority concentration gradient is large and diffusion dominates.

The individual current components in a forward-biased diode are shown in Figure 3.16. Let us first focus on the electron current on the P-side, labeled J_e. We know that there is a concentration gradient of electrons near the depletion region, since these charges are injected from the P-side. This concentration gradient causes a diffusion current to flow. In essence, electrons diffuse away from the junction (to the left in the diagram). This is the current J_e as shown. As the excess carriers diffuse away, they recombine with holes in the p bulk material. This causes a net hole current J_h as shown.

It is assumed that the electron current J_e and the hole current J_h are constant through the depletion region. We assume that the depletion region is so depleted of charges that there is no significant recombination in this region.

Now, what about the exponential diode voltage–current relationship? Since we know that the total diode current is the sum of electron and hole currents and since we assume that electron and hole current is constant through the depletion region, we can write that the diode current is:

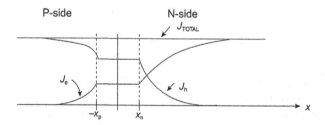

FIGURE 3.16

Various current components in a diode under forward bias. J_e is the current density (ampere per square meter) due to electron flow, J_h is the current density due to hole flow, and J_{total} is the total current density.

---

[30]L_p is the minority carrier diffusion length for holes. It is a measure of how far the holes diffuse into the n-bulk material before recombining. Numerically, $L_p = \sqrt{D_h \tau_h}$, where D_h is the diffusion constant for holes and τ_h is the minority carrier lifetime, or the characteristic lifetime for decay of excess carriers.

$$J_{\text{total}} = J_e(x = -x_p) + J_p(x = x_n) \qquad (3.24)$$

The individual components of current $J_e(x = -x_p)$ and $J_p(x = x_n)$ are dominated by diffusion, and we know that the concentration gradient depends on the junction voltage, or:

$$J_e(x = -x_p) \propto e^{\frac{qV_D}{kT}}$$
$$\qquad (3.25)$$
$$J_h(x = x_n) \propto e^{\frac{qV_D}{kT}}$$

Therefore, we see the derivation of the exponential voltage–current relationship for the ideal diode. The electron current (J_e), the hole current (J_h), and the total current through the diode is shown in Figure 3.16.

Reverse-biased diode

Under reverse bias, the minority carrier concentration at the edges of the depletion region is suppressed below their equilibrium values, as shown in Figure 3.17. As shown before, there is also a large electric field barrier that suppresses current flow across the junction. As we will see in the next section, the diode current under reverse bias is small, but finite.

The ideal diode equation

We showed before that the current densities J_e and J_h, measured at the edge of the depletion region, each have a value that depends exponentially on diode voltage. Since the total current is the sum of these two components, the total diode current has this same functional dependence. This result is summarized by the familiar Shockley equation, which is:

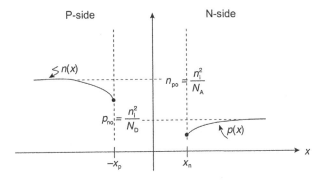

FIGURE 3.17

Minority carrier concentrations under reverse bias.

$$I_D = J_{TOTAL}A = I_s\left(e^{\frac{qV_D}{kT}} - 1\right) \tag{3.26}$$

where J_{TOTAL} is the current density (in amps/m^2), A is the junction area, and I_s is the *saturation current* with typical values of 10^{-15} A for a signal diode. Under forward bias, with $V_D \gg kT/q$, the diode current is approximately:

$$I_D \approx I_s e^{\frac{qV_D}{kT}} \quad \text{for } V_D \gg kT/q \tag{3.27}$$

For reverse voltages $V_D \ll -kT/q$, the diode current saturates at the reverse saturation current $-I_s$.

The total plot of diode current as a function of voltage (I_D vs. V_D) for the ideal diode is shown in Figure 3.18(a). An often-used approximate curve is shown in Figure 3.18(b). Sometimes the more realistic idealized curve of Figure 3.18(c) is used, which incorporates the diode "knee" voltage of approximately 0.6 V. You can also include the effect of any diode ohmic resistance (not accounted for by ideal diode analysis) by including a finite slope in the diode curve, as shown in Figure 3.18(d).

Detailed considerations show that the ideal diode V_D/I_D relationship is given by:

$$I_D = qAn_i^2\left(\frac{D_p}{N_D L_p} + \frac{D_n}{N_A L_n}\right)\left(e^{\frac{qV_D}{kT}} - 1\right) = I_s\left(e^{\frac{qV_D}{kT}} - 1\right) \tag{3.28}$$

where A is the junction area; D_p and D_n are diffusion constants for holes and electrons, respectively; and L_p and L_n are diffusion lengths for holes and electrons., respectively. The term I_s is the reverse saturation current of the diode.

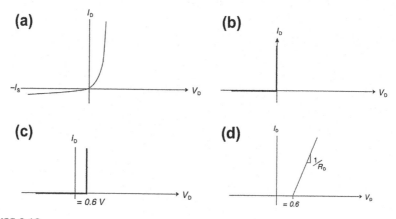

FIGURE 3.18

Plot of I_D vs. V_D for diode, showing various approximations. (a) Result for ideal diode showing exponential $V–I$ curve. (b) Often-used simple approximation where the diode turns on with zero voltage drop. (c) More realistic approximation including diode ON voltage. (d) Including diode ohmic resistance R_D.

Charge storage in diodes

Charge storage in diodes is accomplished via several mechanisms. First, there is charge storage in the depletion regions, under both forward and reverse bias conditions. This charge is stored in the dipole layer near the junction. The depletion layer is sometimes called a "space charge" layer.

Under forward bias, charge is stored in the neutral regions of the P-side and the N-side. We have seen that carriers are injected from one side of the junction to the other and then diffuse away; therefore, there is a "diffusion capacitance", which controls how much charge is stored in the neutral regions. We will discuss each of these charge storage mechanisms next.

Depletion capacitance

We have seen earlier that a variation in applied diode bias results in a change in the depletion width and hence a change in the charge in the depletion region. Let us revisit this from a graphical point of view, as shown in Figure 3.19. When the reverse bias on the diode increases from V_1 to V_2, the depletion region width increases. We can think of this as a change in charge stored in the depletion region. There is a net change in charge of $\Delta Q+$ on the N-side, and a change in charge of $\Delta Q-$ on the P-side. Capacitance is equal to the derivative of charge with respect to voltage, so we can find the junction capacitance as:

$$C_j = \frac{dQ}{dV} \tag{3.29}$$

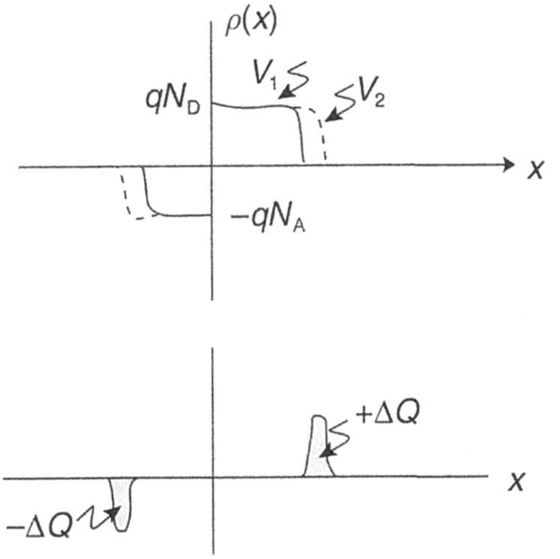

FIGURE 3.19

Illustration of cause of diode junction capacitance. As the reverse bias increases from V_1 to V_2, the depletion width increases, resulting in an increase in depletion charge $+\Delta Q$ and $-\Delta Q$ on either sides of the junction.

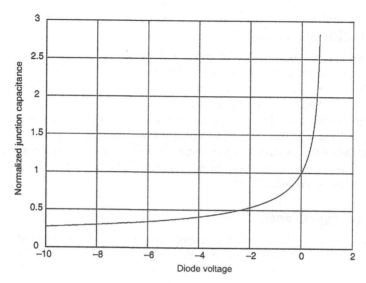

FIGURE 3.20

Nonlinear junction capacitance vs. diode voltage for abrupt junction diode, shown for a hypothetical diode with $\phi_{bi} = 0.8\,V$, $C_{jo} = 1$, and $m = 0.5$.

A detailed analysis shows that the depletion capacitance as a function of applied bias V_j can be expressed as:

$$C_j = \frac{C_{jo}}{\left(1 - \frac{V_J}{\phi_{bi}}\right)^m} \tag{3.30}$$

where $m = 1/2$ for an abrupt junction and $m = 1/3$ for a linearly graded junction. The term C_{jo} is the depletion capacitance at zero diode voltage, V_J is the anode–cathode voltage across the diode junction, and ϕ_{bi} is the built-in voltage. The depletion capacitance for a typical signal diode is plotted in Figure 3.20. Note that as the reverse bias voltage increases, the depletion capacitance decreases.[31]

Charge storage in the diode under forward bias

When the diode is forward biased and carrying forward current, there is charge stored in the neutral P- and N-regions. If the diode current increases, this charge must likewise increase. As the voltage applied to the diode increases, the diode current increases, and the charge stored in the *diffusion capacitance* increases as well. These effects are illustrated in Figure 3.21.

---

[31]This simple junction capacitance equation works well assuming that $V_D < \varphi_{bi}$. Note that this equation shows a junction capacitance that increases without bound as $V_D \to \varphi_{bi}$.

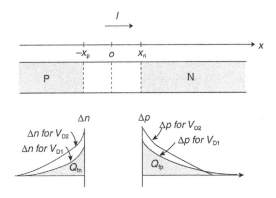

FIGURE 3.21

Excess minority carrier concentrations under forward bias, short diode. As the forward diode voltage increases from V_{D1} to V_{D2}, the electron and hole concentrations increase as shown.

The charge stored in the P-region is the area under the Δn curve and is called Q_{fn}. Likewise, the charge stored in the N-region is the area under the Δp curve, and is called Q_{fp}. Let us say that the diode forward voltage is increased from V_{D1} to V_{D2}. The height curves for the stored charges Δn and Δp also increase because more charges are injected through the depletion region. Correspondingly, the total stored charge in the diode (the sum of Q_{fn} and Q_{fp}) increases as the diode voltage increases from V_{D1} to V_{D2}. A change in charge with a change in diode voltage can be modeled as an equivalent capacitance. This capacitance is sometimes called the *diffusion capacitance* of the forward-biased diode.

These charges diffuse away from the junction and contribute to total diode current I as shown. Note that electrons diffuse from right to left and holes diffuse from left to right, yielding a net current from left to right.

The result of this is that the diffusion capacitance is proportional to the diode forward current, or:

$$C_d \propto I_D \qquad (3.31)$$

The details of the proportionality constant depend on the specific construction of the diode.

Reverse recovery in bipolar diodes

When a bipolar diode is ON and you abruptly reverse the voltage across the diode, you might expect the diode current to immediately drop to zero. However, if you perform this experiment and monitor the diode current, you will find that the diode current actually spikes negative for a period, called the *reverse recovery* time. This reverse recovery is a direct result of the charge stored in the diffusion capacitance of the diode. This charge must be removed before the diode can turn off.

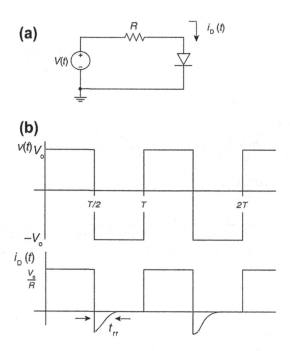

FIGURE 3.22

Test circuit illustrating reverse recovery. (a) Test circuit. (b) Waveforms showing driving voltage $v(t)$ and resultant diode current $i_D(t)$. When the driving voltage transitions negative, the diode current goes negative for a time t_{rr} (the reverse recovery time). During this reverse recovery interval, extra charges are swept out of the diode neutral regions as the diode shuts off.

An experiment illustrating reverse recovery is shown in Figure 3.22. A diode is repetitively pulsed by a voltage source $v(t)$ through a resistor R. Let us assume that the amplitude V_o of the pulse is much larger than the diode ON voltage of ~ 0.7 V. When the input voltage transitions positive, the diode current reaches a limit of $\sim V_o/R$, limited by the resistance.

Reverse recovery occurs after the input voltage transitions from $+V_o$ to $-V_o$ at time $t = T/2$. Looking at the diode current curve $i_D(t)$ we see that the diode current is negative for a time period t_{rr}, called the *reverse recovery time*. During this reverse recovery time, the charge stored in the diffusion capacitance is removed, turning the diode off.

Reverse breakdown

A reverse-biased diode carries a small reverse current. This is true until a large-enough reverse voltage, called the reverse breakdown voltage, is applied

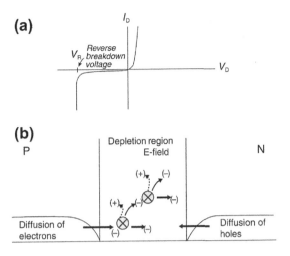

(b)

FIGURE 3.23

Diode reverse breakdown. (a) Diode $V–I$ curve, showing reverse breakdown when the diode reverse voltage reaches V_R. (b) Illustration of reverse breakdown process.

(Figure 3.23(a)). When the reverse breakdown voltage is applied, the reverse current carried by the junction increases significantly. If there is no current-limiting mechanism, this reverse current can destroy the device.

In signal diodes, reverse breakdown is dominated by an effect known as *avalanche breakdown*. In a reverse-biased junction, the applied reverse voltage applies an electric field in the depletion region of the diode. At a critical field E_{max}, the carriers in the depletion region gather sufficient speed to hit the lattice with enough force to cause other electron–hole pairs to be created. These created electron–hole pairs are then accelerated in the electric field, creating other electron–hole pairs. This process results in a rapidly increasing diode current as the reverse voltage is increased.

Taking a look at a diode datasheet

In this section, we will examine the datasheet of a 1N914 diode,[32] an inexpensive high-speed switching diode. The maximum ratings of the diode (Figure 3.24) show that this diode has a reverse voltage rating of 100 V and a forward current of 200 mA. The maximum peak current rating is 500 mA for a short period. These numbers are typical of a small signal switching diode.

Next, let us look at the basic electrical characteristics (Figure 3.25). Again, we see the reverse voltage rating ("reverse breakdown voltage") of 100 V. The reverse leakage current is specified to be a maximum of 5 µA at a reverse voltage of 75 V.

---

[32]The particular datasheet excerpts are from the On Semiconductor MMDL914 diode, the surface mount version of the through-hole 1N914 diode. Reprinted with permission of On Semiconductor.

Maximum ratings			
Rating	Symbol	Value	Unit
Reverse voltage	V_R	100	Vdc
Forward current	I_F	200	mAdc
Peak forward surge current	$I_{FM(surge)}$	500	mAdc

FIGURE 3.24

Maximum ratings of 1N914 diode.

Electrical characteristics (T_A = 25 °C unless otherwise noted)				
Characteristic	Symbol	Min	Max	Unit
OFF characteristics				
Reverse breakdown voltage (I_R = 100 µAdc)	$V_{(BR)}$	100	–	Vdc
Reverse voltage leakage current (V_R = 20 Vdc) (V_R = 75 Vdc)	I_R	– –	25 5.0	nAdc µAdc
Diode capacitance (V_R = 0 V, f = 1.0 MHz)	C_T	–	4.0	pF
Forward voltage (I_F = 10 mAdc)	V_F	–	1.0	Vdc
Reverse recovery time (I_F = I_R = 10 mAdc) (Figure 1)	t_{rr}	–	4.0	ns

FIGURE 3.25

Electrical characteristics of 1N914 diode.

We also see information on the diode capacitance; the capacitance information is better displayed later on in a graph. We also see that this diode has a fast reverse recovery time of $t_{rr} < 4$ ns, when tested with the test circuit shown in Figure 3.26.

The forward and reverse characteristics of the diode at various temperatures are shown in Figure 3.27. The forward voltage characteristic shows a negative temperature coefficient of diode voltage of approximately −2 mV/°C. At constant collector current, the diode voltage decreases approximately this much as the temperature increases. The forward plots also show additional curvature at high diode currents, due to ohmic drops in the diode not accounted for by the ideal diode model. Shown in the reverse leakage plot is the fact that reverse leakage current significantly increases as temperature increases.

The chart of diode capacitance in the reverse region (Figure 3.28) shows that the diode capacitance is nonlinear and decreases as reverse bias voltage increases.

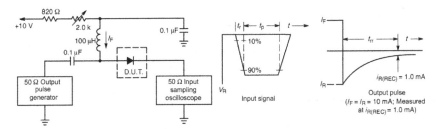

Notes: 1. A 2.0 k Ω variable resistor adjusted for a forward current (I_F) of 10 mA.
2. Input pulse is adjusted so $I_{R(peak)}$ is equal to 10 mA.
3. $t_p \gg t_{rr}$

FIGURE 3.26

Reverse recovery time test circuit for 1N914 diode.

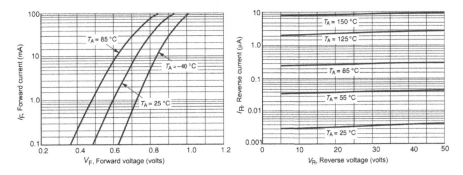

FIGURE 3.27

Forward and reverse characteristics of 1N914 diode.

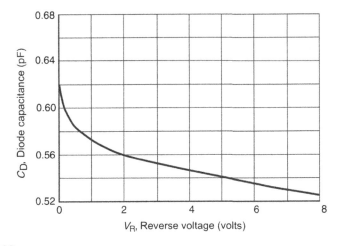

FIGURE 3.28

Diode capacitance in reverse region for 1N914 diode.

Thermal characteristics

Characteristic	Symbol	Max	Unit
Total device dissipation FR-5 Board $T_A = 25°C$ (Note 1) Derate above 25°C	P_D	200 1.57	mW mW/°C
Thermal resistance, Junction-to-ambient	$R_{\theta JA}$	635	°C/W
Junction and storage temperature	T_J, T_{stg}	−55 to 150	°C

FIGURE 3.29

Thermal characteristics of 1N914 diode.

The thermal characteristics (Figure 3.29) show that the maximum device dissipation is 200 mW at an ambient temperature of 25 °C. We need to derate this maximum power dissipation if the ambient temperature is higher than 25 °C.

Some quick comments on Schottky diodes

The analysis that we have done in this chapter has focused exclusively on bipolar diodes composed of a PN junction. Another type of diode, called a Schottky diode, is composed of a metal–semiconductor junction. We will not delve into the details but will leave the reader with a few bullets, summarized in Table 3.1:

- The typical forward voltage of a bipolar diode at operating current is ∼0.6 V or so, while the forward voltage of a Schottky is somewhat less, typically 0.4 V.
- Bipolar diodes, as shown before, exhibit reverse recovery. Schottky diodes, to first order, do not have reverse recovery.

Table 3.1 Comparison of Bipolar and Schottky Diodes

Item	Bipolar	Schottky
Typical forward voltage	>0.6 V	∼0.4 V
Reverse recovery?	Yes	No
Typical voltage rating	<several kilovolts	<100 V

Chapter 3 problems

Problem 3.1

For the ideal diode under forward bias, find the amount of forward voltage increase that causes a 10 times increase in forward current at room temperature.

Problem 3.2

An abrupt PN^+ junction diode has $N_D = 10^{16}/cm^3$ and $N_A = 10^{15}/cm^3$. Find the breakdown voltage if the breakdown electric field $E_{max} = 3 \times 10^5$ V/cm. Assume the abrupt junction approximation, an intrinsic carrier concentration in silicon of $n_i = 1.5 \times 10^{10}/cm^3$, that the diode operates at room temperature, and note that the total depletion region width with an applied bias V_D is:

$$x_n + x_p = \sqrt{\frac{2\varepsilon_{Si}(\phi_{bi} - V_D)}{q}\left(\frac{1}{N_A} + \frac{1}{N_D}\right)} \qquad (3.32)$$

Problem 3.3

An abrupt PN junction has parameters $N_A = 10^{17}/cm^3$, $N_D = 10^{16}/cm^3$, and area $A = 10^{-3}$ cm^2. Find the forward voltage at a forward current $I_D = 10$ mA. For this diode at 300 K, assume that mobilities are: $\mu_n = 1400$ cm^2/V s, $\mu_p = 400$ cm^2/V s, the minority carrier lifetimes are $\tau_n = \tau_p = 10^{-6}$ s, and that the intrinsic carrier concentration is $n_i = 1.4 \times 10^{10}$ cm^{-3}.

Problem 3.4

A diode has a reverse saturation current $I_s = 15$ nA at room temperature (298 K). Plot the diode I_D/V_D curve over the voltage range -10 V $< V_D < +0.5$ V, and at room temperature ($\sim 27\,°C$, or 300 K) and at elevated temperature of 150 °C (423 K). Assume the rule of thumb that the reverse current of a diode doubles for every 10 °C increase in temperature.

Problem 3.5

This problem investigates the use of load lines.[33] The ideal diode of Problem 3.3 above is used in a simple circuit, biased from a $+12$ V source through a 1 kΩ resistor. Assume that the diode operates at room temperature (300 K) and hence $kT/q = 26$ mV. Find the diode current and diode voltage by the following methods:

a. First, assume that the diode ON voltage is a constant 0.6 V, regardless of diode current. Solve for diode current I_D given this assumption.

b. In a more detailed analysis, solve iteratively the diode equation and resistor load-line equation to find diode current and diode voltage, assuming that the diode reverse saturation current $I_s = 10^{-10}$ A.

c. Plot the diode current/voltage load line, indicating the operating point on your plot.

---

[33]The load-line method is a graphical method used to solve nonlinear equations. By plotting the characteristic diode curve and then plotting the curve imposed by an external resistive element and noting the intersection of the two plots, we determine the operating point of the circuit. This method is also very useful for analyzing transistor circuits.

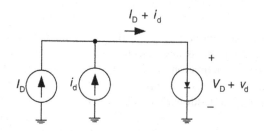

FIGURE 3.30

Diode circuit for Problem 3.6.

Problem 3.6

Consider the diode circuit shown in Figure 3.30. The diode is biased with bias current source I_D and a small signal AC current source is also connected, i_d. The diode voltage has both a DC part, V_D, and a small-signal varying part v_d. The total diode current is given by:

$$i_D = I_s\left(e^{\frac{qv_D}{kT}} - 1\right) \tag{3.33}$$

The diode current and diode voltage can be expressed as a DC part plus an incrementally varying part.

$$\begin{aligned} i_D &= I_D + i_d \\ v_D &= V_D + v_d \end{aligned} \tag{3.34}$$

Assume that the diode is biased in its forward-active region and that the value of the i_d is very small.

a. Find the ratio of small-signal voltage to small-signal current, or v_d/i_d.

b. Draw the low-frequency small-signal model relating v_d to i_d.

Problem 3.7

Shown in Figure 3.31 is a circuit containing a 1N4001 diode. Draw the load line and find the diode current and diode power dissipation.

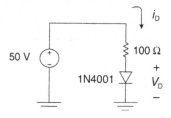

FIGURE 3.31

1N4001 diode circuit for Problem 3.7.

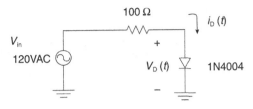

FIGURE 3.32

1N4004 AC diode circuit for Problem 3.8.

Problem 3.8

Shown in Figure 3.32 is a 120-V AC power supply driving a circuit containing a 1N4004 diode. (A 120-V AC waveform is a sine wave at 60 Hz, which has a positive peak at $+170$ V and a negative peak at -170 V). Note that in this case we use the 1N4004, since it has sufficient voltage rating for this application.

a. Plot input voltage $v_{in}(t)$, diode current $i_D(t)$, and diode voltage $v_D(t)$ vs. time. You can ignore any reverse recovery.

b. Calculate average power dissipated in the diode and in the resistor. In order to find power dissipation in the diode, calculate average diode current $\langle i_D \rangle$. The power dissipated in the diode[34] is:

$$P_{DIODE} \approx \langle V_D \rangle \langle i_D \rangle$$

where V_D is the diode ON voltage. Rather than doing a nasty integral, pick some reasonable value of average diode voltage $\langle V_D \rangle$ when the diode is on.

Further reading

[1] Adler RB, Smith AC, Longini RL. *Introduction to semiconductor physics*In *Semiconductor electronics education committee*, vol. 1. John Wiley; 1964.

[2] Bardeen J. Semiconductor research leading to the point contact transistor, December 11, 1956, from Nobel Lectures Physics, 1942–1962, Elsevier Publishing. Available from the Nobel e-museum at: http://www.nobel.se/physics/laureates/1956/.

[3] Brattain WH. Surface properties of semiconductors, December 11, 1956, from Nobel Lectures Physics, 1942–1962, Elsevier Publishing. Available from the Nobel e-museum at: http://www.nobel.se/physics/laureates/1956/.

[4] Feynman R, Leighton R, Sands M. *The Feynman lectures on physics*. Addison-Wesley; 1963.

[5] Gray PE, DeWitt D, Boothroyd AR, Gibbons JF. *Physical electronics and circuit models for transistors* In *SEEC*, vol. 2. John Wiley; 1964.

---

[34]This is a very rough approximation. To get a more accurate answer, we should do the integral $P_{DIODE} = \frac{1}{T} \int_0^T v_d(t) i_d(t) dt$.

[6] Gray PE, DeWitt D, Boothroyd AR, Gibbons JF. *Physical electronics and circuit models of transistors* In *Semiconductor electronics education committee*, vol. 2. John Wiley; 1964. Hannay NB, editor. Semiconductors, American Chemical Society Nomograph Series. Reinhold Publishing Corp.; 1959.

[7] Gray P, Searle C. *Electronic principles physics, models and circuits.* John Wiley; 1967.

[8] Haynes JR, Shockley W. The mobility and life of injected holes and electrons in germanium. *Phys Rev* 1951;**81**:835.

[9] Haynes JR, Shockley W. Investigation of Hole Injection in Transistor Action. *Phys Rev* 1949;**75**:691.

[10] Pierret RF. *Modular series on solid state devices* In *Semiconductor fundamentals*, vol. 1. Addison-Wesley; 1983.

[11] Shockley W. Transistor electronics: imperfections, unipolar and analog transistors. *Proc IRE* November, 1952;**40**(11):1289–313. Reprinted in Proc IEEE December 1997;**85**(12):2055–80.

[12] Shockley W. Circuit element utilizing semiconductive materials, U.S. Patent #2,569,347, issued September 25, 1951 and Semiconductor amplifier, U.S. patent 2,502,488, issued April 5, 1950, found in Semiconductor amplifier patent, Proc IEEE **86**(1):36.

[13] Shockley W. Transistor technology evokes new physics, December 11, 1956, from Nobel Lectures Physics, 1942–1962, Elsevier Publishing. Available from the Nobel e-museum at: http://www.nobel.se/physics/laureates/1956/.

Bipolar Transistor Models

4

"The chemical analogue to an n-type is a base ... similarly a p-type semiconductor is analogous to an acid."

—William Shockley, Nobel lecture, 1956

IN THIS CHAPTER

▶ This chapter builds on the device model work discussed in the previous chapter, and models for the ideal bipolar transistor are derived. Rather than deriving the full transistor equations, we will rely on results from the ideal diode and "talk through" intuitively how the bipolar transistor works. The NPN transistor is considered, but results obtained are germane to analysis of the PNP transistor as well. Most of this discussion will focus on operation of the bipolar transistor in the "forward-active" region, the region of operation where the transistor can be used as an amplifier. In a later chapter, we will extensively discuss operation of the bipolar transistor in the saturation and cutoff regions.

A little bit of history

Pioneering work on the bipolar junction transistor (BJT) was done at Bell Laboratories[1] in the late 1940s, with contributions from William Shockley, William Brattain, John Bardeen, and others. This team (led by Shockley) was challenged by Bell management to invent a solid-state switch to replace mechanical relays. They initially focused their efforts on devices fabricated from germanium. The first device that demonstrated transistor effect and current gain was the "point contact" transistor demonstrated in 1948[2] (Figure 4.1), which had a germanium crystal

---

[1] Bell Telephone Labs, in Murray Hill, New Jersey. The physical layout, performance, and physics of the point contact transistor is described in W. D. Bevitt's book *Transistors Handbook*, Chapter 3, and in Lo et al., *Transistor Electronics*, Chapter 1.14.

[2] It was John Bardeen who coined the terms *emitter, base,* and *collector.* For instance, in an NPN transistor, the emitter emits electrons and the collector collects them. The "base" was the base of a germanium crystal. See, e.g. W. F. Brinkman et al., "A History of the Invention of the Transistor and Where It Will Lead Us." An excellent reference describing the invention and implementation of the point contact transistor is given in John Bardeen's 1956 Nobel lecture.

Intuitive Analog Circuit Design. http://dx.doi.org/10.1016/B978-0-12-405866-8.00004-8

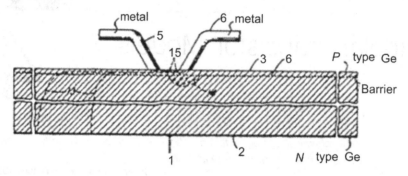

FIGURE 4.1

Point contact transistor from US Patent # 2,524,035 (June 17, 1948) by Brattain and Bardeen of Bell Labs.

with closely spaced gold contacts on the upper surface. With this device, Bardeen and Brattain were able to demonstrate current gain.

Later on, Shockley realized that a device with two semiconductor junctions could also demonstrate current gain. This led to the development of his transistor (Figure 4.2), which was the first BJT and arguably the first practical transistor

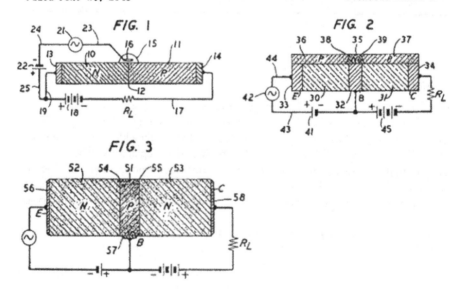

FIGURE 4.2

Shockley's transistor from US Patent # 2,569,347 (September 25, 1951).

design. Shockley patented his device as well as a number of practical amplifying circuits. He had a remarkable ability to simplify and describe the fundamental physics governing carrier motion in diodes and transistors. The results of his pioneering work, coupled with work done by colleagues, led to the semiconductor revolution of the second half of the twentieth century.

Basic NPN transistor

We will discuss some of the physics of the bipolar transistor by considering a basic structure that only varies in one dimension. The structure of this basic NPN transistor (not to scale) is shown in Figure 4.3. It is composed of an N-type emitter region, a P-type base, and an N-type collector region. Metallic, high-conductivity contacts are connected to an external circuit. This picture is cartoonish in that it does not show how the transistor is fabricated. However, it provides a useful framework through which we can analyze the operation. We will assume a one-dimensional geometry (in x) throughout this chapter.

Now, let us talk about the operation of this BJT strike in so-called *thermal equilibrium*, when there are no connections to the device and hence no net current flow. From our work in the previous chapter, we know that there are drift and diffusion components working inside the semiconductor material. If there is no net current flow from any of the terminals, we know that the drift and diffusion currents exactly balance each other out, leaving no currents at the terminals. Furthermore, the hole drift and hole diffusion currents balance each other and the electron drift balances the electron diffusion currents.

Let us first focus our efforts on the base—emitter junction. We know that an electric field barrier to current flow exists at this PN junction. Now, if we connect the NPN transistor as shown in Figure 4.4, we can add a positive voltage to the base—emitter junction. The effect of this forward bias voltage is to reduce the electric field barrier.

Now that we have forward-biased the base—emitter junction and reduced the electric field barrier, there is less of an impediment for electrons to be injected from the emitter region to the collector region. As shown in Figure 4.5, electrons are now injected from the emitter, travel through the base region, and are collected by the collector. Since each electron carries a negative charge, our convention for the direction of current flow in an NPN transistor is as shown.

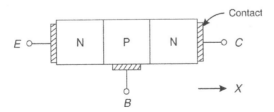

FIGURE 4.3

Basic NPN transistor showing emitter (E), base (B), and collector (C).

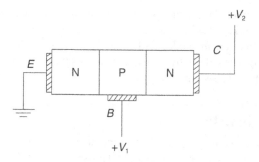

FIGURE 4.4

Basic NPN transistor connected with the base–emitter junction forward biased with $V_2 > V_1$.

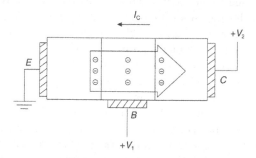

FIGURE 4.5

Basic NPN transistor connected with the base–emitter junction forward biased and the collector–base junction reverse biased.

Let us figure out the cause of base current. We know that a forward-biased transistor requires a small but finite base current I_B to support the collector current I_C. The transistor current gain β_F is defined as:

$$\beta_F = \frac{I_C}{I_B} \tag{4.1}$$

What is the cause of the base current? The individual base current components are shown in Figure 4.6.[3] Total base current is made up of three individual components, labeled I_{B1}, I_{B2}, and I_{B3}.

The emitter injects lots of electrons into the base region, as shown. Most of these electrons diffuse successfully across the base (as the base is very thin[4]). However, a

---

[3] An excellent description of the PNP transistor, showing a similar construction, is given in Gerold Neudeck's *The Bipolar Junction Transistor*, Modular Series on Solid State Devices, volume 3, pp. 8–10.

[4] In fact, transistor action depends on the base being very thin, so that most of the injected electrons will make it across the base. If the base width was very wide, the transistor would behave like back-to-back diodes, and all of the injected electrons would recombine in the base.

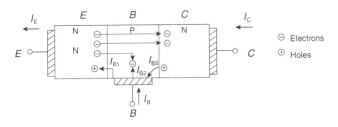

FIGURE 4.6

NPN transistor, forward-active region, showing base current components I_{B1}, I_{B2}, and I_{B3}.

small fraction of these injected electrons recombine with holes. A base current I_{B2} is needed to resupply the hole concentration in the base.

A second current component (I_{B1}) is composed of holes that are injected from the base into the emitter.

There are holes created in the collector N-region by thermal generation. Some of these holes cross through the reverse-biased collector–base junction into the base. A third base current component I_{B3} is composed of these holes that are injected across the reverse-biased base–collector junction. This current component is the reverse leakage current of the reverse-biased collector junction.

The net base current I_B is the sum of these three individual components, and in a well-designed transistor this base current is a small fraction of the total collector current. The DC current gain β_F of a signal transistor can be typically 100 or greater.[5]

A PNP transistor operating in the forward-active region is shown in Figure 4.7, with the base current components shown. The roles of electrons and holes are interchanged in the PNP transistor as compared to the NPN.

The bulk of the collector current is composed of holes that are injected from the emitter into the base and diffuse across the base. Base current I_{B2} is composed of electrons flowing into the base that recombine with holes that are diffusing across the base. Base current I_{B1} consists of electrons that are injected from the base into the emitter. Base current I_{B3} is composed of generated electrons in the collector that travel from the collector to the base.

An NPN transistor, showing terminal variables, is shown in Figure 4.8. In an NPN transistor operating in the forward-active region, the terminal currents (as defined in this diagram) are all positive. The base–emitter junction is forward

---

[5]This holds true for a typical signal transistor. See, e.g. the 2N3904 transistor where at a bias level of $I_C = 10$ mA and $V_{CE} = 1$ V, DC beta β_F (called h_{FE} on the datasheet) shows a range of 100–300. Power transistors, on the other hand, often have β_F less than 10 or so, due to other design tradeoffs. Current gain β_F also varies with collector current bias level. At very low and very high collector currents, β_F falls off.

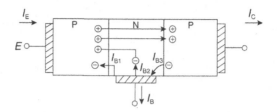

FIGURE 4.7

Basic PNP transistor, operating in the forward-active region, showing current components of the total base current I_B.

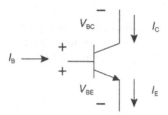

FIGURE 4.8

NPN transistor showing terminal variables.[6]

biased with $V_{BE} \sim 0.7$ V and the base–collector junction is reverse biased with $V_{BC} \leq 0$ V.

Transistor models in different operating regions

When we put an NPN transistor on a curve tracer (Figure 4.9), we find that the transistor has four regions of operation:

- In the cutoff region, $I_B = I_C = 0$ and $V_{BE} << 0.7$ V. The transistor is essentially OFF.
- In the forward-active or linear region, $I_C = \beta_F I_B$ and $V_{BE} = 0.7$ V or so.
- In saturation, V_{CE} is small and both the base–emitter and base–collector junctions are forward biased.
- In the reverse-active region (not shown on the diagram), the roles of collector and emitter are interchanged. Essentially, the transistor is being used "backward".[7]

---

[6]Some texts define I_E as flowing in the opposite direction.
[7]Note that if you interchange the collector and emitter leads in a conventional transistor circuit, the device will operate as a transistor. However, both the β_F and the speed will be low. Transistors are optimized to operate in the forward-active region.

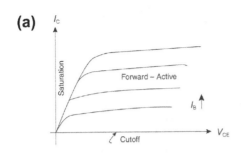

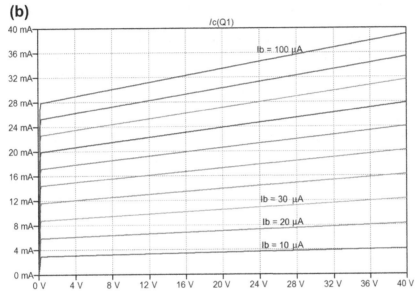

FIGURE 4.9

NPN transistor on a curve tracer. (a) There are three regions of operation in this quadrant: forward-active region, cutoff, and saturation. Not shown is the reverse-active region. (b) LTSPICE-generated I_C–V_{CE} curves for the 2N3904 transistor. (For color version of this figure, the reader is referred to the online version of this book.)

In the forward-active region,[8] the equations describing collector and base currents are:

$$I_C = I_S\left(e^{\frac{qV_{BE}}{kT}} - 1\right)$$
$$I_B = \frac{I_C}{\beta_F} \tag{4.2}$$

---

[8]Remember that the reverse saturation current I_S is the leakage current drawn by the ideal PN junction when it is reverse biased.

In normal operation in transistor amplifiers, the transistor is used in the forward-active region, where the collector current is exponentially related to the base–emitter voltage:

$$I_C \approx I_S e^{\frac{qV_{BE}}{kT}} \qquad (4.3)$$

In order to see how the various regions of operation work, let us do a thought experiment with a simple resistively loaded NPN transistor (Figure 4.10(a)). We will initially assume that the base voltage V_B is zero volts. At $t = 0$, the base voltage starts ramping up slowly as shown. In this thought experiment, we will assume that the ramp is slow enough that bandwidth limitations inside the transistor do not come into play.[9]

In Figure 4.10(b), the transistor characteristic curves are shown with the load line associated with the 1-kΩ resistor superimposed. Before $t = 0$, the operating point is shown at location "A" in the diagram. The transistor is cutoff, with zero collector current and with $V_{CE} = 12$ V as shown.

Some time after $t = 0$, when the base voltage V_B has risen to ~0.5 V or so,[10] the transistor starts turning on and begins to conduct collector current. During this region of operation, the transistor is in the "forward-active" region, labeled "B" in the diagram. When the collector current rises, the transistor collector–emitter voltage falls, due to voltage drop across the 1-kΩ resistor.

As the base voltage continues to rise, the collector voltage falls, until we reach region "C" in the curve. The transistor is now saturated. When the transistor is saturated, there is a relatively low value (less than 0.25 V or so) across the collector–emitter junction. As the base voltage increases, we go further and further into saturation, when the maximum collector current of approximately 12 mA is reached.

This switching profile is shown in Figure 4.10(c), where we plot base voltage and collector current. Note, as expected, that the transistor begins turning on at V_B ~0.4 V or so and that the maximum collector current is approximately 12 mA.

Low-frequency incremental bipolar transistor model

The low-frequency incremental model follows directly from the large-signal model. Assuming that the transistor is operated in the forward-active region, we will linearize about the operating point to get an approximately linear model that is valid for small variations in terminal voltages and currents. Let us assume that we have a resistively loaded NPN transistor properly biased so that the transistor operates

---

[9]In this case, we will use a ramp that transitions from 0 to 1 V in 1 s.

[10]As a rule-of-thumb, we normally assume that a diode forward voltage is approximately 0.7 V; a transistor when it is biased in the forward-active region has a base–emitter drop of approximately 0.7 V. This is the nature of a forward-biased PN junction. However, the transistor does begin to turn-on at a somewhat lower voltage. We will estimate the beginning of the turn-on at V_{BE} ~0.4 V.

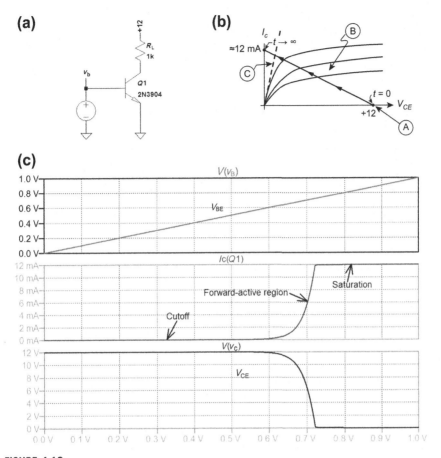

FIGURE 4.10

Resistively loaded NPN transistor. (a) Circuit. (b) Switching trajectory on the transistor I_C–V_{CE} curve in regions A (cutoff), B (forward-active region), and C (saturation). (c) LTSPICE plot showing transistor V_{BE}, I_C, and V_{CE}. (For color version of this figure, the reader is referred to the online version of this book.)

in the middle of the forward-active region.[11] First, we will assume that all transistor terminal variables have a DC part and a small-signal varying part:

$$
\begin{aligned}
i_C &= I_C + i_c \\
i_B &= I_B + i_b \\
v_{BE} &= V_{BE} + v_{be} \\
&\text{etc...}
\end{aligned}
\tag{4.4}
$$

---

[11]Initially, at least, we will ignore the practical problem of how to bias this transistor at this operating point so that the bias point is stable with respect to temperature and component variations.

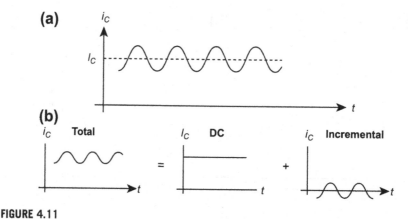

FIGURE 4.11

Diagram illustrating total collector current (i_C), bias level of collector current (I_C), and incremental part of collector current (i_c). (a) Total collector current $i_C(t)$. (b) Total collector current (i_C) broken into individual components of the DC bias level (I_C) and incremental variation (i_c).

where the terms on the left are the total variable, and the terms on the right are the DC (or bias point) and small-signal part, respectively. In Figure 4.11 we have illustrated this for the collector current.

In Figure 4.12(a), we show a transistor circuit biased with a base battery V_{BAT} and an input signal v_{sig}, which has a peak-to-peak amplitude of 1 mV at a frequency of 1 kHz. The level of V_{BIAS} is approximately 0.7 V,[12] but let us assume that we can adjust it so that the transistor collector current is 5 mA. This will bias the transistor in the middle of the forward-active region as shown in Figure 4.12(b). When we wiggle v_{sig}, the collector current varies sinusoidally as shown in Figure 4.12(c).

For the collector current with the transistor biased in the forward-active region, we find again the relationship between base-emitter voltage and collector current:

$$I_C + i_c \approx I_S e^{\frac{q\left(v_{BE} + v_{be}\right)}{kT}} = I_C e^{\frac{q v_{be}}{kT}} \tag{4.5}$$

We can approximately ignore the "−1" term in the exponential transistor voltage–current equation if we assume that the operating point value of V_{BE} is much greater than kT/q. If we restrict *the small-signal variation* of v_{be} to be small compared to kT/q (26 mV at room temperature), we can use the approximation $e^x \approx 1 + x$ to get:

$$I_C + i_c \approx I_S e^{\frac{q\left(v_{BE} + v_{be}\right)}{kT}} \approx I_C \left(1 + \frac{q v_{be}}{kT}\right) \tag{4.6}$$

[12]The actual value of V_{BIAS} used in the LTSPICE simulation is 696 mV, which sets up a bias current of approximately 5 mA.

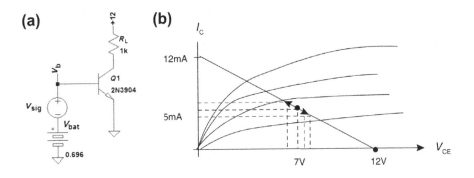

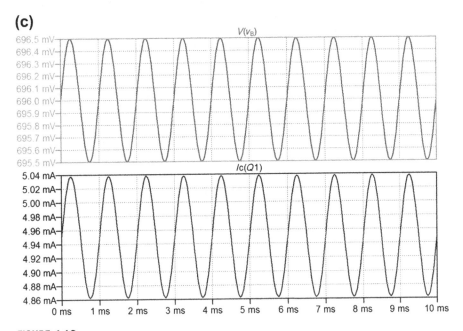

FIGURE 4.12

Resistively loaded NPN transistor biased in the forward-active region. (a) Circuit. (b) Operating point shown on the transistor I_C–V_{CE} curve. (c) LTSPICE plots of the above circuit showing AC output of the collector current riding on a DC bias level of ~4.95 mA. (For color version of this figure, the reader is referred to the online version of this book.)

Solving for the small-signal component i_c of the collector current results in:

$$i_c \approx I_C \left(\frac{q v_{be}}{kT}\right) = g_m v_{be} \qquad (4.7)$$

We have defined a proportionality constant g_m, which relates *small-signal variation* in collector current (i_c) to *small-signal variation* in v_{be}. This term g_m is called the *transconductance* of the transistor and has units of amps per volt.

Now, let us focus on small-signal variations in base current. We know that a small-signal variation in collector current means a proportional small-signal variation in base current. The base current can be described, using this reasoning, as:

$$i_b \approx \frac{i_c}{h_{fe}} = \frac{g_m}{h_{fe}} v_{be} = \frac{v_{be}}{r_\pi} \tag{4.8}$$

where h_{fe} is the small-signal current gain[13] of the transistor (often denoted h_{fe} on the transistor datasheet, and often denoted β_0 in textbooks). The model in Figure 4.13 captures the functional dependence of the collector current and base current equations. This so-called hybrid-pi model is useful for finding midband gain of a transistor amplifier, and we will use it in subsequent chapters extensively.

To summarize, the low-frequency incremental model parameters for the bipolar transistor are:

$$g_m = \frac{|I_C|}{kT/q} = \frac{|I_C|}{V_{TH}}$$

$$r_\pi = \frac{h_{fe}}{g_m} \tag{4.9}$$

High-frequency incremental model

We know that the previous low-frequency model is incomplete, because it does not have any bandwidth-limiting mechanisms. Let us now consider some of the sources of bandwidth limitation in bipolar transistors.

First, we know that a junction has depletion capacitance associated with its depletion layer, as we showed in the diode discussion of the previous chapter.

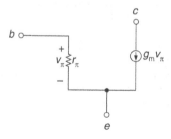

FIGURE 4.13

Bipolar transistor low-frequency incremental model in the forward-active region, showing base (b), collector (c), and emitter (e) terminals.

[13]Do not get small-signal beta (β_0 or h_{fe}) confused with DC or operating point beta (β_F or h_{FE}).

Therefore, we know that there are depletion capacitances in the bipolar transistor at the base–emitter and base–collector junctions. These depletion capacitances[14] show a functional dependence on junction voltage as shown in the previous chapter.

Secondly, when the transistor is biased in the forward-active region, there is charge stored in the base (Figure 4.14), as shown in this one-dimensional depiction of an NPN transistor. In the previous chapter discussion on diodes, this was termed the "diffusion capacitance". As v_{BE} varies, the concentration of extra electrons (n') stored in the base varies as well, supporting the collector current. In this formulation, we will assume that the variation in v_{BE} occurs slowly enough so that the concentration of n' in the base can be modeled as a series of static, triangular distributions. This is the so-called quasi-static approximation and will be used in later chapters when we discuss transistor large-signal switching.

This stored base charge can be modeled as a capacitance that depends on the bias level of the transistor. In the case of the NPN transistor, as V_{BE} increases, the excess minority carrier concentration stored in the base increases as well, as shown.

A circuit model showing the various charge storage mechanisms in a bipolar transistor is shown in Figure 4.15. There are the following capacitances in the model:

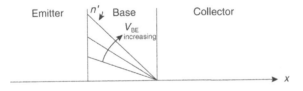

FIGURE 4.14

Charge stored in the base region of an NPN transistor in the forward-active region. As transistor base–emitter voltage V_{BE} increases, the extra electrons in the base n' increase as shown.

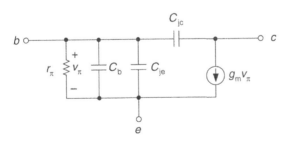

FIGURE 4.15

Transistor model in the forward-active region showing charge storage components C_b (base diffusion capacitance), C_{je} (base–emitter depletion capacitance), and C_{jc} (base–collector depletion capacitance).

---

[14]The depletion capacitance is sometimes called *space charge* capacitance as well.

- C_{je}: base–emitter depletion capacitance (junction voltage dependent)
- C_b: base diffusion capacitance (linearly proportional to collector current)
- C_{jc}: base–collector depletion capacitance (junction voltage dependent).

To this model, we make several modifications (Figure 4.16). First, we have combined C_{je} and C_b into a single capacitance, which we will call C_π. Second, let us rename C_{jc} as C_μ.

Last, let us add the "base spreading" resistance r_x. The base spreading resistance models the resistance between the ohmic base contact and is due in part to two-dimensional base current flow effects. For high-frequency bandwidth calculations, it is important to include r_x in the model because the transistor capacitances C_π and C_μ must be charged through it. If you omit r_x, in certain circuit topologies your model will be overly optimistic as to the bandwidth. For typical transistors, this base resistance is on the order of 50 Ω to 500 Ω.

Now, how do we determine C_π and C_μ from the datasheet? C_μ is relatively easy. We know that C_μ is just the base–collector depletion capacitance. Remember from Chapter 3 that depletion capacitances are junction voltage dependent, as:

$$C_j = \frac{C_{jo}}{\left(1 - \frac{V_J}{\phi_{bi}}\right)^m} \tag{4.10}$$

where $m = 1/2$ for an abrupt junction and $m = 1/3$ for a linearly graded junction, C_{jo} is the depletion capacitance at zero junction voltage, V_J is the junction voltage, and ϕ_{bi} is the built-in voltage. Also remember the polarity of V_J; when the junction is more reverse biased, the junction capacitance goes down.

First, we need to determine the operating point value of the base–collector junction voltage V_{CB}. Then, we can just read the depletion capacitance off the datasheet at the given collector–base bias voltage.

Finding C_π is a little bit more involved. Recall that C_π includes the depletion capacitance part (C_{je}) added to the base diffusion capacitance. The base diffusion capacitance is proportional to the transistor collector current. In order to find

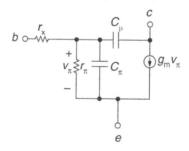

FIGURE 4.16

High-frequency incremental model of a transistor in the forward-active region. In this model we have combined C_{je} and C_b into a single capacitor C_π, have renamed C_{jc} as C_μ, and have added base resistance r_x.

C_π, we need to make use of some information from the datasheet. If you look at the datasheet, there is a number listed, sometimes called the *transition frequency* or *current gain—bandwidth product, or "f_T"*. If we look at a plot of transistor current gain vs. frequency, we will see a plot something like Figure 4.17. The current gain—bandwidth number is the frequency at which extrapolated small-signal current gain reaches unity.

We can use the simple circuit of Figure 4.18 to help us figure out a methodology for finding C_π. First, for simplicity, we have ignored the effects of r_x. Let us solve for incremental collector i_c when the base is driven by an incremental base current i_b.

Since the right-hand side of C_μ is grounded, the voltage v_π is simply:

$$v_\pi = i_b \frac{r_\pi}{r_\pi(C_\pi + C_\mu)s + 1} \tag{4.11}$$

The collector current i_c is:

$$
\begin{aligned}
i_c &= g_m v_\pi + i_f \\
&= g_m i_b \frac{r_\pi}{r_\pi(C_\pi + C_\mu)s + 1} - i_b \frac{r_\pi C_\mu s}{r_\pi(C_\pi + C_\mu)s + 1} \\
&= \frac{(h_{fe} - r_\pi C_\mu s) i_b}{r_\pi(C_\pi + C_\mu)s + 1}
\end{aligned}
\tag{4.12}
$$

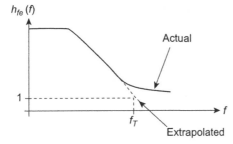

FIGURE 4.17

Plot of bipolar transistor incremental current gain $h_{fe}(f)$ vs. frequency. At frequency f_T, the extrapolated curve reaches a current gain of 1.

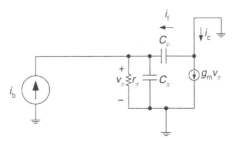

FIGURE 4.18

Transistor incremental model for finding current gain—bandwidth product f_T. The current fed-back through collector-base capacitance C is i_f.

Note that the feedback through C_μ gives a zero in the right-half plane at $\omega_z = +1/r_\pi C_\mu$, which is at a frequency higher than ω_T. Since the zero frequency is higher than the frequency range that we are interested in,[15] we will ignore it, approximating the small-signal transfer function by:

$$\frac{i_c}{i_b} \approx \frac{h_{fe}}{r_\pi (C_\pi + C_\mu)s + 1} \tag{4.13}$$

For frequencies much higher than the breakpoint, and making use of the fact that $h_{fe} = g_m r_\pi$:

$$\frac{i_c}{i_b} \approx \frac{g_m r_\pi}{r_\pi (C_\pi + C_\mu)s} = \frac{g_m}{(C_\pi + C_\mu)s} \tag{4.14}$$

Therefore, the magnitude where this gain drops to 1 is approximately:

$$f_T \approx \frac{g_m}{2\pi (C_\pi + C_\mu)} \tag{4.15}$$

Therefore, our recipe for finding C_π and C_μ from the datasheet and bias information is:

$$C_\mu \approx C_{jc} \text{ (found on datasheet at } V_{BC} \text{ bias)}$$
$$C_\pi = \frac{g_m}{2\pi f_T} - C_\mu \tag{4.16}$$

For the 2N3904 transistor with $I_C = 1$ mA, small-signal parameters are $h_{fe} \approx 100$ and $f_T = 300$ MHz. Transistor small-signal current gain h_{fe} begins dropping at $f \cong 3$ MHz as shown in Figure 4.19.

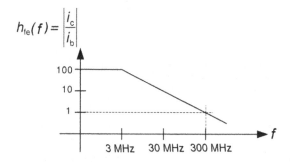

FIGURE 4.19

Idealized small-signal current gain $h_{fe}(f)$ vs. frequency for 2N3904 transistor, assuming low-frequency $h_{fe} = 100$ and $f_T = 300$ MHz. At 300 MHz the small-signal current gain is approximately 1.

---

[15]This frequency is higher than the frequency where the transistor current gain drops to 1. Therefore, in the interest of mathematical simplicity, we simply throw it away.

Reading a transistor datasheet

We will now go through the exercise of reading sections of a commercially available signal transistor datasheet (for the workhorse 2N3904[16]) and extracting important information. The 2N3904 is an inexpensive signal transistor with an approximate f_T of 300 MHz available in a plastic TO-92 package (Figure 4.20) with pins 1, 2 and 3 being the emitter, base, and collector, respectively.

Large-signal parameters (h_{FE}, $V_{CE, SAT}$)

The large-signal transistor parameter that we will consider first is DC current gain (Figure 4.21). On the datasheet,[17] this is often called "h_{FE}". The charts indicate that the 2N3904 has a DC current gain h_{FE} that peaks for intermediate levels of collector current. For instance, at 10 mA collector current, a typical number for h_{FE} is 300, with a minimum of 100. Figure 4.21(b) also shows this variation of h_{FE} with collector current, as well as the variation with temperature. Note that h_{FE} increases as temperature increases, at least for modest collector current. h_{FE} also drops at low collector currents due to recombination of carriers in the base—emitter depletion region. At high collector currents, h_{FE} falls off primarily due to emitter current crowding.

The value of saturation voltage for the 2N3904 is shown in Figure 4.22. Note that in hard saturation the value of $V_{CE, SAT}$ can be as low as 0.1 V or less, depending on how hard the base is driven with base current.

Small-signal parameters (h_{fe}, C_μ, C_π, and r_x)

Small-signal current gain (sometimes called β_0 in textbooks, and usually h_{fe} on datasheets) is not the same as large-signal β_F. Large-signal β_F tells you what the bias value of the base current (I_B) will be if you know the bias value of the collector current (I_C). Small-signal current gain tells you what the variation will be about those operating points, or:

TO-92
CASE 29
STYLE 1

1
2
3

FIGURE 4.20

TO-92 plastic package.

---

[16]I will use the datasheet from On Semiconductor (www.onsemi.com), but this same device is made by Philips, Fairchild, Vishay, National Semiconductor, and other companies. We will extract information from the datasheet in this chapter germane to the topics of transistor biasing and small-signal modeling. In a later chapter on the charge control model and transistor switching, we will extract further information. Reprinted with permission of On Semiconductor.

[17]For historical reasons, DC beta is usually termed β_F in textbooks and journal articles, while in datasheets it is h_{FE}.

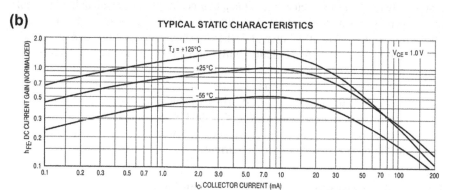

(a)

ON CHARACTERISTICS

DC Current Gain (Note 2)		hFE			–
(I_C = 0.1 mAdc, V_CE = 1.0 Vdc)	2N3903		20	–	
	2N3904		40	–	
(I_C = 1.0 mAdc, V_CE = 1.0 Vdc)	2N3903		35	–	
	2N3904		70	–	
(I_C = 10 mAdc, V_CE = 1.0 Vdc)	2N3903		50	150	
	2N3904		100	300	
(I_C = 50 mAdc, V_CE = 1.0 Vdc)	2N3903		30	–	
	2N3904		60	–	
(I_C = 100 mAdc, V_CE = 1.0 Vdc)	2N3903		15	–	
	2N3904		30	–	
Collector–Emitter Saturation Voltage (Note 2)		V_CE(sat)			Vdc
(I_C = 10 mAdc, I_B = 1.0 mAdc)			–	0.2	
(I_C = 50 mAdc, I_B = 5.0 mAdc			–	0.3	
Base–Emitter Saturation Voltage (Note 2)		V_BE(sat)			Vdc
(I_C = 10 mAdc, I_B = 1.0 mAdc)			0.65	0.85	
(I_C = 50 mAdc, I_B = 5.0 mAdc)			–	0.95	

(b) TYPICAL STATIC CHARACTERISTICS

FIGURE 4.21

Datasheet information for 2N3904 for DC current gain h_{FE}. (a) Datasheet values. (b) Curves. Note the variation in DC current gain with temperature, as well as variation with bias collector current value.

Reprinted with permission of On Semiconductor.

$$h_{fe} = \frac{\partial i_c}{\partial i_b}\bigg|_{I_C} \tag{4.17}$$

If you put an NPN transistor on a curve tracer, you will get a plot something like Figure 4.23. Let us assume that the point "Q" is the operating point where we will be using the transistor in our circuit. Using the curve tracer, we can estimate the small-signal beta h_{fe} near this operating point as:

$$h_{fe} \approx \frac{\Delta i_C}{\Delta i_b}\bigg|_{I_C} = \frac{I_{C3} - I_{C1}}{I_{B3} - I_{B1}} \tag{4.18}$$

Datasheet information for h_{fe} for the 2N3904 is shown in Figure 4.24 where we see that at a collector current of 1 mA, the small-signal current gain is roughly 100 or a little higher. If we know the bias level of the collector current we can find

(a)

Collector–Emitter Saturation Voltage (Note 2)	$V_{CE(sat)}$			Vdc
(I_C = 10 mAdc, I_B = 1.0 mAdc)		–	0.2	
(I_C = 50 mAdc, I_B = 5.0 mAdc		–	0.3	

(b)

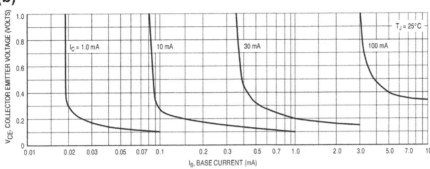

FIGURE 4.22

Datasheet information for 2N3904 on saturation characteristics ($V_{CE,\ SAT}$). Note that as you increase the base current overdrive, the saturation voltage decreases.

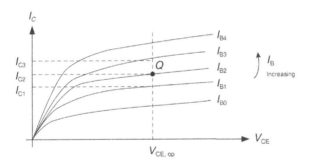

FIGURE 4.23

Curve tracer method for finding small-signal current gain h_{fe}, measured around your operating point "Q".

transconductance g_m. If we know transconductance g_m and small-signal current gain h_{fe}, we can find r_π:

$$r_\pi = \frac{h_{fe}}{g_m} \tag{4.19}$$

Depletion capacitance information for the 2N3904 is shown in Figure 4.25. The curve C_{obo} is the base–collector depletion capacitance and C_{ibo} is the base–emitter depletion capacitance. Both values are dependent on the junction voltage.

Information on base spreading resistance r_x is usually not found directly on transistor datasheets. Furthermore, the hybrid-pi incremental model is the result of an

(a)

Small-Signal Current Gain (I_C = 1.0 mAdc, V_{CE} = 10 Vdc, f = 1.0 kHz)		h_{fe}			–
	2N3903		50	200	
	2N3904		100	400	

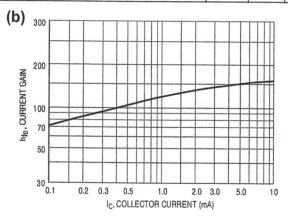

(b)

FIGURE 4.24

Datasheet information for 2N3904 for small-signal current gain (or h_{fe}). (a) Datasheet table. (b) Curve. Note that h_{fe} varies with the bias level of the collector current.

idealized physical model where second-order effects such as base current crowding are not considered. Unfortunately, r_x is difficult to measure accurately and the resistance is collector current dependent (Figure 4.26). The base resistance is also a source of noise.

Some transistor datasheets do provide other information, such as the *collector–base time constant* shown in Figure 4.27(a) for the 2N2222 transistor. We see in this example that the time constant may be used to find r_x, as follows. The collector–base time constant is specified to be 150 ps:

$$r_b C_c \approx r_x C_\mu = \tau_{cb} = 150 \times 10^{-12} \tag{4.20}$$

We note that this value of collector–base time constant is specified at a collector–base reverse bias voltage $V_{CB} = 20$ V. We now look at the chart of C_{cb} vs. frequency (Figure 4.27(a)) and find that at this reverse voltage $C_{cb} = C_\mu \approx 3.4$ pF. Next, we can estimate base spreading resistance as:

$$r_x \approx \frac{\tau_{cb}}{C_\mu} = \frac{150 \times 10^{-12}}{3.4 \times 10^{-12}} \approx 44 \; \Omega \tag{4.21}$$

Note that this value of r_x is approximate at an emitter current (and hence a collector current) of 20 mA. The value varies significantly with collector current due to current crowding. At lower collector currents, the value of r_x will be somewhat higher, but the functional dependence depends on details of the device fabrication.

(a)

Output Capacitance (V_{CB} = 5.0 Vdc, I_E = 0, f = 1.0 MHz)	C_{obo}	–	4.0	pF
Input Capacitance (V_{EB} = 0.5 Vdc, I_C = 0, f = 1.0 MHz)	C_{ibo}	–	8.0	pF

(a)

(b)

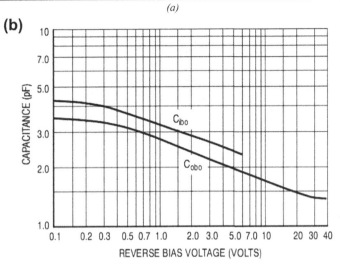

FIGURE 4.25

Datasheet information for 2N3904 on depletion capacitances. (a) Table value at specific values of junction voltages. (b) Curve. Note that as the reverse bias voltage on the junction increases, the depletion capacitances decrease.

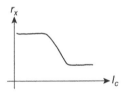

FIGURE 4.26

Base spreading resistance for a generic bipolar transistor showing its dependence on collector current. Base spreading resistance is maximum at low collector current.

Summarizing, in order to find appropriate small-signal models, this is a good design process:

- Find operating point values for collector current I_C and collector–base voltage V_{CB}. If the transistor is in the forward-active region, move on to find small-signal parameters.
- Find transistor transconductance g_m (which depends on transistor collector current).
- From the datasheet, find h_{fe} at your bias collector current.
- Calculate r_π.

(a)

Collector Base Time Constant (I_E = 20 mAdc, V_{CB} = 20 Vdc, f = 31.8 MHz)	rb'C$_c$	—	150	ps

(b)

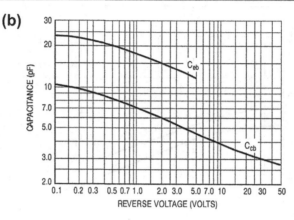

FIGURE 4.27

Datasheet information for 2N2222[18] on base–collector time constant. (a) Time constant found on the datasheet. (b) Information on junction capacitances.

- From the datasheet, find output capacitance C_{obo} (which is the base–collector depletion capacitance[19]). This capacitance is a function of transistor base–collector junction voltage. Small-signal capacitance $C_\mu = C_{obo}$.
- From the datasheet, find f_T. Note that f_T is collector current dependent.
- Calculate C_π using the relationship $C_\pi = \frac{g_m}{2\pi f_T} - C_\mu$.
- In the absence of any other additional information, make an educated guess[20] as to the value of base spreading resistance r_x. Do not necessarily rely on SPICE device models, as they are sometimes incorrect.

Limitations of the hybrid-pi model

A fundamental limitation of the hybrid-pi model is that it is valid for frequencies much less than the f_T of the transistor. We need to be careful in using and relying

---

[18]Datasheet from On Semiconductor, www.onsemi.com, reprinted with permission of On Semiconductor.

[19]Some manufacturers use different names. For instance, in the 2N2222 datasheet, the collector–base depletion capacitance is named "C_{cb}".

[20]The value that I use for the 2N3904 is an estimate from an old laboratory handout from my undergraduate days at MIT.

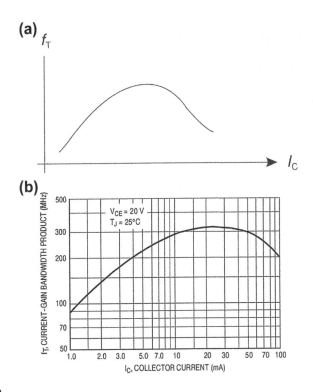

FIGURE 4.28

Variation in f_T with collector current for a generic transistor. (a) For hypothetical small-signal transistor. (b) For 2N2222 NPN transistor.

Reprinted with permission from On Semiconductor.

on the accuracy of the hybrid-pi model for frequencies much higher than $f_T/5$ or so. Let us see what this means for our 2N3904 example.

The datasheet value for f_T of the 2N3904 is 300 MHz. (Note that this value of f_T varies with collector current level; unfortunately, this current variation is not to be found on this particular datasheet.) The variation in f_T with collector current level for a typical transistor is shown in Figure 4.28(a). At low values of collector current, depletion capacitances C_{je} and C_{jc} dominate and f_T is approximately linearly increasing with collector current. At very high collector current, f_T drops off due to high-level injection effects.[21] The variation in f_T vs. collector current for the 2N2222 transistor is shown in Figure 4.28(b).

---

[21]See, e.g. John Choma, Jr., "A Curve-Fitted Circuits Model for Bipolar f_T Roll-Off at High Injection Levels", *IEEE Journal of Solid-State Circuits*, April 1976, pp. 346–348.

2N3904 datasheet excerpts[22]

2N3903, 2N3904

2N3903 is a Preferred Device

General Purpose Transistors

NPN Silicon

Features
- Pb–Free Package May be Available. The G–Suffix Denotes a Pb–Free Lead Finish

ON Semiconductor®

http://onsemi.com

TO–92
CASE 29
STYLE 1

MAXIMUM RATINGS

Rating	Symbol	Value	Unit
Collector–Emitter Voltage	V_{CEO}	40	Vdc
Collector–Base Voltage	V_{CBO}	60	Vdc
Emitter–Base Voltage	V_{EBO}	6.0	Vdc
Collector Current – Continuous	I_C	200	mAdc
Total Device Dissipation @ T_A = 25°C Derate above 25°C	P_D	625 5.0	mW mW/°C
Total Device Dissipation @ T_C = 25°C Derate above 25°C	P_D	1.5 12	W mW/°C
Operating and Storage Junction Temperature Range	T_J, T_{stg}	−55 to +150	°C

THERMAL CHARACTERISTICS (Note 1)

Characteristic	Symbol	Max	Unit
Thermal Resistance, Junction–to–Ambient	$R_{\theta JA}$	200	°C/W
Thermal Resistance, Junction–to–Case	$R_{\theta JC}$	83.3	°C/W

1. Indicates Data in addition to JEDEC Requirements.

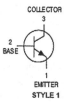

COLLECTOR
3

2
BASE

1
EMITTER
STYLE 1

MARKING DIAGRAMS

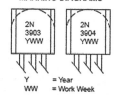

2N
3903
YWW

2N
3904
YWW

Y   = Year
WW = Work Week

ORDERING INFORMATION

Device	Package	Shipping†
2N3903	TO–92	5000 Units/Box
2N3903RLRM	TO–92	2000/Ammo Pack
2N3904	TO–92	5000 Units/Box
2N3904RLRA	TO–92	2000/Tape & Reel
2N3904RLRE	TO–92	2000/Tape & Reel
2N3904RLRM	TO–92	2000/Ammo Pack
2N3904RLRMG	TO–92	2000/Ammo Pack
2N3904RLRP	TO–92	2000/Ammo Pack
2N3904RL1	TO–92	2000/Tape & Reel
2N3904ZL1	TO–92	2000/Ammo Pack

†For information on tape and reel specifications, including part orientation and tape sizes, please refer to our Tape and Reel Packaging Specifications Brochure, BRD8011/D.

*For additional information on our Pb–Free strategy and soldering details, please download the ON Semiconductor Soldering and Mounting Techniques Reference Manual, SOLDERRM/D.

Preferred devices are recommended choices for future use and best overall value.

© Semiconductor Components Industries, LLC, 2003
December, 2003 – Rev. 4

1

Publication Order Number:
2N3903/D

[22]Reprinted with permission of On Semiconductor.

2N3903, 2N3904

ELECTRICAL CHARACTERISTICS (T_A = 25°C unless otherwise noted)

Characteristic		Symbol	Min	Max	Unit
OFF CHARACTERISTICS					
Collector–Emitter Breakdown Voltage (Note 2) (I_C = 1.0 mAdc, I_B = 0)		$V_{(BR)CEO}$	40	–	Vdc
Collector–Base Breakdown Voltage (I_C = 10 μAdc, I_E = 0)		$V_{(BR)CBO}$	60	–	Vdc
Emitter–Base Breakdown Voltage (I_E = 10 μAdc, I_C = 0)		$V_{(BR)EBO}$	6.0	–	Vdc
Base Cutoff Current (V_{CE} = 30 Vdc, V_{EB} = 3.0 Vdc)		I_{BL}	–	50	nAdc
Collector Cutoff Current (V_{CE} = 30 Vdc, V_{EB} = 3.0 Vdc)		I_{CEX}	–	50	nAdc
ON CHARACTERISTICS					
DC Current Gain (Note 2)		h_{FE}			–
(I_C = 0.1 mAdc, V_{CE} = 1.0 Vdc)	2N3903		20	–	
	2N3904		40	–	
(I_C = 1.0 mAdc, V_{CE} = 1.0 Vdc)	2N3903		35	–	
	2N3904		70	–	
(I_C = 10 mAdc, V_{CE} = 1.0 Vdc)	2N3903		50	150	
	2N3904		100	300	
(I_C = 50 mAdc, V_{CE} = 1.0 Vdc)	2N3903		30	–	
	2N3904		60	–	
(I_C = 100 mAdc, V_{CE} = 1.0 Vdc)	2N3903		15	–	
	2N3904		30	–	
Collector–Emitter Saturation Voltage (Note 2)		$V_{CE(sat)}$			Vdc
(I_C = 10 mAdc, I_B = 1.0 mAdc)			–	0.2	
(I_C = 50 mAdc, I_B = 5.0 mAdc)			–	0.3	
Base–Emitter Saturation Voltage (Note 2)		$V_{BE(sat)}$			Vdc
(I_C = 10 mAdc, I_B = 1.0 mAdc)			0.65	0.85	
(I_C = 50 mAdc, I_B = 5.0 mAdc)			–	0.95	
SMALL–SIGNAL CHARACTERISTICS					
Current–Gain–Bandwidth Product		f_T			MHz
(I_C = 10 mAdc, V_{CE} = 20 Vdc, f = 100 MHz)	2N3903		250	–	
	2N3904		300	–	
Output Capacitance (V_{CB} = 5.0 Vdc, I_E = 0, f = 1.0 MHz)		C_{obo}	–	4.0	pF
Input Capacitance (V_{EB} = 0.5 Vdc, I_C = 0, f = 1.0 MHz)		C_{ibo}	–	8.0	pF
Input Impedance		h_{ie}			kΩ
(I_C = 1.0 mAdc, V_{CE} = 10 Vdc, f = 1.0 kHz)	2N3903		1.0	8.0	
	2N3904		1.0	10	
Voltage Feedback Ratio		h_{re}			X 10^{-4}
(I_C = 1.0 mAdc, V_{CE} = 10 Vdc, f = 1.0 kHz)	2N3903		0.1	5.0	
	2N3904		0.5	8.0	
Small–Signal Current Gain		h_{fe}			–
(I_C = 1.0 mAdc, V_{CE} = 10 Vdc, f = 1.0 kHz)	2N3903		50	200	
	2N3904		100	400	
Output Admittance (I_C = 1.0 mAdc, V_{CE} = 10 Vdc, f = 1.0 kHz)		h_{oe}	1.0	40	μmhos
Noise Figure		NF			dB
(I_C = 100 μAdc, V_{CE} = 5.0 Vdc, R_S = 1.0 kΩ, f = 1.0 kHz)	2N3903		–	6.0	
	2N3904		–	5.0	

SWITCHING CHARACTERISTICS

			Symbol	Min	Max	Unit
Delay Time	(V_{CC} = 3.0 Vdc, V_{BE} = 0.5 Vdc,		t_d	–	35	ns
Rise Time	I_C = 10 mAdc, I_{B1} = 1.0 mAdc)		t_r	–	35	ns
Storage Time	(V_{CC} = 3.0 Vdc, I_C = 10 mAdc,	2N3903	t_s	–	175	ns
	I_{B1} = I_{B2} = 1.0 mAdc)	2N3904		–	200	
Fall Time			t_f	–	50	ns

2. Pulse Test: Pulse Width ≤ 300 μs; Duty Cycle ≤ 2%.

2N3903, 2N3904

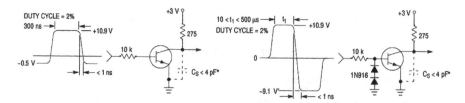

* Total shunt capacitance of test jig and connectors

Figure 1. Delay and Rise Time Equivalent Test Circuit

Figure 2. Storage and Fall Time Equivalent Test Circuit

2N3903, 2N3904

TYPICAL TRANSIENT CHARACTERISTICS

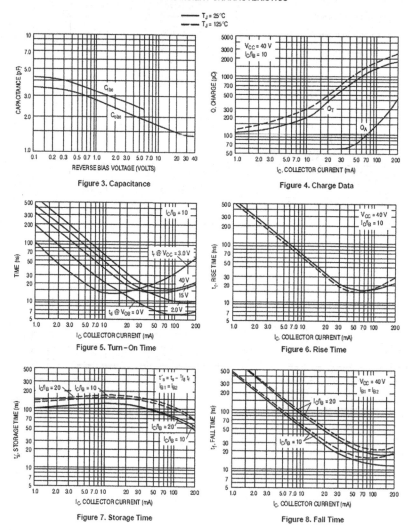

Figure 3. Capacitance

Figure 4. Charge Data

Figure 5. Turn–On Time

Figure 6. Rise Time

Figure 7. Storage Time

Figure 8. Fall Time

2N3903, 2N3904

TYPICAL AUDIO SMALL–SIGNAL CHARACTERISTICS
NOISE FIGURE VARIATIONS
(V_{CE} = 5.0 Vdc, T_A = 25°C, Bandwidth = 1.0 Hz)

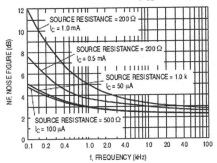

Figure 9.

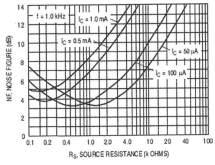

Figure 10.

h PARAMETERS
(V_{CE} = 10 Vdc, f = 1.0 kHz, T_A = 25°C)

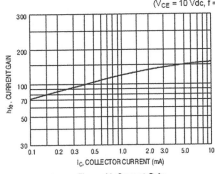

Figure 11. Current Gain

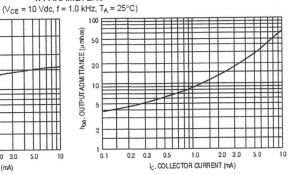

Figure 12. Output Admittance

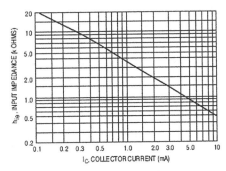

Figure 13. Input Impedance

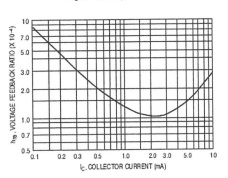

Figure 14. Voltage Feedback Ratio

2N3903, 2N3904

TYPICAL STATIC CHARACTERISTICS

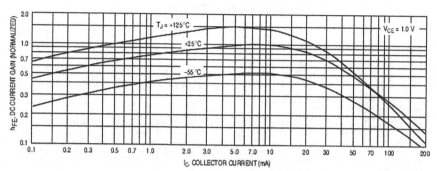

Figure 15. DC Current Gain

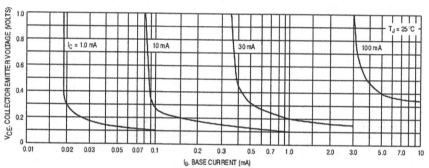

Figure 16. Collector Saturation Region

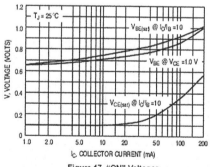

Figure 17. "ON" Voltages

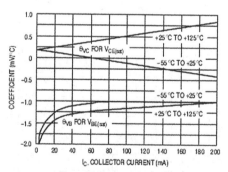

Figure 18. Temperature Coefficients

Chapter 4 problems

Problem 4.1

For the transistor circuit in Figure 4.29, find transistor collector current and small-signal parameters g_m and r_π and draw the low-frequency small-signal model. Assume that large-signal current gain $\beta_F = 100$, small-signal current gain $h_{fe} = 150$, and base resistance $r_x = 200\ \Omega$. Assume room temperature operation of the transistor.

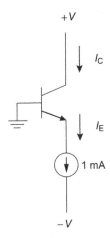

FIGURE 4.29

Transistor circuit for Problem 4.1.

Problem 4.2

Find and sketch the low-frequency incremental model for a transistor with $I_c = 10$ mA and $h_{fe} = 200$. Assume that the transistor operates at room temperature. Assume that the transistor base resistance $r_x = 0$.

Problem 4.3

A transistor has parameters $f_T = 500$ MHz, $C_{jc} = 1$ pF, and $h_{fe} = 100$. Find and sketch the high-frequency incremental model when this transistor is biased with a collector current of 1 mA. Assume that transistor base resistance $r_x = 50\ \Omega$.

Problem 4.4

Using datasheet information from the 2N3904 transistor, find the small-signal model assuming the transistor is biased at a collector current of 5 mA and a collector–base voltage of 5 V.

Problem 4.5

Consider the transistor circuit shown in Figure 4.30. The base of the transistor is biased with bias current source I_B, and a small-signal AC current source is also connected, i_b. The resultant collector current has both a DC part, I_C, and a small-signal varying part i_c.

a. Draw the small-signal model using the small-signal high-frequency model (i.e. including C_π and C_μ), but ignoring the base resistance r_x (i.e. assume r_x is very small).

b. Calculate the small-signal current transfer ratio $i_c(s)/i_b(s)$ in terms of g_m, C_π, and C_μ.

c. Using a reasonable approximation and the result from (2), calculate the frequency ω_T (in radians/second) where the magnitude of the small-signal current gain $i_c(s)/i_b(s)$ drops to 1, in terms of g_m, C_π, and C_μ (*Hint: You should be able to do this graphically without significant calculations.*)

d. For the 2N3904 transistor, biased at $I_C = 10$ mA, plot the magnitude of i_c/i_b using *minimum* numbers from the datasheet, and on your plot, denote the value of ω_T. Label all breakpoints, etc.

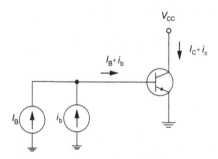

FIGURE 4.30

Transistor circuit for Problem 4.5.

Further reading

[1] Bardeen J. Semiconductor research leading to the point contact transistor, December 11, 1956, from Nobel Lectures Physics, 1942−1962, Elsevier Publishing. Available from the Nobel e-museum at: http://www.nobel.se/physics/laureates/1956/.

[2] Bardeen J, Brattain WH. Physical principles involved in transistor action. *Phys Rev* 1949;**75**:1208.

[3] Bevitt WD. *Transistors handbook*. Prentice-Hall; 1956.

[4] Brinkman WF, Haggan DE, Troutman WW. A history of the invention of the transistor and where it will lead us. *IEEE J Solid-State Circuits* 1997;**32**(12):1858−65.

[5] Choma Jr J. A curve-fitted circuits model for bipolar f_T roll-off at high injection levels. *IEEE J Solid-State Circuits* 1976;**11**(2):346−8.

[6] Fair RB. History of some early developments in ion-implantation technology leading to silicon transistor manufacturing. *Proc IEEE* January 1998;**86**(1):111−37.

[7] Gough R. High-frequency transistor modeling for circuit simulation. *IEEE J Solid-State Circuits* August 1982;**SC-17**(4):666−70.

[8] Gray PE, DeWitt D, Boothroyd AR, Gibbons JF. *Physical electronics and circuit models of transistors*In *Semiconductor electronics education committee*, vol. 2. John Wiley; 1964.

[9] Gray PE, Searle CL. *Electronic principles: physics, models and circuits*. John Wiley; 1969.

[10] Gray PR, Meyer RG. *Analysis and design of analog integrated circuits*. 2nd ed. John Wiley; 1984.

[11] Greeneich EW. An appropriate device figure of merit for bipolar CML. *IEEE Electron Device Lett* 1991;**12**:18.

[12] Hurkx GAM. The relevance of f_T and f_{max} for the speed of a bipolar CE amplifier stage. *IEEE Trans Electron Devices* 1997;**44**(5):775−81.

[13] Lo AW, Endres R, Zawels J, Waldhauer F, Cheng C. *Transistor electronics*. Prentice-Hall; 1955.

[14] Muller RS, Kamins TI. *Device electronics for integrated circuits*. 2nd ed. John Wiley; 1986.

[15] Neudeck GW. *Modular series on solid state devices, the bipolar junction transistor*. Addison-Wesley; 1983. On Semiconductor website: http://www.onsemi.com.

[16] Pritchard RL. Transistor equivalent circuits. *Proc IEEE* January 1998;**86**(1):150−62.

[17] Sansen W, Meyer R. Characterization and measurement of the base and emitter resistances of bipolar transistors. *IEEE J Solid-State Circuits* December 1972;**SC-7**(6):492−8.

[18] Searle CL, Boothroyd AR, Angelo Jr EJ, Gray PE, Pederson DO. *Elementary circuit properties of transistors*In *Semiconductor electronics education committee*, vol. 3. John Wiley; 1964.

[19] Shockley W. Transistor technology evokes new physics, December 11, 1956, from Nobel Lectures Physics, 1942−1962, Elsevier Publishing. Available from the Nobel e-museum at: http://www.nobel.se/physics/laureates/1956/.

[20] Sze SM. *Physics of semiconductor devices*. 2nd ed. John Wiley; 1981.

[21] Warner RM. Microelectronics: its unusual origin and personality. *IEEE Trans Electron Devices* 2001;**48**(11):2457−67.

Basic Bipolar Transistor Amplifiers and Biasing

5

"Amplifier bias networks must be carefully designed to overcome two deficiencies in transistors. First, … transistor incremental parameters vary significantly with temperature. Second, substantial variations in parameters occur among supposedly identical transistors bearing the same type number."

—Paul E. Gray and Campbell L. Searle, "Electronic Principles Physics, Models and Circuits", 1969

IN THIS CHAPTER

▶ In this chapter, we cover some basic transistor topologies, including the common-emitter amplifier, emitter follower, as well as common-base and differential amplifiers. The important issue of biasing—connecting the transistor so that it operates in the forward-active region (FAR) where it can be used for amplification—is also discussed in detail.

The issue of transistor biasing

In order to achieve useful amplification from a transistor, the transistor is generally biased in the FAR. The FAR is the region of operation of the transistor where amplification can occur because the transistor provides incremental current gain. Biasing is the process by which one sets the DC operating point of a transistor amplifier to a known and repeatable point in this FAR. If you do your biasing correctly, the amplifier output bias level will not drift significantly with time, temperature, or component variations.

Let us consider the simple bias circuit shown in Figure 5.1(a). The bias voltage $V_{BB} = 5$ V sets the base current as follows:

$$I_B = \frac{V_{BB} - V_{BE}}{R_B} \approx \frac{5 - 0.7}{28.6 \text{ k}\Omega} \approx 150 \text{ μA} \tag{5.1}$$

This initial calculation assumes that the base–emitter voltage $V_{BE} = 0.7$ V, which is approximately correct if the transistor is biased in the FAR and the

Intuitive Analog Circuit Design. http://dx.doi.org/10.1016/B978-0-12-405866-8.00005-X

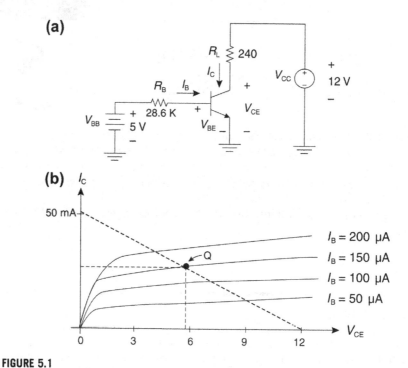

FIGURE 5.1

Bias design example. (a) Circuit. (b) Load line imposed on the transistor characteristic curves by the 240-Ω load resistor. The operating point is labeled "Q" and the operating point is $I_C = 26.25$ mA and $V_{CE} = 5.7$ V.

base–emitter junction is forward biased. Let us assume that this transistor has a DC current gain $\beta_F = 175$. This means that the collector current is:

$$I_C = \beta_F I_B = \beta_F \left(\frac{V_{BB} - V_{BE}}{R_B} \right) \approx 175 \times 150 \ \mu A \approx 26.25 \ mA \qquad (5.2)$$

With a load resistor $R_L = 240 \ \Omega$, there is a 6.3-V drop across this load resistor and hence $V_{CE} = 5.7$ V. The operating point, labeled "Q" for quiescent operating point, is shown in Figure 5.1(b). The operating point is the intersection of the transistor characteristic curves with the load line imposed by the 240-Ω resistor.

This form of simple biasing has several practical design challenges. First, how do you generate the bias supply voltage V_{BB}? In this example, the main supply $V_{CC} = 12$ V, so a second power supply would be needed to provide the 5 V for the base voltage bias.

Second, there is the issue of variation in collector current with transistor current gain variations. As shown in the collector current bias current equation, the collector current is proportional to transistor β_F. If β_F varies, the collector current varies proportionally. Figure 5.2 shows the current–gain curves for a 2N3904 transistor at

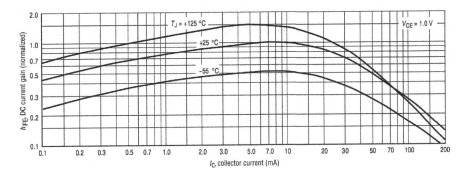

FIGURE 5.2

Variation in DC current gain β_F (called h_{FE} in this datasheet) for the 2N3904 transistor.

Reprinted with permission of On Semiconductor.

three different operating temperatures. Therefore, the DC current gain for a transistor varies with temperature and collector current level as shown.

Third, there is variation in collector current due to transistor V_{BE} variation. For a typical transistor, the base−emitter voltage, measured with a constant collector current bias, decreases approximately 2 mV/°C temperature increase.[1]

A slightly more complicated biasing arrangement can be implemented to significantly reduce bias variations due to transistor V_{BE} and β_F variations. Consider the circuit of Figure 5.3(a), where a base bias resistor divider R_{B1} and R_{B2} set the bias

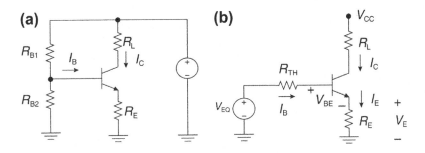

FIGURE 5.3

Alternative bias circuit. (a) Circuit. (b) Thévenin equivalent circuit.

---

[1]See, e.g. Robert J. Widlar, "An Exact Expression for the Thermal Variation of the Emitter Base Voltage of Bi-Polar Transistors", *Proceedings of the IEEE*, January 1967, pp. 96−97. Widlar shows that the expression for transistor, V_{BE} as a function of temperature is:

$V_{BE}(T) = V_{Go}\left(1 - \frac{T}{T_o}\right) + V_{BEo}\left(\frac{T}{T_o}\right) + \frac{nkT}{q}\ln\left(\frac{T_o}{T}\right) + \frac{kT}{q}\ln\left(\frac{I_C}{I_{Co}}\right)$ where V_{Go} = extrapolated bandgap voltage (~ 1.2 V @ 0 K), T = temperature in K, T_o = reference temperature at V_{BEo} and I_{Co}, and n is a process-dependent constant between 1.5 and 3. At room temperature, the V_{BE} temperature coefficient is approximately -2 mV/K, or approximately -3000 ppm/°C.

voltage at the base of the transistor and emitter resistor R_E has been added. As we will see, this combination of components, if properly designed, results in a bias point that is much less dependent on temperature and process variations of β_F and V_{BE} than in the previous circuit.

The Thévenin equivalent of this circuit is shown in Figure 5.3(b) where components are:

$$R_{TH} = R_{B1} \| R_{B2}$$

$$V_{EQ} = V_{CC} \left(\frac{R_{B2}}{R_{B1} + R_{B2}} \right) \tag{5.3}$$

Using the Thévenin circuit, let us solve for the base current:

$$I_B = \frac{V_{EQ} - V_{BE} - V_E}{R_{TH}} \tag{5.4}$$

The voltage across the emitter resistor V_E is found by:[2]

$$V_E = I_E R_E = \left(\frac{\beta_F + 1}{\beta_F} \right) I_C R_E \tag{5.5}$$

Combining these equations results in:

$$I_C = \beta_F \left[\frac{V_{EQ} - V_{BE} - \left(\frac{\beta_F+1}{\beta_F} \right) I_C R_E}{R_{TH}} \right] \approx \beta_F \left[\frac{V_{EQ} - V_{BE} - I_C R_E}{R_{TH}} \right] \tag{5.6}$$

Solving for collector current results in:

$$I_C \approx \beta_F \left(\frac{V_{EQ} - V_{BE}}{R_{TH}} \right) \left(\frac{1}{1 + \frac{\beta_F R_E}{R_{TH}}} \right) \approx \frac{(V_{EQ} - V_{BE})}{\left(R_E + \frac{R_{TH}}{\beta_F} \right)} \tag{5.7}$$

Since we have extra resistors in this bias arrangement, we have the degrees of freedom in our design available to make $V_{EQ} \gg V_{BE}$ and $R_E \gg R_{TH}/\beta_F$. Under this set of assumptions, the collector current bias level is approximately:

$$I_C \approx \frac{V_{EQ}}{R_E} \tag{5.8}$$

Let us examine intuitively how this circuit provides stabilization of the collector bias current. If collector current I_C increases due to an increase in DC current gain β_F or due to a decrease in V_{BE}, emitter current I_E increases as well. As we note from the base–emitter Kirchoff Voltage Law (KVL) equation, an increase in I_E results in a decrease in base current. This reduction in base current will in turn stabilize the collector current.

---

[2]Note the polarities I have given to the currents, which differ from those found in some texts. We can see that if β_F is large, the collector and emitter currents are approximately equal. The actual relationship is: $I_B + I_C = I_E \Rightarrow \frac{I_C}{\beta_F} + I_C = I_E \Rightarrow I_E = \left(\frac{\beta_F+1}{\beta_F} \right) I_C$.

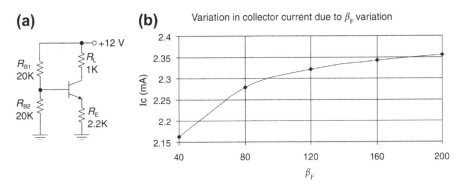

(a)

(b) Variation in collector current due to β_F variation

FIGURE 5.4

Bias example. (a) Circuit. (b) Variation in collector current when β_F varies from 40 to 200.

Example 5.1: Biasing example

For the circuit of Figure 5.4, let us find the variation in collector current if β_F varies from 40 to 200. Assume that V_{BE} remains constant at 0.7 V.

For this circuit, $V_{EQ} = 6$ V and $R_{TH} = 10$ kΩ. The transistor current gain varies by a factor of 5 as shown, while the collector current shows only a small variation ($\sim 20\%$).

Some transistor amplifiers
The common-emitter amplifier

The common-emitter amplifier is a widely used gain block. Using the hybrid-pi model, we will find the bias point, gain, and small-signal bandwidth of a common-emitter amplifier, built with a 2N3904 transistor (Figure 5.5). First we will do the bandwidth calculations the difficult way, by solving the node equations. For the time being, we will gloss over the details of the biasing circuit and assume that the device has the following parameters:

- Transistor $f_T = 300$ MHz
- Base spreading resistance $r_x = 100$ Ω
- Small-signal current gain $h_{fe} = 150$
- Collector–base junction capacitance $C_\mu = 2$ pF
- Collector current bias $I_C = 2$ mA
- Ambient temperature $T_A = 300$ K
- $R_s = R_L = 1$ kΩ

The amplifier is driven from a source resistance $R_s = 1$ kΩ and has a collector resistor $R_L = 1$ kΩ. Our goal in this and the next chapter is to find the gain and the bandwidth by a variety of techniques.

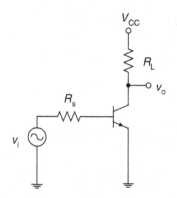

FIGURE 5.5

Common-emitter amplifier (omitting biasing details). We assume that extra components not shown bias this transistor in the FAR. Note that node voltages such as V_{CC} and v_o are measured with respect to ground unless otherwise stated.

We note that the ambient temperature is 300 K and hence the thermal voltage $V_T = kT/q \approx 26$ mV. The transistor is biased on with a collector current of 2 mA, resulting in the following small-signal parameters:

$$g_m = \frac{|I_C|}{V_{TH}} = \frac{2\,\text{mA}}{26\,\text{mV}} = 0.077/\Omega$$

$$r_\pi = \frac{h_{fe}}{g_m} = \frac{150}{0.077/\Omega} = 1950\,\Omega \tag{5.9}$$

$$C_\pi = \frac{g_m}{2\pi f_T} - C_\mu = \frac{0.077}{2\pi(300 \times 10^6)} - 2\,\text{pF} = 38.8\,\text{pF}$$

The small-signal model is shown in Figure 5.6. The simpler circuit of Figure 5.6(a) will be used to find the low-frequency gain of this amplifier. In order to find the high-frequency bandwidth limit, we will use the circuit in Figure 5.6(b).

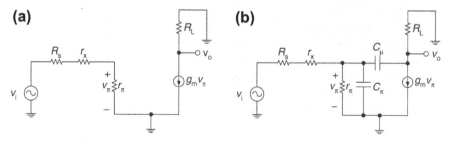

(a)   **(b)**

FIGURE 5.6

Common-emitter amplifier small-signal model. (a) Model valid for low frequencies. (b) Model valid for low and high frequencies.

Low-frequency gain of the common-emitter amplifier

Using the low-frequency small-signal model given in Figure 5.6(a), let us find the low-frequency gain (at frequencies low enough so that the transistor internal capacitors have no effect). The output voltage is given by:

$$v_o = -g_m v_\pi R_L \tag{5.10}$$

The node voltage v_π is given by the voltage divider:

$$v_\pi = v_i \frac{r_\pi}{R_s + r_x + r_\pi} \tag{5.11}$$

Putting this all together, we can easily find the gain:

$$A_V = \frac{v_o}{v_i} = -(g_m R_L)\left(\frac{r_\pi}{R_s + r_x + r_\pi}\right)$$

$$= -(0.077)(1000)\left(\frac{1950}{1000 + 100 + 1950}\right) \approx -49 \tag{5.12}$$

High-frequency bandwidth limit of common-emitter amplifier

It is considerably more work to find the amplifier response including the effects of transistor internal capacitances C_π and C_μ, shown in Figure 5.6(b). The node equations at the v_π and v_o nodes are:[3]

$$1.\ (v_i - v_\pi)G_s' - v_\pi(g_\pi + C_\pi s) + (v_o - v_\pi)C_\mu s = 0 \tag{5.13}$$

$$2.\ (v_\pi - v_o)C_\mu s - v_o G_L - g_m v_\pi = 0$$

where

$$G_s' = \frac{1}{R_s + r_x}$$

$$g_\pi = \frac{1}{r_\pi} \tag{5.14}$$

$$G_L = \frac{1}{R_L}$$

Putting this into matrix form results in:

$$1.\ -v_\pi\left[G_s' + g_\pi + (C_\pi + C_\mu)s\right] + v_o C_\mu s = -v_i G_s' \tag{5.15}$$

$$2.\ v_\pi(C_\mu s - g_m) - v_o(G_L + C_\mu s) = 0$$

or

$$\begin{bmatrix} -\left[G_s' + g_\pi + (C_\pi + C_\mu)s\right] & C_\mu s \\ C_\mu s - g_m & -(G_L + C_\mu s) \end{bmatrix} \begin{Bmatrix} v_\pi \\ v_o \end{Bmatrix} = \begin{Bmatrix} -v_i G_s' \\ 0 \end{Bmatrix} \tag{5.16}$$

---

[3] In many instances, the mathematics is a little easier if you use conductances instead of resistance; the conductance of a resistance of value R is given by $G = 1/R$.

Cramer's rule[4] is used to solve for the output voltage as follows:

$$
v_o = \frac{\det \begin{bmatrix} -[G'_s + g_\pi + (C_\pi + C_\mu)s] & -v_i G'_s \\ C_\mu s - g_m & 0 \end{bmatrix}}{\det \begin{bmatrix} -[G'_s + g_\pi + (C_\pi + C_\mu)s] & C_\mu s \\ C_\mu s - g_m & -(G_L + C_\mu s) \end{bmatrix}}
\tag{5.17}
$$

where the notation "det" denotes the determinant of the matrix. Solving for the transfer function v_o/v_i results in:

$$
\frac{v_o(s)}{v_i(s)} = -(g_m R_L)\left(\frac{G'_s}{G'_s + g_\pi}\right)
$$

$$
\times \left[\frac{1 - \frac{C_\mu}{g_m} s}{\left(\frac{R_L C_\pi C_\mu}{G'_s + g_\pi}\right) s^2 + \left(\frac{1}{G'_s + g_\pi}\right)\left[R_L\left(g_m + g_\pi + G'_s\right)C_\mu + C_\pi + C_\mu\right]s + 1} \right]
$$

$$
\tag{5.18}
$$

It is instructive to break this complicated expression down into various terms:[5]

- The $-g_m R_L$ term is what the DC gain would be if the source resistance and base resistance were zero. This is the maximum gain that you can get from a resistively loaded common-emitter amplifier.
- The second gain term is a degradation term due to the effects of r_π loading down the base resistance r_x and source resistance R_s.
- There is a zero in the right-half plane at frequency $+g_m/C_\mu$, which is higher than the ω_T of the transistor. As we will see later, this zero is at such a high frequency that we can ignore it.
- The denominator shows that there are two poles.

The gain for this amplifier is -49. MATLAB (Figure 5.7) shows that the pole and zero locations are at frequencies:

$$
\omega_z = +3.84 \times 10^{10} \text{ rad/s}
$$

$$
\omega_{p1} = -7.2 \times 10^7 \text{ rad/s}
$$

$$
\omega_{p2} = -2.5 \times 10^9 \text{ rad/s}
$$

Note that both the zero frequency ω_z and the high-frequency pole ω_{p2} are at frequencies higher than the ω_T of the transistor.

---

[4]Cramer's rule is discussed in Chapter 16.

[5]In breaking up the expression into grouped terms, I attempt to show terms grouped in a logical fashion. This follows along with R. D. Middlebrook's concept of "low entropy expressions" (reference at the end of this chapter). We attempt to group terms so that the reader can see the functional dependence of each term in a simple and logical fashion.

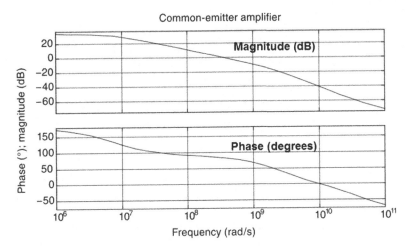

FIGURE 5.7

Frequency response of common-emitter amplifier.

One-pole approximation for estimating bandwidth of common-emitter amplifier

A fundamental limitation of the hybrid-pi model is that it is valid for frequencies significantly lower than the ω_T of the transistor. Given this restriction, the s^2 term in the denominator of the gain expression can be ignored, since:

$$\left| C_\pi C_\mu R_L \omega \right| \ll g_m R_L C_\mu$$

$$\Downarrow \qquad\qquad (5.19)$$

$$\omega \ll \frac{g_m}{C_\pi}$$

Furthermore, we will ignore the zero since it is at a frequency higher than ω_T. Since $g_m \gg g_\pi$, we will ignore g_π wherever possible, resulting in the one-pole approximation:

$$\frac{v_o(s)}{v_i(s)} \approx -(g_m R_L)\left(\frac{G_s'}{G_s' + g_\pi}\right)\left[\frac{1}{\left(\frac{1}{G_s' + g_\pi}\right)\left[\left[1 + (g_m + G_s')R_L\right]C_\mu + C_\pi\right]s + 1}\right]$$

$$(5.20)$$

Note the insight afforded by this approximation. The denominator term has a single pole, with the C_μ term multiplied by $1 + (g_m + G_s')R_L$. This term is, in part, equal to the midband gain of the amplifier—hence the multiplication of the feedback

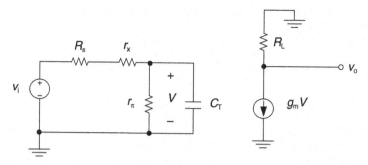

FIGURE 5.8

Circuit illustrating capacitance multiplication caused by the Miller effect.

capacitance due to the *Miller effect*.[6] A circuit model illustrating this approximation is shown in Figure 5.8. We lumped the effects of C_π plus the effects of the Miller effect-modified C_μ into a single equivalent capacitance that we call C_T, with value:

$$C_T = C_\pi + \left[1 + \left(g_m + G_s'\right)R_L\right]C_\mu \tag{5.21}$$

This approximation predicts a dominant pole at:

$$\omega_{dom} = -\frac{G_s' + g_\pi}{\left[1 + \left(g_m + G_s'\right)R_L\right]C_\mu + C_\pi} = \frac{G_s' + g_\pi}{C_T} = -7.29 \times 10^6 \text{ rad/s} \tag{5.22}$$

A plot showing a comparison between the exact result and the Miller approximation is shown in Figure 5.9. Note that at frequencies below $\sim 10^9$ rad/s the magnitude responses track very well.

More discussion of the Miller effect

Using Figure 5.10, we can further illustrate the Miller effect.[7] A capacitor C_f is wrapped around a negative gain of $-A$. By using the circuit of Figure 5.10(b), we can find the input resistance looking into the input of the amplifier. We have applied a voltage source v_t. The amplifier forces the output voltage to be $-Av_t$.

$$i_t = \frac{v_t - (-Av_t)}{\left(\frac{1}{C_f s}\right)} = v_t(1 + A)C_f s \tag{5.23}$$

The input impedance to the amplifier is the ratio of v_t to i_t, or:

$$Z_{in} = \frac{v_t}{i_t} = \frac{1}{(1 + A)C_f s} \tag{5.24}$$

---

[6]As we will see later on, the Miller effect comes into play if you have a negative-gain stage with a capacitance across the input–output terminals.

[7]See Miller's original 1920 paper, referenced at the end of this chapter.

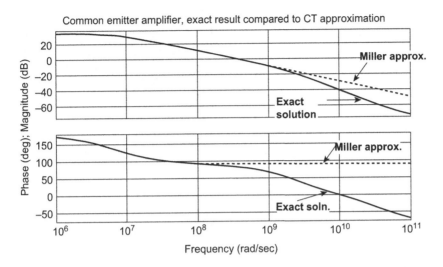

FIGURE 5.9

Comparison of one-pole (C_T or Miller) approximation to exact solution.

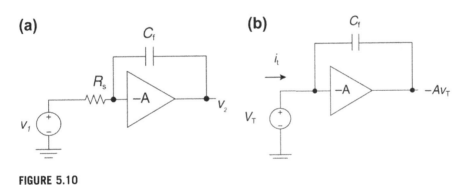

FIGURE 5.10

Illustration of Miller effect. (a) Circuit. (b) Circuit for determining input capacitance.

Therefore, the effect of negative feedback is to make the input capacitance $(1 + A)$ times the feedback capacitance. This effect is known as the *Miller effect* or *Miller multiplication*.

Emitter follower low-frequency gain, input impedance, and output impedance

The emitter follower (Figure 5.11(a)) is a buffer stage with high input impedance, low output impedance, and a gain of approximately unity. Using the small-signal

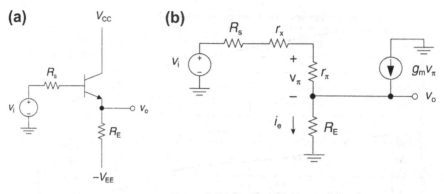

FIGURE 5.11

Emitter follower. (a) Circuit. (b) Small-signal model.

low-frequency circuit of Figure 5.11(b), we will find the gain, input resistance seen at the base, and output resistance seen at the emitter.[8]

Let us first find the input resistance r_{in} looking into the base of the transistor, using Figure 5.12. We have applied a test current source i_t and we will find the test voltage v_t generated by this current source.

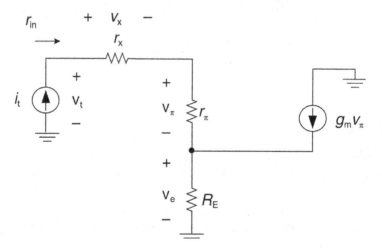

FIGURE 5.12

Circuit for finding small-signal, low-frequency input resistance of emitter follower.

---

[8]Note that the gain, input impedance, and output impedance calculations are valid only for low frequencies. In Chapter 7, we will examine in detail the output impedance of the emitter follower.

The test voltage v_t is:

$$v_t = v_x + v_\pi + v_e = i_t r_x + i_t r_\pi + (i_t + g_m v_\pi)R_E$$

$$\Downarrow$$

$$v_t = i_t r_x + i_t r_\pi + (i_t + g_m(r_\pi i_t))R_E \qquad (5.25)$$

$$\Downarrow$$

$$v_t = i_t \left(r_x + r_\pi + (1 + h_{fe})R_E \right)$$

Making note that $g_m r_\pi = h_{fe}$, we find that the input resistance looking into the base of the emitter follower is:

$$r_{in} = \frac{v_t}{i_t} = r_x + r_\pi + (1 + h_{fe})R_E \qquad (5.26)$$

Note that this input resistance can be large for reasonable values of R_E, since R_E is multiplied by small-signal current gain h_{fe} in finding the input resistance.

Having found the input resistance looking into the base, we are now at a position to easily find the gain of the emitter follower. We will find the gain using the circuit of Figure 5.11(b). We see that the output voltage that we want to find is given by:

$$v_o = i_e R_E \qquad (5.27)$$

The incremental emitter current is given by:

$$i_e = \frac{v_\pi}{r_\pi} + g_m v_\pi = (g_\pi + g_m)v_\pi \qquad (5.28)$$

The voltage v_π is easy to find, since we have previously done the hard work of finding the input resistance to the emitter follower:

$$v_\pi = v_i \left(\frac{r_\pi}{R_s + r_x + r_\pi + (1 + h_{fe})R_E} \right) \qquad (5.29)$$

We now find that the low-frequency voltage gain of the emitter follower is:

$$A_v = \frac{v_o}{v_i} = (g_m + g_\pi) \left(\frac{r_\pi}{R_s + r_x + r_\pi + (1 + h_{fe})R_E} \right) R_E$$

$$\approx \frac{h_{fe}R_E}{R_s + r_x + r_\pi + (1 + h_{fe})R_E} \qquad (5.30)$$

Note that this gain is very close to unity if $h_{fe} R_E \gg R_s$, r_x, and r_π.

Next, we will find the output resistance r_{out} of the emitter follower using the circuit in Figure 5.13. Initially, we will ignore the emitter resistance R_E (and put it back in parallel later on). We apply a test voltage source v_t and calculate the resultant test current i_t. The voltage v_π is given by:

$$v_\pi = -v_t \left(\frac{r_\pi}{R_s + r_x + r_\pi} \right) \qquad (5.31)$$

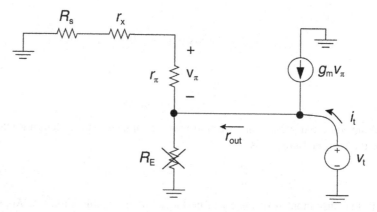

FIGURE 5.13

Circuit for finding small-signal, low-frequency output resistance of an emitter follower. We have omitted R_E for the initial calculation. We will add it back in parallel later on.

The test current i_t is given by:

$$i_t = -g_m v_\pi - \frac{v_\pi}{r_\pi} = -(g_m + g_\pi)v_\pi \tag{5.32}$$

Finally, the test current is:

$$i_t = \left(\frac{g_m r_\pi + 1}{R_s + r_x + r_\pi}\right)v_t \tag{5.33}$$

Therefore, the output resistance of the emitter follower (remembering to put emitter resistance R_E back into the mix[9]) is:

$$r_{out} = \frac{v_t}{i_t} = R_E \left\| \left(\frac{R_s + r_x + r_\pi}{1 + h_{fe}}\right) \right. \tag{5.34}$$

Note that this output resistance is generally small, since the resistors in the base leg are divided by the incremental current gain of the transistor.

Example 5.2: Emitter follower gain and bandwidth

Using the hybrid-pi model, we will find the small-signal gain and bandwidth of an emitter follower, built with a transistor (Figure 5.14(a)) with the following small-signal parameters:

- $f_T = 300$ MHz
- $r_x = 100 \ \Omega$

---

[9]In most cases, we can ignore R_E altogether, since R_E is usually a large bias resistor or a current source, with a resistance much larger than the resistance looking into the emitter of the transistor.

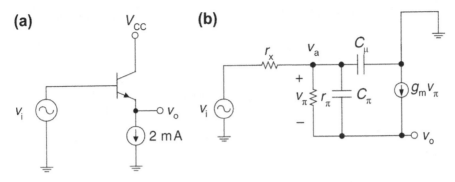

FIGURE 5.14

Emitter follower example. (a) Circuit. (b) High-frequency small-signal model.

- $h_{fe} = 150$
- $C_\mu = 2 \text{ pF}$

The emitter follower is biased with a 2 mA current source, resulting in the following small-signal parameters:

$$g_m = \frac{|I_C|}{V_{TH}} = \frac{2 \text{ mA}}{26 \text{ mV}} = 0.077/\Omega$$

$$r_\pi = \frac{h_{fe}}{g_m} = \frac{150}{0.077/\Omega} = 1950 \ \Omega \tag{5.35}$$

$$C_\pi = \frac{g_m}{2\pi f_T} - C_\mu = \frac{0.077}{2\pi(300 \times 10^6)} - 2 \text{ pF} = 38.8 \text{ pF}$$

The small-signal model for the emitter follower is shown in Figure 5.14(b). Using this small-signal model, the node equations at the v_a and v_o nodes are:

1. $(v_i - v_a)g_x + (v_o - v_a)(g_\pi + C_\pi s) - v_a C_\mu s = 0$

2. $(v_o - v_a)(g_\pi + C_\pi s + g_m) = 0$

(5.36)

Putting this into matrix form, and making the approximation that $g_m = h_{fe}g_\pi \gg g_\pi$ results in:

1. $-v_a[g_x + g_\pi + (C_\pi + C_\mu)s] + v_o(g_\pi + C_\pi s) = -v_i g_x$

2. $v_a(g_\pi + C_\pi s) - v_o(g_\pi + C_\pi s) = 0$

or   (5.37)

$$\begin{bmatrix} -[g_x + g_\pi + (C_\pi + C_\mu)s] & g_\pi + C_\pi s \\ g_\pi + C_\pi s & -(g_\pi + C_\pi s) \end{bmatrix} \begin{Bmatrix} v_a \\ v_o \end{Bmatrix} = \begin{Bmatrix} -v_i g_x \\ 0 \end{Bmatrix}$$

Using Cramer's rule (see Chapter 16) to solve for the transfer function v_o/v_i results in:

$$v_o = \frac{\det\begin{bmatrix} -[g_x + g_\pi + (C_\pi + C_\mu)s] & -v_i g_x \\ g_\pi + C_\pi s & 0 \end{bmatrix}}{\det\begin{bmatrix} -[g_x + g_\pi + (C_\pi + C_\mu)s] & g_\pi + C_\pi s \\ g_\pi + C_\pi s & -(g_\pi + C_\pi s) \end{bmatrix}} \qquad (5.38)$$

Solving for v_o/v_i:

$$\frac{v_o(s)}{v_i(s)} \approx \frac{\frac{C_\pi}{g_m}s + 1}{\frac{r_x C_\pi C_\mu}{g_m}s^2 + \left[r_x C_\mu + \frac{C_\pi}{g_m}\right]s + 1} \qquad (5.39)$$

Note that the denominator is exactly in the form:

$$(\tau_1 s + 1)(\tau_2 + 1) = \tau_1 \tau_2 s^2 + (\tau_1 + \tau_2)s + 1 \qquad (5.40)$$

Therefore, the two poles are at frequencies:

$$\omega_{p1} = \frac{1}{\tau_1} = -\frac{g_m}{C_\pi} = -1.95 \times 10^9 \text{ rad/s}$$

$$(5.41)$$

$$\omega_{p2} = -\frac{1}{\tau_2} = -\frac{1}{r_x C_\mu} = -5 \times 10^9 \text{ rad/s}$$

Note that there is also a zero at $-C_\pi/g_m$, which in this simplified model cancels the pole at $-\omega_{p1}$. (Actually, the pole and zero do not exactly cancel. This is an artifact of our ignoring g_π as compared to g_m.)

The bandwidth for this circuit is:

$$\omega_h \approx -\omega_{p2} = 5 \times 10^9 \text{ rad/s}$$

$$(5.42)$$

$$f_h \approx \frac{\omega_h}{2\pi} = 795 \text{ MHz}$$

which matches closely the LTSPICE result (Figure 5.15). This result shows that the gain $= 1$ (as expected) and the -3 dB bandwidth is approximately 800 MHz.

This result for 800 MHz bandwidth is in serious question, especially in the frequency range higher than 100 MHz. First, this model predicts a bandwidth in excess of the f_T of the transistor, where the hybrid-pi model is only valid to $f_T/5$ or so. Second, in this simplified model, the inductance of the lead connections was not included. In Figure 5.16, we see a circuit where 0–20 nH of lead inductance is included in the base and collector leads. The following LTSPICE plot shows that the parasitic inductance does indeed degrade bandwidth in the high-frequency range. In a later chapter in this book, we discuss in more detail the high-frequency behavior of the emitter follower.

(a)

Emitter follower small-signal model
Parameters chosen given f_T = 300 MHz, C_μ = 2 pF, h_{fe} = 150, r_x = 100
MTT 9/10/12

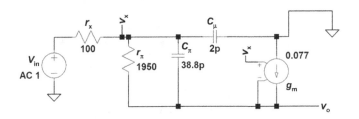

.ac dec 100 10e6 1000e6
file: Emitter follower small signal.asc

(b)

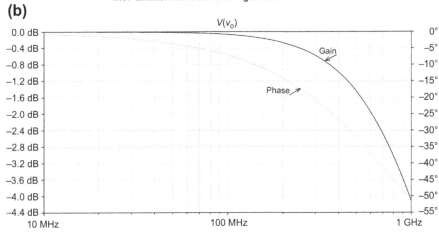

FIGURE 5.15

LTSPICE result for emitter follower, showing bandwidth of approximately 800 MHz and a gain
of about 1 (0 dB). (a) Circuit. (b) Gain magnitude and phase. $V(v_0)$ is the voltage at the v_0
node. (For color version of this figure, the reader is referred to the online version of this book.)

Differential amplifier

The transistor differential amplifier (Figure 5.17) is a ubiquitous building block used
commonly as the front-end of an operational amplifier. The differential amplifier is
used to amplify the difference between its two inputs while rejecting the DC value
common to the two inputs.[10]

If we do KVL around the voltage source and base–emitter junctions, we find:

$$-V_1 + V_{BE1} - V_{BE2} + V_2 = 0 \qquad (5.43)$$

---

[10]Said another way, the differential amplifier is designed to have high "differential-mode" gain and
very low "common-mode" gain.

(a)

Emitter follower with parasitics
Added small lead inductance in the base and collector leads
MTT 9/10/12

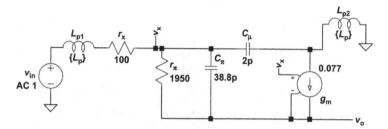

.ac dec 100 10e6 1000e6
.step param Lp list 0.0001n 10n 20n
file: Emitter follower small signal with parasitics.asc

(b)

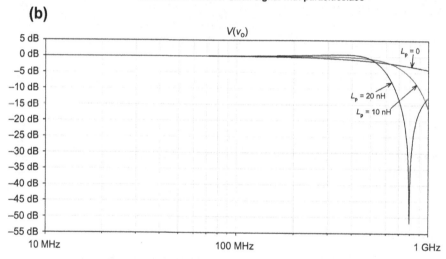

FIGURE 5.16

LTSPICE result for emitter follower, with lead inductance 0–20 nH. (a) Circuit. (b) Gain magnitude for parasitic inductance of 0, 10 and 20 nH. (For color version of this figure, the reader is referred to the online version of this book.)

From the ideal transistor relationships and with $V_{BE} \gg kT/q$ (~ 26 mV at room temperature), we find:

$$I_{C1} \approx I_s e^{\frac{qV_{BE1}}{kT}} \Rightarrow V_{BE1} \approx \frac{kT}{q} \ln\left(\frac{I_{C1}}{I_s}\right)$$

$$I_{C2} \approx I_s e^{\frac{qV_{BE2}}{KT}} \Rightarrow V_{BE2} \approx \frac{kT}{q} \ln\left(\frac{I_{C2}}{I_s}\right)$$

(5.44)

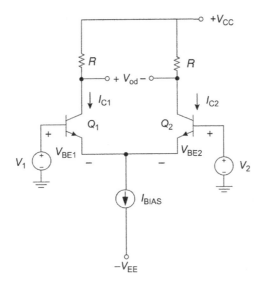

FIGURE 5.17

Ideal transistor differential amplifier. Note that the differential input voltage V_{id} is the difference between V_1 and V_2, or $V_{id} = V_1 - V_2$.

Combining the equations above results in an equation relating the ratio of the two collector currents:

$$\frac{I_{C1}}{I_{C2}} = e^{\frac{q(V_1 - V_2)}{kT}} = e^{\frac{qV_{id}}{kT}} \tag{5.45}$$

We have defined the differential input voltage as the difference between the two inputs, or $V_{id} = V_1 - V_2$. Now, if $\beta_F \gg 1$ then $I_E \approx I_C$ for each transistor and $I_{C1} + I_{C2} \approx I_{BIAS}$, and we can write:

$$I_{C1} \approx \frac{I_{BIAS}}{1 + e^{\frac{-qV_{id}}{kT}}}$$

$$I_{C2} \approx \frac{I_{BIAS}}{1 + e^{\frac{qV_{id}}{kT}}} \tag{5.46}$$

The exponential terms mean that the range of V_{id} over which I_{C1} and I_{C2} vary is only a few kT/q. A plot of collector currents is shown in Figure 5.18(a). We can also find the differential output voltage V_{od} by recognizing that V_{od} is the difference between the two transistor collector voltages. In Figure 5.18(b), we implicitly assume that the load resistors R are small enough so that Q_1 and Q_2 do not saturate. The range of voltages over which approximately linear amplification occurs is on the order of $|V_{id}| < 25\ \text{mV}$. Note that the differential output voltage depends only on

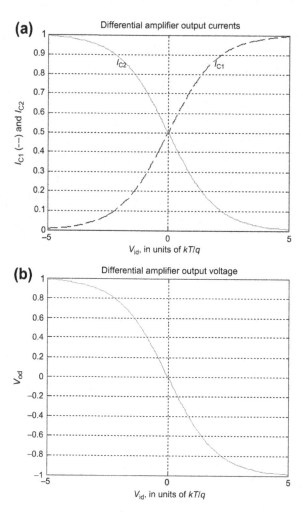

FIGURE 5.18

Ideal transistor differential amplifier outputs. (a) Output currents I_{C1} and I_{C2} in units of I_{BIAS} variation as V_{id} varies from -5 kT/q to $+5$ kT/q. (b) Differential output voltage V_{od} in units of $I_{BIAS}R$. (For color version of this figure, the reader is referred to the online version of this book.)

the differential input voltage V_{id} and *not* on other design parameters such as bias current level I_{BIAS} or power supply voltage V_{CC}. This is one of the advantages of the fully differential amplifier topology.

To analyze the differential gain, we make use of a *half-circuit* technique. Let us assume that we are operating in a mode when both Q_1 and Q_2 are biased ON, and hence $V_1 - V_2$ is a small voltage, less than a few kT/q. In this mode, the differential input voltage is small enough so that both transistors carry approximately the same

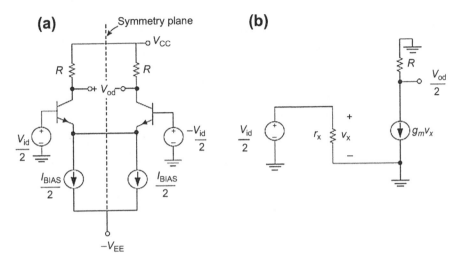

FIGURE 5.19

Illustration of differential-mode operation. (a) Differential-mode operation showing symmetry plane. (b) Small-signal model of differential-mode half circuit.

current (or $\sim I_{BIAS}/2$). In the differential mode of operation, when V_1 moves up, V_2 moves down, and vice versa. Therefore, the emitters of the coupled pair remain at incremental ground during this operation. Therefore, we can analyze the left and right sides of the circuits independently. A circuit valid for the differential mode is shown in Figure 5.19(a).[11] Since the voltage at the symmetry plane does not move when we are in the differential mode, we can ground the symmetry plane, resulting in the incremental circuit of Figure 5.19(b)[12] valid for differential-mode operation. From this circuit, we find that the differential-mode incremental gain ($A_{V,DM}$) is:

$$\frac{v_{od}}{v_{id}} = A_{V,DM} = -g_m R$$

$$g_m = \frac{I_{BIAS}}{2V_T} = \frac{I_{BIAS}}{2\left(\dfrac{kT}{q}\right)} \tag{5.47}$$

A real-world differential amplifier has an important parasitic element—the finite incremental output resistance of the current source. This is shown in Figure 5.20(a), where we see that the I_{BIAS} current source has a finite output resistance R_{CS}. We all

---

[11]Note that in order to further illustrate the symmetry of the circuit, we have broken the bias generator into two equal current sources of value $I_{BIAS}/2$.
[12]For simplicity, we assume that base spreading resistance $r_x = 0$.

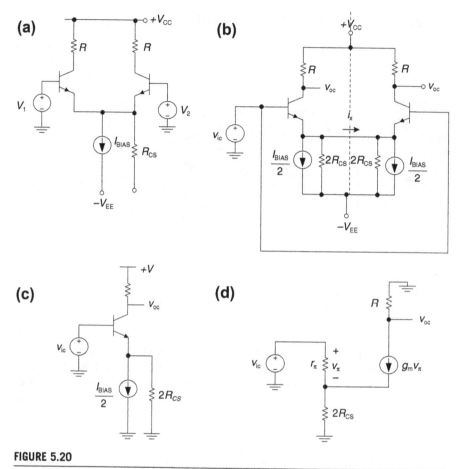

FIGURE 5.20

Differential amplifier showing a parasitic element that affects common-mode gain. (a) Bias current source with finite output resistance R_{CS}. (b) Circuit redrawn showing common-mode symmetry plane. (c) Common-mode half circuit. (d) Incremental circuit valid for common-mode analysis.

see that this finite resistance affects how well the amplifier rejects *common-mode* signals. Common-mode signals are signals that are common to both inputs, as shown in Figure 5.20(b). The same input v_{ic} is applied to the two transistor inputs. Note that in this figure we have redrawn the circuit showing the symmetry plane.

Again, we can exploit symmetry in order to simplify circuit analysis. Note that the same input is applied to the bases of both transistors and that the circuit is symmetric. Therefore, there is no current across the symmetry plane (i.e. current $i_x = 0$). Therefore, we are free to cut the circuit apart at the symmetry plane. This results in the common-mode model of Figure 5.20(c).

Using Figure 5.20(d), the incremental model for the common-mode half circuit, we can find the common-mode gain $A_{V,CM}$, skipping over the mathematical manipulations:[13]

$$\frac{v_{oc}}{v_{ic}} = A_{V,DM} = -(g_m R)\left(\frac{1}{1 + 2g_m R_{CS}}\right)$$

$$g_m = \frac{I_{BIAS}}{2V_T}$$

(5.48)

An important figure of merit is the ratio of differential-mode gain to common-mode gain. In a differential amplifier, we want a high differential-mode gain and a very low (ideally zero) common-mode gain. The common-mode rejection ratio (CMRR) of a differential amplifier is given by:

$$CMMR = \frac{|A_{V,DM}|}{|A_{V,CM}|} = 1 + 2g_m R_{CS}$$

(5.49)

Therefore, to achieve high CMRR we need to bias the differential pair with a current source with high output resistance.

An analogy to help explain differential and common modes of operation

A simple mechanical analogy (Figure 5.21) can illustrate differential and common-mode operation.[14] Let us assume that we have a teeter-totter, with the motion of the ends of the teeter-totter being analogous to the voltage drive levels at the inputs of the differential amplifier. In the differential mode of operation (Figure 5.21(a)), the center of the teeter-totter is fixed, and one input goes UP and the other goes DOWN

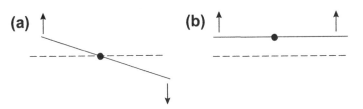

FIGURE 5.21

Teeter-totter analogy to differential- and common-mode operation. (a) Differential mode. (b) Common mode.

[13]We can use the previous work we did on the input resistor of the emitter follower to arrive at this result. The input resistance at the base of the transistor is: $r_\pi + (1 + h_{fe})(2R_{CS})$. We can now find v_π:

$v_\pi = \dfrac{r_\pi}{r_\pi + (1 + h_{fe})(2R_{CS})}$. Next, we find the common-mode gain: $\dfrac{v_{oc}}{v_{ic}} = \dfrac{-g_m R r_\pi}{r_\pi + (1 + h_{fe})(2R_{CS})} =$

$\dfrac{-g_m R}{1 + \frac{(1 + h_{fe})}{r_\pi}(2R_{CS})} \approx \dfrac{-g_m R}{1 + -2g_m R_{CS}}$.

[14]Thanks go to Prof. John McNeill at Worcester Polytechnic Institute, whose course notes on undergraduate IC design use this analogy.

the same amount. This motion of the ends of the teeter-totter is analogous to the variation in the inputs to the differential amplifier in the differential mode.

In the common mode of operation (Figure 5.21(b)), the center of the teeter-totter is allowed to move up and down, but both ends go UP or DOWN equally. This is analogous to the common mode of operation when a common voltage drives the differential amplifier.

Example 5.3: Shunt peaked amplifier

Shown in Figure 5.22(a) is an inductively peaked amplifier. Shunt peaking was used in the days of tube amplifiers to extend bandwidth. Assume that the transistor has parameters: $f_T = 800$ MHz, $h_{fe} = 100$, $C_\mu = 0.7$ pF, and $r_x \sim 0.$[15] As we have seen earlier, the gain of the common-emitter amplifier rolls-off at a finite frequency

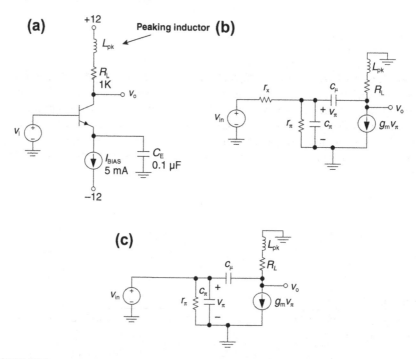

FIGURE 5.22

Shunt peaked amplifier. (a) Circuit with peaking inductor L_{pk}. (b) Small-signal model. (c) Simplified small-signal model assuming that r_x is very small.

---

[15]Ignoring r_x leads to a mathematical result that is relatively easy to understand. As we will see in a future chapter (Chapter 7), r_x can often be a dominant bandwidth limitation in circuits that have a low source resistance. So, ignore r_x at your peril!

due to the bandwidth limiting effects of C_π and C_μ. One way to counteract this bandwidth limit is to use an inductor in series with the load resistor. We add a zero to the transfer function to enhance the bandwidth; at high frequencies, the impedance of the load increases. Using this technique, the bandwidth can be extended if the peaking circuit is properly designed.

The small-signal model of this amplifier is shown in Figure 5.22(b), assuming operation at a frequency high enough where the emitter bypass capacitor behaves as a short circuit. If we assume that r_x is very small, we can further simplify the small-signal circuit as in Figure 5.21(c). The small-signal parameters are:

$$g_m = \frac{|I_C|}{V_{TH}} = \frac{5\ \text{mA}}{26\ \text{mV}} \approx 0.19/\Omega$$

$$r_\pi = \frac{h_{fe}}{g_m} = \frac{100}{0.19} = 526\ \Omega \tag{5.50}$$

$$C_\pi = \frac{g_m}{2\pi f_T} - C_\mu = \frac{0.19}{(2\pi \times 800 \times 10^6)} - 0.7\ \text{pF} = 37\ \text{pF}$$

Using the small-signal model, we can find the midband gain as:

$$A_v = g_m R_L = -190 \tag{5.51}$$

Using the small-signal model of Figure 5.22(c), we can find the overall gain of this amplifier by writing node equations. Let us sum currents at node v_o:

$$(v_i - v_o)C_\mu s - g_m v_i - \frac{v_o}{R_L + L_{pk}s} = 0 \tag{5.52}$$

Solving for the input–output transfer function results in:

$$\frac{v_o(s)}{v_i(s)} = -g_m R_T \frac{\left(1 + \frac{L_{pk}}{R_L}s\right)\left(1 - \frac{C_\mu}{g_m}s\right)}{L_{pk}C_\mu s^2 + R_L C_\mu s + 1} \approx -g_m R_L \frac{\left(1 + \frac{L_{pk}}{R_L}s\right)}{L_{pk}C_\mu s^2 + R_L C_\mu s + 1} \tag{5.53}$$

Note that the addition of an inductor results in a zero at $\omega_z = -R_L/L_{pk}$ rad/s, as well as a second-order denominator polynomial. For $L_{pk} = 0$, this transfer function predicts a bandwidth of $1/(R_L C_\mu)$, or $2\pi \times 227 \times 10^6$ rad/s (227 MHz). This is confirmed by LTSPICE (Figure 5.23), which also shows that the midband voltage gain is -190. With an inductor $L_{pk} = 250$ nH, the -3 dB bandwidth extends to approximately 390 MHz, with a little bit of gain peaking.[16] With larger values of L_{pk}, the response becomes very peaky, as shown in the LTSPICE plots.

---

[16]Detailed mathematical analyses of similar circuits are given in Tom Lee's *The Design of CMOS Radio-Frequency Circuits*, pp. 178–184.

(a)   **Shunt peaked amplifier**

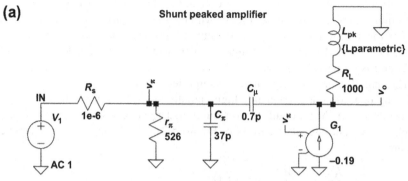

.ac dec 100 1e6 1000e6
.step param Lparametric 0.1n 1000.1n 250n
file: Shunt peaked amplifier.asc

(b)

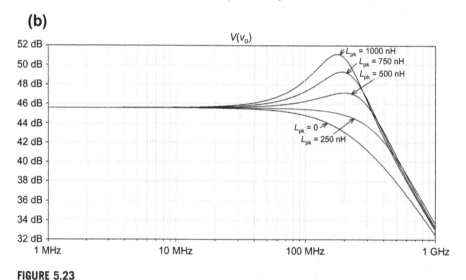

FIGURE 5.23

LTSPICE analysis of shunt peaked amplifier. (a) LTSPICE circuit. (b) Magnitude plot, for $L_{pk} = 0 \ldots 1000$ nH. (For color version of this figure, the reader is referred to the online version of this book.)

Example 5.4: Simple tuned amplifier

Inductors and capacitors can be used in the collector load to design a frequency-selective amplifier. Shown in Figure 5.24(a) is a tuned amplifier. The collector current is approximately 2 mA corresponding to a transconductance of $g_m = 0.077/\Omega$. Maximum gain will be achieved at the resonant frequency of the LC network:

$$f_n = \frac{1}{2\pi\sqrt{L_L C_L}} = 159 \text{ kHz} \tag{5.54}$$

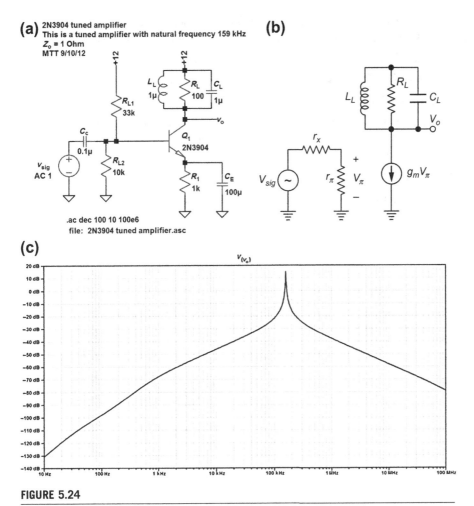

FIGURE 5.24

Simple tuned amplifier. (a) Circuit. (b) Small-signal model assuming that coupling capacitor C_c and emitter bypass capacitor C_E act as short circuits and assuming low enough frequency that transistor internal capacitances C_π and C_μ do not come into play. (c) LTSPICE plot of amplifier gain (in dB) vs. frequency. (For color version of this figure, the reader is referred to the online version of this book.)

Looking at the small-signal model (Figure 5.24(b), with assumptions that the bypass and coupling capacitors act as short circuits and that the transistor internal capacitances have negligible effect on the transfer function), we can figure out the maximum gain of this amplifier. At resonance (159 kHz), the inductor and capacitor load impedances cancel, and the gain of the amplifier at resonance is approximately:

$$A_{\text{v}}|_{159 \text{ kHz}} \approx -\left(\frac{r_\pi}{r_x + r_\pi}\right)(g_m R_L) \tag{5.55}$$

The magnitude of the gain at 159 kHz is approximately $g_m R_L = -7.7$, or 17.7 dB. This is confirmed by the LTSPICE simulation (Figure 5.24(c)).

Chapter 5 problems

Problem 5.1

For the circuit in Figure 5.4(a), find the variation in collector current if V_{BE} varies from 0.6 to 0.8 V. Assume that current gain remains constant at $\beta_F = 100$.

Problem 5.2

For the circuit in Figure 5.25, calculate bias point values:

a. I_{C1}

b. I_{C2}

c. V_{E1}

d. V_{E2}

Throughout, assume that $V_{BE} = 0.7$ V and $\beta_F = 100$.

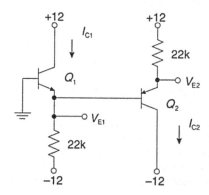

FIGURE 5.25

Transistor circuit for Problem 5.2.

Problem 5.3

The emitter follower circuit of Figure 5.26 is biased at a constant emitter current of 1 mA. Using LTSPICE, find the variation in transistor V_{BE} as the ambient temperature rises from 25 to 75 °C.

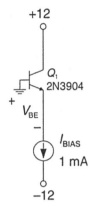

FIGURE 5.26

Circuit for Problem 5.3.

Problem 5.4

a. Using the incremental model calculate the "midband" gain of the transistor amplifier in Figure 5.27, v_o/v_{in}. Midband is the frequency range high enough so that C_c and C_E behave as short circuits but low enough so that the transistor internal capacitances have little or no effect.

b. Using the one-pole approximation (Miller approximation), calculate the high-frequency -3 dB bandwidth f_H (in Hz).

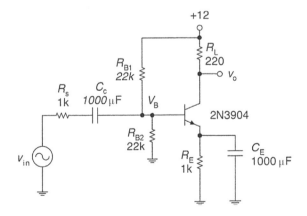

FIGURE 5.27

Common-emitter amplifier with biasing network.

c. Sketch the Bode plot of the magnitude response of v_o/v_{in}, but do not calculate the low-frequency breakpoint that depends on coupling and bypass capacitors C_c and C_E (more on this later). Plot f_h in Hz.

d. Simulate your circuit using LTSPICE, and compare your results to calculations. In your LTSPICE calculation, do not use the LTSPICE built-in 2N3904 model; instead, directly input the hybrid-pi model you derive above.

Problem 5.5

For the transistor circuit of Figure 5.28, calculate:

a. The operating point current I_C and the DC value of V_{out}.

b. Small-signal parameters r_π and g_m.

c. Draw the midfrequency small-signal model, assuming C_{in} is a short circuit. Assume that the frequency is low enough so that the transistor internal capacitances have no effect.

d. Calculate the AC input resistance looking into the base terminal of the transistor (assuming C_{in} acts as a short circuit at signal frequencies).

e. Find AC voltage gain $A_v = |v_{out}/v_{in}|$ (again, assuming that C_{in} acts as a short circuit). (*Hint: there is a clever way to use the results of part (4) to greatly simplify this result.*)

f. Calculate the low-frequency breakpoint of A_v. (*Hint: at DC, the gain of this circuit is zero, due to C_{in}. At high frequency, the gain is what you calculate in part (5). The breakpoint is where the gain has risen to −3 dB from maximum value*). Use the result from part (4) to find this.

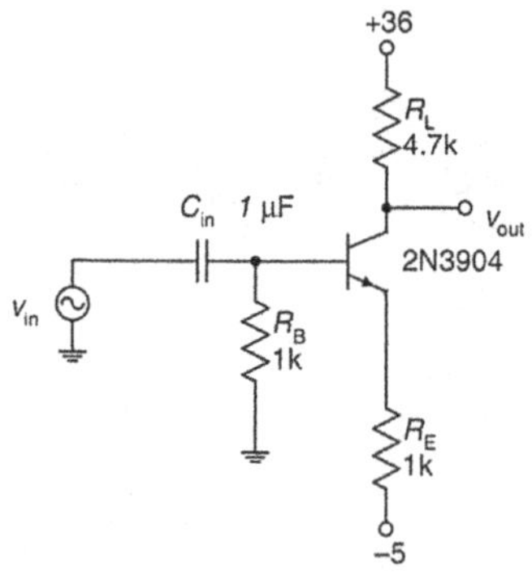

FIGURE 5.28

Circuit for Problem 5.6.

g. Sketch (but do not calculate a high-frequency breakpoint) a Bode plot (magnitude only) of the gain.

Throughout the problem, make reasonable approximations and justify them.

Problem 5.6

For the transistor circuit of Figure 5.29,

a. Estimate collector current at −55, 25, and 125 °C, using information from the 2N2222 datasheet as a guide and a closed-form solution for collector current. Assume that $V_{BE} = 0.7$ V at 25 °C and that the transistor V_{BE} has a temperature coefficient of −2 mV/°C.

b. Simulate using LTSPICE at the three different ambient temperatures. Comment on how well the LTSPICE result matches up with what you calculated in part (1). Also, comment on the bias stability of this circuit and how you might improve the bias stability.

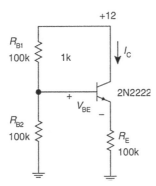

FIGURE 5.29

Circuit for Problem 5.7.

Problem 5.7

For the circuit of Figure 5.30, assume that $r_x = 20\ \Omega$ and $V_{BE} = 0.7$ V.

a. Using the 2N2222 datasheet, find reasonable values for DC current gain h_{FE} and small-signal current gain h_{fe}.

b. Find collector current I_C and base current I_B.

c. Find small-signal parameters g_m and r_π.

d. Draw the low-frequency incremental circuit.

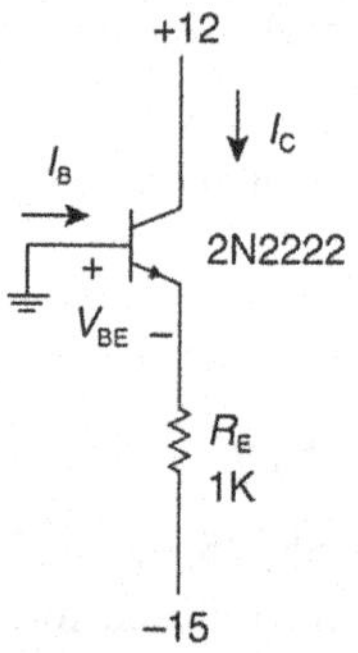

FIGURE 5.30

NPN transistor circuit built with 2N2222. (*For color version of this figure, the reader is referred to the online version of this book.*)

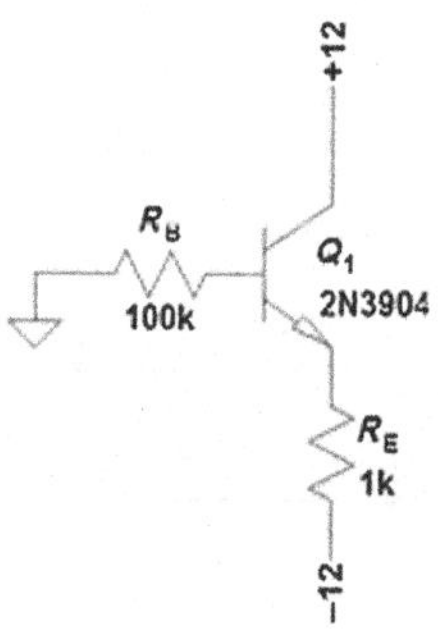

FIGURE 5.31

Transistor circuit for Problem 5.9. (For color version of this figure, the reader is referred to the online version of this book.)

Problem 5.8

In the 2N3904 circuit in Figure 5.31, the collector current will vary with temperature due to the dependencies of V_{BE} and β_F calculated above.

a. Calculate the operating point values of collector current I_C and base–emitter voltage V_{BE} at 25, 50, and 75 °C. You may assume that $V_{BE} = 725$ mV at $T_A = 25$ °C. Assume that the temperature dependence of V_{BE} in a transistor at constant current is approximately -2.2 mV/°C and the temperature dependence of β_F is in the neighborhood of $+7000$ ppm/°C.

b. Now, simulate using LTSPICE and compare your values to that calculated in part (1) above. Can you draw any assumptions about the relative bias stability of this circuit? How might you improve the bias stability?

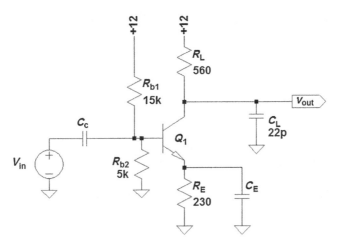

FIGURE 5.32

Circuit for Problem 5.10. (For color version of this figure, the reader is referred to the online version of this book.)

Problem 5.9

The common-emitter amplifier in Figure 5.32 is driven by a small-signal voltage source v_{in}. The transistor has the following parameters: $f_T = 500$ MHz, $\beta_F = 250$, $h_{fe} = 200$, and $r_x = 50\ \Omega$. The circuit operates at ambient temperature 25 °C.

a. Assume that capacitors C_c and C_E are infinitely large. Estimate the low-frequency small-signal parameters for this amplifier. Make reasonable estimates and justify them.

b. Estimate the midband gain of this amplifier.

Further reading

[1] Chuang CT. Analysis of the settling behavior of an operational amplifier. *IEEE J Solid-State Circuits* 1982;**17**(1):74−80.
[2] Filipkowski A. Poles and zeros in transistor amplifiers introduced by Miller effect. *IEEE Trans Educ* 1999;**42**(4):349−51.
[3] Gilbert B, All you ever need to know about bandgaps, lecture 1978.
[4] Gray PE, Searle CL. *Electronic principles physics, models and circuits*. John Wiley; 1969.
[5] Gray PR, Meyer RG. *Analysis and design of analog integrated circuits*. 2nd ed. John Wiley; 1984.
[6] Knapp R. Selection criteria assist in choice of optimum reference. *EDN* Feb 18, 1988:183−92.
[7] Knapp R. Back-to-basics approach yields stable references. *EDN* Jun 9, 1988:193−8. Detailed description of bandgap and Zener voltage references.

[8]   Lee TH. *The design of CMOS radio-frequency integrated circuits*. Cambridge University Press; 1998.

[9]   Lindmayer J, North W. The inductive effect in transistors. *Solid-State Electron* 1965;**8**:409–15.

[10] Mercer MJ, Burns SG. High-frequency broadband amplifier ASIC design optimization using pole-zero compensation techniques. In: *IEEE international symposium on circuits and systems* 1990. p. 3225–9.

[11] Middlebrook RD. Low-entropy expressions: the key to design-oriented analysis. In: *Proceedings of the twenty-first annual conference "Engineering education in a new world order"* 1991. p. 399–403.

[12] Miller JM. Dependence of the input impedance of a three-electrode vacuum tube upon the load on the plate circuit. *Sci Pap Bur Stand* 1920;**15**(351):367–85.

[13] Muller RS, Kamins TI. *Device electronics for integrated circuits*. 2nd ed. John Wiley; 1986.

[14] Neudeck GW. *Modular series on solid state devices, the bipolar junction transistor*. Addison-Wesley; 1983.

[15] Pease B. The design of band-gap reference circuits: trials and tribulations. In: *IEEE proceedings of the 1990 bipolar circuits and technology meeting* September 17–18, 1990. Minneapolis, Minnesota.

[16] Rincón-Mora GA. *Voltage references*. IEEE; 2002.

[17] Searle CL, Boothroyd AR, Angelo Jr EJ, Gray PE, Peterson DO. *Elementary circuit properties of transistors*In *SEEC*, vol. 3. John Wiley; 1964.

[18] Sze SM. *Physics of semiconductor devices*. 2nd ed. John Wiley; 1981.

[19] Thompson MT. Design linear circuits using OCTC calculations. *Electron Des (Special Analog Issue)* June 24, 1993:41–7.

[20] Thompson MT. SCTC analysis estimates low-frequency −3-dB point. *Electron Des* 1993:65–8.

[21] Thompson MT. Network tricks aid in OCTC. *Electron Des* 1993:67–70.

[22] Thompson MT. Tips for designing high-gain amplifiers. *Electron Des* 1994:83–90.

[23] Thornton RD, DeWitt D, Chenette ER, Gray PE. *Characteristics and limitations of transistors*In *SEEC*, vol. 4. John Wiley; 1966.

[24] Thornton RD, Searle CL, Pederson DO, Adler RB, Angelo Jr EJ. *Multistage transistor circuits*In *SEEC*, vol. 5. John Wiley; 1965.

[25] Widlar Robert J. New developments in IC voltage regulators. *IEEE J Solid-State Circuits* 1971;**SC-6**(1):2–7.

[26] Widlar RJ. An exact expression for the thermal variation of the emitter base voltage of bipolar transistors. *Proc IEEE* 1967:96–7 (Widlar was the inventor of the National Semiconductor LM113 reference diode, which is a monolithic band-gap voltage reference introduced in 1971).

[27] Wing-Hung Ki, Der L, Lam S. Re-examination of pole splitting of a generic single stage amplifier. *IEEE Trans Circuits Syst I: Fundamental Theory and Applications* 1997;**44**(1):70–4.

[28] Yahya CB. Design of wideband low noise transimpedance amplifiers for optical communications. In: *Proceedings of the 43rd IEEE midwest symposium on circuits and systems* 2000. p. 804–80.

Amplifier Bandwidth Estimation Techniques

"... the lowest natural frequency of the network can be estimated simply by forming the sum of the open-circuit time constants obtained by examining each capacitor with the others open-circuited."

—R. D. Thornton, C. L. Searle, D. O. Pederson, R. B. Adler and E. J. Angelo, Jr., "Multistage Transistor Amplifiers", SEEC volume 5, 1965

IN THIS CHAPTER

▶ It is possible to use SPICE or other circuit simulators to determine the gain and bandwidth of transistor amplifiers, or for that matter any circuit with lumped elements and dependent sources. However, while SPICE is a useful tool for analyzing circuits, it does not give that much insight and intuition into amplifier design. This chapter covers "back-of-the-envelope" techniques for estimating the bandwidth of transistor amplifiers. The techniques in this chapter are employed on bipolar transistor amplifiers, but can also be used for CMOS.

Introduction to open-circuit time constants

The method of open-circuit time constants (OCTCs) is a powerful approximate analysis tool that allows estimation of the high-frequency −3 dB bandwidth of circuits containing resistors, capacitors, and dependent sources. The method was developed by R. B. Adler and others at MIT[1] and is a valuable design tool in finding the root cause of bandwidth limitation in circuits. The usefulness of the technique lies in the fact that it allows identification of local bandwidth limitations in amplifiers.

Following a brief mathematical discussion of the method of OCTCs, the method is applied to several different amplifier topologies. The OCTC results are compared to closed-form solutions and LTSPICE simulations.

---

[1]See, e.g. the Semiconductor Electronics Education Committee's *Multistage Transistor Circuits*, volume 5, by Thornton, Searle, Pederson, Adler, and Angelo.

Intuitive Analog Circuit Design. http://dx.doi.org/10.1016/B978-0-12-405866-8.00006-1

At first, let us consider a system with n real-axis poles (Figure 6.1). The transfer function for this system is:

$$H(s) = \frac{1}{(\tau_1 s + 1)(\tau_2 s + 1)\cdots(\tau_n s + 1)} \tag{6.1}$$

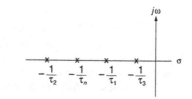

FIGURE 6.1

System with multiple negative real-axis poles.

Expanding the denominator results in:

$$H(s) = \frac{1}{(\tau_1 \tau_2 \cdots \tau_n)s^n + \cdots + (\tau_1 + \tau_2 + \cdots + \tau_n)s + 1} \tag{6.2}$$

We note that the first-order s term has as its coefficient the sum of reciprocal pole locations. In a simpler case, let us look at a system with two negative real-axis poles, with transfer function:

$$H(s) = \frac{1}{\tau_1 \tau_2 s^2 + (\tau_1 + \tau_2)s + 1} \tag{6.3}$$

In using the method of OCTCs, we throw away the higher order s terms. The method of OCTCs approximates the second-order transfer function as:

$$H(s) \approx \frac{1}{(\tau_1 + \tau_2)s + 1} \tag{6.4}$$

Using the method of OCTCs, our estimate of bandwidth is:

$$\omega_h \approx \frac{1}{(\tau_1 + \tau_2)} \tag{6.5}$$

When is this approximation justified? We are ignoring the s^2 term in comparison to the s term in the denominator, or:

$$\left| \tau_1 \tau_2 \omega^2 \right| \ll \left| (\tau_1 + \tau_2)\omega \right| \tag{6.6}$$

This approximation is justified when:

$$\omega \ll \frac{(\tau_1 + \tau_2)}{\tau_1 \tau_2} \ll \frac{1}{\tau_1} + \frac{1}{\tau_2} \tag{6.7}$$

Assume that there is a high-frequency (ω_h) and a low-frequency (ω_l) pole, given by:

$$\omega_l = \frac{1}{\tau_1}$$
$$\omega_h = \frac{1}{\tau_2}$$

(6.8)

This approximation says that our estimate is valid if we are interested in frequencies well below the high-frequency pole. This is surely valid if the amplifier response is dominated by the low-frequency pole ω_l.

How accurate is the estimate? Let us take a look at the second-order system transfer function:

$$H(j\omega) = \frac{1}{-\tau_1\tau_2\omega^2 + \cdots + (\tau_1 + \tau_2)(j\omega) + 1}$$

(6.9)

Let us assume that the sum of the pole time constants $\tau_1 + \tau_2$ is 1 s. Therefore, our estimate for -3 dB bandwidth of this amplifier is 1 rad/s. We will rewrite the transfer function using $\omega_{h,est}$ as our estimate for -3 dB bandwidth:

$$H(j\omega) = \frac{1}{-\tau_1\tau_2\omega_{h,est}^2 + \cdots + (\tau_1 + \tau_2)(j\omega_{h,est}) + 1}$$

(6.10)

Are we justified in neglecting the ω^2 term in the denominator of the polynomial? We note that the magnitude of the ω term is one, since $\omega_{h,est} = 1$ is our estimate for bandwidth. The maximum value of the ω^2 term is 0.25. So, for estimating bandwidth this approximation is justified.

Similarly, we can estimate the bandwidth of a higher order system with n poles as the reciprocal of the sum of the pole time constants τ_{pn}:

$$H(s) \approx \frac{1}{(\tau_1 + \tau_2 + \cdots + \tau_n)s + 1}$$
$$\omega_h \approx \frac{1}{(\tau_1 + \tau_2 + \cdots + \tau_n)} = \frac{1}{\sum\limits^{n} \tau_{pn}}$$

(6.11)

How do we find the sum of pole time constants $(\tau_1 + \tau_2 + \ldots \tau_n)$? A proof by Adler and others at MIT shows that the *sum of the pole time constants* is exactly equal to the *sum of OCTCs*, which is[2] given as:

$$\tau_1 + \tau_2 + \cdots + \tau_n = \sum\limits_{i=1}^{n} \tau_{oi}$$

(6.12)

---

[2]Each individual pole time constant is not the same as each OCTC. However, the *sum* of the pole time constants is equal to the *sum* of the OCTCs.

The approximate bandwidth of the amplifier is:

$$\omega_h \approx \frac{1}{\sum_{i=1}^{n} \tau_{oi}} \tag{6.13}$$

This approximation is often surprisingly accurate and the OCTCs are often *easy* to calculate, while actually the pole locations are usually very *difficult* to calculate. To calculate each OCTC τ_{oi}, do the following steps:

- Open circuit all capacitances in the system that contribute to high-frequency bandwidth limitations.
- Find the resistance facing each capacitor's terminals, one by one.
- Each individual time constant is found by:

$$\tau_{oi} = R_{oi} C_i \tag{6.14}$$

- Sum the individual OCTCs and find the estimate of bandwidth.

$$\omega_h \approx \frac{1}{\sum_{i=1}^{n} \tau_{oi}} \tag{6.15}$$

Example 6.1: Elementary OCTC example

For the circuit in Figure 6.2 we will estimate the -3 dB bandwidth using the method of OCTCs and then find the actual -3 dB bandwidth by calculation and by using LTSPICE.

There are three capacitors, each of which contributes to bandwidth limitation in this circuit. We will perform the method of OCTCs using these three bandwidth-limiting capacitances.

For capacitor C_1, we open-circuit C_2 and C_3 and find the resistance R_{o1} across the C_1 terminals (Figure 6.3(a)). Note that we also short the input generator, as an input voltage generator provides an incremental short. The open-circuit resistance for C_1 is 1 kΩ and hence the OCTC for C_1, which we will call τ_{oi}, is 0.1 ms.

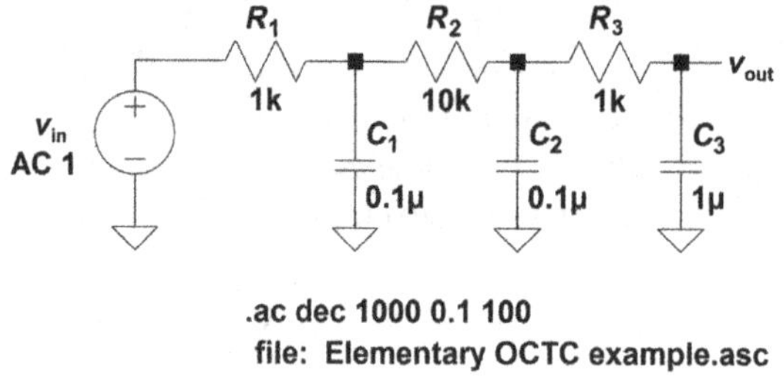

.ac dec 1000 0.1 100
file: **Elementary OCTC example.asc**

FIGURE 6.2

Elementary open-circuit time constants example. (For color version of this figure, the reader is referred to the online version of this book.)

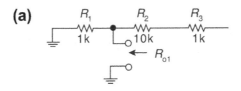

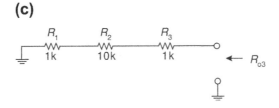

FIGURE 6.3

Individual circuits for finding open-circuit resistances. (a) For C_1. (b) For C_2. (c) For C_3.

For C_2 and C_3, we follow a similar procedure, with the result:

- For C_2: open-circuit resistance $R_{o2} = 11$ kΩ, $\tau_{o2} = 1.1$ ms
- For C_3: open-circuit resistance $R_{o3} = 12$ kΩ, $\tau_{o3} = 12$ ms

Our estimate for bandwidth of this circuit is as follows:

$$\sum \tau_{oc} = \tau_{o1} + \tau_{o2} + \tau_{o3} = 1.32 \times 10^{-2} \text{ s}$$

$$\omega_h \approx \frac{1}{\sum \tau_{oc}} \approx 75.75 \text{ rad/s} \tag{6.16}$$

$$f_h = \frac{\omega_h}{2\pi} \approx 12 \text{ Hz}$$

LTSPICE shows that the -3 dB bandwidth $f_h \approx 12.2$ Hz (Figure 6.4). We also see that the dominant bandwidth-limiting capacitor is C_3, which makes sense since C_3 is larger than the other capacitors.

If we want to find the poles and zeros numerically, first we write the node equations. The node equations for this system, in matrix form, are:

$$\begin{bmatrix} -(G_1 + G_2 + C_1 s) & G_2 & 0 \\ G_2 & -(G_2 + G_3 + C_2 s) & G_3 \\ 0 & G_3 & -(G_3 + C_3 s) \end{bmatrix} \begin{Bmatrix} v_1 \\ v_2 \\ v_{\text{out}} \end{Bmatrix} = \begin{Bmatrix} -v_{\text{in}} G_1 \\ 0 \\ 0 \end{Bmatrix}$$

$$\tag{6.17}$$

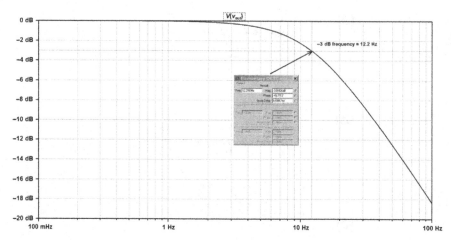

FIGURE 6.4

LTSPICE result of AC analysis for simple *RC* circuit example, showing −3 dB bandwidth of 12.2 Hz.

This is easily solved with a numerical solver. A MATLAB script for finding the poles is as follows:

```
function rcrc
% rcrc network for AC&I notes example
R1=1000; G1=1/R1;
R2=10,000; G2=1/R2;
R3=1000; G3=1/R3;
C1=10^-7;
C2=10^-7;
C3=10^-6;
G=[-(G1+G2) G2 0;
   G2 - (G2+G3) G3;
   0 G3 -G3];
C=[-C1 0 0;
   0 -C2 0;
   0 0 -C3];
poles=-eig(G/C)
poles_in_Hz=poles/(2*pi)
```

The MATLAB result shows a dominant pole at 12.2 Hz and two other higher frequency poles at 1.65 kHz and 1.99 kHz.

Transistor amplifier examples
Example 6.2: Common-emitter amplifier (revisited)

Let us revisit the common-emitter amplifier from the previous chapter (Figure 6.5) and estimate its -3 dB bandwidth using OCTCs. We will assume that the transistor is driven from a source resistance $R_s = 1$ kΩ, and that the transistor base spreading resistance $r_x = 100$ Ω, collector-base capacitance $C_\mu = 2$ pF, and unity gain frequency $f_T = 300$ MHz.

Let us assume that the transistor is biased with a collector current of 2 mA, resulting in the following small-signal parameters:

$$g_m = \frac{|I_C|}{V_{TH}} = \frac{2\text{ mA}}{26\text{ mV}} = 0.077/\Omega$$

$$r_\pi = \frac{h_{fe}}{g_m} = \frac{150}{0.077/\Omega} = 1950\ \Omega \tag{6.18}$$

$$C_\pi = \frac{g_m}{2\pi f_T} - C_\mu = \frac{0.077}{2\pi(300 \times 10^6)} - 2\text{ pF} = 38.8\text{ pF}$$

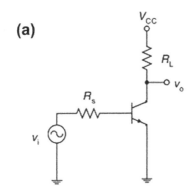

(a)

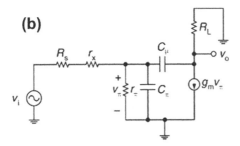

(b)

FIGURE 6.5

Common-emitter amplifier. (a) Circuit, omitting bias details. (b) Small-signal high-frequency model.

Now, we will estimate the bandwidth using OCTCs and the small high-frequency model of Figure 6.5. For C_π, we add a test voltage source v_t and calculate the test current i_t as follows (Figure 6.6(a)). The open-circuit resistance seen across the C_π terminals with C_μ an open circuit is:

$$R_{o\pi} = \frac{v_t}{i_t} \tag{6.19}$$

By inspection, the open-circuit resistance facing C_π is the parallel combination of r_π and the source resistance:

$$R_{o\pi} = r_\pi \| R_s' = r_\pi \| (R_s + r_x) = 1950\ \Omega \| 1100\ \Omega = 703\ \Omega \tag{6.20}$$

The resultant OCTC is:

$$\tau_{o\pi} = R_{o\pi} C_\pi = (703)(38.8\ \text{pF}) = 27.3\ \text{ns} \tag{6.21}$$

For C_μ, we use the circuit of Figure 6.6(b). In order to find the test voltage v_t, we find the two node voltages to ground, v_a and v_b. Voltage v_a is easy to find, since the resistance on the left-hand side of the current source is just the open-circuit resistance we found earlier for C_π, or $R_{o\pi}$. The voltage v_b at the right of the current source is:

$$v_b = -(i_t + g_m v_\pi)R_L = -(i_t + g_m i_t R_{o\pi})R_L \tag{6.22}$$

(a)

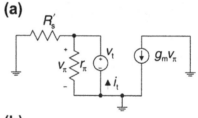

(b)

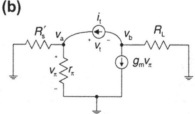

FIGURE 6.6

Circuits for finding open-circuit resistances. (a) Circuit for finding C_π time constant of common-emitter amplifier. (b) Circuit for finding C_μ time constant of common-emitter amplifier.

Hence, the test voltage is:

$$v_t = v_a - v_b = i_t R_{o\pi} + (i_t + g_m i_t R_{o\pi})R_L \tag{6.23}$$

Solving for v_t/i_t results in:

$$R_{o\mu} = \frac{v_t}{i_t} = R_{o\pi} + (1 + g_m R_{o\pi})R_L = 5.58 \times 10^4 \ \Omega \quad (6.24)$$

The resultant OCTC for C_μ is:

$$\tau_{o\mu} = R_{o\mu}C_\mu = (5.58 \times 10^4)(2 \text{ pF}) = 111.7 \text{ ns} \quad (6.25)$$

The sum of the OCTCs is:

$$\sum \tau_{oc} = \tau_{o\pi} + \tau_{o\mu} = 27.3 \text{ ns} + 111.7 \text{ ns} = 139 \text{ ns} \quad (6.26)$$

Our estimation for the -3 dB bandwidth is:

$$\omega_h \approx \frac{1}{\sum \tau_{oc}} \approx \frac{1}{139 \times 10^{-9}} \approx 7.19 \times 10^6 \text{ rad/s}$$

$$f_h \approx \frac{\omega_h}{2\pi} \approx 1.14 \text{ MHz} \quad (6.27)$$

A comparison of the four solution methods (Table 6.1) shows good agreement between the closed-form solution, the one-pole (Miller) approximation, the method of OCTCs, and the LTSPICE simulation (Figure 6.7). Recall that in Chapter 5 we calculated the low-frequency gain of this amplifier to be -49 (or 33.8 dB).

Using the common-emitter amplifier result as an OCTC sanity check

We are now in a position to use the closed-form result of the previous chapter as a sanity check for the OCTCs method. The closed-form transfer function for the common-emitter amplifier, repeated here for the reader's convenience, is:

$$\frac{v_o(s)}{v_i(s)} = -(g_m R_L)\left(\frac{G_s'}{G_s' + g_\pi}\right)$$

$$\times \left[\frac{1 - \frac{C_\mu}{g_m}s}{\left(\frac{R_L C_\pi C_\mu}{G_s' + g_\pi}\right)s^2 + \left(\frac{1}{G_s' + g_\pi}\right)\left[R_L(g_m + g_\pi + G_s')C_\mu + C_\pi + C_\mu\right]s + 1}\right]$$

$$(6.28)$$

Table 6.1 Comparison of Results for Common-Emitter Amplifier

Method	Bandwidth Calculation
Closed form (from Chapter 5)	7.2 Mrad/s (1.15 MHz)
One-pole (Miller) approximation (from Chapter 5)	7.29 Mrad/s (1.16 MHz)
Open-circuit time constants	7.19 Mrad/s (1.14 MHz)
LTSPICE result	7.29 Mrad/s ($\sim$1.16 MHz)

(a)

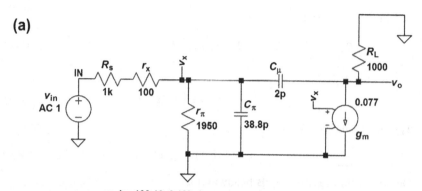

.ac dec 100 10e3 100e6
file: Common emitter amplifier.asc

(b)

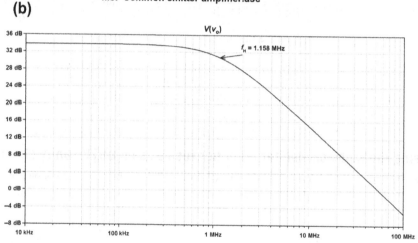

FIGURE 6.7

LTSPICE result for common-emitter amplifier. (a) Circuit. (b) LTSPICE result showing a DC gain of 33.8 dB and a −3 dB bandwidth of 1.16 MHz. (For color version of this figure, the reader is referred to the online version of this book.)

Note that the coefficient of the denominator s term is:

$$\frac{1}{G'_s + g_\pi} \left[R_L \left(g_m + g_\pi + G'_s \right) C_\mu + C_\pi + C_\mu \right] \tag{6.29}$$

We will regroup the terms of the "s" coefficient so that we can see the terms associated with C_π and C_μ individually, and we note that $g_m \gg g_\pi$.

$$\frac{C_\pi}{G'_s + g_\pi} + \frac{C_\mu}{G'_s + g_\pi} + \frac{1}{G'_s + g_\pi} \left[R_L \left(g_m + g_\pi + G'_s \right) C_\mu \right]$$

$$\approx \frac{C_\pi}{G'_s + g_\pi} + \left(\frac{1 + g_m R_L}{G'_s + g_\pi} \right) C_\mu + R_L C_\mu \tag{6.30}$$

The OCTC method allows us to calculate the coefficient of the s term, so the OCTC result should give us the same result as the closed-form equation above. The individual OCTCs for C_π and C_μ are:

$$\tau_{o\pi} = \left(r_\pi \| R_s'\right)C_\pi = \frac{C_\pi}{g_\pi + G_s'}$$

$$\tau_{o\mu} = \left(r_\pi \| R_s' + R_L + g_m\left(r_\pi \| R_s'\right)R_L\right)C_\pi = \left(\frac{1 + g_m R_L}{G_s' + g_\pi} + R_L\right)C_\mu \qquad (6.31)$$

$$= \left(\frac{1 + g_m R_L}{G_s' + g_\pi}\right)C_\mu + R_L C_\mu$$

Note that the two methods give the same answer for the coefficient of the s term, as expected.

Example 6.3: Emitter follower bandwidth estimate using OCTCs

Using OCTCs, let us estimate the bandwidth of the emitter follower (Figure 6.8) from the previous chapter. We will assume small-signal parameters: base spreading resistance $r_x = 100\ \Omega$; $C_\mu = 2$ pF; and the f_T of the transistor $= 300$ MHz.

The emitter follower is biased with a 2 mA current source, resulting in the following small-signal parameters:

$$g_m = \frac{|I_C|}{V_{TH}} = \frac{2\text{ mA}}{26\text{ mV}} = 0.077/\Omega$$

$$r_\pi = \frac{h_{fe}}{g_m} = \frac{150}{0.077/\Omega} = 1950\ \Omega \qquad (6.32)$$

$$C_\pi = \frac{g_m}{2\pi f_T} - C_\mu = \frac{0.077}{2\pi(300 \times 10^6)} - 2\text{ pF} = 38.8\text{ pF}$$

The small-signal model of the emitter follower is shown in Figure 6.9. Note that incrementally the 2 mA current source behaves as an open circuit.

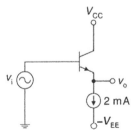

FIGURE 6.8

Emitter follower biased with a current source.

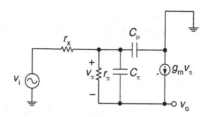

FIGURE 6.9

Emitter follower small-signal model.

Next, we will estimate the bandwidth using OCTCs. For C_π, add a test voltage source v_t, and measure the test current i_t as shown in Figure 6.10.

The test current is:

$$i_t = \frac{v_t}{r_\pi} + g_m v_t \tag{6.33}$$

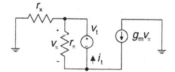

FIGURE 6.10

Circuit for finding C_π time constant for emitter follower.

Hence, the open-circuit resistance facing C_π is (noting that $g_m \gg 1/r_\pi$):

$$R_{o\pi} = \frac{v_t}{i_t} = \frac{1}{g_m + g_\pi} \approx \frac{1}{g_m} \approx 13.07 \ \Omega \tag{6.34}$$

and the resultant OCTC for C_π is:

$$\tau_{o\pi} = R_{o\pi} C_\pi = (13.07)(38.8 \ \text{pF}) = 0.51 \ \text{ns} \tag{6.35}$$

For C_μ, we use the circuit in Figure 6.11(a).

Now, we can make an important simplification that greatly eases the analysis for this case. Note that there is a current loop containing r_π and the $g_m v_\pi$-dependent generator. If we sum currents at the emitter node, the following holds:

$$g_m v_\pi = -\frac{v_\pi}{r_\pi} \tag{6.36}$$

Recognizing that this can only be true when $v_\pi = 0$, we note that the current generator must have zero current and hence we can eliminate it, resulting in the simplified circuit of Figure 6.11(b). Therefore, the open-circuit resistance for C_μ is:

$$R_{o\mu} = r_x = 100 \ \Omega \tag{6.37}$$

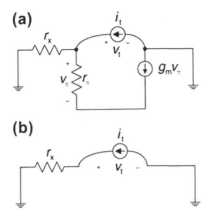

FIGURE 6.11

Circuit for finding C_μ time constant for emitter follower. (a) Original circuit. (b) Simplified circuit acknowledging the fact that $g_m V_\pi = 0$.

and the resultant time constant for capacitor C_μ is:

$$\tau_{o\mu} = R_{o\mu} C_\mu = (100)(2 \text{ pF}) = 0.2 \text{ ns} \tag{6.38}$$

The sum of the OCTCs is:

$$\sum \tau_{oc} = \tau_{o\pi} + \tau_{o\mu} = 0.71 \text{ ns} \tag{6.39}$$

and our estimation for the -3 dB bandwidth is:

$$\omega_h \approx \frac{1}{\sum \tau_{oc}} \approx 1.41 \times 10^9 \text{ rad/s}$$

$$\tag{6.40}$$

$$f_h \approx \frac{\omega_h}{2\pi} \approx 224 \text{ MHz}$$

An LTSPICE result showing the gain and bandwidth is shown in Figure 6.12. This result shows that the gain is approximately unity (as expected) but the -3 dB bandwidth is approximately 794 MHz. What is going on here?

It is time to look at the closed-form solution from the previous chapter in more detail. Previously, we found the transfer function of the emitter follower to be:[3]

$$\frac{v_o(s)}{v_i(s)} = -\frac{\frac{C_\mu}{g_m} s + 1}{\frac{r_x C_\pi C_\mu}{g_\pi} s^2 + \left[r_x C_\mu + \frac{C_\pi}{g_m} \right] s + 1} \tag{6.41}$$

---

[3]Note that the sum of OCTCs for the emitter follower equals the coefficient of the s term in the closed-form solution.

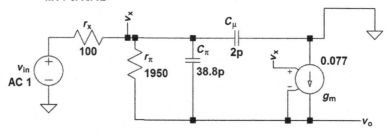

(a)   **Emitter follower small signal model**
Parameters chosen given f_T = 300 MHz, C_μ = 2 pF, h_{fe} = 150, r_x = 100
MTT 9/10/12

.ac dec 100 10e6 1000e6

file: **Emitter follower small signal.asc**

(b)

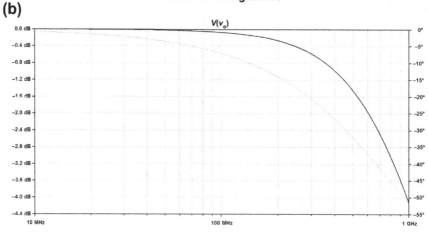

FIGURE 6.12

LTSPICE result for emitter follower. (a) Circuit. (b) Result showing −3 dB bandwidth of approximately 800 MHz. (For color version of this figure, the reader is referred to the online version of this book.)

There are two poles and a zero, and the zero nearly[4] cancels the high-frequency pole (Figure 6.13). The method of OCTCs does not account for this zero and hence greatly underestimates the bandwidth in this specialized case.

Furthermore, the preceding result must be taken with a grain of salt, as the model predicts a bandwidth in excess of the unity current-gain bandwidth product f_T of the transistor. These are caveats to keep in mind when using the OCTC method.

---

[4]The pole and zero nearly cancel out; in the closed-form solution we made an approximation or two, and the pole and zero do not exactly cancel.

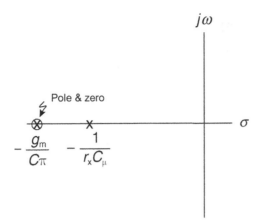

FIGURE 6.13

Emitter follower pole-zero plot.

Example 6.4: Differential amplifier

A differential amplifier is shown in Figure 6.14(a). For differential-mode excitation, the small-signal model of Figure 6.14(b) is valid. We can use this model and the method of OCTCs to estimate the bandwidth of this amplifier. Note that the differential half circuit is identical to the common-emitter amplifier we worked with previously. Hence, the circuits for finding the OCTCs (Figure 6.15) are the same as the common-emitter amplifier.

Let us find the differential gain and bandwidth of a differential amplifier with the following parameters:

- $R_C = 3.3\ \text{k}\Omega$
- $I_{BIAS} = 4\ \text{mA}$
- $h_{fe} = 100$
- $f_T = 300\ \text{MHz}$
- $r_x = 100\ \Omega$
- $C_\mu = 2\ \text{pF}$

The small-signal parameters for this amplifier are:

$$g_m = \frac{|I_C|}{V_{TH}} = \frac{2\ \text{mA}}{26\ \text{mV}} = 0.077/\Omega$$

$$r_\pi = \frac{h_{fe}}{g_m} = \frac{100}{0.077/\Omega} = 1299\ \Omega \qquad (6.42)$$

$$C_\pi = \frac{g_m}{2\pi f_T} - C_\mu = \frac{0.077}{2\pi(300 \times 10^6)} - 2\ \text{pF} = 39\ \text{pF}$$

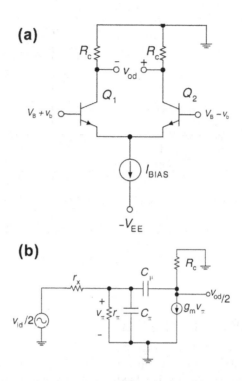

FIGURE 6.14

Differential amplifier. (a) Circuit. (b) Differential-mode half circuit.

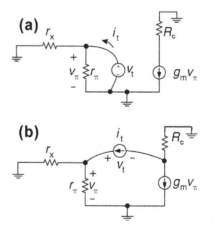

FIGURE 6.15

Circuits for finding OCTCs for differential-mode half circuit. (a) Circuit for C_π. (b) Circuit for C_μ.

The small-signal differential-mode gain is:

$$\frac{v_{od}}{v_{id}} = -g_m R_C \left(\frac{r_\pi}{r_x + r_\pi} \right) \approx -236 \qquad (6.43)$$

Our OCTC calculations are as follows. For C_π, we use the circuit of Figure 6.15(a):

$$R_{o\pi} = r_x \| r_\pi = 100\,\Omega \| 1299\,\Omega \approx 93\,\Omega$$

$$\tau_{o\pi} = R_{o\pi} C_\pi = (93)(39\text{ pF}) = 3.6\text{ ns} \qquad (6.44)$$

For C_μ, we use the circuit of Figure 6.15(b) and we find:

$$R_{o\mu} = r_x \| r_\pi + R_C + g_m(r_x \| r_\pi) R_C$$

$$= 93\,\Omega + 3300\,\Omega + (0.077)(93\,\Omega)(3300\,\Omega) = 27{,}024\,\Omega \qquad (6.45)$$

$$\tau_{o\mu} = R_{o\mu} C_\mu = (27{,}024)(2\text{ pF}) = 54\text{ ns}$$

The sum of OCTCs for this amplifier is 57.6 ns, and our estimate for bandwidth is:

$$\omega_h \approx \frac{1}{57.6 \times 10^{-9}} \approx 17.36\text{ Mrad/s} \qquad (6.46)$$

$$f_h = \frac{\omega_h}{2\pi} \approx 2.76\text{ MHz}$$

LTSPICE (Figure 6.16) shows that the gain is -236 (47.5 dB) as expected and the -3 dB bandwidth is ~ 2.76 MHz, which is in good agreement with the bandwidth predicted by OCTCs.

Example 6.5: Iterative design study using OCTCs

In the previous examples, we used OCTCs to analyze the bandwidth of some circuits. In the following example, we will use OCTCs as a tool in an iterative design exercise. Let us design a transistor amplifier to meet the following gain and bandwidth specifications:

- Magnitude of midband gain: $|A_v|$: >100
- Input source resistance: $R_s = 2$ kΩ (high-impedance source, such as a microphone)
- Load capacitance: $C_L = 10$ pF (scope probe, stray capacitance, etc.)
- -3 dB bandwidth: $f_h > 10$ MHz
- Assume transistor parameters: $h_{fe} = 150$; $f_T = 300$ MHz; $C_\mu = 2$ pF; $r_x = 100\,\Omega$.

As an initial guess at a circuit topology, let us consider the AC-coupled common-emitter amplifier of Figure 6.17(a). The Bode plot of this amplifier looks qualitatively like that in Figure 6.17(b). The high-frequency roll-off is due to the bandwidth-limiting effects of transistor internal capacitances C_π and C_μ. The

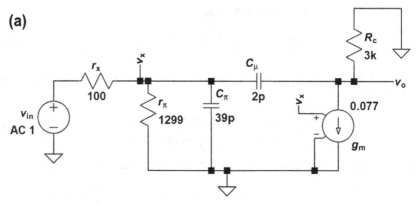

(a)

.ac dec 100 1000e3 1000e6
file: Differential amplifier half circuit small signal.asc

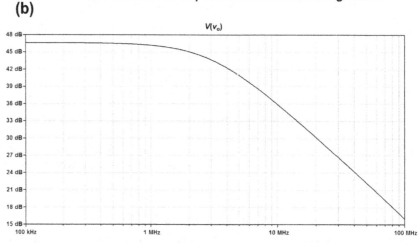

(b)

FIGURE 6.16

LTSPICE result for common-base amplifier of Example 6.4. (a) LTSPICE circuit. (b) Frequency response of common-base amplifier showing low-frequency gain of -236 (47.5 dB) and -3 dB bandwidth of 2.76 MHz. (For color version of this figure, the reader is referred to the online version of this book.)

low-frequency roll-off is due to the coupling capacitor C_C and the emitter bypass capacitor C_E.[5] Later on in this chapter, we will estimate the low-frequency breakpoint f_L using the method of *short-circuit time constants*. In this section, we will estimate the high-frequency roll-off f_h using OCTCs.

---

[5]We will assume that C_C and C_E are so large that the low-frequency breakpoint f_L is well below the high-frequency breakpoint f_h.

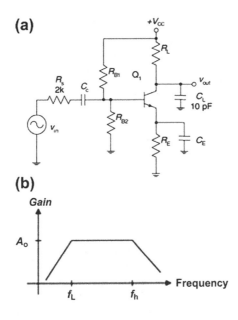

FIGURE 6.17

Initial guess at topology (Try #1). (a) Circuit. (b) Qualitative view of gain vs. frequency curve for amplifier.

TRY #1: Common-emitter amplifier

First, let us arbitrarily assume that the collector current of the transistor is 2 mA.[6] Other small-signal parameters are:

$$g_{m1} = \frac{|I_{C1}|}{V_{TH}} = \frac{2 \text{ mA}}{26 \text{ mV}} = 0.077/\Omega$$

$$r_{\pi 1} = \frac{h_{fe1}}{g_{m1}} = \frac{150}{0.077/\Omega} = 1948 \ \Omega \tag{6.47}$$

$$C_{\pi 1} = \frac{g_{m1}}{2\pi f_T} - C_{\mu 1} = \frac{0.077}{2\pi(300 \times 10^6)} - 2 \text{ pF} = 39 \text{ pF}$$

In order to find the midband gain, we use the small-signal model in Figure 6.18.[7] We find that the small-signal low-frequency gain is:

$$A_V = -g_{m1}R_L\left(\frac{r_{\pi 1}}{r_{\pi 1} + r_{x1} + R_s}\right) \tag{6.48}$$

---

[6]We can easily bias this transistor at a collector current of 2 mA by carefully designing the values of R_{B1}, R_{B2}, and R_E as shown in the previous discussions on transistor biasing.

[7]I have assumed that R_{B1} and R_{B2} are so big that they do not affect the gain and bandwidth calculations. In order for this to be true, R_{B1} and R_{B2} need to be large compared to R_s, r_x, and r_π. If this is not the case, we need to include R_{B1} and R_{B2} in the gain and bandwidth calculations. We also assume that we are operating at frequencies high enough so that the coupling and emitter bypass capacitors act as short circuits.

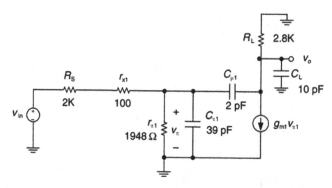

FIGURE 6.18

Small-signal model for Try #1.

Note that $-g_m R_L$ is what the gain would be if the source resistor (R_s) and transistor base spreading resistance (r_x) were zero. Using this gain equation, we find that the minimum value of R_L that achieves our necessary gain of 100 is $R_L = 2.7 \text{ k}\Omega$. We will use $R_L = 2.8 \text{ k}\Omega$ to ensure a little bit of extra gain.

Next, in order to estimate the bandwidth we apply the method of OCTCs. In order to find the open-circuit resistance across $C_{\pi1}$, we use the circuit in Figure 6.19(a). The open-circuit resistance and time constant are:

$$R_{\pi1} = r_{\pi1} \| (R_s + r_{x1}) = 1948 \, \Omega \| 2100 \, \Omega = 1010 \, \Omega$$

$$\tau_{\pi1} = R_{\pi1} C_{\pi1} = (1010 \, \Omega)(39 \text{ pF}) = 39 \text{ ns} \tag{6.49}$$

To find the open-circuit resistance for $C_{\mu1}$, we use the circuit of Figure 6.19(b). We find that the open-circuit resistance across the $C_{\mu1}$ terminals is the impedance to the left ($R_{\pi1}$), the impedance to the right (R_L), and a third term due to the $g_m v_{\pi}$-dependent generator. The result is:

$$R_{\mu1} = R_{\pi1} + R_L + g_{m1} R_{\pi1} R_L$$
$$= 1010 \, \Omega + 2700 \, \Omega + (0.077)(1010 \, \Omega)(2800 \, \Omega) = 221.5 \text{ k}\Omega \tag{6.50}$$
$$\tau_{\mu1} = R_{\mu1} C_{\mu1} = (221.5 \text{ k}\Omega)(2 \text{ pF}) = 443 \text{ ns}$$

The open-circuit resistance for the output load capacitor (Figure 6.19(c)) is easy to find by inspection, because the dependent current source is disabled. The result is:

$$R_{o,L} = R_L = 2.8 \text{ k}\Omega$$
$$\tau_{LOAD} = R_{o,L} C_L = (2800 \, \Omega)(10 \text{ pF}) = 28 \text{ ns} \tag{6.51}$$

The sum of OCTCs for this amplifier is 510 ns and our estimate of bandwidth is:

$$\omega_h \approx \frac{1}{510 \text{ ns}} \approx 1.96 \text{ Mrad/s}$$

$$f_h = \frac{\omega_h}{2\pi} \approx 312 \text{ kHz} \tag{6.52}$$

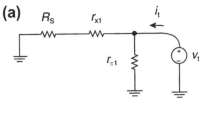

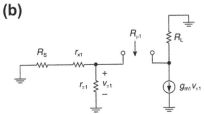

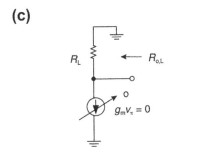

FIGURE 6.19

Circuits for determining open-circuit resistances for Try #1. (a) OCTC circuit for $C_{\pi 1}$. (b) OCTC circuit for $C_{\mu 1}$. (c) OCTC circuit for load capacitor C_{LOAD}.

LTSPICE (Figure 6.20) shows that the gain is approximately -103.8 as calculated and the bandwidth is approximately 313 kHz, in good agreement with OCTCs. A summary of the OCTC results for Try #1 is shown in Table 6.2. There is one dominant pole in the amplifier (due to the Miller effect), so in this case, the OCTC estimate is pretty good. We have not met the bandwidth specification, so we need to iterate the design.

TRY #2: Common emitter + cascode

In Try #1, the OCTC method identifies the time constant of $C_{\mu 1}$ as the dominant bandwidth bottleneck for this amplifier. This result is not surprising considering that we are asking a single common-emitter gain stage to provide all the gain and that the Miller effect has a major effect in high-gain common-emitter amplifiers.

(a)

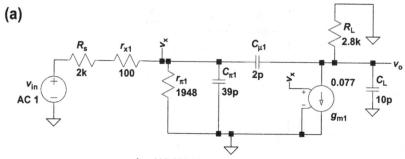

.ac dec 100 100 10e6
file: Amplifier design example try1.asc

(b)

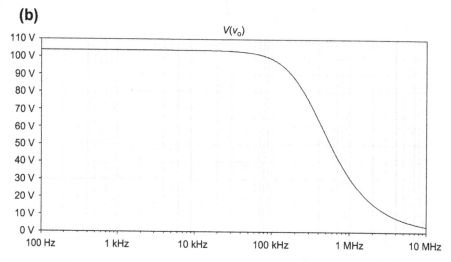

FIGURE 6.20

Frequency response for Try #1. (a) LTSPICE circuit, omitting biasing details. (b) LTSPICE frequency-response result showing gain of −104 and a −3 dB bandwidth of 313 Hz. (For color version of this figure, the reader is referred to the online version of this book.)

Table 6.2 OCTC Summary for Try #1

	TRY #1
$\tau_{\pi 1}$	39 ns
$\tau_{\mu 1}$	443 ns
τ_{LOAD}	28 ns
$\Sigma \tau_{oc}$	510 ns
$\omega_{h,est}$	1.96 Mrad/s
$f_{h,est}$	312 kHz
f_h from LTSPICE	313 kHz

In fact, a *very* rough estimate of bandwidth due to Miller effect (without resorting to the full Miller approximation given before) is:

$$\omega_h \approx \frac{1}{R_s(AC_{\mu 1})} \approx \frac{1}{(2000)(100)(2 \times 10^{-12})} \approx 2.5 \text{ Mrad/s}$$
(6.53)

$$f_h \approx \frac{\omega_h}{2\pi} \approx 398 \text{ kHz}$$

which is a pretty good estimate since it is so simple and intuitive.

In order to reduce the effects of the Miller effect, we can reduce the voltage gain of transistor Q_1 by using a *cascode*[8] transistor, as in Figure 6.21(a). Transistor Q_1 generates a controlled output current that is buffered by common-base transistor Q_2. The resistance looking into the emitter Q_2 is very low, so the voltage gain at the collector of Q_1 is low, hence reducing the Miller effect. Transistor Q_2 provides approximately unity current gain; the load resistor at the collector of Q_2 converts the current to an output voltage. We will assume that the small-signal parameters for Q_1 and Q_2 are identical,[9] resulting in the small-signal model of Figure 6.21(b).

We will now find the five OCTCs for this cascode amplifier. Note that the OCTC for $C_{\pi 1}$ has not changed since Try #1; it is still 39 ns. The OCTC for $C_{\mu 1}$ requires a little bit more work. In Figure 6.22(a), we see the OCTC circuit for $C_{\mu 1}$. We have replaced Q_2 in this circuit with the output resistance r_{out2} seen at the emitter of Q_2. This output resistance at the emitter of Q_2 is:[10]

$$r_{out2} = \frac{r_{x2} + r_{\pi 2}}{1 + h_{fe2}} = \frac{100\,\Omega + 1948\,\Omega}{151} = 13.6\,\Omega$$
(6.54)

We now find the OCTC for $C_{\mu 1}$ as follows:

$$R_{\mu 1} = R_{\pi 1} + r_{out2} + g_{m1}R_{\pi 1}r_{out2}$$
$$= 1010\,\Omega + 13.6\,\Omega + (0.077)(1010\,\Omega)(13.6\,\Omega)$$
$$= 2081\,\Omega \quad \tau_{\mu 1} = R_{\mu 1}C_{\mu 1} = (2081\,\Omega)(2\text{ pF}) = 4.2\text{ ns}$$
(6.55)

Note that we have drastically reduced this OCTC from Try #1, since we have killed the Miller effect. In fact, we have reduced this OCTC by about a factor of 100, consistent with our understanding of the Miller effect.

---

[8]The term *cascode* dates back to vacuum tube circuits. See, e.g., F.V. Hunt and R.W. Hickman, "On Electronic Voltage Stabilizers", *Review of Scientific Instruments*, vol. 10, Jan. 1939, pp. 6–21.
[9]The two transistors operate at the same collector currents so g_m, r_π, and r_x are the same for both. We should actually double-check the values of $C_{\mu 1}$ and $C_{\mu 2}$. With the cascode transistor, Q_1 now operates at a lower V_{CB} than in Try #1, so $C_{\mu 1}$ may go up a little bit. In order to be more rigorous, we should find the V_{CB} of each transistor, than find the C_μ for each transistor individually. However, for this design example, let us assume that $C_{\mu 1} = C_{\mu 1} = 2$ pF.
[10]We can use the results of the previous chapter where we found the output resistance of the emitter follower.

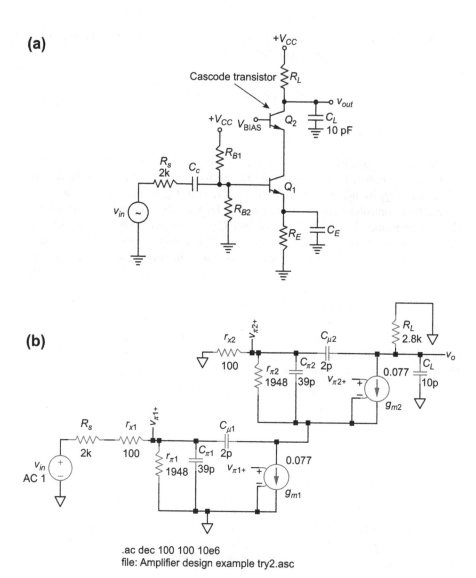

FIGURE 6.21

Second iteration (Try #2) with cascode transistor Q_2 added. (a) Circuit. (b) Small-signal model. (For color version of this figure, the reader is referred to the online version of this book.)

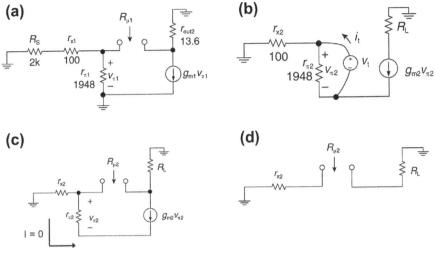

FIGURE 6.22

Incremental circuits for finding OCTCs for Try #2. (a) $C_{\mu 1}$. (b) $C_{\pi 2}$. (c) $C_{\mu 2}$. (d) Simplified circuit for $C_{\mu 2}$.

For $C_{\pi 2}$, we use the circuit in Figure 6.22(b). Note that the impedance looking into the collector of Q_1 is very high, so the emitter of Q_2 looks into an open circuit. In order to find the OCTC for $C_{\pi 2}$, we add a test voltage source v_t and find the test current i_t as follows:

$$i_t = \frac{v_t}{r_{\pi 2}} + g_{m2}v_t \approx g_{m2}v_t \qquad (6.56)$$

The open-circuit resistance and OCTC for $C_{\pi 2}$ are:

$$R_{\pi 2} = \frac{v_t}{i_t} \approx \frac{1}{g_{m2}} \approx 13\ \Omega \qquad (6.57)$$

$$\tau_{\pi 2} = R_{\pi 2}C_{\pi 2} = (13\ \Omega)(39\ \text{pF}) = 0.5\ \text{ns}$$

For $C_{\mu 2}$, we use the circuit in Figure 6.22(c). At first glance this circuit looks like a horrible mess, but a little thought is in order before writing node equations. Note that since the output resistance of Q_1 is so high, all the current from the $g_{m2}v_{\pi 2}$ generator flows through $r_{\pi 2}$. This in turn constrains the following to be true:

$$v_{\pi 2} = -g_{m2}r_{\pi 2}v_{\pi 2} \qquad (6.58)$$

This can only be true if $v_{\pi 2} = 0$, so we can further simplify by noting that with $v_{\pi 2} = 0$, the $g_{m2}v_{\pi 2}$ current generator is also equal to zero. This results in the simplified circuit of Figure 6.22(d). We can now find the open-circuit resistance by inspection and time constant for $C_{\mu 2}$.

$$R_{\mu 2} = r_{x2} + R_L = 2900\ \Omega$$
$$\tau_{\mu 2} = R_{\mu 2}C_{\mu 2} = (2900\ \Omega)(2\ \text{pF}) = 5.8\ \text{ns} \qquad (6.59)$$

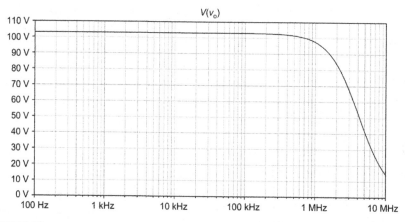

FIGURE 6.23

Try #2. LTSPICE frequency-response result showing gain of -103 and bandwidth of 2.69 MHz.

The OCTC for the load capacitor C_L is still 28 ns. Our sum of OCTCs is now 77.5 ns, and our estimate of bandwidth is:

$$\omega_h \approx \frac{1}{77.5 \text{ ns}} \approx 12.9 \text{ Mrad/s}$$

$$f_h \approx \frac{\omega_h}{2\pi} \approx 2.05 \text{ MHz}$$

(6.60)

LTSPICE (Figure 6.23) shows that the gain is approximately -103.1 as calculated and the bandwidth is approximately 2.69 MHz. A summary of the OCTC results for Try #2 is shown in Table 6.3. We have improved the overall bandwidth significantly by adding the cascode, but we have not yet met the bandwidth specification. We will iterate the design again.

Table 6.3 OCTC Summary for Try #1 and #2		
	TRY #1	**TRY #2**
$\tau_{\pi 1}$	39 ns	39 ns
$\tau_{\mu 1}$	443 ns	4.2 ns
$\tau_{\pi 2}$	–	0.5 ns
$\tau_{\mu 2}$	–	5.8 ns
τ_{LOAD}	28 ns	28 ns
$\Sigma \tau_{oc}$	510 ns	77.5 ns
$\omega_{h,\text{est}}$	1.96 Mrad/s	12.9 Mrad/s
$f_{h,\text{est}}$	312 kHz	2.05 MHz
f_h from LTSPICE	313 kHz	2.69 MHz

TRY #3: Emitter follower + common emitter + cascode

After Try #2, we see that the dominant time constant is due to $C_{\pi 1}$ of Q_1. This is due to the large source resistance R_s interacting with the input capacitance $C_{\pi 1}$ of Q_1. We can reduce this effect by buffering the source resistance from the common-emitter transistor with a unity gain emitter follower buffer, as shown in Figure 6.24(a). Q_3 is an emitter follower that has high input impedance and low output impedance. For simplicity, let us assume that Q_3 is biased at the same collector current as the other transistors and has the same small-signal parameters as Q_1 and Q_2. The small-signal model for Try #3 is shown in Figure 6.24(b).

The addition of Q_3 and its high input impedance also reduces the gain loading effect due to the high source impedance R_s. Therefore, to achieve our desired gain of 100, we can reduce the load resistor R_L to approximately

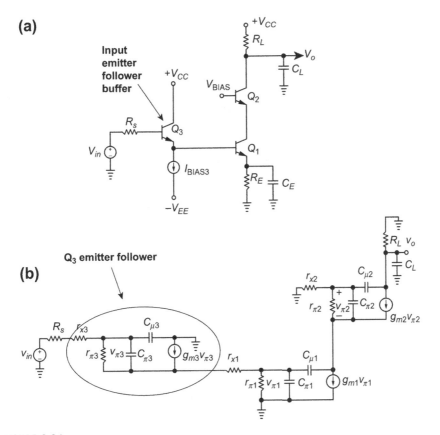

FIGURE 6.24

Third attempt (Try #3). (a) Circuit, with added emitter follower buffer Q_3. (b) Small-signal model.

1.5 kΩ. As we will see later on, this also helps with improving the bandwidth (because reducing the load resistance reduces the OCTCs associated with Q_2 and the load capacitance).

The output impedance seen at the emitter of the emitter follower Q_3 is:

$$r_{out,Q3} \approx \frac{R_s + r_{x3} + r_{\pi3}}{1 + h_{fe3}} = \frac{2000\ \Omega + 100\ \Omega + 1948\ \Omega}{151} = 27\ \Omega \qquad (6.61)$$

We can now work on the OCTCs for Q_1. For $C_{\pi1}$, the calculation is as follows:

$$R_{\pi1} = r_{\pi1} \| (r_{out,Q3} + r_{x1}) = 1948\ \Omega \| 127\ \Omega = 119\ \Omega$$
$$\tau_{\pi1} = R_{\pi1}C_{\pi1} = (119\ \Omega)(39\ \text{pF}) = 4.6\ \text{ns} \qquad (6.62)$$

For $C_{\mu1}$ we find:

$$\begin{aligned}
R_{\mu1} &= R_{\pi1} + r_{out2} + g_{m1}R_{\pi1}r_{out2} \\
&= 119\ \Omega + 13.6\ \Omega + (0.077)(119\ \Omega)(13.6\ \Omega) = 257\ \Omega \qquad (6.63) \\
\tau_{\mu1} &= R_{\mu1}C_{\mu1} = (257\ \Omega)(2\ \text{pF}) = 0.5\ \text{ns}
\end{aligned}$$

For transistor Q_2, the OCTC for $C_{\pi2}$ is unchanged at 0.5 ns. The OCTC for $C_{\mu2}$ is reduced from Try #2, since we have lowered the value of the load resistor:

$$R_{\mu2} = r_{x2} + R_L = 1600\ \Omega$$
$$\tau_{\mu2} = R_{\mu2}C_{\mu1} = (1600\ \Omega)(2\ \text{pF}) = 3.2\ \text{ns} \qquad (6.64)$$

For the emitter follower transistor Q_3, we find for $C_{\pi3}$:

$$R_{\pi3} = r_{\pi3} \left\| \left[\frac{R_s + r_{x3} + r_{x1} + r_{\pi1}}{1 + g_{m3}(r_{x1} + r_{\pi1})} \right] \right. \approx 26\ \Omega$$
$$\tau_{\mu3} = R_{\pi3}C_{\pi3} = (26\ \Omega)(39\ \text{pF}) = 1.0\ \text{ns} \qquad (6.65)$$

Next, for $C_{\mu3}$ we find:

$$R_{\mu3} = [r_{x3} + R_s] \| [r_{\pi3} + (1 + h_{fe3})(r_{x1} + r_{\pi1})] \approx r_{x3} + R_s = 2100\ \Omega$$
$$\tau_{\mu3} = R_{\mu3}C_{\mu3} = (2100\ \Omega)(2\ \text{pF}) = 4.2\ \text{ns} \qquad (6.66)$$

The OCTC due to the load capacitor is 15 ns. The sum of OCTCs for this entire circuit is now 29 ns, and our estimate of bandwidth is:

$$\omega_h \approx \frac{1}{29\ \text{ns}} \approx 34.5\ \text{Mrad/s}$$
$$f_h = \frac{\omega_h}{2\pi} \approx 5.5\ \text{MHz} \qquad (6.67)$$

LTSPICE (Figure 6.25) shows that the gain is approximately -107.7 and the bandwidth is approximately 9.4 MHz. A summary of the OCTC results for Try #3 is shown in Table 6.4. We still have not quite met the 10 MHz bandwidth specification, so we will iterate again.

(a)

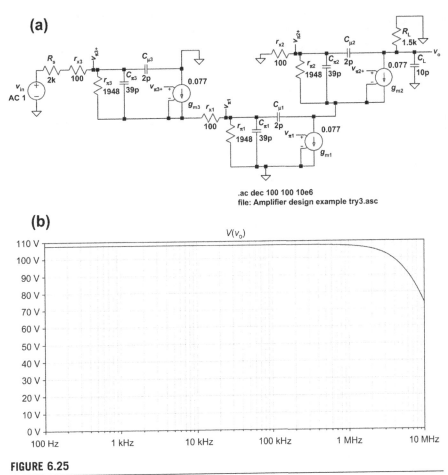

.ac dec 100 100 10e6
file: Amplifier design example try3.asc

(b)

$V(v_o)$

FIGURE 6.25

Frequency response for Try #3. (a) Circuit. (b) LTSPICE frequency-response result showing gain of −108 and −3 dB bandwidth of 9.4 MHz. (For color version of this figure, the reader is referred to the online version of this book.)

TRY #4: Emitter follower + common emitter + cascode + emitter follower

In Try #3, OCTCs identified the interaction of the load capacitor C_L with the load resistor R_L as the major bandwidth bottleneck. We can reduce the effects of C_L by adding a buffer emitter follower Q_4 as shown in Figure 6.26(a) to buffer the load capacitor from the 1.5-kΩ load resistor. The addition of Q_4 does not affect the OCTCs for Q_1, Q_2, and Q_3. We do need to calculate OCTCs for Q_4 as well as a new OCTC for the load capacitor, using the small-signal model of Figure 6.26(b).

Table 6.4 OCTC Summary for Try #1, #2, and #3

	TRY #1	TRY #2	TRY #3
$\tau_{\pi 1}$	39 ns	39 ns	4.6 ns
$\tau_{\mu 1}$	443 ns	4.6 ns	0.5 ns
$\tau_{\pi 2}$	–	0.5 ns	0.5 ns
$\tau_{\mu 2}$	–	5.8 ns	3.2 ns
$\tau_{\pi 3}$	–	–	1.0 ns
$\tau_{\mu 3}$	–	–	4.2 ns
τ_{LOAD}	28 ns	28 ns	15 ns
$\Sigma \tau_{oc}$	510 ns	77.5 ns	29 ns
$\omega_{h,est}$	1.96 Mrad/s	12.9 Mrad/s	34.5 Mrad/s
$f_{h,est}$	312 kHz	2.05 MHz	5.5 MHz
f_h from LTSPICE	313 kHz	2.69 MHz	9.4 MHz

(a)

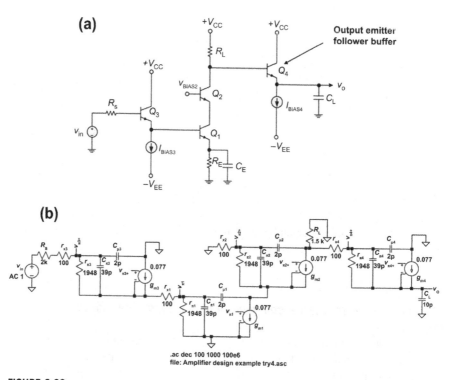

(b)

.ac dec 100 1000 100e6
file: Amplifier design example try4.asc

FIGURE 6.26

Fourth attempt (Try #4) with output emitter follower buffer added. (a) Circuit. (b) LTSPICE model. (For color version of this figure, the reader is referred to the online version of this book.)

For $C_{\pi 4}$, we find:

$$R_{\pi 4} \approx \frac{1}{g_{m4}} = 13\ \Omega$$

$$\tau_{\pi 4} = R_{\pi 4}C_{\pi 4} = (13\ \Omega)(39\ \text{pF}) = 0.5\ \text{ns}$$

(6.68)

Next, for $C_{\mu 4}$:

$$R_{\mu 4} = r_{x4} + R_L = 1600\ \Omega$$

$$\tau_{\mu 4} = R_{\mu 4}C_{\mu 4} = (1600\ \Omega)(2\ \text{pF}) = 3.2\ \text{ns}$$

(6.69)

In order to find the OCTC for the load capacitor C_L, we need to find the output resistance of the Q_4 emitter follower:

$$r_{\text{out},Q4} \approx \frac{R_L + r_{x4} + r_{\pi 4}}{1 + h_{fe4}} = \frac{1500\ \Omega + 100\ \Omega + 1948\ \Omega}{151} = 23.5\ \Omega \qquad (6.70)$$

The resultant OCTC for the load capacitor is:

$$\tau_{\text{LOAD}} = r_{\text{out},Q4}C_L = (23.5\ \Omega)(10\ \text{pF}) = 0.2\ \text{ns} \qquad (6.71)$$

The sum of OCTCs for this circuit is now 17.9 ns, and our estimate of bandwidth is:

$$\omega_h \approx \frac{1}{22.5\ \text{ns}} \approx 55.9\ \text{Mrad/s}$$

$$f_h = \frac{\omega_h}{2\pi} \approx 8.89\ \text{MHz}$$

(6.72)

LTSPICE (Figure 6.27) shows that the gain is approximately -108 and the bandwidth is approximately 19.4 MHz. A summary of the OCTC results for Try #4 is shown in Table 6.5. We have finally met both the gain and bandwidth specifications for this amplifier.

We see that using the method of OCTCs we are able to identify the major bandwidth bottlenecks and attack them. Furthermore, the OCTC method gives a pessimistic approximation of bandwidth, so that the actual bandwidth will be higher. Note the following OCTC caveats:

- The bandwidth estimate is always conservative.
- The estimate is accurate if there is a dominant pole.
- The OCTC estimate can be overpessimistic if there are dominant complex poles, since the method assumes real-axis poles.
- Not all capacitors are used in OCTC analysis (e.g. coupling and bypass capacitors).
  - OCTC applies only to high-frequency models.
- Each OCTC does *not* equate with a single system pole.
  - Remember, the *sum* of pole time constants equals the *sum* of OCTCs, but the individual terms are not necessarily equal.
- System zeros are not accounted for.

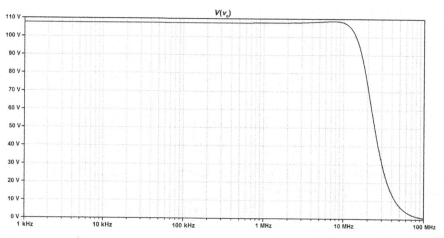

FIGURE 6.27

Frequency response of Try #4 showing low-frequency gain of −107.7 and bandwidth of 19.4 MHz. Note the slight amount of gain peaking near 10 MHz.

Table 6.5 OCTC Summary for Try #1, #2, #3 and #4

	TRY #1	TRY #2	TRY #3	TRY #4
$\tau_{\pi 1}$	39 ns	39 ns	4.6 ns	4.6 ns
$\tau_{\mu 1}$	443 ns	4.6 ns	0.5 ns	0.5 ns
$\tau_{\pi 2}$	–	0.5 ns	0.5 ns	0.5 ns
$\tau_{\mu 2}$	–	5.8 ns	3.2 ns	3.2 ns
$\tau_{\pi 3}$	–	–	1.0 ns	1.0 ns
$\tau_{\mu 3}$	–	–	4.2 ns	4.2 ns
$\tau_{\pi 4}$			–	0.5 ns
$\tau_{\mu 4}$			–	3.2 ns
τ_{LOAD}	28 ns	28 ns	15 ns	0.2 ns
$\Sigma \tau_{oc}$	510 ns	77.5 ns	29 ns	17.9 ns
$\omega_{h,est}$	1.96 Mrad/s	12.9 Mrad/s	34.5 Mrad/s	55.9 Mrad/s
$f_{h,est}$	312 kHz	2.05 MHz	5.5 MHz	8.9 MHz
f_h from LTSPICE	313 kHz	2.69 MHz	9.4 MHz	19.4 MHz

Short-circuit time constants

The method of OCTCs discussed previously allows estimation of the high-frequency breakpoint f_h of a generic amplifier. A similar method, called *short-circuit time constants*, allows estimation of the low-frequency roll-off of a transistor amplifier due to

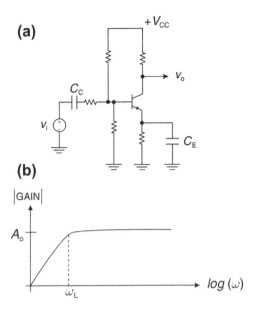

FIGURE 6.28

Typical AC-coupled amplifier. (a) Circuit showing input coupling capacitor C_C and emitter bypass capacitor C_E. (b) Frequency response.

bypass and coupling capacitors. This same methodology can be applied to a generic circuit to determine the low-frequency breakpoint.

A circuit illustrating the use of short-circuit time constants is shown in Figure 6.28(a). We know intuitively that the amplifier has the gain-frequency curve of Figure 6.28(b), for frequencies low enough so that C_π and C_μ do not yet come into play. The method of short-circuit time constants gives us an estimate for the low-frequency -3 dB point ω_L.

Let us examine the transfer function of this amplifier, qualitatively at first. If we are concerned with the low-frequency -3 dB point, we need not be concerned with C_π and C_μ in the transistors. So, we will focus on the coupling capacitor C_C and the emitter bypass capacitor C_E.

We know that there is a zero at zero frequency due to the coupling capacitor C_C. This makes sense, since C_C does not pass any DC component of the input signal. Secondly, there is a zero at a finite frequency due to the emitter bypass capacitor C_E.[11] A transfer function that captures these functional requirements is:

$$H(s) = K \frac{s(\tau_z s + 1)}{(\tau_1 s + 1)(\tau_2 s + 1)} \qquad (6.73)$$

---

[11] As C_E comes into play, the gain increases because C_E shorts out the emitter resistor R_E. This means C_E contributes a zero to the transfer function at a finite frequency.

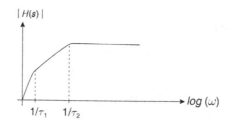

FIGURE 6.29

Frequency response of approximate transfer function $H(s)$.

In order to simplify the mathematics,[12] let us assume that the zero is at a very low frequency, so low that we can approximate it as being at zero frequency. This results in the transfer function:

$$H(s) \approx K' \frac{s^2}{(\tau_1 s + 1)(\tau_2 s + 1)} \tag{6.74}$$

If we multiply out the denominator, we find:

$$H(s) \approx K' \frac{s^2}{\tau_1 \tau_2 s^2 + (\tau_1 + \tau_2)s + 1} \tag{6.75}$$

The Bode plot of this approximate transfer function is shown in Figure 6.29. We note that the -3 dB point is dominated by the higher frequency pole; in this case, the -3 dB point is approximately at frequency $-1/\tau_2$. Using a widely spaced pole approximation (which we developed in Chapter 2) we find that the higher frequency pole location can roughly be approximated as:

$$p_{\text{high}} \approx -\left(\frac{\tau_1 + \tau_2}{\tau_1 \tau_2}\right) = -\left(\frac{1}{\tau_1} + \frac{1}{\tau_2}\right) \tag{6.76}$$

In the general case with n low-frequency poles (and hence n zeros in the transfer function), we find that the approximation for low-frequency bandwidth ω_L is:

$$\omega_L \approx \sum_{j=1}^{n} \frac{1}{\tau_{pj}} \tag{6.77}$$

Said in words, an estimate for low-frequency bandwidth is equal to the *sum of the reciprocal pole time constants*. Unfortunately, in the general case, it is difficult to calculate these pole time constants.

Fortunately, a method for finding the sum of the reciprocal time constants exists. It is the method of short-circuit time constants, which was developed at MIT

---

[12]This methodology is also done in Gray and Searle's *Electronic Principles*, pp. 542–547. The method is also covered in Gray et al.'s *Analysis and Design of Analog Integrated Circuits*.

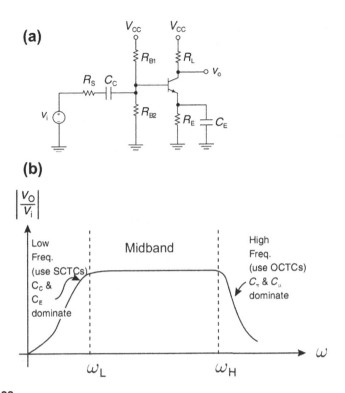

FIGURE 6.30

Common-emitter amplifier for short-circuit time constants analysis. (a) Circuit. (b) Transfer function, showing low-frequency and high-frequency bandwidth limits. We use OCTCs in the high-frequency model to calculate the upper −3 dB point ω_h.

by Adler and others. Using the method of short-circuit time constants, we find that the sum of the reciprocal time constants is exactly equal to the reciprocal sum of relatively easily calculated short-circuit time constants. Mathematically, we find:

$$\omega_L \approx \sum_{j=1}^{N} \frac{1}{\tau_{pj}} \approx \sum_{j=1}^{N} \frac{1}{\tau_{scj}} \tag{6.78}$$

where τ_{sc} are the individual short-circuit time constants. In a method analogous to the method of OCTCs, here is the short-circuit time constant recipe:

- Identify each capacitor that contributes to low-frequency roll-off in your circuit.
- For each capacitor, find the resistance facing this capacitance with other capacitors shorted. (The shorting of the other capacitors is why this is called "short-circuit time constants".)

- Find the time constant associated with each capacitor.
- Our bandwidth estimate is found by:

$$\omega_L \approx \sum_{j=1}^{N} \frac{1}{\tau_{scj}} \tag{6.79}$$

Note that we have to be careful in what capacitors we choose in the method of short-circuit-time constants. For instance, in a generic common-emitter amplifier (Figure 6.30), we will not use C_π and C_μ in our short-circuit time constant calculation; C_π and C_μ contribute to high-frequency bandwidth limit, but have no contribution to the low-frequency limit.

Example 6.6: Short-circuit time constants design example

We will now discuss a short-circuit time constants example and estimate the low-frequency breakpoint and the midband gain of the circuit of Figure 6.31(a). Let us assume that the transistor has the following parameters: DC current gain $\beta_F = 100$, base spreading resistance $r_x = 100\,\Omega$, and small-signal current gain $h_{fe} = 150$. We will assume that the transistor has $V_{BE} = 0.7$ V.

The base and collector currents are found by using the circuit of Figure 6.31(b):

$$6V - I_B(R_{B1} \| R_{B2}) - V_{BE} - I_E R_E = 0$$

$$6V - \left(\frac{I_C}{\beta_F}\right)(R_{B1} \| R_{B2}) - V_{BE} - I_C R_E \approx 0 \tag{6.80}$$

$$I_C \approx \frac{6 - V_{BE}}{\left(R_E + \frac{R_{B1} \| R_{B2}}{\beta_F}\right)} \approx \frac{5.3}{1\,k\Omega + \frac{5\,k\Omega}{100}} \approx 5\text{ mA}$$

Knowing the bias point value, we can now calculate the small-signal parameters:

$$g_m = \frac{|I_C|}{V_{TH}} = \frac{0.005}{0.026} \approx 0.19/\Omega \tag{6.81}$$

$$r_\pi = \frac{h_{fe}}{g_m} = \frac{150}{0.19} = 780\,\Omega$$

A small-signal model for this circuit, valid for low and midband frequencies, is shown in Figure 6.32(a). At midband frequencies, the coupling capacitor C_C and the emitter bypass capacitor C_E short out, resulting in the circuit in Figure 6.32(b). We will now find the midband gain using this circuit:

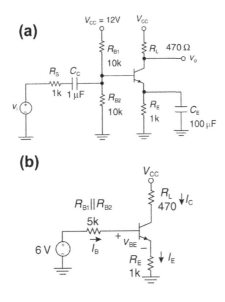

FIGURE 6.31

Common-emitter amplifier for short-circuit time constants bandwidth estimate. (a) Original circuit, with $V_{CC} = 12$ V. (b) Circuit for finding collector current.

$$v_x = v_i \left(\frac{(R_{B1}\|R_{B2})\|(r_x + r_\pi)}{R_s + (R_{B1}\|R_{B2})\|(r_x + r_\pi)} \right) = v_i \left(\frac{5000\,\Omega\|880\,\Omega}{1000\,\Omega + 5000\,\Omega\|880\,\Omega} \right)$$

$$= v_i \left(\frac{748}{1000 + 748} \right) = 0.428 v_i$$

$$v_\pi = v_x \left(\frac{r_\pi}{r_x + r_\pi} \right) = 0.886 v_x = 0.379 v_i$$

$$v_o = -g_m v_\pi R_L = -(0.19)(0.379 v_i)(470\,\Omega) = -33.9 v_i \Rightarrow \frac{v_o}{v_i} = -33.9$$

$$(6.82)$$

Having found the gain, we will next estimate the low-frequency breakpoint using short-circuit time constants. The short-circuit time constant circuit for the coupling capacitor is shown in Figure 6.33(a). The short-circuit resistance with the emitter bypass capacitor acting as a short is:

$$R_{SC1} = R_s + (R_{B1}\|R_{B2})\|(r_x + r_\pi) = 1000\,\Omega + 5000\,\Omega\|880\,\Omega = 1748\,\Omega$$

$$(6.83)$$

The resultant short-circuit time constant for the coupling capacitor is:

$$\tau_{SC1} = R_{SC1} C_C = (1748\,\Omega)(10^{-6}) = 1.75 \times 10^{-3}\,\text{s} \qquad (6.84)$$

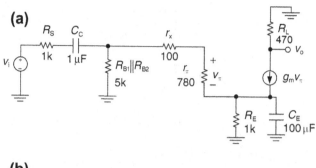

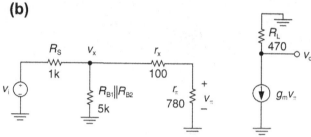

FIGURE 6.32

Small-signal models of common-emitter amplifier. (a) Circuit valid for low and midband frequencies. (b) Midband circuit for finding midband gain.

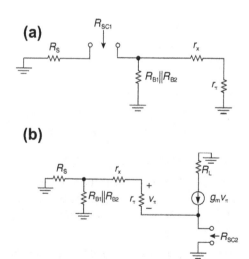

FIGURE 6.33

Circuits for finding short-circuit time constants. (a) Short-circuit time constant circuit for coupling capacitor C_C. (b) Short-circuit time constant circuit for emitter bypass capacitor C_E. Note that R_E has been omitted here for clarity; it needs to be added in parallel.

(a)

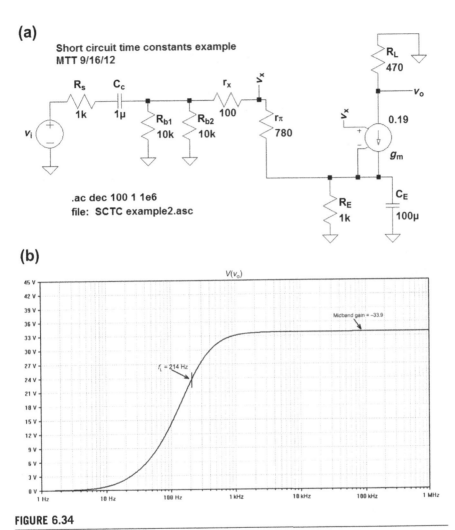

Short circuit time constants example
MTT 9/16/12

.ac dec 100 1 1e6
file: SCTC example2.asc

(b)

FIGURE 6.34

LTSPICE frequency response for short-circuit time constants example showing a midband gain of −34 and a low-frequency −3 dB breakpoint of ∼214 Hz. (a) Circuit. (b) Magnitude of frequency response.

We will next find the short-circuit resistance for the emitter bypass capacitor with the coupling capacitor acting as a short, using the circuit of Figure 6.33(b). This short-circuit resistance is:

$$R_{SC2} = R_E \left\| \left[\frac{R_S \| R_{B1} \| R_{B2} + r_x + r_\pi}{1 + h_{fe}} \right] = 1000\,\Omega \right\| \left[\frac{833\,\Omega + 100\,\Omega + 780\,\Omega}{151} \right]$$

$$= 11.3\,\Omega$$

(6.85)

The resultant short-circuit time constant for the emitter bypass capacitor is:

$$\tau_{SC2} = R_{SC2}C_E = (11.3\,\Omega)(100 \times 10^{-6}) = 1.1 \times 10^{-3}\,\text{s} \qquad (6.86)$$

Our estimate for low-frequency breakpoint is:

$$\omega_L \approx \sum \frac{1}{\tau_{SC}} \approx \frac{1}{\tau_{SC1}} + \frac{1}{\tau_{SC2}} \approx \frac{1}{1.75 \times 10^{-3}} + \frac{1}{1.1 \times 10^{-3}} \approx 1480\ \text{rad/s}$$

$$f_L \approx \frac{\omega_L}{2\pi} \approx 235\ \text{Hz}$$

(6.87)

An LTSPICE analysis (Figure 6.34) shows that the gain is as calculated and that the −3 dB bandwidth is approximately 220 Hz. As in the method of OCTCs, short-circuit time constants give results that are conservative; your actual −3 dB breakpoint will be lower than that predicted by the method.

Chapter 6 problems

Problem 6.1

Using the method of OCTCs, estimate the bandwidth (in Hz) of the RC ladder in Figure 6.35.

Problem 6.2

For the circuit in Figure 6.36:

a. Estimate the −3 dB bandwidth using the method of OCTCs.

b. Find the −3 dB bandwidth exactly by calculation or by using LTSPICE.

Problem 6.3

The transistor amplifier shown (Figure 6.37) is a high-bandwidth, common-emitter configuration used to amplify a video signal. A small-signal video input drives from a 75-Ω source impedance and is terminated with a 75-Ω termination resistor. The transistor is biased with a DC power supply $V_{DC} = 5.7$ V. For the MPSH20 transistor you may assume $r_x = 20\,\Omega$, AC beta $h_{fe} \approx$ DC beta $h_{FE} = 25$, $C_\mu = 0.9$ pF, and

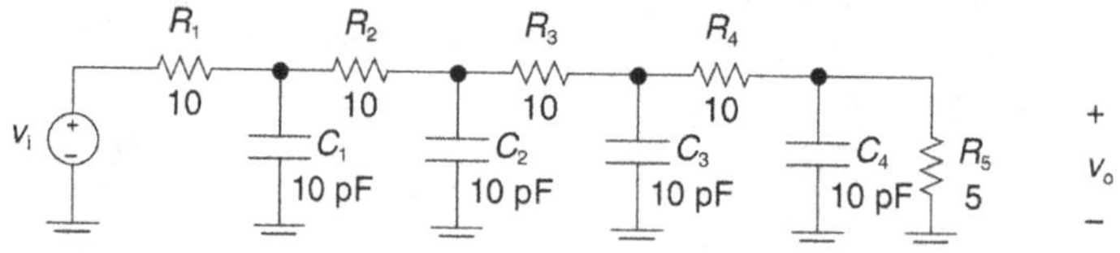

FIGURE 6.35

RC ladder for Problem 6.1.

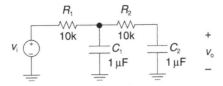

FIGURE 6.36

Circuit for Problem 6.2.

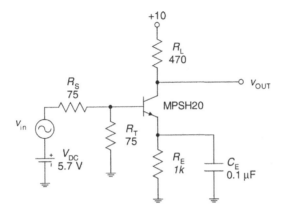

FIGURE 6.37

Common-emitter amplifier for Problems 6.3 and 6.4.

$f_T = 630$ MHz. Make reasonable approximations, state them, and justify them. Assume that room temperature is 25 °C and hence $kT/q = 26$ mV.

a. Estimate the DC bias point values of collector current I_C and collector-base voltage V_{CB}. Find the power dissipated in the transistor.

b. Calculate small-signal parameters g_m, r_π, C_π, and C_μ and draw the small-signal model. Include the emitter bypass capacitor C_E in your model.

c. Now, assume that the operating frequency is high enough so that C_E acts as a short circuit, but low enough so that C_π and C_μ are open circuits. Redraw the small-signal model given these conditions and calculate the midband voltage gain A_V.

d. Using the method of OCTCs, estimate the approximate high-frequency bandwidth f_h, in hertz.

Problem 6.4

When you build the circuit of Problem 6.3 and attach an oscilloscope to the output, you find that the measured bandwidth f_h is significantly lower than you expected based on your OCTC analysis. This effect is especially pronounced, since you know that the method of OCTC gives answers that are conservative. You trace down two causes:

- There is 2 pF parasitic capacitance across the collector–base junction, due to the way you wired the circuit on the protoboard.

- There is 10 pF capacitance from v_{out} to ground, due to the loading effect of the scope probe.

  - Given these two parasitic effects, what bandwidth f_h does the method of OCTCs predict for your circuit?
  - In order to improve the bandwidth of your circuit, you fix the parasitic capacitance problem at the base-collector junction by directly soldering the collector lead to the load resistor without plugging the node into the protoboard. You may assume that this reduces the 2 pF parasitic capacitance to zero at Q_1's base-collector node. (Remember that you still have $C_{\mu 1}$ across this junction.) You also add an emitter follower to isolate the load resistor from the output capacitive load due to the scope (Figure 6.38). Assume that the C_π time constant for Q_2 is 0.25 ns and the C_μ time constant for Q_2 is 0.4 ns. What value of f_h does OCTCs predict?

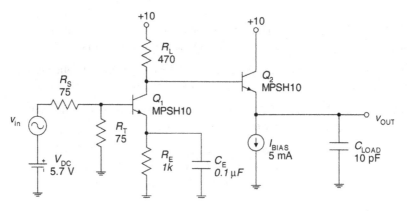

FIGURE 6.38

Common-emitter amplifier with output emitter follower (Problem 6.4).

Problem 6.5

As we have seen, there is a relationship between bandwidth and risetime. The method of OCTCs, since it enables us to estimate bandwidth, may also allow us to estimate the risetime of a single system, or of multiple systems in cascade. Consider a cascade of N amplifiers, each with a single pole with time constant τ.

a. Using OCTCs, estimate the bandwidth of the overall amplifier. From this bandwidth, estimate the risetime of the overall system.

b. Using the risetime addition rule for systems in cascade,[13] i.e. estimate the risetime. Comment on how this compares with the OCTC-derived estimate.

$$\tau_{R,total} \approx \sqrt{\tau_{R1}^2 + \tau_{R2}^2 + \cdots + \tau_{RN}^2}$$

c. Using LTSPICE for the case where $N=5$ and $\tau=1$, find the actual bandwidth and risetime. Compare the results to your previous calculations. Comment on the accuracy of the OCTC estimate.

d. Now, repeat your calculations for the case with five systems in cascade, with four pole time constants $\tau=1$ and one dominant time constant $\tau=10$. Comment on the accuracy of OCTC estimate for risetime.

Problem 6.6

a. Assume that the circuit in Figure 6.39 is driven by an AC voltage source. Using an intuitive approach, thought experiments, etc. sketch the shape of the Bode magnitude response of v_{out}/v_{in}.

b. Using the method of OCTCs, find the high-frequency breakpoint f_h. Fill in f_L and f_h in your Bode plot.

FIGURE 6.39

Circuit for Problem 6.6.

---

[13]Note that this expression is not exact.

Problem 6.7

The circuit in Figure 6.40 is a video amplifier that uses a common-base voltage amplifier driving an emitter follower. The source resistance R_s may be, for instance, the resistance of a 50 Ω transmission line used to input the signal to the amplifier.

You may assume that, under the operating conditions in the circuit, $h_{FE} = 50$, $h_{fe} = 50$, and $r_x = 50 \, \Omega$ for both transistors.

a. What is the operating point value of the output voltage V_{out}?

b. What is the midband voltage gain v_o/v_{in}?

c. Assume that the design specifies a lower cutoff frequency of 50 Hz. What value of C_E is required?

d. Verify your results using LTSPICE.

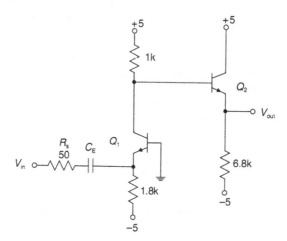

FIGURE 6.40

Circuit for Problem 6.7.

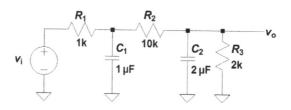

FIGURE 6.41

Circuit for Problem 6.8.

Problem 6.8

a. Using the method of OCTCs, estimate the bandwidth of the circuit in Figure 6.41, (f_h, in hertz).

b. Sketch the Bode plot magnitude of v_o/v_i vs. frequency (in hertz). Plot the magnitude (not in dB).

c. Estimate the 10−90% risetime of the output of this circuit in response to an input unit step. Sketch the step response, labeling axes, etc.

Further reading

[1] Davis A, Moustakas E. Analysis of active RC networks by decomposition. *IEEE Trans Circuits Syst* 1980;**27**(5):417−9.

[2] Fox RM, Lee SG. Extension of the open-circuit time-constant method to allow for transcapacitances. *IEEE Trans Circuits Syst* 1990;**37**(9):1167−71.

[3] Gray PE, Searle CL. *Electronic principles physics, models and circuits*. John Wiley; 1969.

[4] Gray PR, Meyer RG. *Analysis and design of analog integrated circuits*. 2nd ed. John Wiley; 1984.

[5] Hunt FV, Hickman RW. On electronic voltage stabilizers. *Rev Sci Instrum* 1939;**10**:6−21.

[6] Rathore TS. Generalized Miller theorem and its applications. *IEEE Trans Educ* 1989;**32**(3):386−90.

[7] Thompson MT. Design linear circuits using OCTC calculations. *Electron Des (Special Analog Issue)* 1993:41−7.

[8] Thompson MT. Network tricks aid in OCTC. *Electron Des* 1993:67−70.

[9] Thornton RD, Searle CL, Pederson DO, Adler RB, Angelo Jr EJ. *Multistage transistor circuits*In *SEEC*, vol. 5. John Wiley; 1965.

Advanced Amplifier Topics and Design Examples

"With the advent of television and radar during the Second World War, the behavior of wideband amplifiers in the time domain has become very important. In today's digital world this is even more the case. It is a paradox that designers and troubleshooters of digital equipment still depend on oscilloscopes, which — at least in their fast and low level input part — consist of analog wideband amplifiers."

—Peter Starič and Erik Margan, "Wideband Amplifiers," 2007

IN THIS CHAPTER

▶ Various and sundry issues related to advanced amplifier design techniques are discussed.

Note on cascaded gain stages and the effects of loading

Let us take a quick look at the two-stage amplifier in Figure 7.1(a) (with biasing details omitted). When calculating the overall gain of the amplifier, we need to be very careful about the effects of loading of the second gain stage on the first. The gain at the collector of Q_1 (denoted "v_x" in the small-signal model of Figure 7.1(b)) depends not only on the collector resistor R_{L1} but also on the input resistance of the second gain stage looking into the base of transistor Q_2.

We can break up this problem into smaller pieces by recognizing the relationship between the output voltage v_o, the intermediate voltage v_x, and the input voltage v_i:

$$\frac{v_o}{v_i} = \left(\frac{v_x}{v_i}\right)\left(\frac{v_o}{v_x}\right) \tag{7.1}$$

The voltage at node v_x (including the effects of second-stage loading) is:

$$v_x = -v_i\left(\frac{r_{\pi 1}}{R_s + r_{x1} + r_{\pi 1}}\right)(g_{m1}(R_{L1}\|(r_{x2} + r_{\pi 2}))) \tag{7.2}$$

Intuitive Analog Circuit Design. http://dx.doi.org/10.1016/B978-0-12-405866-8.00007-3

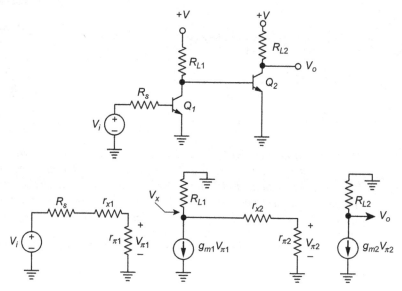

FIGURE 7.1

Cascaded common-emitter gain stages. (a) Circuit, biasing details omitted. (b) Midband incremental model.

Next, the relationship between v_o and v_x is:

$$v_o = -v_x \left(\frac{r_{\pi 2}}{r_{x2} + r_{\pi 2}} \right) (g_{m2} R_{L1}) \tag{7.3}$$

Worst-case open-circuit time constants calculations

The most complicated open-circuit time constant (OCTC) calculation is found for a transistor amplifier that has a source resistance (R_s), emitter resistance (R_E), and collector load resistance (R_L), as shown in Figure 7.2. In the following section, we will figure out the open-circuit resistances for C_π and C_μ for this worst-case circuit.

The worst-case OCTCs circuit for C_π is shown in Figure 7.3(a). The mathematics rapidly gets difficult if we blindly begin writing node equations. First, let us recognize that the test voltage source v_t sets the value of v_π. Since the v_t generator sets the value of the dependent[1] current source, we *can* use superposition to find the test current.

---

[1] Normally, we cannot use superposition in circuits with dependent sources. However, in this special case we can, since the test voltage source v_t directly sets the dependent current source value as well.

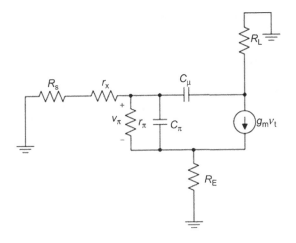

FIGURE 7.2

Worst-case open-circuit time constant circuit with source resistance R_s, emitter resistor R_E, and collector load resistor R_L.

Using the circuit of Figure 7.3(b), we will find the test current due to the test voltage source, with the $g_m v_t$-dependent generator set to zero as an open circuit. By inspection, we see that the resistance across the v_t test source is r_π in parallel with $R_s + r_x + R_E$. Therefore, the test current i_{t1} due to the v_t generator only is:

$$i_{t1} = \frac{v_t}{r_\pi} + \frac{v_t}{R_s + r_x + R_E} \tag{7.4}$$

To find the current due to the $g_m v_t$ test generator only, we use the circuit in Figure 7.3(c). Using this circuit and recognizing the current divider we find:

$$i_{t2} = g_m v_t \frac{R_E}{R_s + r_x + R_E} = v_t \frac{g_m R_E}{R_s + r_x + R_E} \tag{7.5}$$

The total test current i_t is the sum of i_{t1} and i_{t2}, and is:

$$i_t = i_{t1} + i_{t2} = v_t \left(\frac{1}{r_\pi} + \frac{1 + g_m R_E}{R_s + r_x + R_E} \right) \tag{7.6}$$

We find that the open-circuit resistance for C_π is:

$$R_{o\pi} = \frac{v_t}{i_t} = r_\pi \left\| \left(\frac{R_s + r_x + R_E}{1 + g_m R_E} \right) \right. \tag{7.7}$$

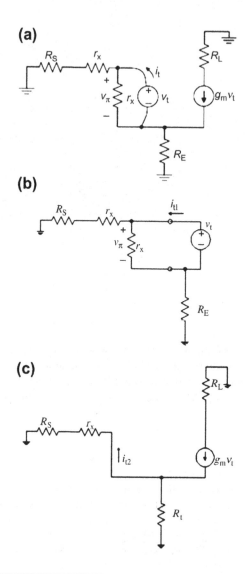

FIGURE 7.3

Circuits for finding worst-case OCTC for C_π. (a) Original circuit. (b) Circuit for finding contribution due to test voltage source v_t only. (c) Circuit for finding contribution due to $g_m v_t$ generator only.

Let us note some important limiting cases. In the case of a large emitter resistor with $R_E \gg r_x$ and R_s, we find:

$$R_{o\pi} \approx \frac{1}{g_m} \quad \text{if } R_E \gg R_s, r_x \tag{7.8}$$

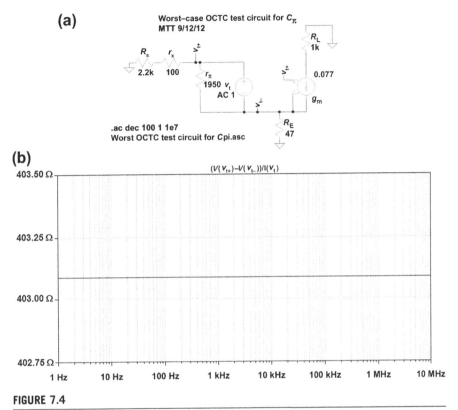

FIGURE 7.4

"Sanity check" circuit for C_π. (a) LTSPICE circuit where a test voltage source v_t is applied across the C_π terminals with capacitor C_μ open-circuited. (b) LTSPICE result of AC sweep, showing resistance across C_π terminals of 403 Ω. (For color version of this figure, the reader is referred to the online version of this book.)

In the case of a small emitter resistor[2] with $R_E \ll r_x$ and R_s and with $g_m R_E \ll 1$, we find:

$$R_{o\pi} \approx r_\pi \| (R_s + r_x) \tag{7.9}$$

We will do a "sanity check" on this calculation using LTSPICE. Assume that we have a small-signal circuit (Figure 7.4(a)) with $R_s = 2.2$ kΩ, $r_x = 100$ Ω, $r_\pi = 1950$ Ω, $h_{fe} = 150$, $R_L = 1$ kΩ, $g_m = 0.077$ A/V, and $R_E = 47$ Ω. We put a test voltage source v_t across the C_π terminals with C_μ open-circuited and find the test current i_t and resistance v_t/i_t at these terminals. Our theoretical calculation and LTSPICE both find the open-circuit resistance facing C_π to be $R_{o\pi} = 403$ Ω (so we did it correctly).

---

[2]We worked out this case in the previous chapter for the common-emitter amplifier with $R_E = 0$.

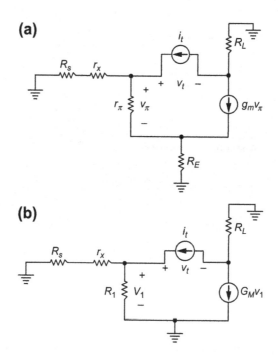

FIGURE 7.5

Circuit for finding worst-case OCTC for C_μ. (a) Original circuit with test current source i_t applied across the C_μ terminals with C_π open-circuited. (b) Circuit redrawn in equivalent form, which is easier to solve.

The worst-case OCTC circuit for C_μ (Figure 7.5(a)) looks like a difficult problem as well. However, let us transform the circuit to a simpler form (Figure 7.5(b)). The circuit of Figure 7.5(b) is equivalent to the circuit of Figure 7.5(a), provided we choose the correct values[3] of R_1 and G_M. The value of R_1 is easy to find; it is just the input resistance looking into the top of r_π so we find that:

$$R_1 = r_\pi + (1 + h_{fe})R_E \qquad (7.10)$$

Next, we need to find the value G_M of the new dependent generator. We can make use of the fact that the output currents from the old and transformed circuits are the same, or:

$$g_m v_\pi = G_M v_1 \qquad (7.11)$$

---

[3]A similar development is given in Gray, Hurst, Lewis, and Meyer, *Analysis and Design of Analog Integrated Circuits*, 4th edition, pp. 197–200.

Let us imagine using test current sources[4] i_t to solve for v_π in Figure 7.5(a) and for v_1 in Figure 7.5(b). We find[5] for v_π:

$$v_\pi = i_t \left(\frac{R_s + r_x}{R_s + r_x + r_\pi + (1 + h_{fe})R_E} \right) r_\pi \qquad (7.12)$$

Next, we solve for v_1:

$$v_1 = i_t \left(\frac{(R_s + r_x)R_1}{R_s + r_x + R_1} \right) = \frac{(R_s + r_x)(r_\pi + (1 + h_{fe})R_E)}{R_s + r_x + r_\pi + (1 + h_{fe})R_E} \qquad (7.13)$$

Next, we solve[6] for G_M:

$$G_M = \frac{g_m v_\pi}{v_1} = g_m \frac{\left(\frac{(R_s + r_x)}{R_s + r_x + r_\pi + (1 + h_{fe})R_E} \right) r_\pi}{\left(\frac{(R_s + r_x)(r_\pi + (1 + h_{fe})R_E)}{R_s + r_x + r_\pi + (1 + h_{fe})R_E} \right)} = \frac{g_m}{1 + \frac{R_E}{r_\pi} + g_m R_E} \approx \frac{g_m}{1 + g_m R_E} \qquad (7.14)$$

To summarize, the worst-case open-circuit resistance for C_μ is:

$$R_{o\mu} = R_{EQ1} + R_L + G_M R_{EQ1} R_L$$

$$R_{EQ1} = (R_s + r_x)\|R_1 = (R_s + r_x)\|(r_\pi + (1 + h_{fe})R_E) \qquad (7.15)$$

$$G_M \approx \frac{g_m}{1 + g_m R_E}$$

We will do another "sanity check" on this calculation using LTSPICE (Figure 7.6). We put a test voltage source v_t across the C_μ terminals with C_π open-circuited and find the test current i_t and resistance v_t/i_t at these terminals. Our calculation and LTSPICE both calculate the open-circuit resistance facing C_μ to be about 33 kΩ.

Example 7.1: Estimating gain and bandwidth of common-emitter amplifier with emitter degeneration

Shown in Figure 7.7 is a common-emitter amplifier with "emitter degeneration". Note that there is a finite resistor R_E in the emitter leg of the transistor. Emitter degeneration lowers the gain of the common-emitter amplifier but extends the bandwidth by partially bootstrapping the base–emitter capacitance and by lowering the C_μ time constant by lowering the gain.

---

[4]You could use a test voltage source v_t; the choice of whether to use a test voltage or a test current is up to the designer, based primarily on calculation simplicity.

[5]This is done by calculating a current divider. The fraction of the current that travels through r_π is:

$$\left(\frac{R_s + r_x}{R_s + r_x + r_\pi + (1 + h_{fe})R_E} \right)$$

[6]Yes, I know it is a mess, but follow it through. It arrives at a nice simple result. Note that $g_m \gg 1/r_\pi$.

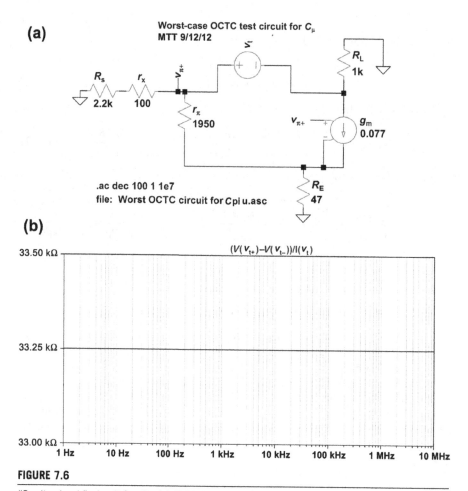

(a)

FIGURE 7.6

"Sanity check" circuit for C_μ. (a) LTSPICE circuit with test voltage source v_t applied across C_μ terminals with C_π open-circuited. (b) LTSPICE result of AC sweep, showing resistance across C_μ terminals of 33.25 kΩ. (For color version of this figure, the reader is referred to the online version of this book.)

The midband gain equation for this amplifier, left as an exercise for the reader, is:

$$A_v = \frac{v_o}{v_i} = (-g_m R_L)\left(\frac{r_\pi}{R_s + r_x + r_\pi + (1 + h_{fe})R_E}\right) \qquad (7.16)$$

Important limiting cases are the standard common-emitter amplifier, with $R_E = 0$:

$$A_v = (-g_m R_L)\left(\frac{r_\pi}{R_s + r_x + r_\pi}\right) \qquad (7.17)$$

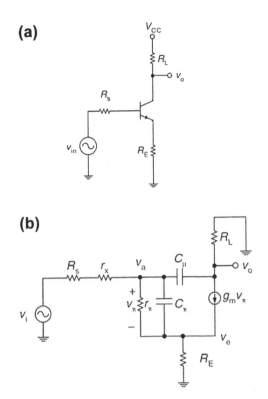

FIGURE 7.7

Common-emitter amplifier with emitter degeneration, biasing details omitted. (a) Circuit.
(b) High-frequency incremental model.

In the special case where $h_{fe}R_E \gg R_s$, r_x, and r_π, the gain is approximately:

$$A_v \approx -\frac{h_{fe}R_L}{(1 + h_{fe})R_E} \approx -\frac{R_L}{R_E} \tag{7.18}$$

We can use the previously derived worst-case open-circuit resistance calculations to estimate the bandwidth of this amplifier.

Assume that the transistor is biased at a collector current $I_C = 2$ mA and that
the transistor has small-signal current gain $h_{fe} = 100$, base resistance $r_x = 100\ \Omega$,
collector–base capacitance $C_\mu = 2$ pF, and unity-gain frequency $f_T = 300$ MHz.
The load resistor is $R_L = 1000\ \Omega$ and the emitter resistor R_E is variable. The gain
and OCTCs estimate of bandwidth for this circuit is shown in Table 7.1. Note that
as the emitter resistor is increased, the gain decreases, while the bandwidth
increases. This gives the designer another degree of freedom in setting gain and
bandwidth in common-emitter amplifiers. These results are confirmed by an
LTSPICE simulation (Figure 7.8).

Table 7.1 Common-Emitter Amplifier with Emitter Degeneration Results. The Computation $f_{h,est}$ is the Estimated −3 dB Bandwidth Using OCTCs

R_E (Ω)	Gain (Calculated)	OCTC for C_π (ns)	OCTC for C_μ (ns)	$\omega_{h,est}$ (Mrad/s)	$f_{h,est}$ (MHz)	−3 dB from LTSPICE (MHz)
0	−41.7	23.1	94.9	8.48	1.35	1.35
22	−21.6	12.2	51.6	15.7	2.49	2.52
47	−14.0	8.1	34.9	23.3	3.7	3.75
100	−8.0	4.8	21.8	37.6	6.0	6.10

(a)

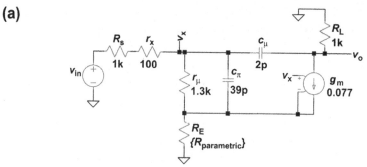

(b)

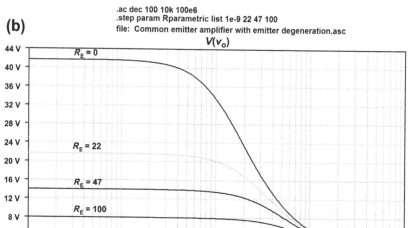

FIGURE 7.8

Common-emitter amplifier with emitter degeneration (Example 7.1). (a) LTSPICE circuit. Note that the value of R_E is variable. (b) Circuit simulations showing voltage gain for $R_E = 0$, 22, 47, and 100 Ω. (For color version of this figure, the reader is referred to the online version of this book.)

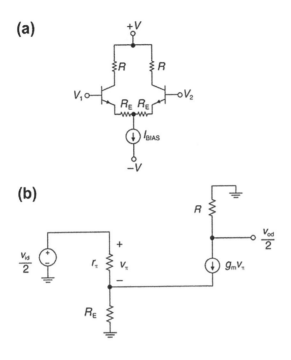

FIGURE 7.9

Differential amplifier with emitter degeneration. (a) Circuit. (b) Low-frequency small-signal model for the differential mode of operation.

Example 7.2: Differential amplifier with emitter degeneration

The model of the common-emitter amplifier with emitter degeneration can be used to analyze the gain of the differential amplifier with emitter degeneration (Figure 7.9(a)) when operated in the differential mode. By half-circuit techniques, we see that in the differential mode,[7] the circuit is equivalent to the common-emitter amplifier with emitter degeneration (Figure 7.9(b)). Inclusion of emitter resistors R_E increases the range of differential input voltages over which the differential amplifier provides approximately linear gain.[8]

Shown in Figure 7.10 is an LTSPICE simulation of a differential amplifier with emitter degeneration. The base of transistor Q_2 is grounded, and the voltage at the base of transistor Q_1 is swept from -500 to $+500$ mV and the voltage at the base of Q_2 is swept simultaneously from $+500$ to -500 mV. Results are shown in the

---

[7]In the differential mode, the input to one of the transistors is $v_{id}/2$ and the output of one of the transistors is $v_{od}/2$.

[8]However, there are tradeoffs—an increase in noise due to the resistors, as well as a lowering of differential gain and CMRR. A nice discussion is given in Grebene's *Bipolar and MOS Analog Integrated Circuit Design*, pp. 221–224.

(a)

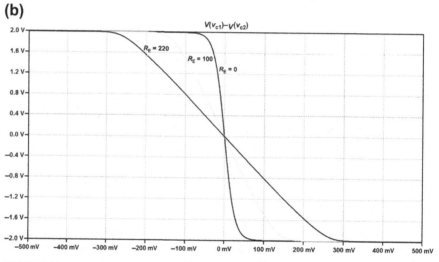

file: Differential amplifier with emitter degeneration.asc

(b)

FIGURE 7.10

Differential amplifier with emitter degeneration (Example 7.2). (a) LTSPICE circuit. (b)
Simulation result showing differential output voltage ($v_{c1} - v_{c2}$) with $R_E = 0$, 100, and 220 Ω
when differential input ($v_+ - v_-$) is swept from -500 to $+500$ mV. (For color version of this
figure, the reader is referred to the online version of this book.)

simulation of Figure 7.10(b). With $R_E = 0$, the amplifier is a standard differential
amplifier, and the output swings over its full range when the input varies a few
kT/q. With $R_E = 100$ and 220 Ω, the amplifier has linear gain over a much wider
input voltage range.

High-frequency output and input impedance of emitter follower buffers

An ideal buffer has very high input impedance and very low output impedance. The output and input impedance of an emitter follower varies with frequency due to the effects of C_π and C_μ. Using the circuits of Figure 7.11, we will further investigate the high-frequency output impedance of the emitter follower. We will model the case when there is a resistance (R_s) in series with the base. We found in Chapter 5, using the circuit of Figure 7.11(b), that the low-frequency output resistance of the emitter follower is:

$$Z_{\text{out}_{w \to 0}} = \left(\frac{R_s + r_x + r_\pi}{1 + h_{\text{fe}}} \right) \approx \left(\frac{R_s + r_x}{h_{\text{fe}}} \right) + \frac{1}{g_m} \tag{7.19}$$

We saw earlier that the small-signal current gain of the transistor decreases at very high frequencies. At very high frequencies, C_π shorts out r_π and the $g_m v_\pi$ generator is turned off as well, resulting in the incremental model of Figure 7.11(c). Ignoring the effects of C_μ, the output impedance at very high frequencies is:

$$Z_{\text{out}_{\omega \to \infty}} = R_s + r_x \tag{7.20}$$

If we wanted, we could find the output impedance in closed form by using the following methodology. We can reuse the previous result for the low-frequency output impedance of the emitter follower. However, if we replace r_π in the original

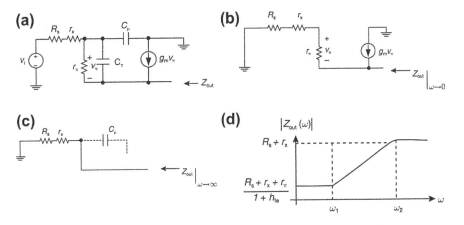

FIGURE 7.11

Analysis of high-frequency output impedance of emitter follower. (a) Incremental circuit. (b) Low-frequency model. (c) High-frequency model. (d) Impedance plot of output impedance showing that output impedance increases over a finite frequency range from ω_1 to ω_2. In this region, the output impedance of the emitter follower is inductive.

equation with an impedance representing r_π in parallel with C_π, we can solve for the output impedance as a function of frequency:

$$Z_{out}(s) = \left(\frac{R_s + r_x + z_\pi(s)}{1 + h_{fe}} \right)$$

$$z_\pi(s) = \frac{r_\pi C_\pi}{r_\pi C_\pi s + 1}$$

(7.21)

However, this gets mathematically complicated, so let us resort to an approximate result, which we can find by inspection with a minimum of calculation.

A plot of emitter follower output impedance magnitude is shown in Figure 7.11(d), assuming[9] that collector current is high enough so that $1/g_m < R_s + r_x$. We note that between frequencies ω_1 and ω_2, the output impedance increases linearly with frequency, and hence is inductive in this frequency range.

The frequency ω_1 when the impedance begins to increase is the frequency at which C_π begins shorting out r_π. This frequency is approximately:

$$\omega_1 \approx \frac{1}{r_\pi C_\pi}$$

(7.22)

The frequency ω_2 is approximately h_{fe} times this frequency:

$$\omega_2 \approx h_{fe}\omega_1 \approx \frac{h_{fe}}{r_\pi C_\pi} \approx \frac{g_m}{C_\pi}$$

(7.23)

An equivalent circuit that mimics this frequency-dependent impedance is shown in Figure 7.12. We can fit the parameters R_1 and R_2:

$$R_1 = R_s + r_x$$

$$R_1 \| R_2 = \frac{R_s + r_x + r_\pi}{1 + h_{fe}}$$

(7.24)

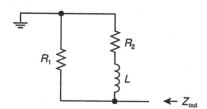

FIGURE 7.12

Model of output impedance $Z_{out}(s)$ of emitter follower showing parameters L, R_1, and R_2. Note that this models the case when there is a resistance in series with the base lead.

---

[9]This assumption is that the high-frequency impedance is higher than the low-frequency impedance. A necessary condition for this to be true is: $\frac{R_s + r_x}{h_{fe}} + \frac{1}{g_m} < R_s + r_x$. This usually occurs at modest collector currents.

In order to find the inductor value L, we note that the time constants of the original transistor circuit and the inductive circuit model must be the same. The open-circuit resistance facing C_π is approximately $1/g_m$. The resistance facing the inductor L is $R_1 + R_2$. Hence we can find L by the following method of equating the RC and L/R time constants of the two circuits:

$$\frac{C_\pi}{g_m} = \frac{L}{R_1 + R_2} \quad \Rightarrow \quad L \approx \left(\frac{C_\pi}{g_m}\right)(R_1 + R_2) \tag{7.25}$$

Example 7.3: Emitter follower output impedance numerical example

Let us do a numerical example. Assume we have an emitter follower with $R_s = 1\ \text{k}\Omega$, $r_x = 200\ \Omega$, $I_C = 2\ \text{mA}$, $h_{fe} = 150$, $g_m = 0.077/\Omega$, $r_\pi = 1950\ \Omega$, $C_\mu = 2\ \text{pF}$, $f_T = 300\ \text{MHz}$, and $C_\pi = 39\ \text{pF}$. The incremental model for this emitter follower is shown in Figure 7.13(a). Our inductive model predicts:

$$R_1 = R_s + r_x = 1200\ \Omega$$

$$R_1 \| R_2 = \frac{R_s + r_x + r_\pi}{1 + h_{fe}} = 20.9\ \Omega \Rightarrow R_2 \approx 20.9\ \Omega \tag{7.26}$$

$$L \approx \left(\frac{C_\pi}{g_m}\right)(R_1 + R_2) = \left(\frac{39\ \text{pF}}{0.077}\right)(1220\ \Omega) \approx 0.61\ \mu\text{H}$$

This model predicts an impedance that starts to rise at a frequency $\sim R_2/L$ or at 34.3 Mrad/s (5.5 MHz).

We can simulate this with LTSPICE by adding an AC test current source i_t and measuring the resultant test voltage v_t, as in Figure 7.13(b), with $Z_{out} = v_t/i_t$. The simulation result (Figure 7.13(c)) shows that the low-frequency impedance is approximately 20 Ω as expected and that the impedance starts to rise at about 5 MHz. However, the high-frequency impedance (at frequencies above approximately 100 MHz) does not reach our expected high-frequency limit of 1200 Ω. This is due to the loading effect of C_μ, which our simplified analysis has ignored. At very high frequencies, the shunting effects of C_μ cause the output impedance to become capacitive and decrease. So, in this design example, the models show that the range over which the output impedance is inductive is approximately 5–100 MHz.

One ramification of the inductive output impedance is that you can have small-signal gain peaking if you drive capacitive loads. Figure 7.14(a) shows an emitter follower driving a capacitive load.[10] The LTSPICE simulation (Figure 7.14(b)) shows gain peaking for modest capacitive loading. Note that the peak value occurs at frequencies in the 5–100 MHz range where we expect the output of the emitter follower to have inductive output impedance.[11]

---

[10]In this simulation, we use the 2N3904 transistor, which has parameters (f_T, h_{fe}, etc.) similar to that of the previous example.

[11]We could damp this by adding a resistor in series with the load capacitor.

(a)

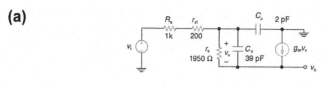

(b)

Emitter follower small-signal model, output impedance
Parameters chosen given f_T = 300 MHz, C_μ = 2 pF, h_{fe} = 150, r_x = 100
MTT 9/11/12

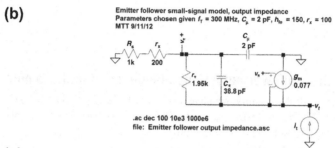

.ac dec 100 10e3 1000e6
file: Emitter follower output impedance.asc

(c)

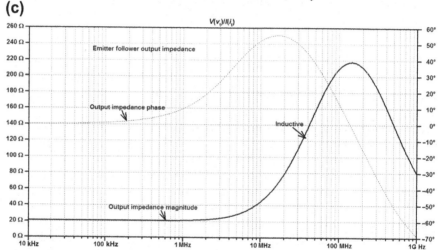

FIGURE 7.13

Analysis of high-frequency output impedance of emitter follower using LTSPICE
(Example 7.3). (a) High-frequency incremental circuit including r_x and C_μ. (b) LTSPICE
circuit, where an AC current source of value $i_t = 1$ A drives the emitter so that we can find the
incremental emitter output impedance. (c) LTSPICE simulation, showing region of inductive
output impedance, followed by impedance rolloff due to effects of C_μ. (For color version of
this figure, the reader is referred to the online version of this book.)

(a)

2N3904 emitter follower driving a capacitive load
This circuit demonstrates the resonant gain peaking that can occur
when an inductive output impedance drives a capacitive load
MTT 9/12/12

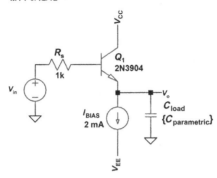

.ac dec 100 100e3 1000e6
.step param Cparametric list 1e-15 50p 100p 220p 330p
file: 2N3904 emitter follower driving a capacitive load.asc

(b)

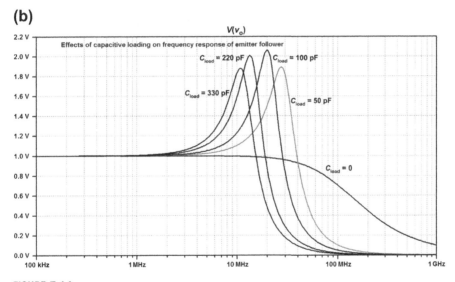

FIGURE 7.14

Emitter follower driving capacitive load (Example 7.3). (a) Circuit using 2N3904 transistor.
(b) LTSPICE simulation of gain of this emitter follower with load capacitor values $C_L = 0$, 50,
100, 220, and 330 pF. $V_{CC} = +12$ V and $V_{EE} = -12$ V. (For color version of this figure, the
reader is referred to the online version of this book.)

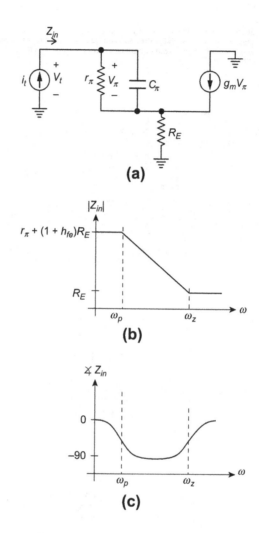

FIGURE 7.15

Simplified model for finding input impedance $Z_{in}(s)$ of emitter follower (Example 7.4).
(a) Original circuit assuming $C_\mu = 0$ and $r_x = 0$. (b) Qualitative plots of magnitude of input impedance. (c) Qualitative plot of angle of input impedance.

Example 7.4: Unloaded emitter follower input impedance

We will next investigate the input impedance of the unloaded emitter follower by using the simplified incremental model[12] of Figure 7.15(a). In this first example, we will assume that the emitter follower is unloaded, or at worst, loaded with a purely resistive load that we can lump into an equivalent emitter resistor. Using a

---

[12]It is simplified because we are ignoring r_x and C_μ throughout the following discussion. This makes the math much easier and gives us some insight into the physical operation of the circuit.

similar methodology as in Chapter 5 when we found the low-frequency input impedance of the emitter follower, we find that the input impedance is:

$$Z_{in}(s) = \frac{v_i(s)}{i_t(s)} = z_\pi(s) + \left(1 + h_{fe}(s)\right)R_E = z_\pi(s) + \left(1 + g_m z_\pi(s)\right)R_E$$

$$(7.27)$$

$$z_\pi(s) = \frac{r_\pi}{r_\pi C_\pi s + 1}$$

We find that at very low frequencies, the input impedance is:

$$Z_{in_{\omega \to 0}} = r_\pi + \left(1 + h_{fe}\right)R_E \qquad (7.28)$$

At very high frequencies, C_π shorts out r_π and disables the dependent current
source, resulting in:

$$Z_{in_{\omega \to \infty}} = R_E \qquad (7.29)$$

Plots of the magnitude and phase of the input impedance of the emitter follower
as a function of frequency are shown in Figure 7.15(b and c). Note that at very high
frequencies, the impedance $Z_{in}(\omega)$ will further decrease due to the shunting effect of
C_μ, which we ignored in this previous analysis.

Example 7.5: Input impedance of capacitively loaded emitter follower

Using the results of Example 7.4, let us see what happens when we have the added
complication of a capacitor loading the emitter follower as in Figure 7.16. The next
logical progression in the algebra is to replace R_E by a load ($Z_E(s)$), which is the
parallel combination of R_E and a load capacitor C_L:

$$Z_{in}(s) = z_\pi(s) + (1 + g_m z_\pi(s))Z_E(s)$$

$$z_\pi(s) = \frac{r_\pi}{r_\pi C_\pi s + 1} \qquad (7.30)$$

$$Z_E(s) = \frac{R_E}{R_E C_L s + 1}$$

Following through with the algebra, we find the messy result:

$$z_\pi(s) = \frac{r_\pi}{r_\pi C_\pi s + 1} + \left(1 + g_m\left(\frac{r_\pi}{r_\pi C_\pi s + 1}\right)\right)\left(\frac{R_E}{R_E C_L s + 1}\right)$$

$$(7.31)$$

$$= \left(r_\pi + \left(1 + h_{fe}\right)R_E\right)\left(\frac{\frac{r_\pi R_E(C_L + C_\pi)s}{(r_\pi + (1 + h_{fe})R_E)} + 1}{(r_\pi C_\pi s + 1)(R_E C_L s + 1)}\right)$$

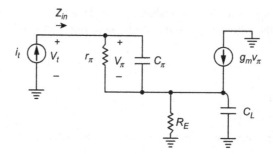

FIGURE 7.16

Incremental model of capacitively loaded emitter follower of Example 7.5 for calculation of input impedance $Z_{in} = v_t/i_t$. For mathematical simplicity, we at first ignore transistor collector–base capacitance C_μ.

Normally, R_E is a large value resistor (or the high output impedance of a bias current source) and $h_{fe}R_E \gg r_\pi$, so we can approximate this input impedance as:

$$Z_{in}(s) \approx \left(r_\pi + (1 + h_{fe})R_E\right) \left(\frac{\frac{(C_L + C_\pi)}{g_m} + 1}{(r_\pi C_\pi s + 1)(R_E C_L s + 1)} \right) \qquad (7.32)$$

Now let us pause and look at this result. First, there is a term out front $(r_\pi + (1 + h_{fe})R_E)$, which is the low-frequency input impedance, as expected. Second, we find that there are two poles and a zero at frequencies:

$$\omega_{p1} = -\frac{1}{R_E C_L}$$

$$\omega_{p2} = -\frac{1}{r_\pi C_\pi} \qquad (7.33)$$

$$\omega_z = -\frac{g_m}{C_L + C_\pi}$$

There are two interesting cases: relatively large C_L and relatively small C_L. Let us plot the pole-zero plot and Bode plot for two different cases. First, for relatively small C_L, with $C_L < h_{fe}C_\pi$ (Figure 7.17(a)), we find that the zero frequency ω_z is at a frequency higher than ω_{p2}. This means that the angle of the input impedance dips below $-90°$ for some frequency range. In the frequency range where the angle is less than $-90°$, the real part of the input impedance is negative.[13] This is a negative resistance. In other words, in this frequency range, the negative real input impedance can help sustain an oscillation.

---

[13] A negative resistance can source power, while a positive resistance can only dissipate power.

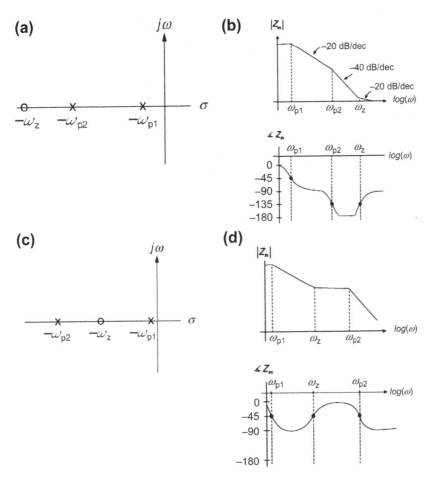

FIGURE 7.17

Input impedance of capacitively loaded emitter follower (Example 7.5). (a) Pole-zero plot and (b) Bode plot for relatively small C_L, with $C_L < h_{fe}C_\pi$. (c) Result for relatively large C_L with $C_L > h_{fe}C_\pi$, pole-zero plot, and (d) Bode plot of magnitude and phase.

Second, for relatively large C_L (Figure 7.17b), we find that the zero frequency ω_z is at a frequency higher than ω_{p2}. Thus, the real part of the input impedance never becomes negative.

This result shows that emitter followers can exhibit strange behavior—circuits containing emitter followers can have gain peaking, or in some cases can even oscillate.[14] One design strategy is sometimes is to put a resistor or ferrite bead in series with the base of the transistor.

---

[14]A negative resistor delivers power to an external circuit. A positive resistor can only dissipate power. If the negative resistor delivers power to an *LC* circuit, an oscillation can result.

Next we will consider an LTSPICE example of a 2N3904 emitter follower driving a capacitive load (Figure 7.18(a)). We have an emitter follower biased with a 2 mA current source and loaded by different capacitor values in the 0–1000 pF range. The resultant AC analysis (Figure 7.18(b)) plots the input impedance as measured at the base. At frequencies when the angle of the input impedance dips below −90°, there is a *negative resistance* at the base. This phenomenon occurs in the frequency range between approximately 1 and 200 MHz, depending on the load capacitor value and on

(a)   Input impedance of 2N3904 emitter follower driving a capacitive load
This circuit demonstrates the possibility of negative input resistance
MTT 9/12/12

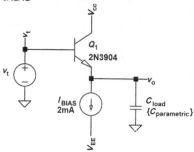

.ac dec 1000 10000 1000e6
.step param Cparametric list 1e-15 50p 100p 220p 330p 1000p
file: Input impedance of 2N3904 emitter follower driving a capacitive load.asc

(b)

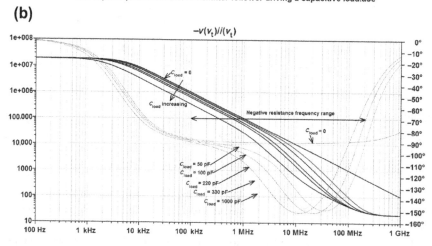

FIGURE 7.18

Emitter follower with capacitive load (Example 7.5). (a) Circuit for finding input impedance of the emitter follower, with emitter follower loaded with 0, 50, 100, 220, 330, and 1000 pF; $V_{CC} = 12$ V; and $V_{EE} = -12$ V (b) Input impedance magnitude and phase. There is the possibility of a negative resistance in the 1–200 MHz range, where the angle dips below −90°. (For color version of this figure, the reader is referred to the online version of this book.)

the collector current bias level of the transistor. This negative resistance means that there is the potential to create unwanted parasitic oscillations in this frequency range. We will investigate the possibility of oscillation next.

Example 7.6: The dreaded emitter follower unintentional high-frequency oscillation

A negative impedance can be a very troublesome thing. If there are reactive components surrounding the transistor (for instance, transistor capacitances and wiring inductance), the possibility of an unintentional oscillator exists if there is a negative resistance looking into the base terminal. A circuit illustrating the possibility of an unintentional oscillator is shown in Figure 7.19. In this case, L_p is the parasitic wiring inductance and C_p is transistor parasitic capacitances attached to the base. The negative resistance $-R$ adds energy to the LC tank circuit, causing an oscillation. In practice, the $-R$ at the base of the transistor will be partially canceled by series-positive resistances. Also, we note that in this simplified model the poles of the system will be in the right-half plane and a constant-amplitude oscillator has poles on the $j\omega$ axis. In practice, if an unintentional oscillator is created, the oscillation amplitude self-limits due to the nonlinear gain of the transistor.

An unintentional oscillator is shown in Figure 7.20, where we have a 2N3904 emitter follower loaded by a 100-pF load capacitor. External parasitic wiring inductances of 100 nH are in the base and collector leads. The resultant transient analysis (Figure 7.20(b)) shows a parasitic oscillation at ~ 120 MHz.

The classic "fix" to the oscillation problem of emitter followers is to put a small-valued resistor in the base (Figure 7.21). In order to prevent oscillation, the external resistance must be larger than the negative resistance $-R$ presented by the base circuit. In the transient analysis shown, it looks like an external resistor of around 100 Ω will do a good job of killing any oscillation, but the value that you will need in your circuit will vary, depending on the type of transistor you use, transistor bias level, and the amount of parasitic inductance surrounding the transistor terminals.

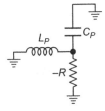

FIGURE 7.19

Circuit for Example 7.6 illustrating the possibility of an unintentional oscillator due to parasitic components L_p and C_p attached to the base of an emitter follower, where the follower presents a negative impedance of value $-R$.

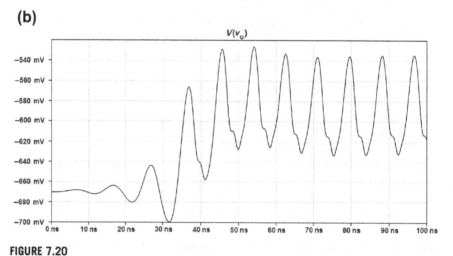

(a)

PULSE(0 0.001 0 1ns 1ns 100ns 200ns 1)

.tran 0 100n 0 0.01n
file: 2N3904 emitter follower driving a capacitive load parasitic oscillation.asc

(b)

FIGURE 7.20

Unintentional oscillator caused by capacitively loading an emitter follower (Example 7.6).
(a) Circuit. (b) Transient LTSPICE analysis, showing oscillation at 120 MHz. (For color version
of this figure, the reader is referred to the online version of this book.)

Example 7.7: Laboratory experiment demonstrating dreaded emitter follower unintentional high-frequency oscillation

A circuit demonstrating the dreaded emitter follower high-frequency unintentional
oscillator is shown in Figure 7.22(a). The circuit was built on a protoboard, with
an intentionally bad high-frequency layout (i.e. no ground plane, no power supply
bypassing, etc.). A scope probe was attached to the emitter (Figure 7.22(b)). The

(a)

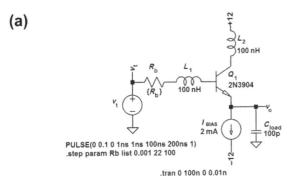

PULSE(0 0.1 0 1ns 1ns 100ns 200ns 1)
.step param Rb list 0.001 22 100

.tran 0 100n 0 0.01n
file: 2N3904 EF with oscillation killing base resistor.asc

(b)

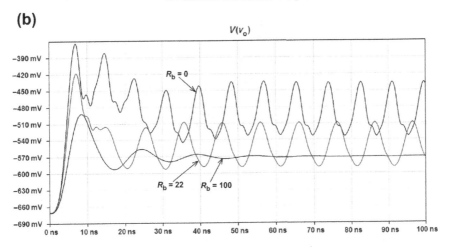

FIGURE 7.21

Classic fix for the unintentional 120 MHz oscillator of Example 7.6, by adding a small-value base resistor. (a) Circuit. (b) Transient LTSPICE analysis, effects of external base resistor $R_b = 0$, 22 and 100 Ω. The 100-Ω resistor effectively kills the oscillation. (For color version of this figure, the reader is referred to the online version of this book.)

resultant output (Figure 7.22(c)) shows the DC value of the emitter voltage to be approximately −0.7 V, but with a ∼ 100-MHz oscillation riding on top of the DC bias.

There are many bad ramifications of the dreaded oscillation. The oscillation can be rectified and cause DC shifts in the bias voltage of the transistor, or surrounding transistors. And, of course, the resultant oscillation can cause electromagnetic interference problems with nearby circuits.

Next, we will discuss a little bit about instrumentation and measurement. The scope photograph of Figure 7.22(c) shows a waveform that is quite sinusoidal.

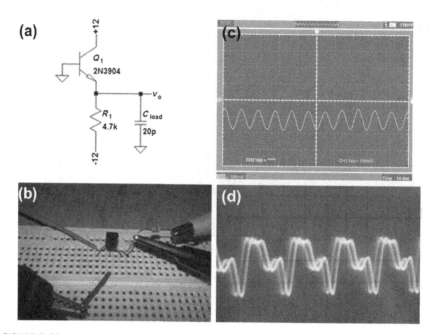

FIGURE 7.22

Circuit demonstrating the dreaded high-frequency emitter follower oscillation of Example 7.7.
(a) Emitter follower circuit as built with a 2N3904 transistor, driving a 20-pF capacitive load.
(b) Photograph of circuit built on a protoboard with an intentionally poor circuit layout.
(c) Scope capture showing DC level of transistor emitter voltage at −0.7 V with a ∼100-MHz
oscillation. This photograph was taken with a digital storage scope; horizontal: 10 ns/division.
(d) Photograph taken with high-bandwidth (Tektronix 465) scope; horizontal: 5 ns/division.
(For color version of this figure, the reader is referred to the online version of this book.)

However, in this case, details of the waveform are masked by the relatively low
bandwidth of the digital storage scope used in the experiment. In Figure 7.22(d),
we see a photograph taken with a higher bandwidth Tektronix 465 analog scope,
showing that the oscillation is not sinusoidal. (The moral is use a scope and probe
with sufficient bandwidth to do the job.)

Bootstrapping

We have seen earlier in our discussion of the Miller effect that the effects of capacitance can be multiplied if the capacitance is wrapped around a negative gain. Referring to Figure 7.23, remember our result from a previous chapter that the capacitance
looking into the input terminal of this amplifier is:

$$C_{in} = C(1 - A) \qquad (7.34)$$

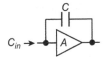

FIGURE 7.23

Circuit illustrating Miller effect.

This result suggests another way to reduce the effects of circuit capacitances. What happens if the amplifier has a gain of +1? With a gain of +1, the capacitance looking into the terminals of the amplifier is exactly zero. This methodology can be used to reduce the effects of capacitances and hence to enhance bandwidth.

Example 7.8: Bootstrapping an emitter follower

We can illustrate how to do bootstrapping by looking at an emitter follower circuit (Figure 7.24). We will assume that the transistor small-signal parameters are $h_{fe} = 150$, $r_x = 200 \, \Omega$, $C_\mu = 2 \, \text{pF}$, and $f_T = 300 \, \text{MHz}$.

Other incremental parameters are:

$$I_C \approx \frac{11.3 \text{ V}}{10 \text{ k}\Omega} \approx 1.13 \text{ mA}$$

$$g_m = \frac{|I_C|}{V_{TH}} = \frac{1.13 \text{ mA}}{26 \text{ mV}} = 0.043/\Omega$$

$$(7.35)$$

$$r_\pi = \frac{h_{fe}}{g_m} = \frac{150}{0.043/\Omega} = 3488 \, \Omega$$

$$C_\pi = \frac{g_m}{\omega_T} - C_\mu = \frac{0.043}{2\pi(300 \times 10^6)} - 2\text{pF} = 20.8 \text{ pF}$$

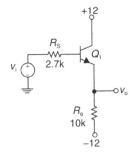

FIGURE 7.24

Emitter follower for Example 7.8.

The OCTC calculations for C_π and C_μ are:

$$R_{o\pi} = \frac{v_t}{i_t} = r_\pi \left\| \left(\frac{R_s + r_x + R_E}{1 + g_m R_E} \right) \right.$$

$$= 3488\,\Omega \left\| \left(\frac{2700\,\Omega + 200\,\Omega + 10,000\,\Omega}{1 + (0.043)(10,000\,\Omega)} \right) \approx 29.7\,\Omega \right.$$

$$\tau_{o\pi} = R_{o\pi} C_\pi = (29.7\,\Omega)(20.8\,\text{pF}) = 0.6\,\text{ns} \tag{7.36}$$

$$R_{o\mu} = (R_s + r_x) \| (r_\pi + (1 + h_{fe}) R_E)$$

$$= (2700\,\Omega + 200\,\Omega) \| (3488 + (1 + 150)(10\,\text{k}\Omega)) = 2894\,\Omega$$

$$\tau_{o\mu} = R_{o\mu} C_\mu = (2894\,\Omega)(2\,\text{pF}) = 5.8\,\text{ns} \tag{7.37}$$

The OCTC calculation shows that the OCTC for C_π is approximately 0.6 ns and that for C_μ is approximately 5.8 ns. An estimate of bandwidth from OCTCs is therefore 24.9 MHz. An LTSPICE simulation (Figure 7.25) shows a bandwidth of ~26.5 MHz.

We can improve the −3 dB bandwidth by bootstrapping the C_μ of the input transistor, as shown in Figure 7.26(a), where we have added a gain of +1 across the collector–base junction of Q_1. The simulation result (Figure 7.27) shows that we have improved the bandwidth significantly, at the expense of some gain peaking. One way to reduce the gain peaking is to add some damping (Figure 7.28). In Figure 7.28(b), we show the results (empirically derived) for a damping resistor in the 1–10 kΩ range. It looks like an optimum value for this circuit is ~5 kΩ, which results in good bandwidth with a minimum of gain peaking.

Example 7.9: Another bootstrapping design example

We will now revisit an amplifier similar to that from Chapter 6 (Figure 7.29(a)) and further extend the bandwidth by using bootstrapping.[15] In this circuit, the dominant time constant is due to the $C_{\mu 3}$ of the input emitter follower Q_3 interacting with the 2.7 kΩ source resistance. The gain of this circuit is over 100 and the bandwidth (Figure 7.29(b)) is approximately 19 MHz.

In order to reduce the effects of $C_{\mu 3}$, let us force a gain of +1 across the collector–base junction of Q_3 by adding bootstrap transistor Q_5 (Figure 7.30(a)). Q_5 is a PNP emitter follower and forces the collector of Q_3 to follow the base of Q_3 incrementally. The LTSPICE analysis (Figure 7.30(b)) shows that we have

---

[15]This example is similar in topology to that previously discussed in this chapter, but some of the details (bias point levels, etc.) are different. This section resorts exclusively to LTSPICE simulations; the OCTC calculations, although illustrative, are somewhat cumbersome.

(a)

Emitter follower example
Parameters chosen given fT = 300 MHz, Cmu = 2 pF, hfe = 150, rx = 200
MTT 9/16/12

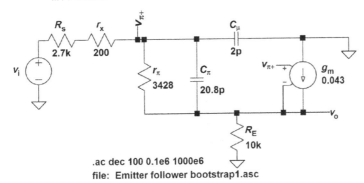

.ac dec 100 0.1e6 1000e6
file: Emitter follower bootstrap1.asc

(b)

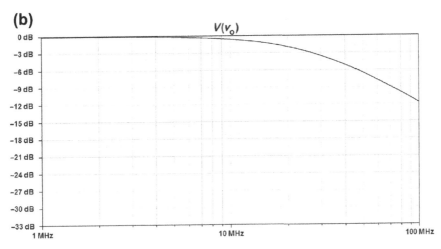

FIGURE 7.25

Emitter follower for Example 7.8, LTSPICE result. (a) Small-signal incremental model.
(b) Gain vs. frequency showing a −3 dB bandwidth of about 26.5 MHz. (For color version of
this figure, the reader is referred to the online version of this book.)

indeed extended the bandwidth of the overall amplifier to approximately 40 MHz,
although at the expense of some gain peaking at the high end.

We can carry this concept one step further by bootstrapping Q_4 (Figure 7.31(a))
by inclusion of bootstrap PNP emitter follower Q_6. Note that we get lots of gain
peaking (Figure 7.31(b)).

We can qualitatively trace back the gain peaking to several possible causes. We
know that gain peaking can occur due to the inductive output impedance of an
emitter follower interacting with any stray capacitance. Some strategies to lower

(a)

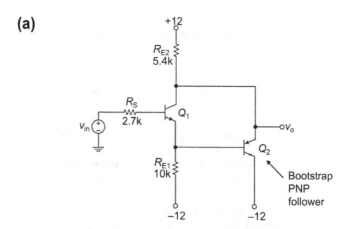

(b)

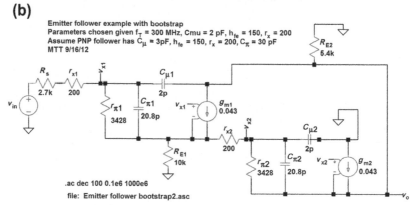

FIGURE 7.26

Bootstrapped emitter follower for Example 7.8. (a) Circuit. (b) Small-signal model. (For color version of this figure, the reader is referred to the online version of this book.)

the effects of inductive peaking are to change the bias levels of the bootstrap transistors and also to include external base resistors. Shown in Figure 7.32 is the response of the amplifier with the collector currents of the bootstrap transistors each run at higher collector currents. (We have also cheated a little bit by adding an extra low-pass filter with $R_4 C_{load}$.) We see that the gain peaking has been significantly reduced, while the bandwidth of the overall amplifier is ~ 50 MHz. We could certainly do better than this by continuing to tweak the design, but you get the idea...

Example 7.10: Shunt peaked amplifier revisited

We will revisit the shunt peaked amplifier of Chapter 5, but this time without making any restrictions as to the value of transistor base resistance r_x. We saw in Chapter 5

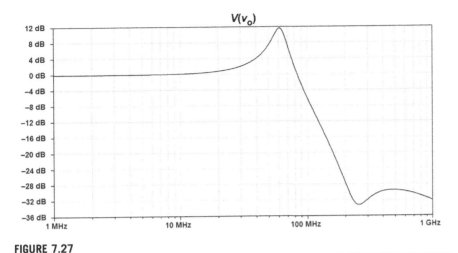

FIGURE 7.27

Bootstrapped emitter follower simulation for Example 7.8 result (magnitude), showing +12 dB of gain peaking around 60 MHz.

that if the value of r_x is very small, the transfer function can easily be found in closed form. However, in many small-signal transistors, the value of r_x can be significant. For instance, for the 2N3904 transistor, the value of r_x is 100–250 Ω, depending on collector current bias level. The resultant circuit is shown in Figure 7.33(a), and the small-signal model is shown in Figure 7.33(b).

In Chapter 5, we were able to increase the bandwidth of the shunt peaked amplifier to approximately 390 MHz, by using a peaking inductor $L_{pk} = 0.25$ μH. However, in that example we assumed that r_x was zero. Let us examine the performance of the peaking amplifier, but with finite values of base resistance.

In Figure 7.34, we see the effect on the inductively peaked amplifier (with $L_{pk} = 0.25$ μH) and with $r_x = 0, 10, 20, 30,$ and 40 Ω. We see that small values of r_x have a significant detrimental effect on both midband gain and bandwidth. Hence, the warning here is to be careful in neglecting r_x in your bandwidth calculations and simulations.[16]

Example 7.11: Common-base amplifier

The common-base amplifier is a topology commonly used in high-frequency circuits. In fact, the cascode amplifier uses a common-base transistor as an output buffer, as we have seen before. A common-base amplifier is shown in Figure 7.35(a) with $R_E = 1$ kΩ and $R_L = 3.3$ kΩ. Let us assume that $V_{CC} = 12$ V and that $V_{BIAS} = 6$ V. The transistor has $I_C = 1$ mA, $h_{fe} = 100$, $f_T = 500$ MHz,

---

[16]Unfortunately, there is very little datasheet information available for most transistors on the value of base spreading resistance. Note that r_x also affects low-frequency gain.

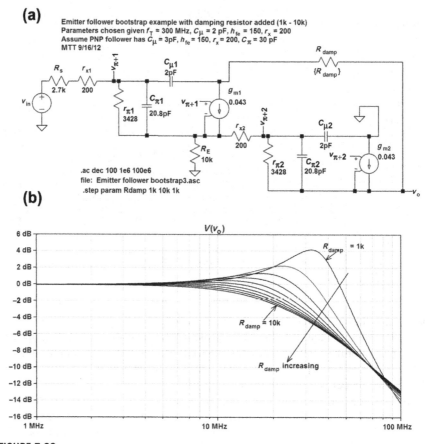

(a)

Emitter follower bootstrap example with damping resistor added (1k - 10k)
Parameters chosen given f_T = 300 MHz, C_μ = 2 pF, h_{fe} = 150, r_x = 200
Assume PNP follower has C_μ = 3pF, h_{fe} = 150, r_x = 200, C_π = 30 pF
MTT 9/16/12

.ac dec 100 1e6 100e6
file: Emitter follower bootstrap3.asc
.step param Rdamp 1k 10k 1k

(b)

FIGURE 7.28

Bootstrapped emitter follower of Example 7.8 with damping added to reduce gain peaking.
(a) Circuit small-signal model. (b) LTSPICE simulation results for a damping resistor of
1, 2, ... 10 kΩ. (For color version of this figure, the reader is referred to the online version of
this book.)

$r_x = 50\ \Omega$, and $C_\mu = 1$ pF. The high-frequency small-signal model for this amplifier
is shown in Figure 7.35(b). We will find the gain and estimate the bandwidth using
OCTCs.

The small-signal parameters for this amplifier are:

$$g_m = \frac{|I_C|}{V_{TH}} = \frac{1\ \text{mA}}{26\ \text{mV}} = 0.038/\Omega$$

$$r_\pi = \frac{h_{fe}}{g_m} = \frac{100}{0.038/\Omega} = 2631\ \Omega \qquad (7.38)$$

$$C_\pi = \frac{g_m}{\omega_T} - C_\mu = \frac{0.038}{2\pi(500 \times 10^6)} - 1\ \text{pF} = 11\ \text{pF}$$

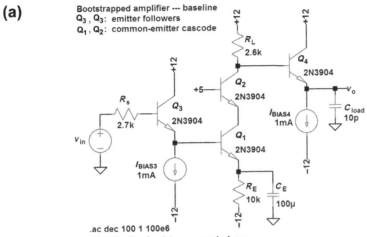

(a)

Bootstrapped amplifier --- baseline
Q_3, Q_3: emitter followers
Q_1, Q_2: common-emitter cascode

.ac dec 100 1 100e6
file: Another bootstrap example 1.asc

(b)

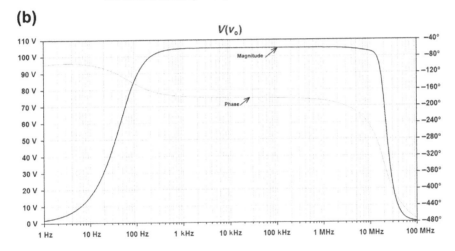

$V(v_o)$

FIGURE 7.29

Another design example (Example 7.9). (a) Circuit. (b) LTSPICE simulation showing a midband gain of −106 and a bandwidth of 19.6 MHz. (For color version of this figure, the reader is referred to the online version of this book.)

By using the low-frequency small-signal model (Figure 7.35(c)), we can find a closed-form solution for the gain of the common-base amplifier that results in:

$$A_v = \frac{v_o}{v_i} \approx \frac{R_L}{R_E + \left(\frac{r_x + r_\pi}{h_{fe}} \right)}$$

(7.39)

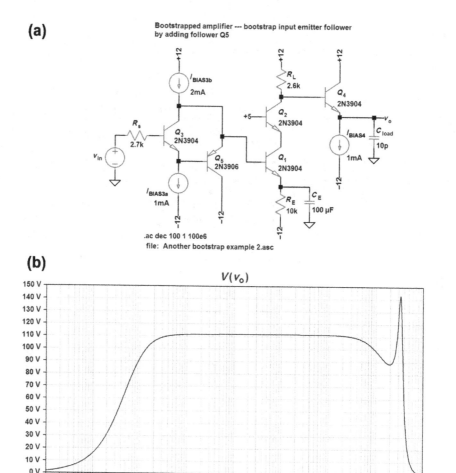

FIGURE 7.30

Circuit of Example 7.9 modified by bootstrapping of input emitter follower Q_3 with bootstrap transistor Q_5. (a) Circuit. (b) Frequency response magnitude from LTSPICE showing significant gain peaking at around 40 MHz. (For color version of this figure, the reader is referred to the online version of this book.)

This equation predicts a gain of +3.2 for this amplifier.

As a side note, this predicted gain makes perfect sense intuitively. The 1-kΩ emitter resistor is much larger than the input resistance looking into the emitter of the transistor ($\sim 1/g_m$). Therefore, the AC current that gets injected into the emitter is approximately v_i/R_E. This current travels through the transistor with a current gain

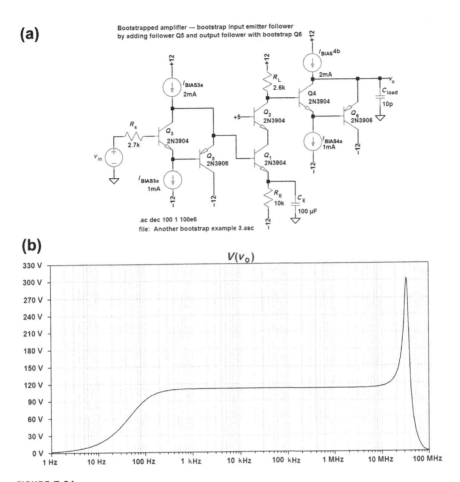

FIGURE 7.31

Circuit of Example 7.9 further modified by bootstrapping of output emitter follower Q_4.
(a) Circuit. (b) Frequency response magnitude from LTSPICE, showing significant gain peaking. (For color version of this figure, the reader is referred to the online version of this book.)

of ~ 1, and the resultant voltage generated at the collector is $(v_i/R_E \times R_L)$, resulting in a voltage gain of ~ 3.3.

We will predict the bandwidth of the amplifier using OCTCs and the circuits of Figure 7.36. For C_π, we find the OCTC:

$$R_{\pi o} = r_\pi \left\| \left(\frac{r_x + R_E}{1 + g_m R_E} \right) \approx \frac{1}{g_m} \approx 26 \, \Omega \right. \tag{7.40}$$

$$\tau_{\pi o} = R_{\pi o} C_\pi = (26 \, \Omega)(11 \, \text{pF}) = 0.3 \, \text{ns}$$

(a)

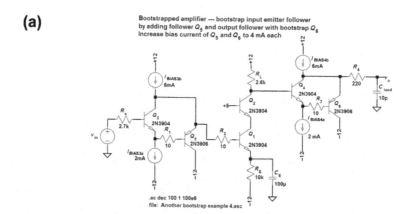

(b)

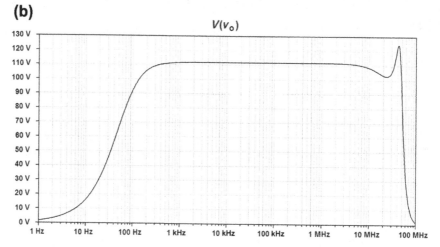

FIGURE 7.32

Response of circuit of Example 7.9 with emitter followers run at a collector current of 4 mA; base resistors and output damping resistor added. (For color version of this figure, the reader is referred to the online version of this book.)

For C_μ, we find:

$$R_{EQ1} = r_x \| (r_\pi + (1 + h_{fe})R_E) \approx r_x \approx 50\ \Omega$$

$$G_M \approx \frac{g_m}{1 + g_m R_E} \approx \frac{1}{R_E} \approx 0.001$$

$$R_{\mu o} = R_{EQ1} + R_L + G_M R_L R_{EQ1} = 50\ \Omega + 3300\ \Omega + (0.001)(3300\ \Omega)(50\ \Omega)$$

$$\approx 3515\ \Omega$$

$$\tau_{\mu o} = R_{\mu o} C_\mu = (3515\ \Omega)(1\ \text{pF}) = 3.5\ \text{ns}$$

$$(7.41)$$

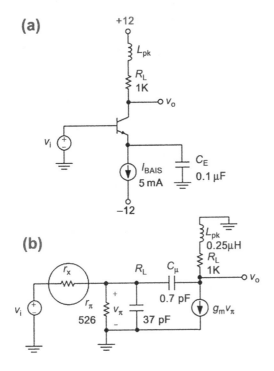

FIGURE 7.33

Inductively peaked amplifier of Example 7.10. (a) Circuit. (b) Small-signal model, including base spreading resistance r_x.

The sum of OCTCs for this amplifier is 3.8 ns, and hence our estimate of bandwidth is:

$$\omega_h \approx \frac{1}{3.8 \text{ ns}} \approx 263 \text{ Mrad/s}$$

$$f_h = \frac{\omega_h}{2\pi} \approx 41.9 \text{ MHz}$$

(7.42)

LTSPICE (Figure 7.37) shows that the gain is +3.18 and the −3 dB bandwidth is 42.3 MHz.

Example 7.12: Current amplifier

An amplifier used for fast current switching is shown in Figure 7.38. Circuits similar to this, where currents are switched and/or amplified, are used extensively in high-frequency circuits such as current-feedback operational amplifiers. We will estimate the bandwidth of this circuit assuming small-signal[17] operation, using OCTCs. For

---

[17]We will see more on current-feedback op-amps in a later chapter.

(a)   Peaked amplifier with finite base spreading resistance r_x

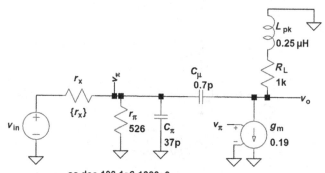

.ac dec 100 1e6 1000e6
.step param r_x 0.001 40.001 10
file: Peaking amplifier with finite base spreading resistance r_x.asc

(b)

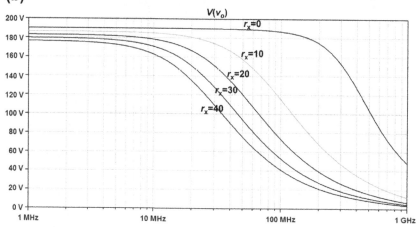

FIGURE 7.34

LTSPICE simulation of inductively peaked ($L_{pk} = 0.25$ µH) amplifier of Example 7.10, with r_x in the range 0–40 Ω. (a) Circuit. (b) Simulation result showing gain and bandwidth degradation with finite r_x. (For color version of this figure, the reader is referred to the online version of this book.)

incremental differential-mode signals, we can ground the emitter connection (as the incremental voltage swing at this node is zero), resulting in the circuits shown. By inspection, the open-circuit resistances for C_π and C_μ are the same, resulting in:

$$R_{o1} = r_x \| r_\pi$$

$$R_{o2} = r_x \| r_\pi$$

$$\sum \tau_{oc} = (r_x \| r_\pi)(C_\pi + C_\mu) \tag{7.43}$$

$$\omega_h \approx \frac{1}{\sum \tau_{oc}} = \frac{1}{(r_x \| r_\pi)(C_\pi + C_\mu)}$$

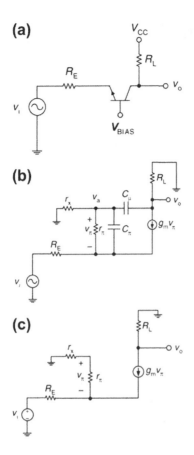

FIGURE 7.35

Common-base amplifier of Example 7.11. (a) Circuit. (b) Small-signal model. (c) Low-frequency small-signal model.

This is likely to be at a very high frequency since there is no Miller effect. For example, for a switch with $f_T = 300$ MHz, $\omega_T = 1.89 \times 10^9$ rad/s, $r_x = 100$ Ω, $h_{fe} = 150$, and $C_\mu = 2$ pF. If we bias the transistors with $I_{BIAS} = 4$ mA, then $I_{C1} = I_{C2} = 2$ mA and $g_{m1} = g_{m2} = 0.077/\Omega$. We then find that $C_\pi = 38.8$ pF and $r_\pi = 1950$ Ω. This results in an OCTC estimate of bandwidth $\omega_h \approx 257$ Mrad/s ($f_h \approx 41$ MHz). LTSPICE (Figure 7.39) shows $f_h = 41$ MHz, so the estimate is pretty good.

Example 7.13: Effects of parasitic inductance on the performance of high-speed circuits

Next, we will investigate the effects of parasitic inductance on the small-signal bandwidth of a high-speed current amplifier. In Figure 7.40(a), we see a current amplifier

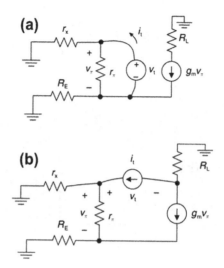

FIGURE 7.36

Circuits for finding OCTCs for common-base amplifier of Example 7.11. (a) Circuit for C_π. (b) Circuit for C_μ.

where we put in a differential input voltage ($v_{base1} - v_{base2}$) and the output is ($i_{c1} - i_{c2}$). In Figure 7.40(b), we see that in the case where there is no parasitic inductance, the small-signal bandwidth is about 80 MHz. With inclusion of only 10 nH of parasitic inductance in each transistor lead, the bandwidth has reduced to ~35 MHz. It does not take much lead length to create a parasitic inductance of 10 nH—a very approximate rule of thumb says that 10 mm of wire is very roughly 10 nH of inductance.[18]

Pole splitting

"Pole splitting" is a technique widely used to tailor the response of amplifiers, especially in operational amplifiers. We will illustrate pole splitting, using the circuit in Figure 7.41. This circuit models an amplifier with gain $-A$, input resistance R_i, input capacitance C_i, output resistance R_o, and output capacitance C_o. We add a feedback capacitor C_f to tailor the pole locations, with the following explanation. It may be counterintuitive, but there are two poles in this system (not three poles).

---

[18]Yes, I know that inductance can only be defined in terms of a loop of wire. But assuming that there is a ground trace or ground plane somewhere around, the inductance of a transistor lead 10 mm long is roughly 10 nH.

(a)

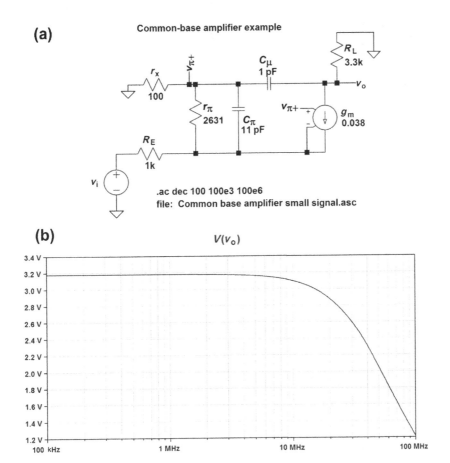

Common-base amplifier example

.ac dec 100 100e3 100e6
file: Common base amplifier small signal.asc

(b)

$V(v_o)$

FIGURE 7.37

Common-base amplifier of Example 7.11. (a) Small-signal model. (b) LTSPICE result (magnitude) showing a gain of +3.18 and a −3 dB bandwidth of 42 MHz. (For color version of this figure, the reader is referred to the online version of this book.)

First, using the circuit, we will write the node equations for nodes v_x and the output node v_o:

$$1.\ (v_i - v_x)G_i - v_xC_is + (v_o - v_x)C_fs = 0$$
$$2.\ (-Av_x - v_o)G_o - v_oC_os + (v_x - v_o)C_fs = 0$$

(7.44)

Let us next group terms in v_x and v_o:

$$1.\ -v_x\left[G_i + (C_i + C_f)s\right] + v_oC_fs = -v_iG_i$$
$$2.\ v_x\left[C_fs - AG_o\right] - v_o\left[G_o + (C_o + C_f)s\right] = 0$$

(7.45)

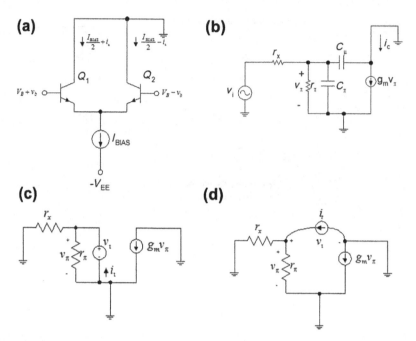

FIGURE 7.38

Current switching amplifier of Example 7.12. (a) Circuit. (b) High-frequency differential-mode half-circuit model. (c) Circuit for finding C_π OCTC. (d) Circuit for finding C_π OCTC.

Next, divide (1) by G_i and divide (2) by G_o:

$$1. \quad -v_x\left[1 + R_i(C_i + C_f)s\right] + v_o R_i C_f s = -v_i$$

$$2. \quad v_x\left[R_o C_f s - A\right] - v_o\left[1 + R_o(C_o + C_f)s\right] = 0 \tag{7.46}$$

We put this into matrix form in anticipation of using Cramer's rule as follows:

$$\begin{bmatrix} [R_i(C_i + C_f)s + 1] & R_i C_f s \\ R_o C_f s - A & [R_o(C_o + C_f)s + 1] \end{bmatrix} \begin{Bmatrix} v_x \\ v_o \end{Bmatrix} = \begin{Bmatrix} -v_i \\ 0 \end{Bmatrix} \tag{7.47}$$

Using Cramer's rule, we find the solution for the output voltage v_o:

$$v_o = \frac{\det\begin{bmatrix} [R_i(C_i + C_f)s + 1] & -v_i \\ R_o C_f s - A & 0 \end{bmatrix}}{\det\begin{bmatrix} [R_i(C_i + C_f)s + 1] & R_i C_f s \\ R_o C_f s - A & [R_o(C_o + C_f)s + 1] \end{bmatrix}} \tag{7.48}$$

(a) Small signal model of current amplifier
Differential-mode half-circuit
Added tiny collector resistor R_c through which we measure collector current

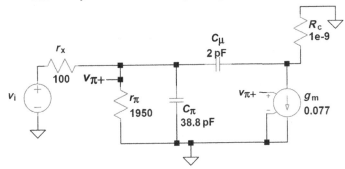

.ac dec 100 1000e3 1000e6
file: Current amplifier small signal.asc

(b)

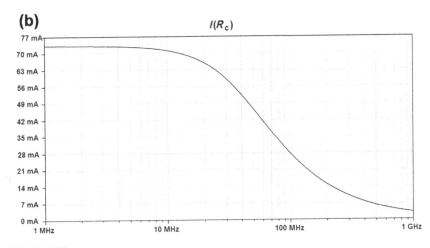

FIGURE 7.39

Current switching amplifier of Example 7.12. (a) LTSPICE circuit, with tiny collector resistor R_c added though which LTSPICE measures current. (b) Frequency response, showing a -3 dB bandwidth of 41 MHz. (For color version of this figure, the reader is referred to the online version of this book.)

After much algebraic manipulation, we find the solution for the input–output transfer function:

$$\frac{v_o(s)}{v_i(s)} = \frac{-A\left(1 - \frac{R_oC_fs}{A}\right)}{R_iR_o\left(C_iC_o + C_iC_f + C_fC_o\right)s^2 + \left(R_i\left(C_i + (1+A)C_f\right) + R_o\left(C_o + C_f\right)\right)s + 1} \tag{7.49}$$

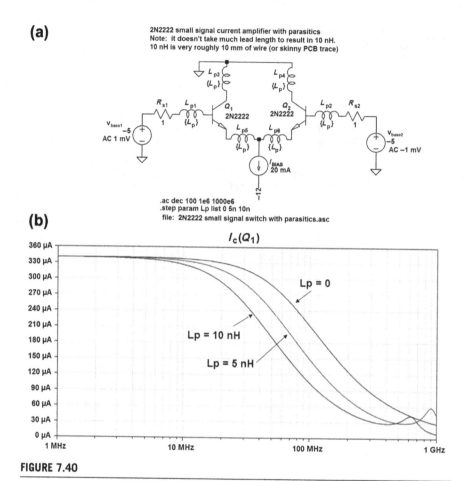

FIGURE 7.40

2N2222 current amplifier of Example 7.13. Each base has a bias voltage of −5 V. (a) Circuit with 0, 5 nH or 10 nH of parasitic inductance in each transistor lead. (b) Small-signal output showing collector current of transistor Q_1 (i_{c1}) for the three cases. The low-frequency small-signal gain is about 330 µA/mV or about 0.33 A/V. (For color version of this figure, the reader is referred to the online version of this book.)

Let us next do a sanity check with $C_f = 0$ and determine if the transfer function looks correct:

$$\frac{v_o}{v_i}\bigg|_{C_f=0} = \frac{-A}{R_i R_o C_i C_o s^2 + (R_i C_i + R_o C_o)s + 1} = \frac{-A}{(R_i C_i s + 1)(R_o C_o s + 1)} \quad (7.50)$$

With $C_f = 0$, the input and output circuits are uncoupled. This is correct, since with $C_f = 0$ we have one pole at $1/(R_i C_i)$ and a second pole at $1/(R_o C_o)$.

In another sanity check, we can apply OCTCs to this amplifier and determine whether the OCTC results sum to the same value as the coefficient of the s term in the gain expression earlier derived.

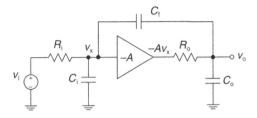

FIGURE 7.41

Circuit used to illustrate pole splitting.

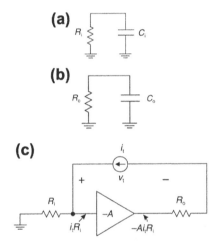

FIGURE 7.42

OCTCs circuits for pole-split amplifier. (a) OCTC circuit for C_i. (b) OCTC circuit for C_o. (c) OCTC circuit for C_f.

For C_i, we open-circuit C_o and C_f and find the resistance facing C_i using the circuit in Figure 7.42(a).

$$R_{o1} = R_i$$

$$\tau_{o1} = R_{o1}C_i = R_iC_i \qquad (7.51)$$

For C_o, we open-circuit C_i and C_f and find the resistance facing C_o using the circuit in Figure 7.42(b).

$$R_{o2} = R_o$$

$$\tau_{o2} = R_{o2}C_o = R_oC_o \qquad (7.52)$$

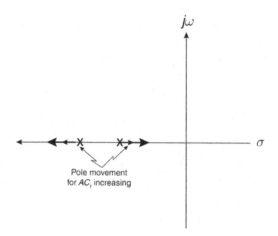

FIGURE 7.43

Movement of poles as AC_f increases.

For C_f, we need to do a little bit more work using the circuit of Figure 7.42(c).

$$v_t = i_t R_i + i_t R_o + A i_t R_i$$

$$R_{o3} = \frac{v_t}{i_t} = R_o + (1+A)R_i \tag{7.53}$$

$$\tau_{o3} = R_{o3} C_f = \left(R_o + (1+A)R_i\right) C_f$$

The sum of the OCTCs is:

$$\sum \tau_{oc} = \tau_{o1} + \tau_{o2} + \tau_{o3} = R_i C_i + R_o C_o + \left(R_o + (1+A)R_i\right) C_f \tag{7.54}$$

This is exactly the same as the coefficient of the s term in the closed-form solution, and this is a further sanity check on the result of Eqn 7.49.

In order to find the pole locations, we will assume that we have two poles and that they are widely spaced and on the negative real axis, a circumstance that occurs often in wideband amplifiers. Our widely spaced pole location approximation for a second-order system is:

$$H(s) = \frac{1}{a_2 s^2 + a_1 s + 1}$$

$$p_{low} \approx -\frac{1}{a_1} \tag{7.55}$$

$$p_{high} \approx -\frac{a_1}{a_2}$$

Using this approximation, we find the approximate low-frequency (p_{low}) and high-frequency pole (p_{high}) locations for this amplifier as:

$$p_{\text{low}} \approx -\frac{1}{\left(R_i\left(C_i + (1+A)C_f\right) + R_o\left(C_o + C_f\right)\right)}$$

$$p_{\text{high}} \approx -\frac{\left(R_i\left(C_i + (1+A)C_f\right) + R_o\left(C_o + C_f\right)\right)}{R_i R_o\left(C_i C_o + C_i C_f + C_f C_o\right)} \tag{7.56}$$

We can further approximate if we assume that $A R_i C_f \gg R_i C_i$, $R_o C_o$, and $R_o C_f$:

$$p_{\text{low}} \approx -\frac{1}{A R_i C_f}$$

$$p_{\text{high}} \approx -\frac{A R_i C_f}{R_i R_o\left(C_i C_o + C_i C_f + C_f C_o\right)} \approx -\frac{A}{R_o\left(\dfrac{C_i C_o}{C_f} + C_i + C_o\right)} \tag{7.57}$$

We note the important functional relationship: as AC_f increases, the low-frequency pole decreases in frequency and the high-frequency pole increases in frequency, as shown in Figure 7.43. In fact, the dominant pole is due to the input resistance and Miller multiplication of the feedback capacitance. In op-amps, it is useful to have a single dominant pole, and pole splitting is used to accomplish this.

Another model of a pole-split amplifier is shown in Figure 7.44. This models a common-emitter amplifier driving a capacitive load. We can start out finding the transfer function by writing the node equations at the v_a and v_o nodes:

$$\begin{aligned}
\text{1. } & (v_i - v_a)G_i - v_a C_i s + (v_o - v_a)C_f s = 0 \\
\text{2. } & -g_m v_a - v_o G_o - v_o C_o s + (v_a - v_o)C_f s = 0
\end{aligned} \tag{7.58}$$

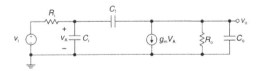

FIGURE 7.44

Another pole-split amplifier. This model is typical of that found in single-stage transistor amplifiers. The equations show that this amplifier is topologically equivalent to the previously shown pole-split amplifier.

We note that these node equations are identical to that found in the previous pole-split amplifier. Therefore, we can use the previous result with the following substitution:

$$A \Rightarrow g_m R_o \tag{7.59}$$

Therefore, the transfer function for this amplifier is:

$$\frac{v_o(s)}{v_i(s)} =$$

$$\frac{-g_m R_o \left(1 - \frac{C_f s}{g_m}\right)}{R_i R_o (C_i C_o + C_i C_f + C_f C_o)s^2 + \left(R_i \left(C_i + (1 + g_m R_o)C_f\right) + R_o(C_o + C_f)\right)s + 1}$$

(7.60)

Example 7.14: Pole splitting

A numerical example will show how pole splitting works. Let us consider the amplifier of Figure 7.44 with the following parameters: $g_m = 0.01$ A/V, $R_o = 10^5$ Ω, $R_i = 10^6$ Ω, and $C_i = C_o = 10$ pF. We will see how the pole locations move as C_f is varied from 0 to 30 pF. With $C_f = 0$, we find two uncoupled poles as follows:

$$p_1 = -\frac{1}{R_i C_i} = -10^5 \text{ rad/s}$$

(7.61)

$$p_2 = -\frac{1}{R_o C_o} = -10^6 \text{ rad/s}$$

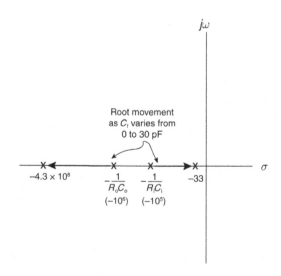

FIGURE 7.45

Movement of poles in pole-split amplifier of Example 7.14 as C_f is varied from 0 to 30 pF.

Table 7.2 Pole and Zero Movement in Pole-Split Amplifier as a Function of Feedback Capacitor C_f

C_f	ω_{p1}	ω_{p2}	ω_z
0	-10^5 rad/s	-10^6 rad/s	$+\infty$
1 pF	-988	-8.4×10^7	$+10^{10}$
5 pF	-199	-2.5×10^8	$+2 \times 10^9$
10 pF	-100	-3.3×10^8	$+10^9$
20 pF	-49	-4×10^8	$+5 \times 10^8$
30 pF	-33	-4.3×10^8	$+3.3 \times 10^8$

When we increase C_f, the low-frequency pole moves down in frequency and the high-frequency pole moves up in frequency. We can approximate the pole locations if we assume that $g_m R_o R_i C_f \gg R_i C_i$, $R_o C_o$, and $R_o C_f$ as follows:

$$p_{low} \approx -\frac{1}{g_m R_o R_i C_f}$$

$$p_{high} \approx -\frac{g_m}{\left(\dfrac{C_i C_o}{C_f} + C_i + C_o\right)} \tag{7.62}$$

The plot of Figure 7.45 shows the pole locations as C_f increases, with results tabulated in Table 7.2.

Chapter 7 problems

Problem 7.1

a. Assume that the circuit in Figure 7.46 is driven by an AC voltage source. Using an intuitive approach, thought experiments, etc., sketch the shape of the Bode magnitude response of v_o/v_i.

b. Using the method of short-circuit time constants, find the low-frequency breakpoint f_L.

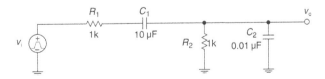

FIGURE 7.46

Circuit for Problem 7.1.

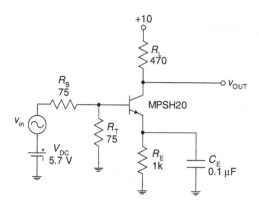

FIGURE 7.47

Common-emitter amplifier for Problem 7.2.

Problem 7.2

For the transistor amplifier shown in Figure 7.47, estimate the low-frequency -3 dB point using the method of short-circuit time constants. For the MPSH20 transistor, you may assume $r_x = 20\,\Omega$, AC beta $h_{fe} \approx$ DC beta $h_{FE} = 25$, $C_\mu = 0.9$ pF, and $f_T = 630$ MHz. Make reasonable approximations, state them, and justify them. Assume room temperature is 25 °C and hence $kT/q = 26$ mV.

Problem 7.3

Shown in Figure 7.48 is a possible model for an amplifier exhibiting the Miller effect. The amplifier has infinite input impedance, zero output impedance, and gain $-A$ with transfer function $v_o = -Av_x$. Using the method of OCTCs, find the -3 dB bandwidth of this amplifier. What is the bandwidth if $A = +1$?

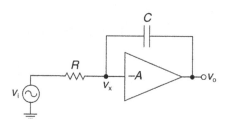

FIGURE 7.48

Circuit for Problem 7.3.

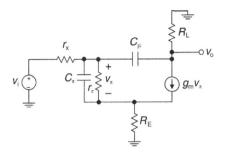

FIGURE 7.49

Common-emitter amplifier with emitter degeneration for Problems 7.4 and 7.5.

Problem 7.4

Shown in Figure 7.49 is the small-signal model of a common-emitter amplifier with emitter degeneration. The circuit has load resistor $R_L = 1$ kΩ and emitter degeneration resistor $R_E = 22$ Ω. Assume that the transistor is biased at $I_C = 10$ mA. Find the gain and estimate the bandwidth using OCTCs. Small-signal parameters are $r_x = 100$ Ω, $r_\pi = 1$ kΩ, $g_m = 0.1$ A/V, $C_\pi = 20$ pF, and $C_\mu = 2$ pF.

Problem 7.5

Simulate the circuit in Figure 7.49 using LTSPICE. In this simulation, comment on the differences between circuit operation for $R_E = 22$, 47, and 100 Ω.

Problem 7.6

The circuit in Figure 7.50 is a video amplifier that uses a common-base voltage amplifier driving an emitter follower. The source resistance R_s is a 50 Ω transmission

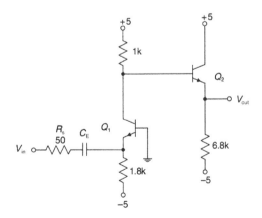

FIGURE 7.50

Circuit for Problem 7.6.

line used to drive the input of the amplifier. You may assume that under the operating conditions in the circuit, the following parameters hold: $h_{FE} = 50$, $h_{fe} = 50$, $C_\mu = 1$ pF, and $f_T = 500$ MHz.

a. What is the operating point value of the output voltage V_{OUT}?

b. What is the midband small-signal voltage gain v_{out}/v_{in}?

c. Assume that the design specifies a low-frequency -3 dB frequency of 50 Hz. What value of emitter capacitor C_E is required?

d. Verify your results using LTSPICE.

Problem 7.7

For the circuit in Figure 7.51, assume that the input voltage source is an AC source with a DC value of 0 V and that the circuit operates at 25 °C.

a. Find the low-frequency and midband gain of the emitter follower, assuming that transistor internal capacitances C_π and C_μ have no effect.

b. Using the method of OCTCs, estimate the bandwidth of the buffer shown below. (In your OCTC estimate, find appropriate values of C_π and C_μ from the 2N3904 datasheet.)

c. Simulate your circuit using LTSPICE and compare your OCTC result with the LTSPICE result. In your LTSPICE model, directly input the small-signal model (i.e. do not use the LTSPICE 2N3904 model; input directly C_π, C_μ, r_π, the $g_m v_\pi$ generator, etc., in a small-signal model).

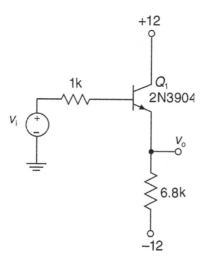

FIGURE 7.51

Circuit for Problem 7.7.

Further reading

[1]   Abidi A. On the operation of cascode gain stages. *IEEE J Solid-State Circuits* 1988;**23**(6):1434–7. An interesting paper illustrating the use of bootstrapping to improve the DC gain of a MOS amplifier.

[2]   Barna A. On the transient response of emitter followers. *IEEE J Solid-State Circuits* 1973:233–5.

[3]   Centurelli F, Luzzi R, Olivieri M, Trifiletti A. A bootstrap technique for wideband amplifiers. *IEEE Trans Circuits Syst I: Fundam Theory Appl* 2002;**48**(10):1474–80.

[4]   Chuang CT. Analysis of the settling behavior of an operational amplifier. *IEEE J Solid-State Circuits* 1982;**17**(1):74–80.

[5]   Choma Jr J. Simplified design guidelines for dominant pole amplifiers peaked actively by emitter or source followers. *IEEE Trans Circuits Syst* 1989;**36**(7):1005–10.

[6]   Eschauzier RGH, Kerklaan LPT, Huijsing JH. A 100 MHz 100 dB operational amplifier with multipath nested miller compensation structure. In: *Proceedings of the 1992 IEEE Solid-State Circuits Conference, (ISSCC)* 1992. p. 196–7.

[7]   Filipkowski A. Poles and zeros in transistor amplifiers introduced by Miller effect. *IEEE Trans Educ* 1999;**42**(4):349–51.

[8]   Fong KL, Meyer RG. High-frequency nonlinearity analysis of common-emitter and differential-pair transconductance stages. *IEEE J Solid-State Circuits* 1998;**33**(4):548–55.

[9]   Grebene A. *Bipolar and MOS analog integrated circuit design*. John Wiley; 1984.

[10] Gray PR, Hurst P, Lewis S, Meyer R. *Analysis and design of analog integrated circuits*. 4th ed. John Wiley; 2001.

[11] Hamilton DK. Use of inductive compensation for improving bandwidth and noise performance of high frequency optical receiver preamplifiers. *IEEE Proc G: Circuits Devices Syst* 1991;**138**(1):52–5.

[12] Ki WH, Der L, Lam S. Re-examination of pole splitting of a generic single stage amplifier. *IEEE Trans Circuits Syst I: Fundam Theory Appl* 1997;**44**(1):70–4.

[13] Kozikowski J. Analysis and design of emitter followers at high frequencies. *IEEE Trans Circuit Theory* 1964:129–36.

[14] Lee T. *The design of CMOS radio-frequency integrated circuits*. Cambridge University Press; 1998.

[15] Makris CA, Toumazou C. Two pole, high speed operational amplifier modelling, methods and techniques. In: *Proceedings of the European conference on circuit theory and design* 1989. p. 304–8.

[16] Oh YH, Lee SG. An inductance enhancement technique and its application to a shunt-peaked 2.5 Gb/s transimpedance amplifier design. *IEEE Trans Circuits Syst II: Express Briefs* 2004;**51**(11):624–8.

[17] Palmisano G, Palumbo G. An optimized Miller compensation based on voltage buffer. In: *Proceedings of the 38th Midwest symposium on circuits and systems* 1995. p. 1034–7.

[18] Solomon JE. The monolithic op amp: a tutorial study. *IEEE J Solid-State Circuits* 1974;**9**(6):314–32.

[19] Thompson MT. Design linear circuits using OCTC calculations. *Electron Des (Special Analog Issue)* 1993:41–7.

[20] Thompson MT. Network tricks aid in OCTC. *Electron Des* 1993:67–70.

[21] Yang HC, Allstot DJ. An equivalent circuit model for two-stage operational amplifiers. In: *Proceedings of the 1988 IEEE international symposium on circuits and systems* 1988. p. 635—8.

[22] Yang HC, Allstot DJ. Modified modeling of Miller compensation for two-stage operational amplifiers. In: *Proceedings of the 1991 IEEE international symposium on circuits and systems* 1991. p. 2557—60.

BJT High-Gain Amplifiers and Current Mirrors

8

"Life is like an onion: you peel it off one layer at a time, and sometimes you weep."

— Carl Sandburg

IN THIS CHAPTER

▶ In this chapter, we discuss a more detailed incremental model of the bipolar transistor that takes into account the base-width modulation effect. The resulting resistive elements resulting from base-width modulation have significant design impact on high-gain amplifiers, emitter-followers, and current mirrors.

The need to augment the hybrid-pi model

The hybrid-pi model considered so far assumes that the incremental output resistance of a bipolar transistor is infinite. However, this is not the case, as anyone who has examined in detail transistor V/I curves on a curve tracer can attest. We need to consider the effects of base-width modulation to account for the fact that the output resistance seen at a transistor collector is finite.

A resistively loaded common-emitter amplifier is shown in Figure 8.1. To maximize output voltage swing, we will set the bias point of V_o to half the positive power supply voltage $V_{CC}/2$ (through means that are not explicitly shown in this schematic). This, in turn, sets the collector current and hence the transconductance of the transistor:

$$g_m = \frac{|I_C|}{kT/q} = \frac{V_{CC}}{2R_L V_{TH}} \tag{8.1}$$

This results in a maximum incremental gain[1] for the resistively loaded common-emitter amplifier as:

$$A_{v,max} = -\frac{V_{CC}}{2V_{TH}} \tag{8.2}$$

---

[1]The incremental gain for the configuration of Figure 8.1(b) is $-g_m R_L$.

Intuitive Analog Circuit Design. http://dx.doi.org/10.1016/B978-0-12-405866-8.00008-5

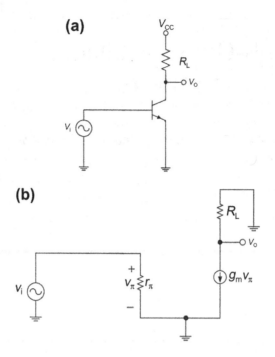

FIGURE 8.1

Common-emitter amplifier with resistive load. (a) Circuit, omitting biasing details. (b) Low-frequency hybrid-pi model developed so far.

For $V_{CC} = +12$, this results in a maximum incremental gain of -230.

In order to get higher gain without arbitrarily high collector voltage, an active load (i.e. a current source) can be used. Let us first consider the ramifications of ignoring base-width modulation and the transistor finite output resistance. Consider the common-emitter amplifier with a current source load (Figure 8.2(a)) where, for simplicity, details of the bias circuit have been omitted. In this case, instead of a collector load resistor we have a collector active load, which is the I_{DC} current source.

The small-signal model is shown in Figure 8.2(b). Let us assume that the incremental output resistance of the I_{DC} current source approaches infinity; this means that the load resistor in the incremental model $R_O \Rightarrow \infty$ as well. Note that in this simplified model, the gain is:

$$A_v = -g_m R_o \Rightarrow \infty \tag{8.3}$$

We know this cannot be the case in a real-world circuit. As we will see later, other transistor internal incremental resistances limit this maximum gain to a more reasonable, finite value.

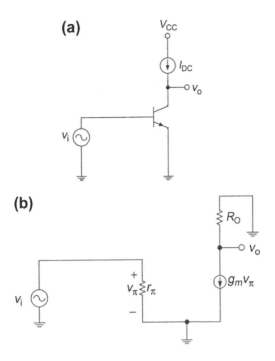

FIGURE 8.2

Common-emitter amplifier with current source load. (a) Circuit, omitting biasing details.
(b) Incremental model, assuming that the I_{DC} current source has incremental output
resistance R_O.

Base-width modulation and the extended hybrid-pi model

The simple hybrid-pi model used so far leads us to believe that the transistor collector current does not change when the collector-emitter voltage changes. In other words, the simple model indicates infinite output impedance at a transistor's collector. If you look closely at transistor V/I curves (for instance, on a curve tracer, as shown in Figure 8.3(a) and (b)) you see that there is a finite slope of the collector current in the linear region of operation. This corresponds to the small-signal output resistance of the transistor when operated in the linear operating region.

If we extend a sloped line corresponding to the transistor curves to negative voltages as shown in Figure 8.3(c), we find that the lines cross the negative voltage axis at the same point V_A. The magnitude of V_A is the *Early voltage*, named after James Early who discovered the phenomenon at Bell Laboratories. We can modify the transistor voltage–current relationship to account for the Early voltage as:

$$I_C = I_S \left(e^{\frac{qV_{BE}}{kT}} - 1 \right) \left(1 + \frac{V_{CE}}{V_A} \right) \tag{8.4}$$

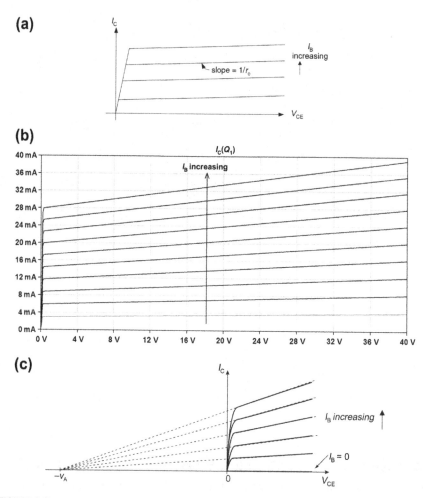

FIGURE 8.3

NPN transistor V–I curves. (a) Generic curves showing effects of finite transistor output resistance r_o, due to base-width modulation. (b) LTSPICE-produced V–I curve for 2N3904 transistor with base current I_B stepped in increments of 10 μA. (c) Definition of the Early voltage V_A. (For color version of this figure, the reader is referred to the online version of this book.)

Equation (8.4) captures the functional dependence of collector current on collector–emitter voltage.

The changing of the effective base width is due to the widening and narrowing of the collector–base depletion region when the transistor collector–base voltage V_{CB} changes (Figure 8.4). When V_{CB} decreases, the width of the collector–base depletion region decreases, increasing the effective base-width from W to $W + \Delta W$. Since

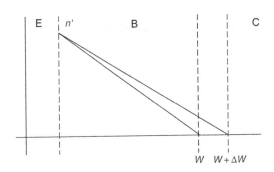

FIGURE 8.4

Illustration of base-width modulation in an NPN transistor. The slope of the minority carrier concentration $n'(x)$ varies as V_{CE} varies, resulting in a change in the collector current.

the collector current is proportional to the *slope* of the minority carrier concentration $n'(x)$ in the base (in this case, electrons), the collector current changes when V_{CB} changes. This process is called *base-width modulation* and was first described by James Early[2] in 1952.

Base-width modulation causes both the collector current and base current of the transistor to change when the collector voltage changes. This small-signal effect means that a transistor current source does not have infinite output impedance. These effects are modeled by adding two more resistors to the hybrid-pi model (Figure 8.5). This model should be used when analyzing transistor stages with gains in excess of a few hundred.

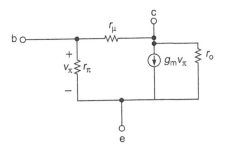

FIGURE 8.5

Transistor extended hybrid-pi model (low frequency) including effects of base-width modulation, resulting in added circuit elements r_o and r_μ. Resistance r_o models the fact that collector current varies as V_{CE} varies. Resistance r_μ models the effects of the extra base current needed to support this extra collector current.

---

[2]See the James Early reference at the end of this chapter.

Calculating small-signal parameters using a transistor datasheet

A detailed analysis shows that the output resistance of the transistor r_o is inversely proportional to collector current,[3] or:

$$r_o = \frac{1}{\eta g_m} = \frac{V_A}{\left(\frac{kT}{q}\right) g_m} \tag{8.5}$$

where η is a constant called the *base-width modulation factor* that has typical values of 10^{-3} to 10^{-4}. Changes in collector–base voltage have a much smaller effect on output current than changes in base–emitter voltage. Therefore, the effects of base-width modulation and output resistance r_o are only significant if the gain of the amplifier approaches $1/\eta$. A rule of thumb is that we need to consider r_o in gain calculations when the gain of our amplifier is greater than a few hundred.

The change in collector current also results in a proportional change of base current, with the proportionality constant being the small-signal current gain h_{fe}, resulting in:

$$r_\mu = h_{fc} r_o = \frac{h_{fe}}{\eta g_m} \tag{8.6}$$

Transistor manufacturers sometimes specify transistors using two-port h-parameters, corresponding to the circuit of Figure 8.6.

The equations describing the operation of this circuit are as follows:

$$v_{be} = i_b h_{ie} + h_{re} v_{ce}$$
$$i_c = h_{fe} i_b + h_{oe} v_{ce} \tag{8.7}$$

Comparing this to the extended hybrid-pi model results in:

$$h_{ie} = r_\pi$$
$$h_{re} = \eta$$
$$h_{fe} = \beta_o \tag{8.8}$$
$$h_{oe} = \eta g_m$$

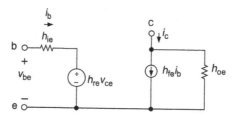

FIGURE 8.6

Transistor model showing two-port h-parameters.

---

[3]For a detailed mathematical derivation, see, e.g. Gray, Hurst, Lewis and Meyer, pp. 14–16 or P. E. Gray, et al., *Physical Electronics and Circuit Models of Transistors*, SEEC volume 2, pp. 149–152.

Shown in Figure 8.7 are the *h*-parameters from the 2N3904 transistor[4] datasheet.

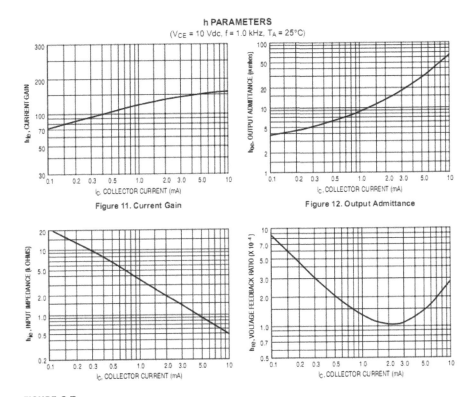

h PARAMETERS
(V_{CE} = 10 Vdc, f = 1.0 kHz, T_A = 25°C)

Figure 11. Current Gain

Figure 12. Output Admittance

FIGURE 8.7

h-Parameters from 2N3904 datasheet.

Example 8.1: Common-emitter amplifier with an ideal current source load

Let us assume that we have a common-emitter amplifier biased with an ideal current source of value $I_{DC} = 1$ mA (Figure 8.8). Our assumption is that the current source is ideal—hence the current source has an infinite incremental output resistance. Let us find the gain of the amplifier, given that the base width modulation factor of the transistor $\eta = 3.3 \times 10^{-4}$.

A small-signal model of this circuit is shown in Figure 8.9, where we have used the transistor extended hybrid-pi model.

---

[4]Found at www.onsemi.com, reprinted with permission of On Semiconductor.

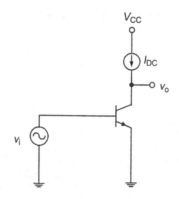

FIGURE 8.8

Common-emitter amplifier of Example 8.1 with active load, many biasing details omitted.

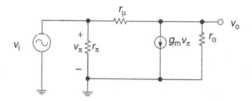

FIGURE 8.9

Small-signal model of the common-emitter amplifier with active load using extended hybrid-pi model for the transistor and assuming that I_{DC} is a magic current source with infinite output resistance.

Applying Kirchoff's Current Law (KCL) at the output (v_o) node results in:[5]

$$(v_i - v_o)g_\mu - v_o g_o - v_i g_m = 0 \tag{8.9}$$

Solving for gain v_o/v_i results in:[6]

$$A_v = \frac{v_o}{v_i} = \frac{g_\mu - g_m}{g_o + g_\mu} \approx \frac{-g_m}{g_o} \approx \frac{-1}{\eta} \tag{8.10}$$

For this example, we predict a gain of -3000. Now, achieving a gain as high as this is contingent on how *ideal* a current source we can build. We will see in the next section methods for building a current source with a high output resistance.

---

[5]Again, we note that the math is a little bit easier if we use conductances instead of resistances. For instance, in this circuit, $g_o = 1/r_o$.

[6]Note that $g_m \gg g_\mu$ and that $g_o \gg g_\mu$.

Building blocks
Incremental output resistance of a bipolar current source

Let us attempt to answer the question of how ideal a current source built with bipolar junction transistors (BJTs) can be. Figure 8.10(a) shows a current source with the transistor having an emitter resistor R_E. The small-signal model is shown in Figure 8.10(b), where we have added a test current source i_t in order to calculate

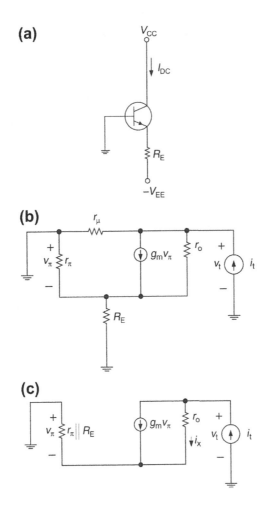

FIGURE 8.10

Current source with finite emitter resistor R_E. (a) Schematic. (b) Small-signal model. (c) Simplified small-signal mode by combining r_π and R_E into a parallel combination and by removing r_μ, to be put back in later.

the incremental output resistance of the current source, as seen at the transistor collector.

The analysis of this circuit is greatly simplified by combining r_π and R_E into a common resistor and by removing r_μ (and adding it in parallel later). Using Figure 8.10(c), the test voltage v_t is given by:

$$v_t = i_t[r_\pi \| R_E] + i_x r_o \tag{8.11}$$

The current i_x is the difference between i_t and $g_m v_\pi$, or:

$$i_x = i_t - g_m v_\pi \tag{8.12}$$

We also know that v_π is related to i_t by the parallel combination of r_π and R_E. So, we can solve for v_t as:

$$v_t = i_t[r_\pi \| R_E] + [i_t + g_m i_t[r_\pi \| R_E]]r_o \tag{8.13}$$

Therefore, the output resistance of the current source is:

$$r_{out} = \frac{v_t}{i_t} = [r_\pi \| R_E] + [1 + g_m[r_\pi \| R_E]]r_o \approx \frac{[1 + g_m[r_\pi \| R_E]]}{g_o} \tag{8.14}$$

For the final result, we have to add r_μ back in parallel, resulting in:

$$r_{out} \approx r_\mu \left\| \frac{[1 + g_m[r_\pi \| R_E]]}{g_o} \right. \tag{8.15}$$

There are several important limiting cases. If the emitter is grounded (i.e. if $R_E = 0$), the output resistance is:

$$r_{out} \approx r_o \quad \text{for} \quad R_E = 0 \tag{8.16}$$

If the emitter resistance is large compared to r_π, the output resistance is much higher and reaches a limit of $r_\mu/2$:

$$r_{out} \approx \frac{r_\mu}{2} \quad \text{for} \quad R_E \gg r_\pi \tag{8.17}$$

The lesson here is that we can build a current source with a high incremental output resistance provided that we have a resistance in the output transistor emitter leg that is large compared to r_π. One way to accomplish this without using large-valued resistors is to use a cascode current source (Figure 8.11). Transistor Q_2 is the output transistor, which carries the desired output current I_{DC}. Transistor Q_1 provides a high incremental output impedance seen by the emitter of Q_2. This in turn ensures that the incremental output resistance at the collector of Q_2 is high. The inclusion of R_{E1} increases the output resistance of the collector of Q_1, further increasing the incremental output resistance seen at the collector of Q_2.

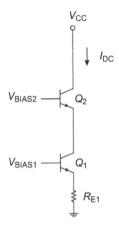

FIGURE 8.11

Cascode current source. The incremental output resistance at the collector of Q_2 is increased by inclusion of current source Q_1 in its emitter.

Emitter-follower incremental input resistance

The incremental input resistance of an emitter-follower (Figure 8.12(a)) can be calculated using the circuit of Figure 8.12(b). As a design example, let us say that you have a common-emitter gain stage followed by an emitter-follower buffer; you would like to know how much the emitter-follower finite input resistance loads decrease the gain stage. In Figure 8.12(b) we see the small-signal model of the emitter-follower, including extended hybrid-pi values r_o and r_μ.

Note that in Figure 8.12(b), incrementally, the transistor output resistance r_o appears in parallel with R_E. Hence, the output resistance (with r_μ taken out to simplify the math as in Figure 8.12(c)) is:

$$r_{in} = r_\pi + (1 + h_{fe})[R_E \| r_o] \tag{8.18}$$

Now, let us put back in r_μ in parallel, to arrive at the final result:

$$r_{in} = r_\mu \| (r_\pi + (1 + h_{fe})[R_E \| r_o]) \tag{8.19}$$

The limit for very big emitter resistance R_E is:

$$r_{in} \approx \frac{r_\mu}{2} \text{ if } R_E \gg r_o \tag{8.20}$$

The limit for very small emitter resistance is:

$$r_{in} \approx r_\pi \quad \text{if} \quad R_E \to 0 \tag{8.21}$$

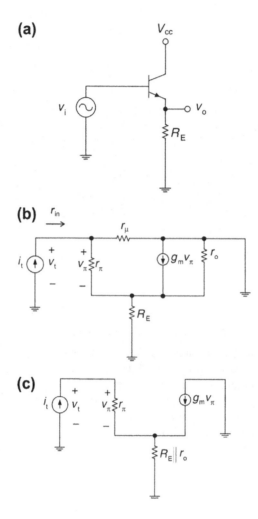

FIGURE 8.12

Emitter-follower (a) Circuit. (b) Small-signal model using extended hybrid-pi transistor model for finding incremental input resistance r_{in} at the base terminal. (c) Circuit simplified by removal of r_μ.

For intermediate values of R_E, we find:

$$r_{in} \approx h_{fe}R_E \quad \text{if} \quad r_\pi \ll R_E \ll r_o \tag{8.22}$$

Example 8.2: Incremental input resistance of emitter-follower

Find the incremental input resistance of an emitter-follower with the following transistor parameters and bias conditions: collector current $I_C = 1$ mA, emitter resistor

$R_E = 10$ kΩ, small-signal current gain $h_{fe} = 100$, and base-width modulation factor $\eta = 10^{-4}$.

Solution: We find the small signal parameters as follows:

$$g_m = \frac{I_C}{V_{TH}} = \frac{10^{-3}}{0.026} = 0.038 \text{ A/V}$$

$$r_\pi = \frac{h_{fe}}{g_m} = \frac{100}{0.038} = 2600 \ \Omega$$

$$r_o = \frac{1}{\eta g_m} = \frac{1}{(10^{-4})(0.038)} = 260 \text{ k}\Omega$$

$$r_\mu = h_{fe} r_o = 260 \text{ M}\Omega$$

(8.23)

The small-signal model for this emitter-follower is shown in Figure 8.13 where we apply a test current source i_t to find the input resistance r_{in}.

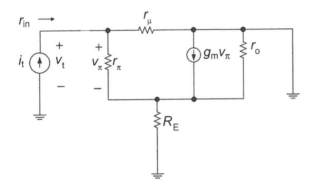

FIGURE 8.13

Emitter-follower small-signal model for Example 8.2.

Bypassing all the algebra, the incremental input resistance is:

$$r_{in} = r_\mu \| (r_\pi + (1 + h_{fe})[R_E \| r_o]) \approx (26 \text{ M}\Omega) \| ((100)[(10 \text{ k}\Omega) \| 260 \text{ k}\Omega]) \approx 1 \text{ M}\Omega$$

(8.24)

Note that this input resistance is approximately $r_{in} \approx h_{fe} R_E$, which is the result we expect for intermediate values of emitter resistance.

Current mirrors

Current mirrors, also called *current repeaters,* are commonly used as biasing elements and as active loads in amplifiers.[7] A basic BJT current mirror is made up

---

[7]See, for instance, J. Roberge, *Operational Amplifiers Theory and Practice*, p. 393; Gray, Hurst, Lewis and Meyer, *Analysis and Design of Analog Integrated Circuits*, 4th edition, p. 255.

of a first diode-connected transistor[8] with an input control current and a second output transistor where the output current is controlled by the voltage generated across the first diode-connected transistor. Hence, in a current mirror, the base-to-emitter voltages of the two transistors are identical, causing collector currents in the two transistors to be proportional.[9] Current mirrors can be constructed from NPN transistors, PNP transistors, or combinations of the two. Some basic current mirror circuits and the equivalent circuits (with the diode-connected transistor replaced by a simple diode in the schematic) are shown in Figure 8.14.

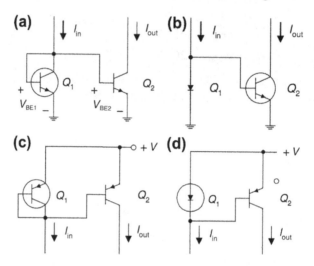

FIGURE 8.14

Basic current mirrors and equivalent circuits. (a) Basic NPN mirror, showing two transistors with equal V_{BE}s. (b) Basic NPN mirror equivalent circuit with diode-connected transistor Q_1 replaced by a diode for illustrative purposes. (c) Basic PNP mirror. (d) Basic PNP mirror equivalent circuit.

Current mirrors are found in many integrated circuit operational amplifiers, including the Fairchild μA709, μA741, and μA776; Analog Devices' OP07; and National Semiconductor's LM301A and LM308, among others. Current mirrors can be fabricated with either BJTs or metal oxide—semiconductor field-effect transistors. A 741 op-amp schematic with current mirrors highlighted is shown in Figure 8.15.

---

[8]A "diode-connected" BJT is one where the collector is shorted to the base and the transistor is operated as a diode utilizing the base—emitter junction. See J. Roberge, *Operational Amplifiers: Theory and Practice*, p. 390.

[9]This analysis assumes that base currents are negligible and assumes that the transistor has high output impedance. This analysis also ignores base-width modulation and the fact that the input and output transistors generally operate at different collector-emitter voltages. For the collector currents of the two transistors to be equal, the transistors must be matched and have the same emitter areas. If the emitter areas of the two transistors are different, there is a scale factor relating the input and output currents.

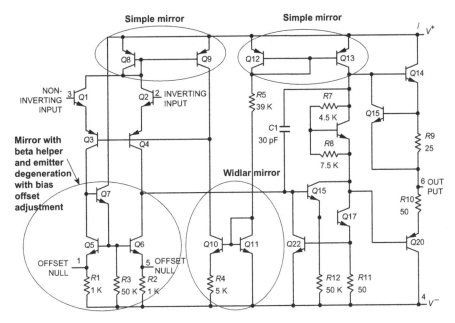

FIGURE 8.15

Schematic of the 741 op-amp, with current mirrors highlighted. The various current mirrors are described in detail later in this chapter.

Reprinted with permission from National Semiconductor.

Basic current mirror accuracy

The basic bipolar current mirror has an input/output gain error due to the finite DC current gain β_F of the transistor. In Figure 8.16, we see that the two transistors are operating at the same V_{BE}, and hence to first order the transistors have the same collector current. However, the input current is not the collector current, because I_{in} also supplies base currents to Q_1 and Q_2. We can express this relationship as:

$$I_{in} = I_{C1} + \frac{I_{C1}}{\beta_F} + \frac{I_{out}}{\beta_F} \tag{8.25}$$

Solving for current mirror gain results in:

$$\frac{I_{out}}{I_{in}} = \frac{1}{1 + \frac{2}{\beta_F}} \tag{8.26}$$

We see that the error is inversely proportional to the value of DC current gain β_F. For current gain $\beta_F = 100$, this results in approximately 2% error in output current.

There is another error due to the fact that the two transistors operate at different collector–emitter voltages V_{CE}, due to base-width modulation and the Early effect. Figure 8.17(a) shows a basic mirror circuit, and we see that the V_{CE} of transistor Q_1 is about 0.7 V due to the diode connection, but the V_{CE} of transistor Q_2 is whatever

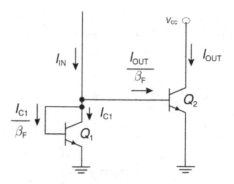

FIGURE 8.16

Circuit for determining gain error of basic current mirror. Intermediate circuit currents are shown for clarity.

the power supply voltage V_{CC} is. Figure 8.17(b) shows the variation in the output current (collector current of Q_2) as V_{CC} varies from 0 to 20 V.

Example 8.3: Speed of ideal, basic current mirror

Next, we will investigate the speed of the basic bipolar current mirror. First, if we consider small-signal bandwidth, we can apply open-circuit time constants (OCTCs). If you do an OCTC analysis of the basic mirror, a bandwidth of a little less than $\sim \omega_T/2$ is predicted if you assume that the effect of base spreading resistance r_x is negligible. For a transistor with $f_T = 300$ MHz and negligible base spreading resistance, the bandwidth is about 150 MHz. An LTSPICE simulation is shown in Figure 8.18 where we see that the bandwidth is indeed greater than 100 MHz.

In Figure 8.19, we apply a pulse of current to the input of the transistor Q_1 and see that the resultant output current risetime is a few nanoseconds. Remembering that risetime $t_r \cong 0.35/f_h$, a bandwidth of 100 MHz is consistent with a small-signal risetime and falltime of a few nanoseconds.

Example 8.4: The effect of parasitic inductance on the speed and transient response of a current mirror

Of course, the accuracy of this model depends on the accuracy of the LTSPICE transistor model, as well as any parasitic inductances that may be surrounding the transistor. Next we will investigate the bandwidth and pulse response degradation when there are a few nanohenries of parasitic inductance in each transistor lead. The circuit model in this case is shown in Figure 8.20(a) where we put 5–10 nH of inductance in each transistor lead.[10] For high-frequency circuits, small values of parasitic inductance can have a significant detrimental effect on frequency and pulse response, as shown in Figure 8.20(b)

---

[10]Off-the-shelf through-hole transistors have a few nanohenries of lead inductance due to the geometry of the package. The parasitic inductance in surface-mount packages is a little less.

(a)

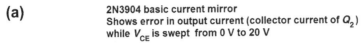

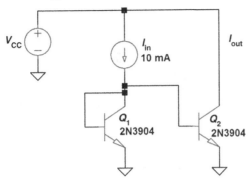

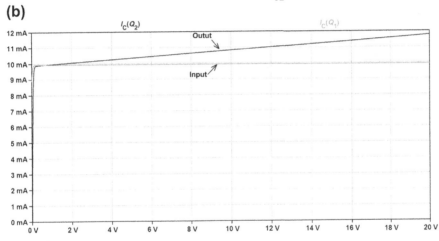

.dc V_{CC} 0 20 0.001
file: Basic 2N3904 current mirror V_{CE} error.asc

(b)

FIGURE 8.17

Circuit for determining mirror error due to base-width modulation. (a) LTSPICE circuit where transistor Q_1 is biased with a 10-mA current source and power supply V_{CC} is varied from 0 to 20 V. (b) Output, showing error between I_{in} and I_{out} due to differing V_{CE} in the two transistors. (For color version of this figure, the reader is referred to the online version of this book.)

and (c). (The moral here is to pick a device with low package inductance if possible and do a good job in your printed circuit board layout.)

Current mirror with emitter degeneration

One way to reduce current mirror errors due to mismatched V_{BES} is to use some emitter degeneration in the current mirror (Figure 8.21) by inclusion of emitter

(a)

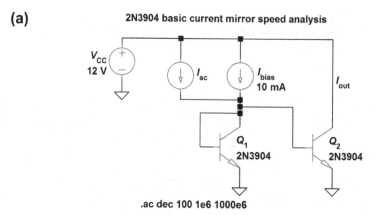

(b)

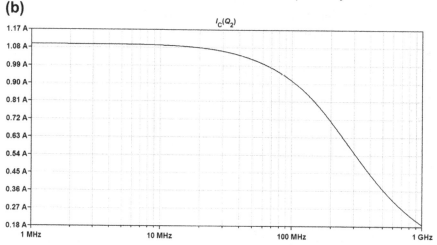

FIGURE 8.18

Circuit for determining small-signal bandwidth of basic current mirror of Example 8.3. (a) LTSPICE circuit where transistor Q_1 is biased with a 10-mA current source, $V_{CC} = 12$ V, and there is a small signal current i_{ac} injected at the collector of Q_1. (b) Output, showing small-signal bandwidth of output current at the collector of Q_2 (i_{out}) greater than 150 MHz. (For color version of this figure, the reader is referred to the online version of this book.)

resistors. Applying KVL around the loop containing the base-emitters of Q_1 and Q_2 results in:

$$I_{E1}R_{E1} + V_{BE1} = I_{E2}R_{E2} + V_{BE2} \tag{8.27}$$

We see that if $I_E R_{E1} \gg V_{BE1}$ and $I_E R_{E2} \gg V_{BE2}$, any effects of V_{BE} mismatches are reduced. Inclusion of the emitter resistors also increases the incremental output

(a)

2N3904 basic current mirror step response
Input current I_{pulse} **is stepped from 1 mA to 10 mA**

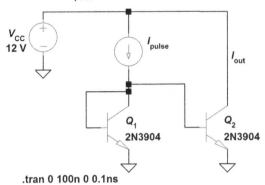

.tran 0 100n 0 0.1ns
file: Basic 2N3904 current mirror step response.asc

(b)

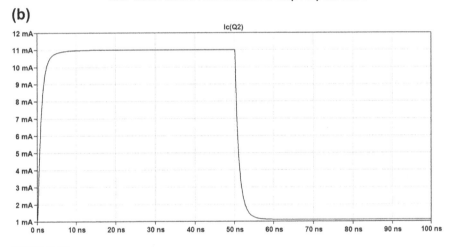

FIGURE 8.19

Circuit for determining step response of basic current mirror of Example 8.3. (a) LTSPICE circuit where transistor Q_1 collector current is pulsed from 1 to 10 mA current source with a 50 ns wide current pulse. (b) Output (collector current of Q_2), showing risetime of a few nanoseconds. (For color version of this figure, the reader is referred to the online version of this book.)

resistance of the current source. We can also adjust the mirror ratio by adjusting the ratio of R_1 to R_2.

Current mirror with "beta helper"

The dependence of output current on transistor DC current gain β_F can be improved by adding another transistor and altering the configuration of the current mirror as in

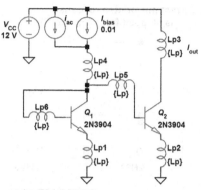

(a) 2N3904 basic current mirror frequency response, with parasitics
Input bias current ≈ 10 mA
Parasitic inductance added, 0, 5 nH, and 10 nH in each transistor lead

.ac dec 100 1e6 1000e6
file: Basic 2N3904 current mirror frequency response with parasitics.asc
.step param Lp list 0.001n 5n 10n

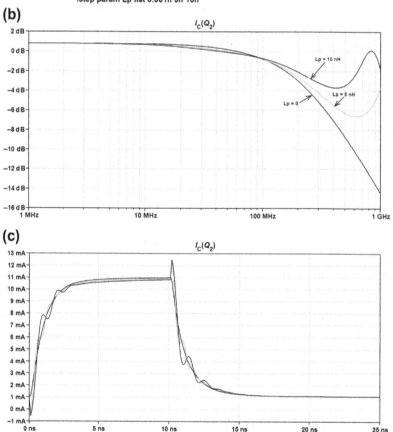

FIGURE 8.20

Circuit for determining effects of parasitic inductance on current mirror bandwidth (Example 8.4). (a) LTSPICE circuit with parasitic inductance of 0, 10, or 20 nH. (b) Frequency response of output current showing bandwidth degradation and high-frequency peaking due to parasitic inductance. (c) Pulse response of output current in response to 10-ns input pulse. (For color version of this figure, the reader is referred to the online version of this book.)

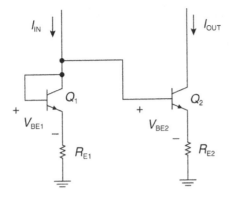

FIGURE 8.21

Basic current mirror with emitter degeneration resistors R_{E1} and R_{E2}.

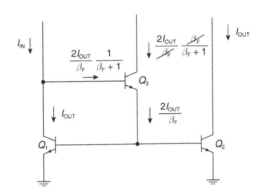

FIGURE 8.22

Current mirror with beta helper, showing intermediate currents to help in calculating the output/input ratio.

Figure 8.22. The values of each current are as shown; transistor Q_3 provides base current to both Q_1 and Q_2 and hence reduces the dependence of the output current on transistor current gain.

In analyzing this circuit, we recognize that Q_1 and Q_2 operate the same V_{BE} and hence their collector currents are the same (I_{OUT}). Transistor Q_3 provides base current for Q_1 and Q_2, and hence its emitter current is $2I_{OUT}/\beta_F$. The input current (I_{IN}) is not exactly the same as the collector current of Q_1, since we have the base current of Q_3 as well. We find I_{OUT} as follows:

$$I_{IN} = I_{OUT} + \left(\frac{2I_{OUT}}{\beta_F}\right)\left(\frac{1}{\beta_F + 1}\right) \qquad (8.28)$$

After some algebraic manipulation we find that the output and input currents are related as:

$$\frac{I_{OUT}}{I_{IN}} = \frac{\beta_F^2 + \beta_F}{\beta_F^2 + \beta_F + 2} = \frac{1}{1 + \frac{2}{\beta_F^2 + \beta_F}} \tag{8.29}$$

We see that the error is inversely proportional to one over the square of β_F, so the error is reduced as compared to the simple current mirror.

Cascode current mirror

A cascode current mirror (Figure 8.23) has an output stack of transistors. We have seen earlier in our work on the extended hybrid-pi model that the output resistance of a transistor with emitter resistance $R_E \gg r_\pi$ results in an output resistance approaching $r_\mu/2$. One way to accomplish this is to stack output transistors. This cascode current mirror has high output impedance, since output transistor Q_4 has a large resistance seen at its emitter, due to the output resistance of transistor Q_2.

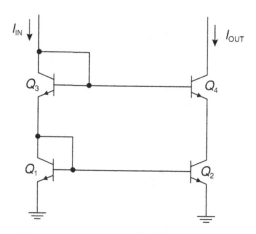

FIGURE 8.23

Cascode current mirror.

Wilson current mirror

The Wilson current mirror (Figure 8.24) reduces errors due to finite β_F by the use of negative feedback. The figure shows intermediate currents, which help us determine the input–output relationship. Let us assume that I_{B3} increases a little bit. This tends to increase the collector current of Q_3 and hence also the collector current Q_2. This increases the collector current of Q_2 and tends to reduce I_{B3} (since an increase in I_{C1}

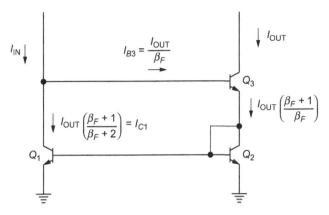

FIGURE 8.24

Wilson current mirror.

steals current away from the base of Q_3). This is negative feedback, which we will investigate in more detail later in this book.

A detailed analysis similar to that we did with the mirror with beta helper results in the input/output current relationship for the Wilson current mirror as:

$$I_{IN} = \frac{I_{OUT}}{\beta_F} + I_{OUT}\left(\frac{\beta_F + 1}{\beta_F + 2}\right) \tag{8.30}$$

After some algebra, we find for the Wilson mirror the transfer function relating input and output current as:

$$\frac{I_{OUT}}{I_{IN}} = \frac{\beta_F^2 + 2\beta_F}{\beta_F^2 + 2\beta_F + 2} = \frac{1}{1 + \frac{2}{\beta_F^2 + 2\beta_F}} \tag{8.31}$$

Example 8.5: Speed of Wilson current mirror

Shown in Figure 8.25(a) is a Wilson mirror. Using LTSPICE we will investigate the bandwidth and pulse response of this mirror, assuming that we use 2N3904 transistors. As shown in Figure 8.25(b), the bandwidth is greater than 100 MHz, but there is some gain peaking. In Figure 8.25(c), we see the pulse response of the Wilson mirror, with some current overshoot, which is consistent with the gain peaking shown in the small-signal simulation. In a later chapter in this book, we will investigate further how systems with feedback can have overshoot.

(a)

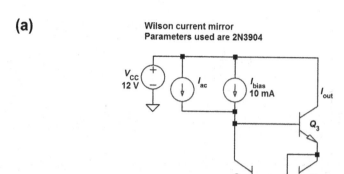

Wilson current mirror
Parameters used are 2N3904

.ac dec 100 1e6 1000e6
file: Wilson current mirror.asc

(b)

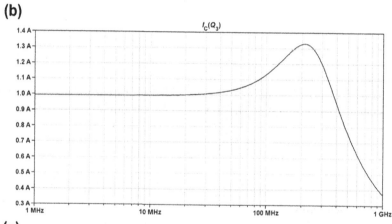

(c)

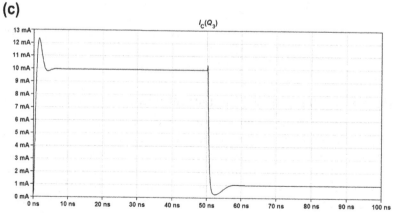

Widlar current mirror

A Widlar mirror (named after Robert Widlar) can be used when low levels of output currents are needed. In the forward-active region, the base—emitter voltage of a transistor is given by:

$$V_{BE} \approx \frac{kT}{q} \ln\left(\frac{I_C}{I_S}\right) \tag{8.32}$$

where kT/q is the thermal voltage and I_S is the reverse saturation current of the transistor. In Figure 8.26, we can solve the Kirchoff Voltage Law (KVL) loop around the Q_1 and Q_2 base—emitter loops, resulting in:

$$V_{BE1} = V_{BE2} + I_{E2}R_E \tag{8.33}$$

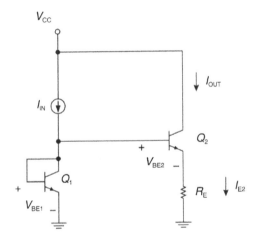

FIGURE 8.26

Widlar current mirror.

This means that the emitter resistor R_E steals away some of the voltage available to drive the base—emitter junction of Q_2. Hence Q_2 runs at a lower collector current than Q_1, but Q_1 still maintains some level of control.

FIGURE 8.25

Circuit for determining speed of Wilson current mirror of example 8.5. (a) LTSPICE circuit with transistor models of 2N3904 used. (b) Small-signal frequency response of output current showing some gain peaking. (c) Pulse response of output current in response to 50-ns input pulse. There is some overshoot in the step response, consistent with some gain peaking in the frequency response. (For color version of this figure, the reader is referred to the online version of this book.)

Example 8.6: Widlar current mirror output current

For the Widlar current mirror of Figure 8.26, find the collector current of transistor Q_2 for an input current $I_{in} = 100\ \mu A$ and emitter resistor $R_E = 10\ k\Omega$. Assume that each transistor has a reverse saturation current $I_S = 10^{-6}$ A, that $\beta_F \gg 1$ for each transistor, and that the transistors operate at a temperature of 300 K.

Solution: Since we assume that $\beta_F \gg 1$, then $I_E \approx I_C$ and we will ignore base currents.

$$V_{BE1} = V_{BE2} + I_{E2}R_E \Rightarrow \frac{kT}{q}\ln\left(\frac{I_{IN}}{I_{S1}}\right) \approx \frac{kT}{q}\ln\left(\frac{I_{OUT}}{I_{S2}}\right) + I_{OUT}R_E \qquad (8.34)$$

Let us simplify this expression by combining the logarithmic terms:

$$\frac{kT}{q}\ln\left(\frac{I_{IN}}{I_{OUT}}\right) \approx I_{OUT}R_E \qquad (8.35)$$

Note that the dependence on the reverse saturation current drops out. Iteratively solving this equation with $I_{IN} = 100\ \mu A$, we find that $I_{OUT} \approx 6.9\ \mu A$.

Example 8.7: Widlar mirror incremental output resistance

For the Widlar current mirror of Figure 8.26, find the incremental output resistance measured at the collector of Q_2. Again, assume that $R_E = 10\ k\Omega$. For Q_2, assume the following parameters: $h_{fe} = 100$; $\eta = 3.3 \times 10^{-4}$.

Solution: We find the small-signal parameters as follows:

$$g_m = \frac{I_{C2}}{V_{TH}} = \frac{6.9 \times 10^{-6}}{0.026} = 2.65 \times 10^{-4}\ A/V$$

$$r_\pi = \frac{h_{fe}}{g_m} = \frac{100}{2.65 \times 10^{-4}} = 377\ k\Omega \qquad (8.36)$$

$$r_o = \frac{1}{\eta g_m} = \frac{1}{(3.3 \times 10^{-4})(2.65 \times 10^{-4})} = 11.4\ M\Omega$$

$$r_\mu = h_{fe}r_o = 1140\ M\Omega$$

The full expression for the output resistance of transistor Q_2 with emitter degeneration is:

$$r_{out} \approx r_\mu \left\|\frac{[1 + g_m[r_\pi\|R_E]]}{g_o}\right\| \qquad (8.37)$$

Note that in this case, $R_E \ll r_\pi$ and hence we can approximate the output resistance as:

$$r_{out} \approx g_m R_E r_o \approx (2.65 \times 10^{-4})(10{,}000\ \Omega)(11.4\ M\Omega) \approx 30.2\ M\Omega \qquad (8.38)$$

Example 8.8: Output "voltage compliance" of current mirrors

In order for a current mirror to operate correctly, we need to be careful to bias the transistors "ON". The "compliance voltage" of a current source is the minimum voltage that the output transistor needs to have across it to ensure that the output mirrors the input. A circuit illustrating voltage compliance of a simple mirror is shown in Figure 8.27, where we see that the simple current mirror needs at least 0.2 V across the output transistor for the mirror to operate correctly.

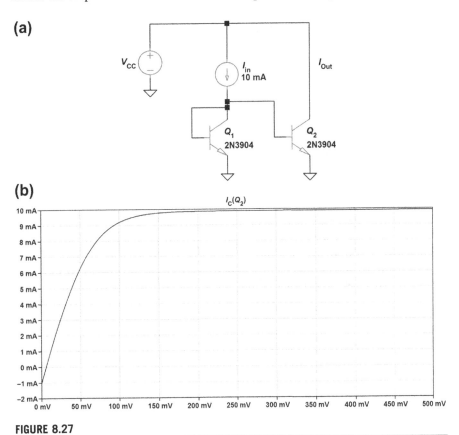

FIGURE 8.27

Circuit illustrating "voltage compliance" in a current mirror (Example 8.8). (a) LTSPICE circuit, where V_{CC} is varied from 0 V to 500 mV. (b) Simulation result showing how output varies as V_{CC} varies, showing the voltage compliance of approximately 200 mV. (For color version of this figure, the reader is referred to the online version of this book.)

Example 8.9: Design example—high-gain amplifier

Consider the high-gain transistor amplifier in Figure 8.28(a). Transistor Q_1 is a standard common-emitter stage, biased with the PNP current source Q_2. The

(a)

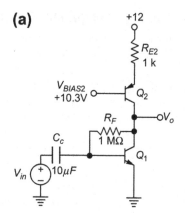

(b)

High gain amplifier small signal model
Cmu2 models the output capacitance of the current source load

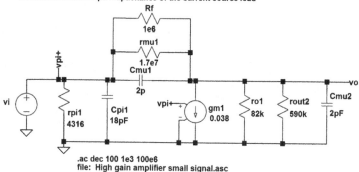

FIGURE 8.28

High-gain amplifier (Example 8.9). (a) Circuit. (b) Small-signal model assuming operation at frequencies high enough so that the coupling capacitor C_c acts as a short-circuit. The PNP current source has been replaced by its incremental output resistance r_{out2}. (For color version of this figure, the reader is referred to the online version of this book.)

1000-kΩ resistor feedback biases the amplifier so that both Q_1 and Q_2 are biased in their linear operating region. Note that we have made the emitter resistor of Q_2 (R_{E2}) large enough so that the incremental output resistance r_{out2} of Q_2 is large. This emitter resistor also sets a collector current of approximately 1 mA in Q_1 and Q_2.

We will find the gain of this amplifier, given that the base-width modulation factors for the NPN and the PNP transistors are $\eta_{npn} = 3.2 \times 10^{-4}$ and $\eta_{pnp} = 1.2 \times 10^{-3}$. Other parameters of interest are $h_{fe,npn} = 203$ and $h_{fe,pnp} = 164$. Each transistor has $f_T = 300$ MHz and $C_\mu = 2$ pF.

In Figure 8.28(b), we draw the incremental model, replacing transistor Q_2 by its equivalent output resistance r_{out2}. We have also included $C_{\mu2}$, which is the capacitance looking into the collector of Q_2. Parameters for Q_2 are as follows:

$$I_{C2} \approx 1 \text{ mA}$$

$$g_{m2} = \frac{1 \text{ mA}}{26 \text{ mV}} \approx 0.038 \text{ A/V}$$

$$r_{\pi2} = \frac{h_{fe2}}{g_{m2}} = \frac{164}{0.038} = 4316 \text{ }\Omega$$

$$r_{o2} = \frac{1}{\eta_{pnp}g_{m2}} = \frac{1}{(1.2 \times 10^{-3})(0.038)} = 2.2 \times 10^4 \text{ }\Omega$$

(8.39)

$$r_{\mu2} = h_{fe2}r_{o2} = (164)(2.2 \times 10^4) = 3.6 \times 10^6 \text{ }\Omega$$

$$r_{out2} = r_{\mu2}\|r_{o2}[1 + g_{m2}[r_{\pi2}\|R_{E2}]] = 5.9 \times 10^5 \text{ }\Omega$$

For transistor Q_1, bias level and incremental parameters are:

$$I_{C1} \approx 1 \text{ mA}$$

$$g_{m1} = \frac{1 \text{ mA}}{26 \text{ mV}} \approx 0.038 \text{ A/V}$$

$$r_{o1} = \frac{1}{\eta_{npn}g_{m1}} = \frac{1}{(3.2 \times 10^{-4})(0.038)} = 8.2 \times 10^4 \text{ }\Omega$$

(8.40)

$$r_{\mu1} = h_{fe1}r_{o1} = (203)(8.2 \times 10^4) = 1.7 \times 10^7 \text{ }\Omega$$

$$C_{\pi1} = \frac{g_{m1}}{\omega_T} - C_{\mu1} = 18 \text{ pF}$$

The node equation for finding the gain is:

$$(v_i - v_o)G_f' - g_{m1}v_i - v_oG_o = 0$$

$$G_f' = \frac{1}{R_f\|r_{\mu1}}$$

(8.41)

$$G_o = \frac{1}{r_{o1}\|r_{out2}}$$

Reducing this results in the equation for gain:

$$v_i(G_f' - g_{m1}) = v_o(G_f' + G_o)$$

$$A_v = \frac{v_o}{v_i} = \frac{(G_f' - g_m)}{(G_f' + G_o)} \approx \frac{-g_{m1}}{(G_f' + G_o)} \approx -g_{m1}(R_f\|r_{\mu1}\|r_{o1}\|r_{out2}) = -2541$$

(8.42)

OCTC calculations for the three capacitors are in Table 8.1:

Table 8.1 Summary of OCTC calculations for Example 8.9

Capacitance	Open-Circuit Resistance	OCTC
$C_{\pi 1}$	$= 0$ since base and emitter are both grounded incrementally	—
$C_{\mu 1}$	$= r_{o1} = 82\ \mathrm{k\Omega}$	164 ns
$C_{\mu 2}$	$= r_{o1} = 82\ \mathrm{k\Omega}$	164 ns

The sum of OCTCs is 328 ns, resulting in a bandwidth estimate $f_h \cong 485$ kHz. An LTSPICE simulation (Figure 8.29) shows that the midband gain and bandwidth are approximately as calculated. The low-frequency rolloff can be adjusted by modifying the value of the coupling capacitor C_c.

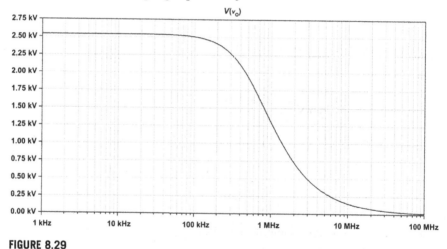

FIGURE 8.29

High-gain amplifier LTSPICE simulation result for Example 8.9, showing a midband gain of approximately -2500 and bandwidth $f_h \sim 600$ kHz.

Example 8.10: Another high-gain amplifier example

Let us design an AC-coupled transistor amplifier with gain magnitude of $|A_v| > 1000$, using high-gain techniques. We are driving a 100-pF capacitive load with the amplifier, so we need an emitter-follower at the output to isolate the high-gain node from the capacitive load.

Assume that you have at your disposal transistors with $\eta_{npn} = 6.7 \times 10^{-4}$, $\eta_{pnp} = 1.8 \times 10^{-4}$, $C_\mu = 2$ pF, $f_T = 300$ MHz, $h_{fe,npn} = 200$, and $h_{fe,pnp} = 175$. In this example, we will not worry much about biasing details; i.e. you can assume that your collector currents will magically run at the correct bias levels. We will find a design that meets the gain specification and then estimate the high- and low-frequency breakpoints using open-circuit and short-circuit time constants.

One solution: An initial iteration on a circuit topology is shown in Figure 8.30(a). Transistor Q_1 is the common-emitter amplifier loaded by current source Q_2. The base bias voltage V_{BIAS2} at the base of Q_2 and the emitter resistor R_{E2} set the quiescent current of Q_1 and Q_2. Feedback resistor R_F biases Q_1 in the forward-active region. We will assume that a value of 1 MΩ suffices for R_F, but the actual value depends on the DC current gain of transistor Q_1 and the desired output voltage at the collector of Q_1. Transistor Q_3 buffers the load capacitor from the high-gain

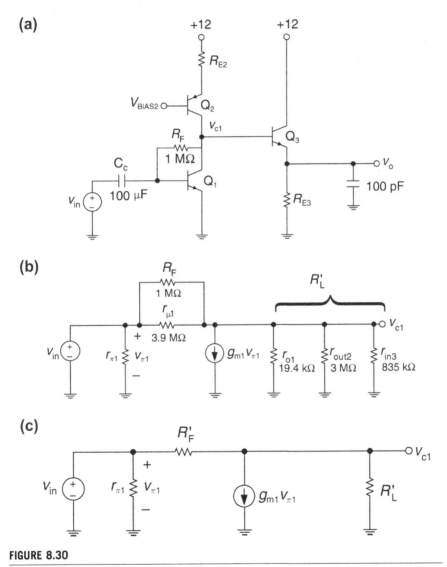

FIGURE 8.30

High-gain amplifier design example for Example 8.10. (a) Circuit (biasing details omitted). (b) Small-signal model. (c) Simplified small-signal model.

node found at the collector of Q_1. In order to achieve high gain, we want current source Q_2 to have a high output resistance and we want emitter-follower Q_3 to have a high input resistance. As a sanity check, we know that the best gain that we can achieve with this topology is:

$$|A_v| < \frac{1}{\eta_{npn}} < \frac{1}{6.7 \times 10^{-4}} < 1492 \tag{8.43}$$

Shown in Figure 8.30(b) is the small-signal model[11] of this amplifier where we can find the input–output relationship between the input v_{in} and the incremental voltage v_{c1} at the collector of Q_1. For this incremental analysis, we have replaced the current source Q_2 by a resistance r_{out2}, which is the output resistance of the current source. Likewise, we have replaced emitter-follower Q_3 by a resistance r_{in3}, which is the input resistance of the emitter-follower. We will assume that the voltage gain from v_{c1} to v_o is ~ 1, an assumption that needs to be checked later on.

To simplify the gain analysis, we will redraw the incremental circuit, resulting in Figure 8.30(c). We lump the parallel combination of r_{o1}, r_{out2}, and r_{in3} into a single resistor, which we will call R'_L. We lump the parallel combination of $r_{\mu1}$ and R_F into a feedback resistance R_F'. Mathematically, we express this as:

$$R'_L = r_{o1} \| r_{out2} \| r_{in3}$$
$$R'_F = R_F \| r_{\mu1} \tag{8.44}$$

If we apply KCL at the output node v_{c1}, we find:

$$(v_{in} - v_{c1})G'_F - g_{m1}v_{in} - v_{c1}G'_L = 0 \tag{8.45}$$

Next, solving for v_{c1}/v_{in} we find:

$$\frac{v_{c1}}{v_{in}} = \frac{G'_F - g_{m1}}{G'_F + G'_L} \approx -\frac{g_{m1}}{G'_F + G'_L} \approx -\frac{g_{m1}}{G_F + g_{\mu1} + g_{o1} + g_{out2} + g_{in3}} \tag{8.46}$$

Let us arbitrarily set the collector currents of Q_1 and Q_2 to be 2 mA. We find the following small-signal parameters for common-emitter amplifier transistor Q_1:

$$g_{m1} = \frac{I_{C1}}{\left(\dfrac{kT}{q}\right)} = \frac{0.002}{0.026} = 0.077 \text{ A/V}$$

$$r_{\pi1} = \frac{h_{fe1}}{g_{m1}} = \frac{200}{0.077} = 2.6 \text{ k}\Omega \tag{8.47}$$

$$r_{o1} = \frac{1}{\eta_{npn}g_{m1}} = \frac{1}{(6.7 \times 10^{-4})(0.077)} = 19.4 \text{ k}\Omega$$

$$r_{\mu1} = h_{fe1}r_{o1} = (200)(19.4 \text{ k}\Omega) = 3.9 \text{ M}\Omega$$

---

[11]In this analysis, we assume that $r_x = 0$ for each transistor. (As we'll see in Example 8.11 following, this is not always a good assumption.)

We now see that the value of r_{o1} is much smaller than R_F and $r_{\mu 1}$, so we can approximate our gain relationship further as:

$$\frac{v_{c1}}{v_{in}} \approx -\frac{g_{m1}}{g_{o1} + g_{out2} + g_{in3}} \tag{8.48}$$

In order to achieve as high a gain as possible, we need to make the output impedance of the current source and the input impedance of the voltage source very large compared to r_{o1}. For the current source, let us set $R_{E2} = 1$ kΩ and $V_{BIAS2} = 9.3$ V, which will give us approximately what we want in terms of bias level. Using these values, we find the incremental parameters of the current source transistor as follows:

$$g_{m2} = 0.077 \text{ A/V}$$

$$r_{\pi 2} = \frac{h_{fe2}}{g_{m2}} = \frac{175}{0.077} = 2.3 \text{ k}\Omega$$

$$\tag{8.49}$$

$$r_{o2} = \frac{1}{\eta_{pnp}g_{m2}} = \frac{1}{(1.8 \times 10^{-4})(0.077)} = 72.1 \text{ k}\Omega$$

$$r_{\mu 2} = h_{fe2}r_{o2} = (175)(72.1 \text{ k}\Omega) = 12.6 \text{ M}\Omega$$

Next, we find the output resistance of the current source:

$$r_{out2} \approx r_{\mu 2} \left\| \frac{[1 + g_{m2}[r_{\pi 2} \| R_{E2}]]}{g_{o2}} \right\| \approx 3 \text{ M}\Omega \tag{8.50}$$

Note that this value of output resistance is so high (compared, for instance, to r_{o1}) that it does not load down the high-gain node significantly.

Next, for the emitter-follower Q_3, let us assume that we have biased the transistor at a collector current of 1 mA. The value of R_{E3} is a little difficult to determine, since we do not know exactly the bias point set at the collector of Q_1. Let us assume that the collector of Q_1 is at approximately 6 V (or half the supply voltage). This means that $R_{E3} = 5.3$ to set a collector current of 1 mA.

We find the incremental parameters of the emitter-follower Q_3 as follows:

$$g_{m3} = \frac{I_{C3}}{\left(\frac{kT}{q}\right)} = 0.038 \text{ A/V}$$

$$r_{\pi 3} = \frac{h_{fe3}}{g_{m3}} = \frac{200}{0.038} = 5.3 \text{ k}\Omega \tag{8.51}$$

$$r_{o3} = \frac{1}{\eta_{npn}g_{m3}} = \frac{1}{(6.7 \times 10^{-4})(0.038)} = 39.3 \text{ k}\Omega$$

$$r_{\mu 3} = h_{fe3}r_{o3} = (200)(39.3 \text{ k}\Omega) = 7.9 \text{ M}\Omega$$

Next, we find the input resistance of the emitter-follower:

$$r_{in3} = r_{\mu3} \| \left(r_{\pi3} + \left(1 + h_{fe3}\right)\left[R_{E3}\|r_{o3}\right]\right)$$

$$\approx (7.9\text{ M}\Omega)\|((200)[(5.3\text{ k})\|39.3\text{ k}]) \approx 835\text{ k}\Omega \tag{8.52}$$

In order to continue with the analysis, we will tabulate the small-signal parameters for all transistors in Table 8.2.

Table 8.2 Small-Signal Parameters for High-Gain Amplifier example 8.10

	Q_1	Q_2	Q_3
h_{fe}	200	175	200
g_m	0.077 A/V	0.077 A/V	0.038 A/V
r_π	2.6 kΩ	2.3 kΩ	5.3 kΩ
r_o	19.4 kΩ	72.1 kΩ	39.3 kΩ
r_μ	3.9 MΩ	12.6 MΩ	7.9 MΩ
C_π	39 pF	39 pF	20 pF
C_μ	2 pF	2 pF	2 pF

Note that the output resistance r_{o1} of Q_1 is very small compared to the output resistance of the Q_2 current source (r_{out2}) and the input resistance of emitter-follower Q_3 (r_{in3}). Therefore, we expect the gain of this amplifier to be approximately $-1/\eta_{npn}$ or approximately -1400.

Next, let us estimate the bandwidth using OCTCs. For $C_{\pi1}$, we find the incremental circuit of Figure 8.31(a). In the following OCTC calculations, we will ignore the r_μ of all transistors and include the effects of transistor output resistance r_o only where needed. We note that the open-circuit resistance for $C_{\pi1}$ is zero, because incrementally both sides of $C_{\pi1}$ are grounded.

Next, we note that incrementally $C_{\mu1}$, $C_{\mu2}$, and $C_{\mu3}$ are all in parallel since all three transistors share the high-gain node v_{c1}. The OCTCs circuit for this case is shown in Figure 8.31(b). The incremental resistance to ground at this node is approximately r_{o1}, which is the incremental output resistance of transistor Q_1. Therefore, this OCTC at the high-gain node is:

$$\tau_{oc1} = r_{o1}\left(C_{\mu1} + C_{\mu2} + C_{\mu3}\right) = (19.4\text{ k}\Omega)(6\text{ pF}) = 116.4 \times 10^{-9}\text{ s} \tag{8.53}$$

The open-circuit resistances for $C_{\pi2}$ and $C_{\pi3}$ are found using the circuits of Figure 8.31(c) and Figure 8.31(d), respectively. In each case, the open-circuit

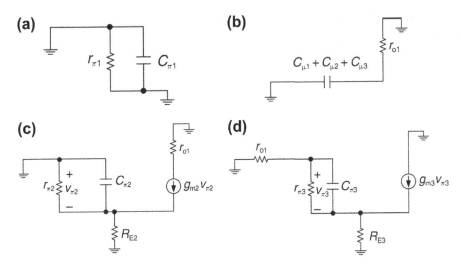

FIGURE 8.31

OCTCs circuits for Example 8.10. (a) OCTC circuit $C_{\pi 1}$. (b) OCTC circuit for $C_{\mu 1}$, $C_{\mu 2}$, and $C_{\mu 3}$, which is the high-gain node of the amplifier. (c) OCTC circuit for $C_{\pi 2}$ (d) OCTC circuit for $C_{\pi 3}$.

resistance across each C_π is approximately $1/g_m$, since the emitter resistors are relatively large.[12] This results in two more time constants that are small compared to the time constant at the high-gain node:

$$\tau_{oc2} \approx \frac{C_{\pi 2}}{g_{m2}} \approx \frac{39 \text{ pF}}{0.077} \approx 0.5 \times 10^{-9} \text{ s}$$

$$\tau_{oc3} \approx \frac{C_{\pi 3}}{g_{m3}} \approx \frac{20 \text{ pF}}{0.038} \approx 0.5 \times 10^{-9} \text{ s}$$

(8.54)

The OCTC due to the load capacitor is the incremental output resistance of the Q_3 emitter-follower multiplied by the load capacitor value. This calculation is as follows:

$$r_{out,Q_3} \approx R_{E3} \left\| \left(\frac{r_{\pi 3} + r_{o1}}{1 + h_{fe3}} \right) \approx 120 \ \Omega \right.$$

$$\tau_{oc4} = r_{out,Q_3} C_L = (120 \ \Omega)(100 \text{ pF}) = 12 \times 10^{-9} \text{ s}$$

(8.55)

The sum of the OCTCs for this amplifier is 129.4 ns, resulting in a bandwidth estimate of:

$$\omega_h \approx \frac{1}{\sum \tau_{oc}} \approx \frac{1}{129.4 \times 10^{-9}} \approx 7.73 \text{ Mrad/s}$$

$$f_h \approx \frac{\omega_h}{2\pi} \approx 1.23 \text{ MHz}$$

(8.56)

---

[12]For details of this calculation, see Chapter 7.

We note that the bandwidth bottleneck is the time constant associated with the capacitance at the high-gain node.

In order to estimate the low-frequency breakpoint of this amplifier due to the coupling capacitor, we note that the incremental resistance seen by the coupling capacitor C_c is simply $r_{\pi 1}$. Therefore, the low-frequency breakpoint is:

$$\omega_L = \frac{1}{r_{\pi 1} C_c} = \frac{1}{(2600)(100 \times 10^{-6})} = 3.8 \text{ rad/s}$$

$$f_L = \frac{\omega_L}{2\pi} = 0.6 \text{ Hz}$$

(8.57)

A PSPICE simulation (Figure 8.32) shows that our estimates of gain and bandwidth (both f_L and f_h) are pretty good. Furthermore, we see that the assumption that the emitter-follower has unity gain is justified in this case.

Example 8.11: Another high-gain amplifier example (revisited)

In the previous example, we made the assumption that $r_x = 0$ for each transistor. As we will soon see, this assumption, although it simplifies the math significantly, gives us gain and bandwidth estimates that need to be revisited. Let us redo these estimates, using a value of $r_x = 100 \ \Omega$ for each transistor.

With respect to gain, we expect the finite r_x to reduce the gain, due to its loading effect with $r_{\pi 1}$ and the input resistance of Q_1. We could march forward and write node equations using the circuit of Figure 8.33, but we can approximate the gain reduction with a little forethought. We recognize that the inclusion of r_{x1} means that the voltage $v_{\pi 1}$ is now set by a voltage divider between r_π and r_{x1}. Although the feedback resistor of Q_1 is R_F (1000 kΩ) in parallel with $r_{\mu 1}$ (3900 kΩ), the input resistance r_{in} is much lower due to the high negative gain (approximately -1400) around this resistance. The input resistance looking into the transistor base, including the effects of R_F' only, is approximately:

$$r_{in} \approx \frac{R_F'}{A_V} \approx \frac{796\text{k}}{1400} \approx 570 \ \Omega$$

(8.58)

Therefore, the voltage divider that reduces the gain is now:

$$\frac{r_{\pi 1} \| r_{in}}{r_{\pi 1} \| r_{in} + r_x} = \frac{467}{467 + 100} = 0.82$$

(8.59)

Hence, we expect the gain to be about 82% of that in the previous example,[13] or approximately -1150. Therefore, the gain reduction due to a finite r_x is significant.

We will now delve into some detail of the OCTCs calculations, with circuits for each C_π and C_μ shown in Figure 8.34.

---

[13]We should iterate the r_{in} calculation with the new value of gain, but you get the idea.

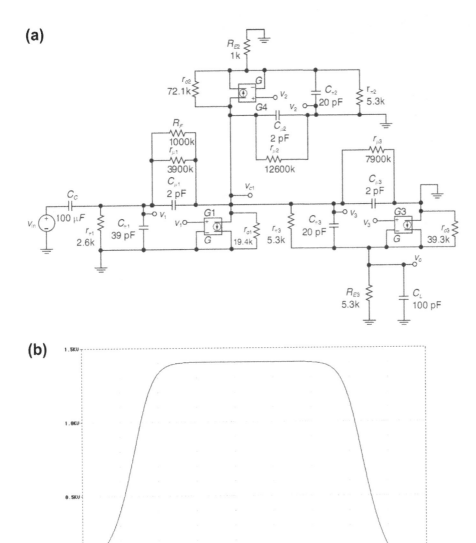

FIGURE 8.32

PSPICE analysis of high-gain amplifier of Example 8.10. (a) Circuit. (b) PSPICE simulation showing frequency response with a high-frequency bandwidth over 1.2 MHz.

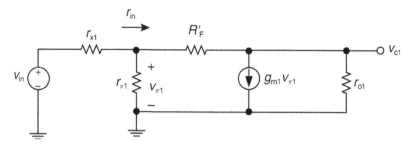

FIGURE 8.33

Circuit for calculating gain reduction in high-gain amplifier due to finite r_x. (Example 8.11)

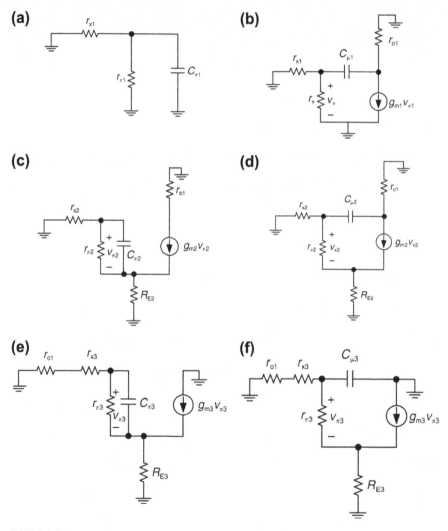

FIGURE 8.34

OCTCs circuits for revisited circuit of Example 8.11 with finite r_x. (a) Circuit for $C_{\pi 1}$, (b) $C_{\mu 1}$, (c) $C_{\pi 2}$, (d) $C_{\mu 2}$, (e) $C_{\pi 3}$, and (f) $C_{\mu 3}$.

Table 8.3 Summary of OCTC calculations for high-gain amplifier with $r_x = 100\ \Omega$			
	Q_1	Q_2	Q_3
R_{oc} for C_π	96.3 Ω	14 Ω	120 Ω
R_{oc} for C_μ	163.3 kΩ	21.4 kΩ	19.1 kΩ
τ_{oc} for C_π	3.8 ns	0.5 ns	2.4 ns
τ_{oc} for C_μ	327 ns	43 ns	38 ns
τ_{oc} for C_L	12 ns		
Sum of OCTCs	427 ns		
Estimate of ω_h	2.34 Mrad/s		
Estimate of f_h	373 kHz		

For $C_{\pi 1}$, we use the circuit of Figure 8.34(a), where we find the open-circuit resistance to be 96 Ω (or approximately the value of r_{x1}) since $r_{x1} << r_{\pi 1}$. The OCTC for $C_{\pi 1}$ is now 3.8 ns.

For $C_{\mu 1}$, we use the circuit of Figure 8.34(b). We note that r_{x1} provides a resistance in the base of transistor Q_1. We may now expect more Miller multiplication of the capacitance $C_{\mu 1}$. Calculations show that the open-circuit resistance facing $C_{\mu 1}$ is 163.3 kΩ and the resultant OCTC is 327 ns (about 2.8× higher than it was before). Note that this OCTC is greater than the total sum of OCTCs for the circuit with $r_x = 0$. Hence, we already know that there is significant bandwidth degradation due to r_x. In fact, since the dominant time constant has increased by about 2.8×, we expect the overall amplifier bandwidth to decrease by the same factor.

We will spare the reader the painful details of further OCTC calculations for the remaining four circuits (Figure 8.34(c)–(f)) and the load capacitance and proceed immediately to the summary in Table 8.3. We note that inclusion of r_x has greatly increased the OCTC for $C_{\mu 1}$ of the common-emitter transistor Q_1. The overall effect of including r_x is that the revised OCTC estimate bandwidth of the amplifier is ~373 kHz, as compared to a bandwidth of greater than 1.2 MHz with $r_x = 0$. This result is verified with the PSPICE simulation of Figure 8.35. We see that the base spreading resistance reduces the midband gain to approximately −1150 and reduces the bandwidth to ~500 kHz. The warning here is to ignore r_x at your own peril!

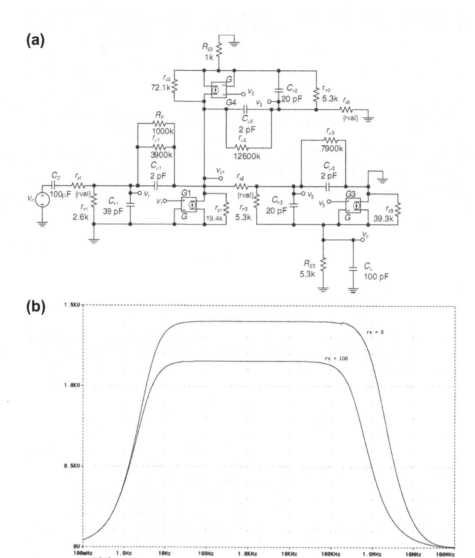

FIGURE 8.35

LTSPICE analysis of high-gain amplifier with $r_x = 100\,\Omega$. (a) Circuit. (b) Comparison of frequency responses of amplifiers with $r_x = 0$ and $r_x = 100\,\Omega$ for each transistor. Note that a finite r_x both decreases the gain and bandwidth significantly as compared to the case with $r_x = 0$.

Chapter 8 problems
Problem 8.1

Using the extended hybrid-pi model, determine the low-frequency input resistance of the emitter-follower connection in Figure 8.36 as a function of transistor parameters and quiescent operating levels. Assume both transistors are identical and that the collector of Q_1 is tied to $+V_{cc}$. When you are all done, you should be able to express the incremental input resistance seen at the base of Q_1 in terms of h_{fe}, r_o, and r_μ. Give an approximate value.

Problem 8.2

Again using the extended hybrid-pi model and the circuit of Figure 8.36, find the output resistance seen at the collector of Q_1, as a function of transistor parameters and quiescent operating conditions.

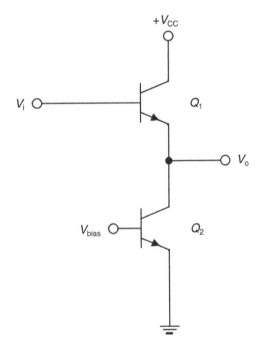

FIGURE 8.36

Circuit for Problems 8.1 and 8.2.

Problem 8.3

Using OCTCs, estimate the small-signal speed of response of the current mirror shown in Figure 8.37. Assume that each transistor is a 2N3906 biased at 10 mA with $V_{CB} = 6$ V for Q_2. Assume that $r_x = 0\ \Omega$ (even though this is a darn lie!) and that incrementally I_{in} is a very high impedance and I_{out} presents a very low impedance.

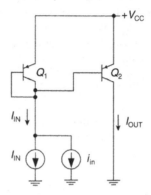

FIGURE 8.37

Circuit for Problems 8.3 and 8.4. Transistor Q_1 is biased with DC current I_{IN}. There is a small signal source current as well (i_{in}).

a. Draw the small-signal model. Identify how many OCTCs calculations there are.

b. Estimate the bandwidth f_h using OCTCs. Relate this f_h to the f_T of the transistor.

c. Simulate using LTSPICE and compare your LTSPICE result to the OCTCs estimate.

Problem 8.4

Let us revisit the current mirror circuit of Problem 8.3, but this time assume that the transistors in question instead of having zero base resistance have base resistance in the range $r_x = 100-200\ \Omega$. Unfortunately, with a finite value of r_x, the OCTCs problem is very difficult (but still solvable!) in this case. Resimulate the circuit using LTSPICE and find the bandwidth with this range of finite value of r_x. Comment on how this solution compares to the solution with $r_x = 0$.

Problem 8.5

Design a transistor amplifier with gain magnitude of $|A_v| = 10,000$, using high-gain techniques. You're driving a capacitive load with your amplifier, and assume that you need an emitter-follower at the output to isolate the high-gain node from the capacitive load. Assume that you have at your disposal transistors with $\eta_{npn} = 6.7 \times 10^{-5}$ and $\eta_{pnp} = 1.8 \times 10^{-4}$ Do not worry about biasing details (you can assume that your collector currents will magically run at the correct bias levels). Ignore r_x in all your calculations and assume that $h_{fe, npn} = 200$ and $h_{fe, pnp} = 175$. Simulate the gain of your circuit using LTSPICE. Assuming that $f_T = 300$ MHz and $C_\mu = 2$ pF for each transistor, estimate the bandwidth of your circuit using OCTCs and confirm using LTSPICE.

Further reading

[1] Early JM. Effects of space-charge layer widening in junction transistors. *Proc IRE* 1952:1401–6.

[2] Gray PE, DeWitt D, Boothroyd AR, Gibbons JF. *Physical electronics and circuit models of transistors*. In: *Semiconductor Electronics Education Committee*, vol. 2. John Wiley; 1964.

[3] Gray PR, Hurst PJ, Lewis SH, Meyer RG. *Analysis and design of analog integrated circuits*. 4th ed. John Wiley; 2001.

[4] Grebene AB. *Bipolar and MOS analog integrated circuit design*. John Wiley; 1984.

[5] Hart BL. Modeling the early effect in bipolar transistors. *IEEE J Solid State Circuits* 1983;**18**(1):139–40.

[6] Huiting C, Whiteside F, Geiger R. Current mirror circuit with accurate mirror gain for low β transistors. In: *Proceedings of the 2001 IEEE international symposium on circuits and systems, (ISCAS 2001)* 2001. p. 536–9.

[7] Joardar K. A new Approach for extracting base width modulation parameters in bipolar transistors. In: *Proceedings of the 1994 bipolar/BiCMOS circuits and technology meeting* 1994. p. 140–3.

[8] Kimura K. Low voltage techniques for bias circuits. *IEEE Trans Circuits Syst I Fundam Theory Appl* 1997;**5**:459–65.

[9] Liou JJ. Comments on 'Early voltage in very-narrow-base bipolar transistors,' by D. J. Roulston. *IEEE Electron Device Lett* 1990;**11**(5):236.

[10] Mahattanakul J, Pookaiyaudom S, Toumazou C. Understanding Wilson current mirror via the negative feedback approach. In: *Proceedings of the 2001 IEEE international symposium on circuits and systems, (ISCAS 2001)* 2001. p. 532–5.

[11] McAndrew CC, Nagel LW. SPICE early modeling [bipolar transistors]. In: *Proceedings of the 1994 bipolar/BiCMOS circuits and technology meeting* 1994. p. 144–7.

[12] Roberge JK. *Operational amplifiers: theory and practice*. John Wiley; 1975.

[13] Roulston DJ. Early voltage in very-Narrow-base bipolar transistors. *IEEE J Solid State Circuits* 1990;**11**(2):88–9.

[14] Rucker LM. Monolithic bipolar diodes and their models. *IEEE Circuits Devices Mag* 1991;**7**(2):26–31.

[15] Soclof S. *Analog integrated circuits*. Prentice-Hall; 1985.

[16] Thompson MT. Tips for designing high-gain amplifiers. *Electron Des* 16, 1994:83–90.

[17] van Kessel J, van de Plassche RJ. Integrated linear basic circuits. *Philips Tech Rev* 1971;**32**(1):1–10.

[18] Yuan JS, Liou JJ. An improved Early voltage model for advanced bipolar transistors. *IEEE Trans Electron Devices* 1991;**38**(1):179–82.

Introduction to Field-Effect Transistors (FETs) and Amplifiers

"Our life is frittered away by detail. Simplify, simplify, simplify!"

—Henry D. Thoreau

IN THIS CHAPTER

▶ We next take a detour from the world of bipolar transistors and enter the world of field-effect transistors (FETs). The basic signal metal oxide–semiconductor (MOS) structure is discussed, followed by a discussion of metal oxide–semiconductor field-effect transistor (MOSFET) amplifiers. The incremental model of the MOS transistor is shown, and it is used in a design example where gain and bandwidth are calculated for an MOS amplifier. We then finish with the junction field-effect transistor (JFET) and several examples of JFET amplifiers.

Early history of field-effect transistors

The invention of the MOSFET predates the bipolar transistor. An excerpt from one US patent granted in 1933 to Dr Julius Lilienfeld is shown in Figure 9.1. In three patents, Dr Lilienfeld gave structures of the MOSFET, MESFET, and other MOS devices, but he was not able to build any working FET, underscoring the difficulty in fabricating practical semiconductor devices at that time. In fact, it was not until the 1960s that the first commercially successful FET devices were manufactured.

Today, there are two broad classes of FET in common use: the MOSFET and the JFET. We will examine each in some detail.

Qualitative discussion of the basic signal MOSFET

The basic N-channel lateral MOS[1] transistor is shown in Figure 9.2. This is a lateral device because current flow is in the $-x$ direction laterally across the surface of the

---

[1] An n-channel MOSFET is sometimes called "NMOS". Likewise, a p-channel device is called "PMOS".

Intuitive Analog Circuit Design. http://dx.doi.org/10.1016/B978-0-12-405866-8.00009-7

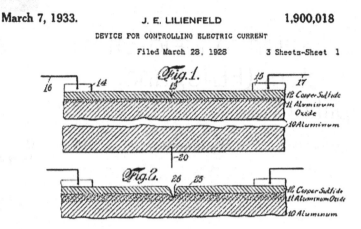

FIGURE 9.1

Excerpt from Lilienfeld's US patent 1,900,018[2] (1933) showing two different versions of the MOSFET. In Figure 1 from the patent, terminal #16 is the source, terminal #15 is the drain, and terminal #20 is the gate connection.

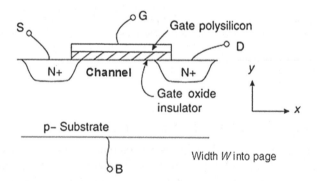

FIGURE 9.2

Basic N-channel MOS transistor. Device has width W into the page. The terminals are source (S), gate (G), drain (D), and the substrate (B, for "bulk").

device.[3] An N-channel MOS device starts with a lightly doped p-substrate. N-type source and drain regions are added at either ends of the channel. Next, an insulating oxide layer is grown on the surface. A gate connection is isolated from the p-substrate by this oxide layer; this oxide is a very good electrical insulator. This

---

[2]Lilienfeld had three patents in succession covering basic MOS transistor structures. They are US patent #1,745,175 (filed 10/8/26, granted 1/18/30); US patent #1,877,140 (filed 12/8/28, granted 9/13/32), and US patent #1,900,018 (filed 3/28/28, granted 3/7/33).

[3]Power MOS devices are generally vertical devices; current flow is vertically through the device. Signal MOS devices have current flow laterally across the surface.

oxide layer is the dielectric of the gate capacitance. A fourth terminal (called bulk or substrate) is connected to the lowest potential in the circuit. The four terminals shown are source (S), gate (G), drain (D), and substrate or bulk (B).

The gate contact forms a capacitance per unit area through the oxide layer, of value:

$$C_{\text{ox}} = \frac{\varepsilon_{\text{ox}}}{t_{\text{ox}}} \tag{9.1}$$

where ε_{ox} is the dielectric permittivity of the oxide layer (in Farads/meter) and t_{ox} is the thickness of the oxide layer. The capacitance C_{ox} has units of Farads per square meter, so to get total oxide capacitance, you multiply by the cross-sectional area of the oxide.

Now, in a thought experiment, let us consider the equivalent circuit of the MOSFET, when it is OFF. If the gate terminal is not connected, the source–channel junction and drain–channel junction path both behave as diodes, and these diodes are back to back as shown in Figure 9.3. No current can flow through this device other than diode leakage current.

In a second thought experiment, let us ground the source, drain, and bulk connections, and apply a positive voltage to the gate (Figure 9.4). This gate voltage produces an electric field in the oxide layer that originates on positive charges in the gate electrode and terminates on negative charges in the channel. If we apply sufficient positive voltage to the gate, we will attract sufficient mobile electrons to the channel underneath the gate to form an *inversion layer.* Basically, at this magic voltage (called the *threshold voltage* V_{T}) we have caused the P-material under the gate to behave as if it were N-material, due to this inversion layer that we have formed. Between the N+ islands, we see an inversion layer that behaves as an N-region. Below a gate-source voltage of V_{T}, the MOSFET is essentially OFF; for $V_{\text{GS}} > V_{\text{T}}$, the MOSFET can support current flow under the gate if a drain-source voltage is applied.

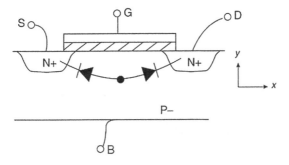

FIGURE 9.3

N-channel MOS device, showing back-to-back diodes. With the gate disconnected there is no current flow in the MOS device.

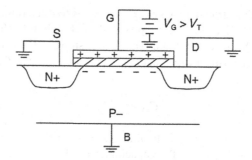

FIGURE 9.4

N-channel MOS (NMOS) device with positive gate potential greater than the threshold voltage (V_T) applied to the gate terminal, with all other MOS terminals grounded. This forms a negative charge "inversion layer" under the gate oxide inside the P-substrate.

In order to figure out the details of the shape of the MOS transistor *V-I* curve, we have to resort to a mathematical model.

Figuring out the *V-I* curve of a MOS device

Let us consider a more detailed model of an N-channel MOSFET (Figure 9.5) in the hopes of generating the ideal MOS *V-I* curve. If there is current flowing in the MOSFET channel, we would expect a varying voltage along the length of the channel. Let us call this channel voltage $V_c(x)$. Remember that no current will flow for gate to source (V_{GS}) voltages less than the threshold voltage V_T. We can find the charge per unit area under the gate in the channel as:

$$Q_c(x) = -C_{ox}(V_{GS} - V_T - V_c(x)) \qquad (9.2)$$

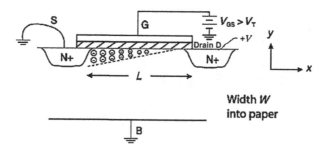

FIGURE 9.5

N-channel MOS device with positive gate potential ($> V_T$) applied, showing an inversion layer under the gate. The width of the inversion layer reduces as we approach the drain terminal due to voltage drop along the channel.

Note that a mobile gate charge exists only if the gate-source voltage exceeds the threshold voltage. The gate charge $Q_c(x)$ is the charge density (Coulombs/m^2) at position x along the gate.

The drain current is equal to:[4]

$$I_D = -WQ_c(x)v(x) \tag{9.3}$$

where W is the width of the MOS device into the paper and $v(x)$ is the "drift velocity" of the charges across the channel.

If we assume a low electric field (and hence a long length device), the drift velocity is linearly related to the electric field with particle mobility as a proportionality constant, as:

$$v(x) = \mu_n E_x = \mu_n \frac{dV_c(x)}{dx} \tag{9.4}$$

We can then express the drain current as:

$$I_D = WC_{ox}[V_{GS} - V_c(x) - V_T]\mu_n \frac{dV_c(x)}{dx} \tag{9.5}$$

We can bring the derivative dx to the left-hand side, resulting in:

$$I_D dx = W\mu_n C_{ox}[V_{GS} - V_c(x) - V_T]dV_c(x) \tag{9.6}$$

Next, let us integrate both sides. We integrate x over the length of the channel from 0 to L, and we integrate the channel voltage $V_c(x)$ from 0 to V_{GS}.

$$\int_0^L I_D dx = \int_0^{V_{GS}} W\mu_n C_{ox}[V_{GS} - V_c(x) - V_T]dV_c(x) \tag{9.7}$$

Following through with the integral results in an equation relating drain current to device voltages:

$$I_D L = \mu_n WC_{ox}\left[(V_{GS} - V_T)V_{DS} - \frac{V_{DS}^2}{2}\right]$$

$$\Downarrow \tag{9.8}$$

$$I_D = \mu_n C_{ox}\left(\frac{W}{L}\right)\left[(V_{GS} - V_T)V_{DS} - \frac{V_{DS}^2}{2}\right]$$

This is the drain current equation for the ideal MOSFET, valid for gate-source voltage greater than the threshold voltage, or $V_{GS} > V_T$.

---

[4]Note that the units work out. Current has units of coulombs/second, W has units of meters, $Q_c(x)$ has units of coulombs/m^2, and velocity has units of meter/second.

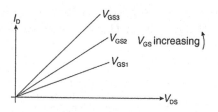

FIGURE 9.6

NMOS *V-I* curve for $V_{DS} \ll V_{GS}$ and $V_{GS} > V_T$. The drain–source connection behaves as a voltage-controlled resistor, controlled by the value of V_{GS}.

For very small V_{DS} ($V_{DS} \ll V_{GS}$) and with gate-source voltage greater than the threshold voltage ($V_{GS} > V_T$), this device behaves like a voltage-controlled resistor:

$$I_D \approx \mu_n C_{ox} \left(\frac{W}{L}\right)((V_{GS} - V_T)V_{DS}) \quad V_{DS} \ll V_{GS} \tag{9.9}$$

As V_{DS} increases, you get higher current because there is a higher electric field sweeping carriers through the channel. Current also increases with higher $V_{GS} - V_T$ since this increases the charge under the gate available to support current. This region of operation is shown in Figure 9.6.

If V_{DS} is equal to or higher than $V_{GS} - V_T$, there is sufficient drain–source voltage to "pinchoff" the channel at the drain end of the channel. This means that the gate charge at the end of the channel approaches zero as shown in Figure 9.7, and we achieve maximum drain current supported by the available V_{GS}. If V_{DS} increases, to first order, there is no further increase in drain current. In the MOS device, this is the so-called saturation[5] region.

For $V_{DS} > (V_{GS} - V_T)$, we find that the drain current saturates. Indeed, we can find the drain current in the saturation region by setting $V_{DS} = (V_{GS} - V_T)$ in the drain current equation. This results in the drain current in saturation:

$$I_D = \frac{\mu_n C_{ox}}{2} \frac{W}{L}(V_{GS} - V_T)^2 \tag{9.10}$$

The ideal MOS *V-I* curves are shown in Figure 9.8, showing cutoff, linear and saturation regions of operation.

In a real-world MOSFET, we note that the drain current vs. V_{DS} curves in saturation are not perfectly horizontal. The MOS transistor has finite output impedance due to *channel-length modulation*, which is analogous to base-width modulation (the "Early effect") in the bipolar transistor. The width of the drain depletion region varies as V_{DS} varies, with the effect that drain current also varies with V_{DS}. Analogous to our development of the bipolar transistor, we can modify the large-signal

---

[5]Do not get this confused with the bipolar transistor, where the saturation region occurs for V_{CE} less than $V_{CE,sat}$.

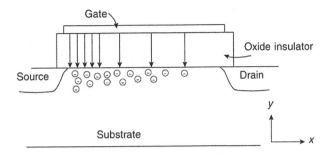

FIGURE 9.7

Qualitative view of gate charge and electric field inside oxide layer for MOS device operating in the "pinchoff" region with $V_{DS} > V_{GS} - V_T$. At the drain end of the channel, there is no mobile gate charge and the drain is pinched off. (Dimensions distorted for clarity).

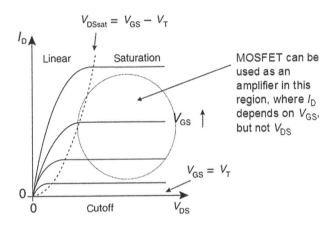

FIGURE 9.8

Ideal MOS *V-I* curve for an N-channel (NMOS) device showing linear, saturation, and cutoff regions of operation. In the "linear" region, the MOSFET can be used as an amplifier to produce voltage gain.

V–I characteristic of the MOSFET by including a channel-length modulation parameter λ that models the finite output resistance of the transistor:

$$I_D = \frac{\mu_n C_{ox}}{2} \frac{W}{L}(V_{GS} - V_T)^2(1 + \lambda V_{DS}) \tag{9.11}$$

The factor λ is process dependent and is inversely proportional to channel length (or $\propto 1/L$). Typical values for λ are 0.005–0.02/V. Figures 9.9 and 9.10 show sets of LTSPICE-produced I_D vs. V_{DS} curves for hypothetical N-channel (NMOS) and P-channel (PMOS) devices. The SPICE parameter KP is equal to (μC_{ox}) for the MOSFET.

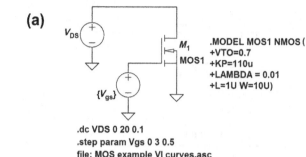

(a)

```
.MODEL MOS1 NMOS (
+VTO=0.7
+KP=110u
+LAMBDA = 0.01
+L=1U W=10U)
```

.dc VDS 0 20 0.1
.step param Vgs 0 3 0.5
file: MOS example VI curves.asc

(b)

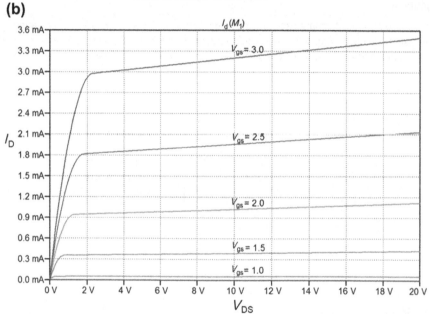

FIGURE 9.9

MOS V-I curve for a hypothetical N-channel MOS device with device length $L = 1$ μm, device width $W = 10$ μm, threshold voltage $V_T = 0.7$ V, and channel-length modulation factor $\lambda = 0.01/V$. (a) LTSPICE model. (b) V-I curves. (For color version of this figure, the reader is referred to the online version of this book.)

MOS small-signal model (low frequency)

A MOS device is useful as an amplifier only if operated in the saturation region with drain—source voltage higher than the pinchoff voltage (i.e. $V_{DS} > (V_{GS} - V_T)$) and with the gate—source voltage higher than the threshold voltage (or $V_{GS} > V_T$). In the saturation region of operation, amplification can occur since the drain current

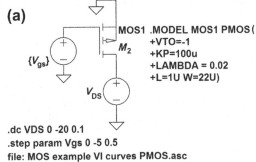

(a)

MOS1 .MODEL MOS1 PMOS(
+VTO=-1
+KP=100u
+LAMBDA = 0.02
+L=1U W=22U)

.dc VDS 0 -20 0.1
.step param Vgs 0 -5 0.5
file: MOS example VI curves PMOS.asc

(b)

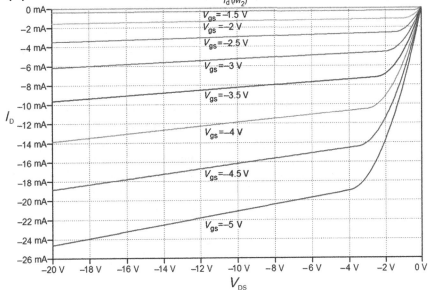

FIGURE 9.10

MOS V-I curve for a hypothetical P-channel MOS (PMOS) device with device length $L = 1$ μm, device width $W = 220$ μm, threshold voltage $V_T = -1$ V, and channel-length modulation factor $\lambda = 0.02/V$. (a) LTSPICE model. (b) V-I curves. (For color version of this figure, the reader is referred to the online version of this book.)

varies with V_{GS} but (to first-order) does not vary with V_{DS}. This region of operation is analogous to the forward-active region in bipolar transistors, where amplification can occur.

If we are operating in the MOS saturation region, in order to find small-signal variation of drain current with gate–source voltage, we need to solve for transconductance. Similar to our derivation for the bipolar transistor, we linearize the current–voltage equation about an operating point, to find transconductance:

$$g_m = \frac{\partial I_D}{\partial V_{GS}} = \mu_n C_{ox} \frac{W}{L} (V_{GS} - V_T) = \sqrt{2\mu_n C_{ox} \frac{W}{L} I_D} \qquad (9.12)$$

This expression means that the transconductance of a MOS device scales as the square-root of drain current I_D. Compare this expression to that of the bipolar transistor, where transconductance scales linearly with collector bias current. The result is that MOS transistors have lower gain than comparably-sized and biased bipolar junction transistors (BJTs).

Next, we need to model the small-signal change in output current due to a change in V_{DS}, due to channel-length modulation. Incrementally, we can model this effect as a small-signal output resistance r_o across the drain–source terminals. The output resistance r_o in the small-signal model can be found as:

$$r_o = \frac{1}{\dfrac{\partial I_D}{\partial V_{DS}}} = \frac{1}{\lambda I_D} \qquad (9.13)$$

Finally, we note that a MOSFET is a four-terminal device; the source–bulk terminals act as a second set of input terminals. In some cases, we will tie the source and bulk terminals together, and therefore we can ignore the effects of this second input. However, if the bulk is *not* tied to the same potential as the source, we need to consider the effect of changing bulk-to-source voltage. The result is to include in our small-signal model a transconductance to model this *back-gate effect*, or:

$$g_{mb} = \frac{\partial I_D}{\partial V_{BS}} \qquad (9.14)$$

Typically, the back-gate transconductance g_{mb} is much smaller than the ideal MOS transconductance g_m. Putting this all together, we find the low-frequency small-signal model of the MOS transistor, in the saturation region (Figure 9.11).

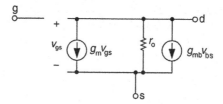

FIGURE 9.11

MOS low-frequency small-signal model in the saturation region, showing gate (g), drain (d), and source (s) terminals. The bulk (b), or substrate, connection is also assumed, but not shown. The $g_{mb}v_{sb}$ generator source causes a current flow proportional to bulk–source voltage v_{bs}.

MOS small-signal model (high frequency)

The MOS small-signal model at high frequencies has capacitive elements modeling gate–source (C_{gs}), gate–drain (C_{gd}), source–substrate (C_{sb}), and drain–substrate (C_{db}) capacitance. Capacitance C_{gs} models the effect of the charge under the gate. C_{gd} models the effect of the gate–oxide overlap over the drain region. C_{db} and C_{sb} are depletion capacitances between drain–substrate and source–substrate, respectively. We should note that many of these capacitances are nonlinear in that the capacitance varies with the voltage across the capacitance. A MOS transistor high-frequency, small-signal model is shown in Figure 9.12.

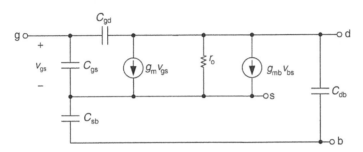

FIGURE 9.12

MOS high-frequency small-signal model showing gate–source capacitance (C_{gs}), gate–drain capacitance (C_{gd}), source–bulk capacitance (C_{sb}), and drain–bulk capacitance (C_{db}). The terminals are gate (g), source (s), drain (d), and bulk (b).

Basic MOS amplifiers

The topologies of MOS amplifiers are similar to those of bipolar transistors, as shown in Table 9.1.

When analyzing MOS amplifiers, there are several differences to consider, as compared to bipolar amplifier analysis:

- The incremental input resistance of a MOS device is very high. In the BJT world, the input resistance at the base is limited by r_π and $(1 + h_{fe})R_E$.

Table 9.1 Comparison of BJT and MOS Amplifiers

BJT	MOS
Emitter follower	Source follower
Common emitter	Common source
Common base	Common gate
BJT differential amplifier	MOS differential amplifier

- A MOSFET is a four-terminal device. In the MOS world, the connection to the substrate (or bulk) must also be considered. Furthermore, there are extra parasitic capacitances to the substrate (C_{sb} and C_{db}) that must be considered.
- The MOS device has another dependent current generator that must be considered, due to the back-gate effect. In effect, the gate–bulk voltage acts as an extra gate, with drain current depending on the bulk–source voltage V_{BS}.

Source follower

A source-follower buffer is shown in Figure 9.13(a). This circuit has high input impedance, low output impedance, and a gain less than, but hopefully close to, 1.0. The small-signal low-frequency model is shown in Figure 9.13(b). Using this model, we can find the small-signal gain of this buffer by doing KCL at the output node (noting that $G_S = 1/R_S$ and assuming the bulk terminal is incrementally grounded) as:

$$g_m(v_i - v_o) - v_o g_o - v_o g_{mb} - v_o G_S = 0 \qquad (9.15)$$

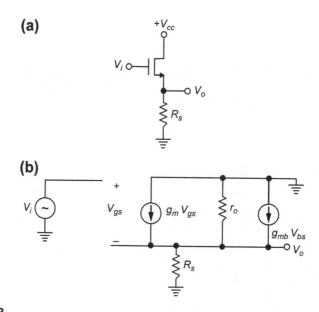

FIGURE 9.13

Source follower. (a) Circuit, biasing details omitted. (b) Small-signal low-frequency model, including back-gate effect.

Simplifying, we find:

$$A_v = \frac{v_o}{v_i} = \frac{g_m}{g_m + g_o + g_{mb} + G_S} \qquad (9.16)$$

This gain is close to $+1$ if the transconductance of the MOS device is large compared to $1/r_o$, g_{mb}, and $1/R_S$.

Common-source amplifier

A common-source amplifier is shown in Figure 9.14(a). This circuit has high input impedance and provides voltage gain with a voltage inversion. The small-signal low-frequency model is shown in Figure 9.14(b). There is no back-gate effect in this case, since the bulk terminal is usually grounded and the source terminal is grounded; hence $v_{bs} = 0$. Using this model, we find the small-signal gain:

$$A_v = \frac{v_o}{v_i} = -g_m(R_L \| r_o) \tag{9.17}$$

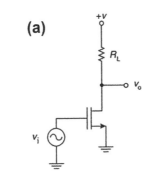

(a)

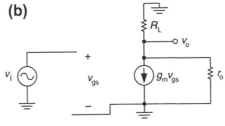

(b)

FIGURE 9.14

Common-source amplifier with resistive load. (a) Circuit, omitting biasing details for simplicity. (b) Small-signal low-frequency model.

If we load the common-source amplifier with an ideal current source instead of a resistor, we have the circuit of Figure 9.15. The gain of this amplifier is:

$$A_v = \frac{v_o}{v_i} = -g_m r_o \tag{9.18}$$

This is the maximum gain per stage we can get with a MOS common-source amplifier.

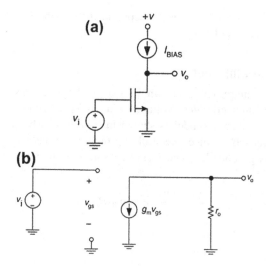

FIGURE 9.15

Common-source amplifier with current source load. (a) Circuit, omitting biasing details for simplicity and assuming an ideal current source with infinite incremental output resistance. (b) Small-signal low-frequency model assuming negligible back-gate effect (i.e. assuming $g_{mb} = 0$).

Common-gate amplifier

A common-gate amplifier is shown in Figure 9.16. We can solve for v_{gs} by using the incremental circuit of Figure 9.16(b) and by summing currents at the source:

$$\frac{v_i + v_{gs}}{R_i} + g_m v_{gs} = 0 \implies v_{gs} = -v_i \left(\frac{1}{1 + g_m R_i} \right) \tag{9.19}$$

The gain is found as follows:

$$v_o = -g_m v_{gs} R_L = v_i \left(\frac{g_m R_L}{1 + g_m R_i} \right)$$

$$\therefore A_v = \frac{v_o}{v_i} = \frac{g_m R_L}{1 + g_m R_i} \tag{9.20}$$

Note that the gain is approximately R_L/R_i if $g_m R_i \gg 1$.

Common-source amplifier with cascode

We can use the common-gate amplifier to augment the common-source amplifier in a cascode configuration (Figure 9.17(a)). In this widely used configuration, common-source transistor M_1 is loaded by common-gate buffer M_2. The voltage-controlled current generated at the drain of M_1 is buffered by M_2 and flows through

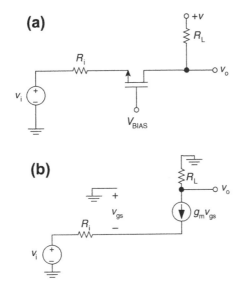

(a)

(b)

FIGURE 9.16

Common-gate amplifier with resistive load. (a) Circuit, omitting biasing details for simplicity. (b) Small-signal low-frequency model assuming negligible back-gate effect (i.e. assuming $g_{mb} = 0$).

the load resistor R_L. The source of M_2 presents a low impedance to the drain of M_1, hence eliminating the Miller effect.

Another variation in the theme is the *folded cascode* of Figure 9.17(b). The folded cascode is topologically similar to the basic cascode in that it consists of a common-source transistor feeding a common-gate buffer transistor with current. The P-channel common-gate provides buffering between M_1 and the current source load.

MOS current mirrors

A basic CMOS current mirror is shown in Figure 9.18(a). This building block works on the same basic premise as a bipolar mirror—to first order, CMOS gates with the same V_{GS} will have the same drain current I_D, provided that both devices operate in the saturation region.

In order to estimate the small-signal bandwidth of the MOS mirror, we refer to Figure 9.18(b). Note that incrementally, all capacitances appear in parallel to one another. The incremental resistance across the gate–source node is $\sim 1/(2g_m)$. Therefore, using open-circuit time constants, an estimate of the small-signal bandwidth of this circuit is:

$$\omega_h \approx \frac{2g_m}{C_{gs1} + C_{db1} + C_{gs2} + C_{gd2}} \tag{9.21}$$

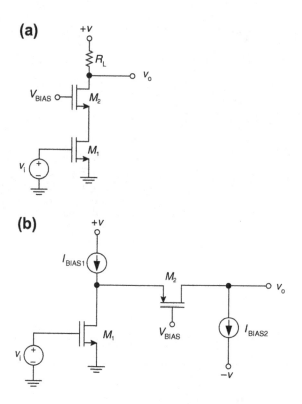

FIGURE 9.17

MOS cascode amplifiers. (a) MOSFET M_1 is a common-source amplifier stage. MOSFET M_2 is a common-gate connection providing buffering between the common-source amplifier and load resistor R_L. (b) Folded cascode where M_1 is an N-channel MOSFET and M_2 is a P-channel MOSFET.

We also note that the low-frequency incremental output resistance of the simple mirror can be quite modest, approximately r_o.

The MOS Wilson current mirror (Figure 9.19(a)) uses negative feedback to increase the output resistance of the mirror. We can see how this operates using the low-frequency small-signal model of Figure 9.19(b). We will assume that all the MOS devices are identical and biased at the same drain currents, and hence $g_{m1} = g_{m2} = g_{m3}$ and $r_{o1} = r_{o2} = r_{o3}$. The diode-connected MOS device M_2 has been replaced by its output resistance $(1/g_{m2})$. We have added a test current source i_t and we will now calculate the test voltage v_t in order to find the incremental output resistance at the M_3 drain node.

Applying KVL around the loop containing the test voltage source results in:

$$v_t = i_x r_o + v_b = i_x r_{o3} + \frac{i_t}{g_{m2}} \qquad (9.22)$$

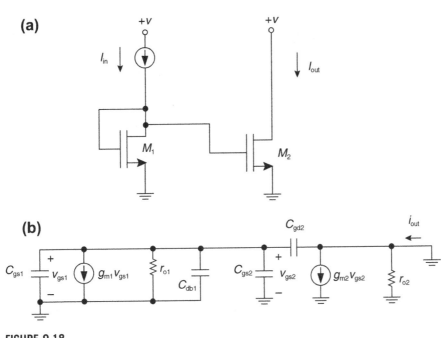

FIGURE 9.18

Basic MOS current mirror. (a) Circuit. (b) High-frequency incremental model, ignoring back-gate effect. Note that in this topology $v_{gs1} = v_{gs2}$.

The unknown current i_x is found by:

$$i_x = i_t - g_{m3}v_a \tag{9.23}$$

The unknown voltage v_a is found by:

$$v_a = -g_{m1}v_b r_{o1} = -g_{m1}\left(\frac{i_t}{g_{m2}}\right)r_{o1} \tag{9.24}$$

Solving for the output resistance, we find:

$$r_{out} = \frac{v_t}{i_t} \approx \left(1 + \frac{g_{m1}g_{m3}}{g_{m2}}r_{o1}\right)r_{o3} \tag{9.25}$$

Note that this output resistance is significantly higher than that of the simple current mirror.

Another variation on the Wilson mirror is the "improved" Wilson mirror (Figure 9.20). MOS device M_4 is added, which ensures that the drain–source voltages of M_1 and M_2 are equal. This reduces gain errors due to the finite output resistance of M_1 and M_2.

Yet another MOS mirror that provides high output impedance is the cascode current mirror (Figure 9.21). Note that output transistor M_3 drives into the drain terminal of M_2. This ensures that the output resistance seen at the drain of M_3 is high.

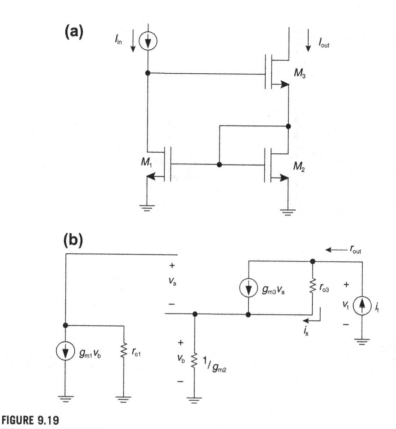

FIGURE 9.19

Wilson current mirror. (a) Circuit. (b) Low-frequency incremental model used to determine incremental output resistance r_{out} of the Wilson mirror.

There are many other "variations on the theme" of MOS current mirrors.

Example 9.1: MOS amplifier design example

Let us work out a MOS design example illustrating the use of MOS models and open-circuit time constants for bandwidth analysis. Let us assume that we need a gain magnitude of at least 10 (i.e. +20 dB) and a −3 dB bandwidth in excess of 350 MHz. The input source has a series resistance of 1 kΩ and the amplifier drives a 1-pF load capacitance. Ignore the back-gate effect (i.e. assume that $g_{mb} = 0$). Assume that the bulk (i.e. substrate) connections of the transistors are tied to ground. The hypothetical MOS transistors have the following parameters:

- $g_m = 0.01$ A/V
- $r_o = 3$ kΩ
- Gate−source capacitance $C_{gs} = 0.2$ pF

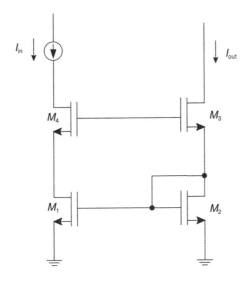

FIGURE 9.20

"Improved" Wilson current mirror.

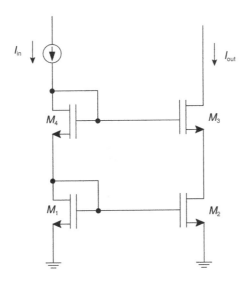

FIGURE 9.21

Cascode current mirror.

- Gate–drain capacitance $C_{gd} = 0.05$ pF
- Source–substrate capacitance $C_{sb} = 0.02$ pF
- Drain–substrate capacitance $C_{db} = 0.02$ pF

TRY #1: Common-source amplifier

A first-cut design for this amplifier is shown in Figure 9.22(a), with low-frequency small-signal model shown in Figure 9.22(b). We can use the low-frequency

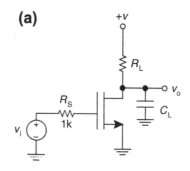

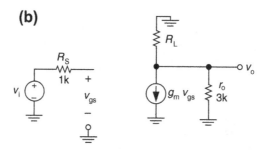

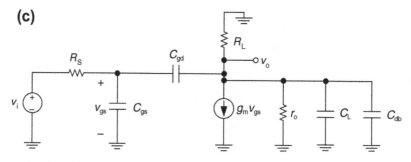

FIGURE 9.22

Try #1, MOS amplifier. (a) Amplifier, omitting biasing details for clarity. (b) Small-signal (low-frequency) model assuming no back-gate effect, or $g_{mb} = 0$. (c) High-frequency small-signal model.

small-signal model to find the minimum value of the load resistor R_L in order to meet the gain specification, as follows:

$$A_v = \frac{v_o}{v_i} = -g_m R_L \| r_o \Rightarrow R_L > 1500 \,\Omega \tag{9.26}$$

We will use $R_L = 1.6 \, \text{k}\Omega$ so that there is a little bit of extra gain and also to account for the fact that we have ignored the back-gate effect.

We will find the bandwidth using the high-frequency small-signal model of Figure 9.22(c). Note that the load (C_L) and drain–substrate (C_{db}) capacitances are in parallel. The source–substrate (C_{sb}) capacitance is not included since the source and substrate are both incrementally grounded.

The circuits for finding the open-circuit time constants are shown in Figure 9.23. For the gate–source capacitance, we use the circuit of Figure 9.23(a) to find:

$$R_{o1} = R_s = 1 \, \text{k}\Omega$$

$$\tau_{o1} = R_{o1} C_{gs} = (1000 \,\Omega)(0.2 \, \text{pF}) = 0.2 \, \text{ns} \tag{9.27}$$

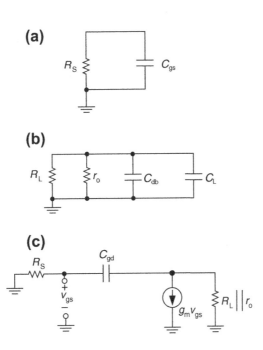

(a)

(b)

(c)

FIGURE 9.23

Circuits for finding open-circuit time constants for Try #1. (a) Circuit for C_{gs}. (b) Circuit for $C_{db} + C_L$. (c) Circuit for C_{gd}.

For the load and drain–substrate capacitances, we use the circuit of Figure 9.23(b) to find:

$$R_{o2} = R_L \| r_o = 1043 \ \Omega$$

$$\tau_{o2} = R_{o2}(C_L + C_{db}) = (1043 \ \Omega)(1.02 \ \text{pF}) = 1.06 \ \text{ns}$$

(9.28)

For the gate–drain capacitances, we use the circuit of Figure 9.23(c) to find:

$$R_{o3} = R_S + R_L \| r_o + g_m R_S(R_L \| r_o) = 12473 \ \Omega$$

$$\tau_{o3} = R_{o3}C_{gd} = (12473 \ \Omega)(0.05 \ \text{pF}) = 0.62 \ \text{ns}$$

(9.29)

The sum of open-circuit time constants and estimate of bandwidth are as follows:

$$\sum \tau_{oi} = \tau_{o1} + \tau_{o2} + \tau_{o3} = 1.88 \ \text{ns}$$

$$\omega_h \approx \frac{1}{\sum \tau_{oi}} \approx \frac{1}{1.98 \times 10^{-9}} = 532 \times 10^6 \ \text{rad/s}$$

(9.30)

$$f_h \approx \frac{\omega_h}{2\pi} \approx 84.7 \ \text{MHz}$$

We note that we will not meet the 250-MHz bandwidth specification yet, which is confirmed by an LTSPICE simulation (Figure 9.24) which shows a gain of -10.4 and a bandwidth of ~ 92 MHz. The method of open-circuit time constants identifies the output load capacitance as the dominant bandwidth limitation. In order to improve the bandwidth, we will isolate the load capacitance from the high-gain node by using a source follower.

TRY #2: Add output source follower M_2

In Try #2, we have added device M_2, which is a source-follower buffer (Figure 9.25) to isolate the load capacitor from the high-gain node. We will assume that the current source I_{BIAS2} biases M_2 at the same drain current as M_1 and hence $g_{m2} = g_{m1}$. We will also assume that all other MOS capacitances are the same for both devices. Next, we perform an open-circuit time constant analysis using the small-signal model of Figure 9.25(b).

For M_1, we note that the open-circuit time constants for C_{gs1} and C_{gd1} are unchanged from the previous design iteration.[6]

$$\tau_{o1} = 0.2 \ \text{ns}$$

$$\tau_{o2} = 0.62 \ \text{ns}$$

(9.31)

---

[6]Note that I am renumbering the "τ_s" with this iteration.

(a)

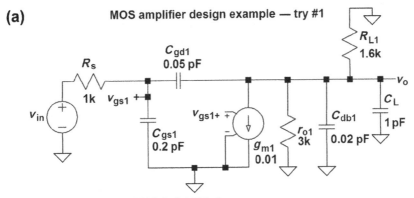

MOS amplifier design example — try #1

.ac dec 1000 1e6 1000e6
file: MOS amplifier design example try1.asc

(b)

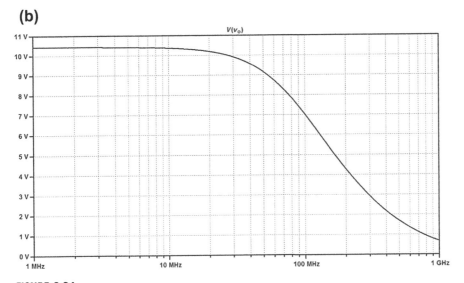

FIGURE 9.24

LTSPICE result showing gain and bandwidth of MOS amplifier Try #1. (a) Circuit which was simulated. (b) LTSPICE simulation result, showing a gain of -10.4 and a -3 dB bandwidth of approximately 92 MHz. (For color version of this figure, the reader is referred to the online version of this book.)

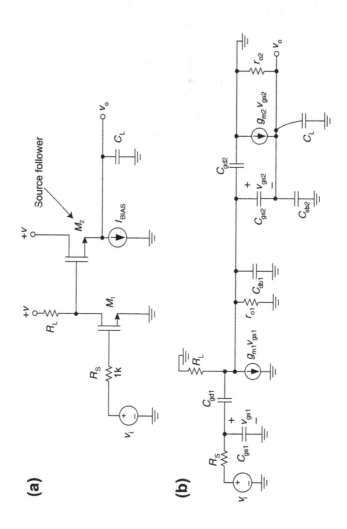

FIGURE 9.25

Common-source amplifier M_1 with source follower M_2 added (Try #2). (a) Circuit. (b) Small-signal model.

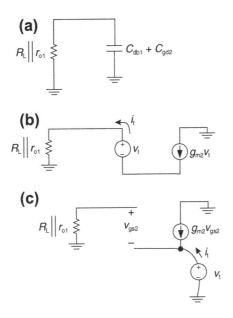

FIGURE 9.26

Open-circuit time constants circuits for Try #2. (a) Circuit for finding open-circuit time constants for $C_{db1} + C_{gd2}$. (b) Circuit for finding OCTCs for C_{gs2}. (c) Circuit for finding OCTCs for $C_L + C_{sb2}$.

In the previous iteration, the high-gain node was loaded by $(C_L + C_{db})$. In this iteration, we have C_{db1} in parallel with C_{gd2} at the high-gain node as shown in Figure 9.26(a). The open-circuit time constant for these capacitors are found as follows:

$$R_{o3} = R_L \| r_{o1} = 1043 \ \Omega$$

$$\tau_{o3} = R_{o3}\left(C_{db1} + C_{gd2}\right) = (1043)(0.07\text{pF}) = 0.07\text{ns} \tag{9.32}$$

For the gate–source capacitance of M_2, we use the test circuit of Figure 9.26(b), where we have added a test voltage source across the C_{gs2} terminals. By inspection,[7] we find that the test current $i_t = g_{m2}v_t$ and hence the open-circuit time constant for C_{gs2}, is as follows:

$$R_{o4} = \frac{1}{g_{m2}} = 100 \ \Omega$$

$$\tau_{o4} = R_{o4}C_{gs2} = (100)(0.2 \text{ pF}) = 0.02 \text{ ns} \tag{9.33}$$

---

[7]I have ignored r_{o2}, which is OK in this case.

Table 9.2 Summary of Open-Circuit Time Constant Results for Try#2

τ_{o1} (from C_{gs1})	0.20 ns
τ_{o2} (C_{gd1})	0.62 ns
τ_{o3} ($C_{db1} + C_{gd2}$)	0.07 ns
τ_{o4} (C_{gs2})	0.02 ns
τ_{o5} ($C_L + C_{sb2}$)	0.10 ns
Sum of OCTCs	1.01 ns
Estimate of ω_h	990 Mrad/s
Estimate of f_h	157.6 MHz

For the $C_L + C_{sb2}$ combination at the source terminal of M_2, we use the test circuit of Figure 9.26(c), where we have added a test voltage source across the source-follower output terminals. By inspection,[8] we find that the test current $i_t = g_{m2}v_t$, and hence the open-circuit time constant for $C_L + C_{sb2}$ is as follows:

$$R_{o5} = \frac{1}{g_{m2}} = 100\ \Omega$$

$$\tau_{o5} = R_{o5}(C_L + C_{sb2}) = (100)(1.02\ \text{pF}) = 0.1\ \text{ns} \tag{9.34}$$

A summary of the open-circuit time constants for Try #2 is shown in Table 9.2. We expect not to meet the bandwidth specification yet, since our open-circuit time constant estimate is only 157.6 MHz. LTSPICE (Figure 9.27) confirms this; the simulated bandwidth is approximately 185 MHz.

TRY #3: Add cascode transistor M_3

In the last open-circuit time constants calculation, we found that the dominant time constant is due to the Miller effect of device M_1. We will add a common-gate (cascode) transistor M_3 in order to try to reduce the Miller effect, resulting in the circuit of Figure 9.28(a).

The low-frequency small-signal model is shown in Figure 9.28(b). We note from this model that the gain of this amplifier configuration is now roughly:

$$A_v = \frac{v_o}{v_i} \approx -g_{m1}R_L \tag{9.35}$$

This assumes that the loading effect of r_{o3} is minimized at the high-gain node and that the gain of the M_3 source follower is close to unity. We can argue that r_{o3} does not significantly affect the gain by noting that the incremental output resistance seen at the drain of M_3 is increased due to the fact that the source of M_3 is loaded with a relatively high resistance (i.e. r_{o1}). Thus, the effects of loading due to the output

---

[8]Again, ignoring r_{o2}.

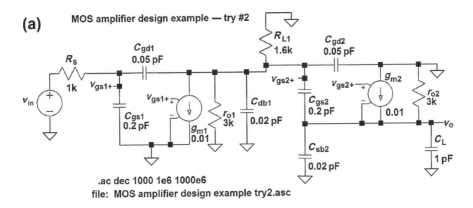

(a) MOS amplifier design example — try #2

.ac dec 1000 1e6 1000e6
file: MOS amplifier design example try2.asc

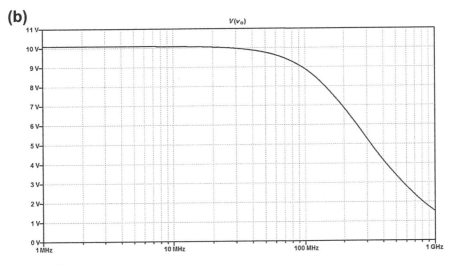

(b)

FIGURE 9.27

LTSPICE result showing gain and bandwidth of MOS amplifier Try #2. (a) Circuit. (b) LTSPICE simulation, showing Try #2 has a higher bandwidth ($\sim$185 MHz) and a slightly lower gain (-10.1). (For color version of this figure, the reader is referred to the online version of this book.)

resistance r_{o3} is minimized. We can then reduce the load resistor in this iteration to perhaps 1.1 kΩ (reduced from 1.6 kΩ) and still meet the gain specification. This will also help us meet the bandwidth specification.

Referring to the small-signal circuit of Figure 9.28(c), we need to do some book-keeping to keep track of all the various capacitances and open-circuit time constant calculations. There are 13 capacitances that contribute to open-circuit time constants: four for each transistor plus the load capacitance. However, we find that some capacitances are shorted out since all transistors have the bulk terminal

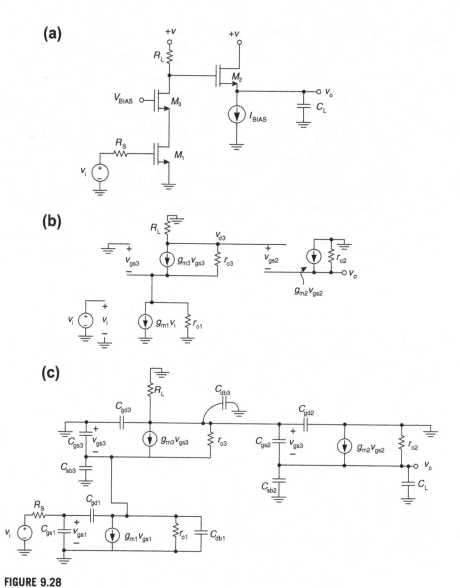

FIGURE 9.28

Try #3, with cascode transistor M_3 added. (a) Circuit. (b) Low-frequency small-signal model. (b) High-frequency small-signal model.

grounded (i.e. C_{sb1} and C_{db2}) and that there are many capacitances that are in parallel (i.e. $C_{db1} + C_{gs3} + C_{sb3}$, $C_{sb2} + C_L$, and $C_{gd2} + C_{gd3} + C_{db3}$). The open-circuit time constant calculations for the various capacitances are shown in Table 9.3.

The sum of the open-circuit time constants is now 0.58 ns, resulting in a bandwidth estimate of 1721 Mrad/s, or 274 MHz. This circuit looks like a good candidate

Table 9.3 Summary of OCTC Calculations for Try#3

Capacitance(s)	Open-Circuit Resistance	Open-Circuit Time Constant Calculation
$C_{gs1} = 0.2$ pF	Unchanged from previous iteration at 1 kΩ	τ_{o1} unchanged at 0.2 ns
$C_{gd1} = 0.05$ pF	Open-circuit resistance is greatly reduced due to low output resistance $\approx r_{out3} = 100\ \Omega$ at source of M_3. $R_{o2} = R_s + r_{out3} + g_{m1}R_s r_{out3} = 2100\ \Omega$	$\tau_{o2} = (2100\ \Omega)(0.05$ pF$) = 0.105$ ns
$C_{db1} + C_{sb3} + C_{gs3} = 0.24$ pF	Open-circuit resistance = $r_{out3} \approx 1/g_{m3} = 100\ \Omega$	$\tau_{o3} = (100\ \Omega)(0.24$ pF$) = 0.024$ ns
$C_{gs2} = 0.2$ pF	Open-circuit resistance unchanged at $1/g_{m2} = 100\ \Omega$	$\tau_{o4} = (100\ \Omega)(0.2$ pF$) = 0.02$ ns
$C_{gd2} + C_{db3} + C_{gd3} = 0.12$ pF	Resistance from this common node to ground $\approx R_L = 1100\ \Omega$	$\tau_{o5} = (1100\ \Omega)(0.12$ pF$) = 0.132$ ns
$C_{sb2} + C_L = 1.02$ pF	Open-circuit resistance = $1/g_{m2} = 100\ \Omega$	$\tau_{o5} = (100\ \Omega)(1.02$ pF$) = 0.10$ ns

for final simulation, since we know the method of open-circuit time constants is always conservative in estimating bandwidth. A detailed LTSPICE simulation shows that the bandwidth is 460 MHz (Figure 9.29), and hence we meet our 350-MHz bandwidth specification.

(a)

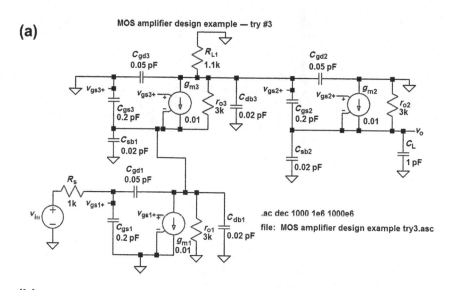

(b)

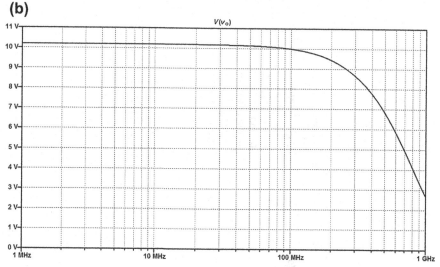

FIGURE 9.29

LTSPICE result showing gain and bandwidth of MOS amplifier Try #3. (a) Circuit, with input common-source amplifier (M_1), cascode transistor (M_3), and output source follower (M_2). (b) LTSPICE result, showing low-frequency gain of -10.2 and a -3 dB bandwidth of 460 MHz. (For color version of this figure, the reader is referred to the online version of this book.)

Example 9.2: MOS amplifier design example, even more bandwidth

If further bandwidth enhancement was needed, we could add an input source follower M_4 to isolate the source resistance from the input capacitances of M_1, as shown in Figure 9.30(a). This attacks the largest open-circuit time constant that is associated with C_{gs1}. Adding the source follower increases the bandwidth to over 800 MHz with some gain peaking, as shown in Figure 9.30(b). We also need to tweak the value of R_{L1} up to 1.15 kΩ to offset the fact that the gain of the source follower buffer M_4 is a little less than $+1$.

(a)

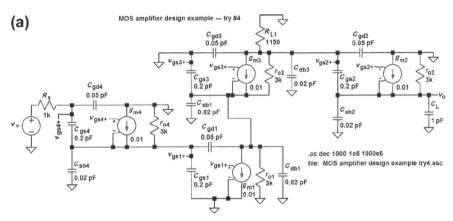

(b)

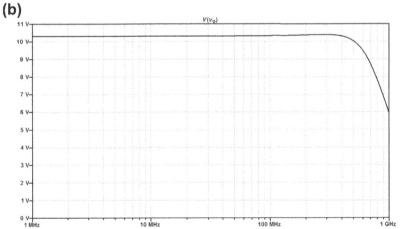

FIGURE 9.30

LTSPICE result showing gain and bandwidth of MOS amplifier Try #4 where an input source follower M_4 was added to isolate M_1's input capacitances from the 1-kΩ source resistance. (a) Circuit. Note that load resistor was slightly increased to 1150 Ω to meet the gain specification. (b) LTSPICE result, showing low-frequency gain of -10.3 and a -3 dB bandwidth of 855 MHz. (For color version of this figure, the reader is referred to the online version of this book.)

Example 9.3: MOS differential amplifier

A MOS differential amplifier is shown in Figure 9.31(a). The amplifier is biased with a 1 mA current source and has differential inputs of ± 1 mV at 1 kHz. We note that the bias current splits equally between M_1 and M_2, so each of the four transistors is

(a)

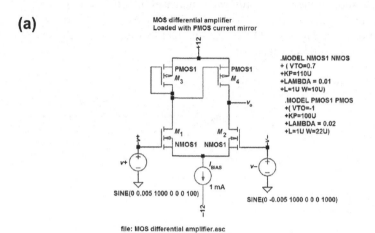

MOS differential amplifier
Loaded with PMOS current mirror

.MODEL NMOS1 NMOS
+ (VTO=0.7
+KP=110U
+LAMBDA = 0.01
+L=1U W=10U)

.MODEL PMOS1 PMOS
+(VTO=-1
+KP=100U
+LAMBDA = 0.02
+L=1U W=22U)

SINE(0 0.005 1000 0 0 0 100)

SINE(0 -0.005 1000 0 0 0 1000)

file: MOS differential amplifier.asc

(b)

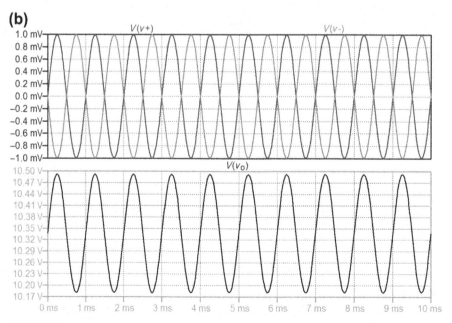

FIGURE 9.31

MOS differential amplifier with current mirror load. (a) Circuit. (b) LTSPICE result for a differential input of ± 1 mV at 1 kHz showing a gain of $+150$. (For color version of this figure, the reader is referred to the online version of this book.)

biased at a drain current of 0.5 mA. Small-signal parameters of interest for this amplifier are:

$$g_{m2} = \sqrt{2(\mu_n C_{ox})\left(\frac{W}{L}\right)I_{D2}} = \sqrt{2(110 \times 10^{-6})\left(\frac{10\mu}{1\mu}\right)(5 \times 10^{-4})}$$

$$= 1.04 \times 10^{-3} \text{ A/V} \tag{9.36}$$

$$r_{o4} = \frac{1}{\lambda_{PMOS}I_{D4}} = \frac{1}{(0.02)(5 \times 10^{-4})} = 200 \text{ k}\Omega$$

$$r_{o2} = \frac{1}{\lambda_{NMOS}I_{D2}} = \frac{1}{(0.01)(5 \times 10^{-4})} = 100 \text{ k}\Omega$$

The input transistors are loaded with PMOS current mirror M_3-M_4. The gain of the amplifier is:

$$A_v = \frac{v_o}{(v_+ - v_-)} \approx +2g_{m2}(r_{o2}\|r_{o4}) = (2)(1.04 \times 10^{-3})(66.7 \text{ k}\Omega) = +139 \tag{9.37}$$

The resultant output (Figure 9.31(b)) shows a DC level of $+10.35$ V and an output sine wave of 300 mV pp. The amplifier has a gain of about $+150$.

Example 9.4: MOS shunt-peaked amplifier

The small-signal model of a single-stage MOS common-source amplifier with shunt inductance peaking is shown in Figure 9.32(a). This amplifier uses the common-source amplifier from Example 9.1, try #1, as a baseline. We recall that the original amplifier had a gain of about 10 ($+20$ dB) and a bandwidth of about 92 MHz. We have added an inductance L_{pk} in series with the load resistor. We see that with a peaking inductor $L_{pk} = 2$ μH the bandwidth is extended to about 140 MHz with some slight amount of gain peaking. Inductors can be useful.

Basic JFETs

The JFET is a voltage-controlled device, where a gate—source voltage on a reverse-biased gate controls the current flow through a drain—source channel. The basic structure of an N-channel JFET is shown in Figure 9.33. There are ohmic contacts at the source (S), drain (D) and gate (G) terminals.

The main current flow path through the JFET is the drain—source channel. Remembering our basic device physics from Chapters 3 and 4, we note that if a negative voltage is applied to the gate, (relative to the source and drain) there is an increase in the depletion width near the P—N junction. If the depletion width gets wide enough, the current through the JFET is limited, or completely shut off.

(a) MOS peaked amplifier
This example uses the design example, try #1 as a baseline
Shows bandwidth improvement possible with inductance shunt peaking

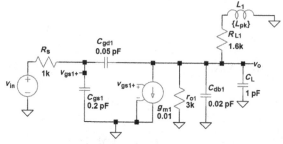

.step param Lpk list 0 1u 2u 3u
.ac dec 1000 10e6 300e6
file: MOS peaked amplifier

(b)

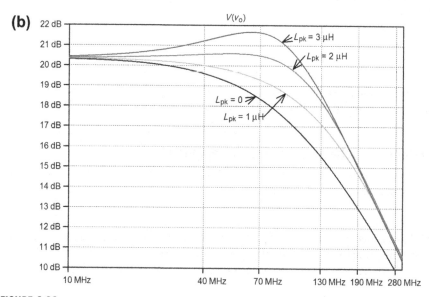

FIGURE 9.32

Shunt-peaked amplifier (using Example 9.1, Try #1 amplifier as a baseline design). (a) Circuit. (b) LTSPICE result showing frequency response magnitude, for inductance of $L_{pk} = 0$, 1, 2, and 3 μH. (For color version of this figure, the reader is referred to the online version of this book.)

We will qualitatively analyze operation of the N-channel JFET by considering Figure 9.34. In Figure 9.34(a), we have the source tied to the gate ($V_{GS} = 0$), and a small voltage V_{DS} is applied between drain and source. There is a depletion layer that extends into the N-channel cross-sectional area, but opposite sides of the depletion layer do not meet.

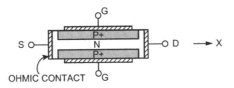

FIGURE 9.33

Basic N-channel JFET structure.

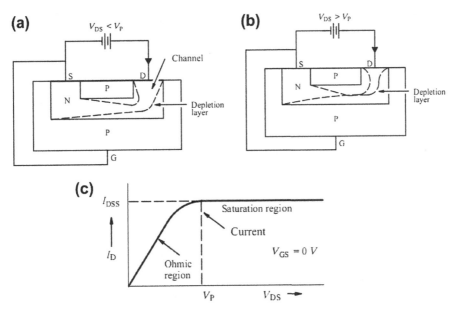

FIGURE 9.34

N-channel JFET, under bias. (a) JFET biased ON with source grounded, a small positive voltage ($< V_P$) applied to drain, gate grounded. (b) JFET biased on with large positive voltage on the drain, "pinching off" the channel resulting in constant current even if V_{DS} increases more. (c) Resultant V-I curve for N-channel JFET, with $V_{GS} = 0$.

Excerpt from application note courtesy of Vishay Intertechnology, Inc.

In Figure 9.34(b), we keep $V_{GS} = 0$, and increase the drain–source voltage so that it is high enough so that opposite sides of the depletion layer meet and the channel is "pinched off". In the pinchoff region (with $V_{DS} > V_P$), we cannot force any more current through the channel even if the drain–source voltage increases.

In Figure 9.34(c), we see qualitatively the V-I curve for this device in the special case when $V_{GS} = 0$. The low-voltage region is called the "ohmic region" where the drain–source channel looks resistive. In the "saturation region", the channel is

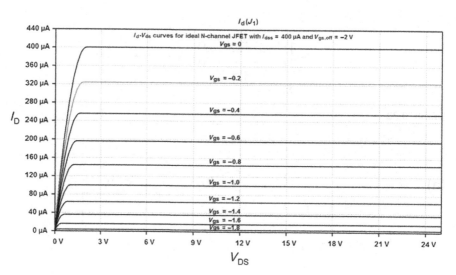

FIGURE 9.35

Ideal N-channel JFET curves with $I_{DSS} = 400\,\mu A$ and $V_{GS,OFF} = -2.0\,V$. (For color version of this figure, the reader is referred to the online version of this book.)

pinched off and the drain current is constant even if V_{DS} increases. The constant value of drain current with $V_{GS} = 0$ is called the "I_{DSS}" of the device, which is the drain saturation current at a gate–source voltage of zero.

We can control the drain current by changing the V_{GS} of the device. If we allow V_{GS} to go negative, we generate a set of curves (Figure 9.35) where we control the drain current with negative values of V_{GS}. (We need to make sure that V_{GS} is always less than or equal to zero volts in the N-channel JFET, so we never forward bias the gate P-N junction). At a negative voltage sufficiently negative ($V_{GS,OFF}$), the JFET is fully off and there is no drain current.

A detailed analysis for a JFET operating in the saturation region (where amplification can occur) shows that the drain current to gate–source voltage relationship is:

$$I_D = I_{DSS}\left(1 - \frac{V_{GS}}{V_{GS,OFF}}\right)^2 \qquad (9.38)$$

A set of curves for an ideal N-channel JFET with $I_{DSS} = 400\,\mu A$ and $V_{GS,OFF} = -2\,V$ is shown in Figure 9.35.

For low values of V_{DS}, the channel looks like a voltage-controlled resistance, under control of V_{GS}. A set of curves for the same ideal N-channel JFET with V_{DS} less than 1 V is shown in Figure 9.36.

A set of V-I curves for a 2N5484 N-channel JFET (as generated by LTSPICE) is shown in Figure 9.37. We note that there is a finite slope to the curves in the

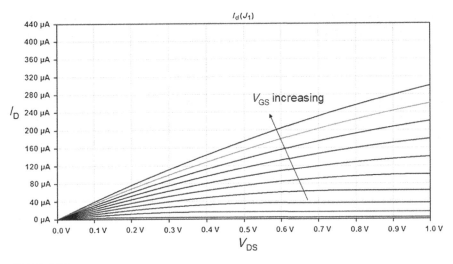

FIGURE 9.36

Ideal JFET curves in the ohmic region, for $V_{DS} < 1$ V, showing that the channel behaves like a voltage-controlled resistance under control of V_{GS}. (For color version of this figure, the reader is referred to the online version of this book.)

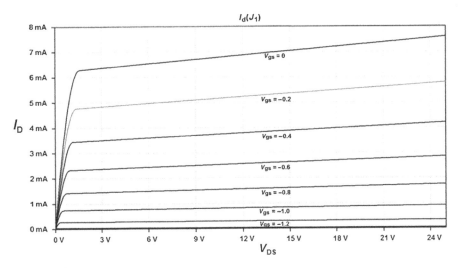

FIGURE 9.37

JFET curve for 2N5484 N-channel JFET, showing channel-length modulation. We also note that $I_{DSS} \sim 7$ mA and $V_{GS,OFF} \sim -1.2$ V. (For color version of this figure, the reader is referred to the online version of this book.)

saturation region of operation (due to an analogous channel-length modulation effect as in MOSFETs), which results in a finite incremental output resistance. We also note that JFET parameters are: $I_{DSS} \sim 7$ mA and $V_{GS,OFF} \sim -1.2$ V.

Next, we will investigate some common JFET amplifiers.

Example 9.5: AC-coupled JFET source follower

The JFET source follower, commonly used as a high-input-impedance buffer, is shown in Figure 9.38. We note that in the simple biasing scheme of Figure 9.38(a), the gate–source voltage is about -1 V. This circuit will provide buffering as long as the -1 V bias across V_{GS} is not as negative as the $V_{GS,OFF}$ of the JFET. We see in Figure 9.38(b) that the midband voltage gain of this buffer is about 0.85.

(a)

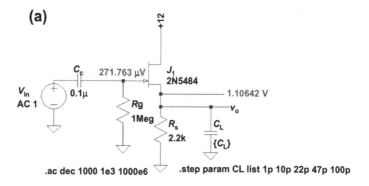

(b)

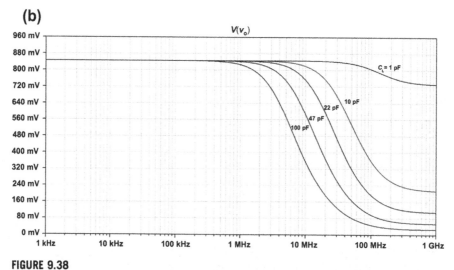

FIGURE 9.38

AC-coupled JFET source follower. (a) Circuit. (b) LTSPICE simulation for various capacitive loads showing frequency response and a midband gain of about 0.85. (For color version of this figure, the reader is referred to the online version of this book.)

Example 9.6: JFET common-source amplifier with source degeneration

The JFET common-source amplifier, with source degeneration provided by the resistor R_s, is shown in Figure 9.39. This amplifier provides modest gain and wide bandwidth.

(a)

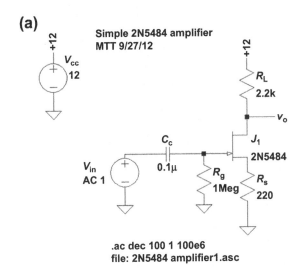

(b)

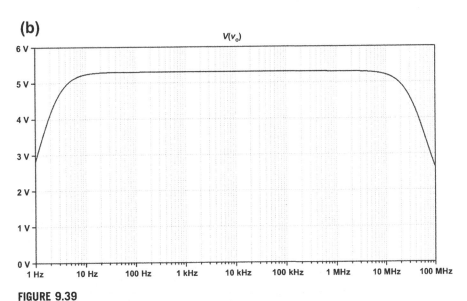

FIGURE 9.39

JFET common-source amplifier with source degeneration. (a) Circuit. (b) LTSPICE simulation showing frequency response and a midband gain of −5.3. (For color version of this figure, the reader is referred to the online version of this book.)

Example 9.7: JFET common-source amplifier

The JFET common-source amplifier, with source degeneration provided by the resistor R_s, is shown in Figure 9.40. This amplifier provides a gain of around -11.

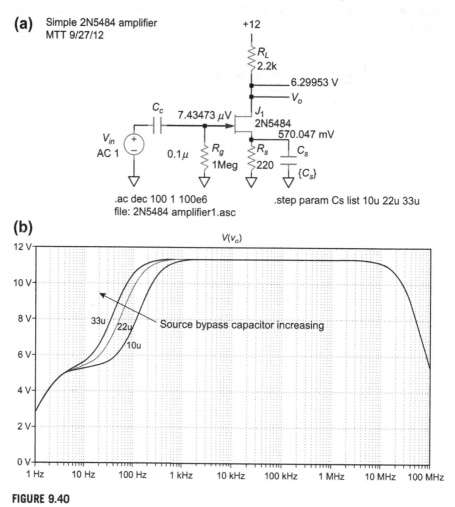

FIGURE 9.40

JFET common-source amplifier. (a) Circuit. (b) LTSPICE simulation showing frequency response with three different values of source bypassing capacitor. (For color version of this figure, the reader is referred to the online version of this book.)

Example 9.8: JFET shunt-peaked common-source amplifier

The JFET amplifier with shunt peaking is shown in Figure 9.41. We see significant improvement in the bandwidth when an inductor of $0.78\ \mu H$ is in series with the drain load resistor.

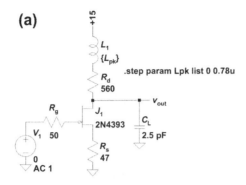

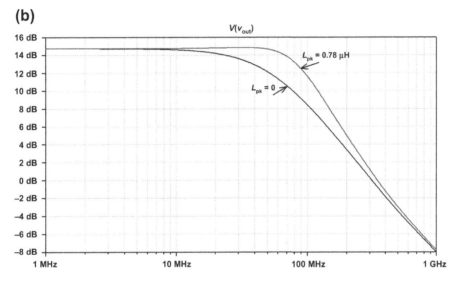

FIGURE 9.41

Shunt-peaked JFET-peaked amplifier. (a) Circuit. (b) LTSPICE simulation showing frequency response for $L_{pk} = 0$ and 0.78 µH. (For color version of this figure, the reader is referred to the online version of this book.)

Chapter 9 problems

Problem 9.1

A MOS current mirror is shown Figure 9.42. Assume that each transistor has $g_m = 0.01$ A/V, $C_{gs} = 1$ pF, and $C_{gd} = 0.5$ pF and that all other transistor parasitic elements are negligible.

a. Draw the small-signal incremental model of the circuit.

b. Estimate the bandwidth of the circuit using open-circuit time constants.

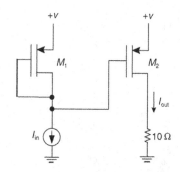

FIGURE 9.42

MOS current mirror for Problem 9.1.

Problem 9.2

A MOS common-source amplifier is shown in Figure 9.43. Assume that the transistor has $g_m = 0.02$ A/V, $C_{gs} = 1$ pF, and $C_{gd} = 0.5$ pF and transistor incremental output resistance $r_o = 5$ kΩ.

a. Draw the small-signal incremental model of the circuit, ignoring the back-gate effect.

b. Find the low-frequency gain.

c. Estimate the bandwidth of the circuit using open-circuit time constants.

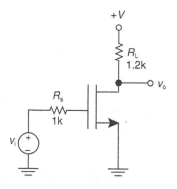

FIGURE 9.43

MOS common-source amplifier for Problem 9.2, omitting biasing details.

Problem 9.3

Using the four-transistor MOS amplifier (Figure 9.30(a)) from the CMOS amplifier design example in this chapter as a basis, suggest possible circuit modifications that would further improve bandwidth. Verify your design modifications using open-circuit time constants and LTSPICE simulations.

Further reading

[1] Arns RG. The other transistor: early history of the metal-oxide semiconductor field-effect transistor. *Eng Sci Educ J* 1998;**7**(5):233—40.

[2] Comer DJ, Comer DT. Teaching MOS integrated circuit amplifier design to undergraduates. *IEEE Trans Educ* 2001;**44**(3):232—8.

[3] Gray PR, Meyer RG. MOS operational amplifier design-a tutorial overview. *IEEE J Solid State Circuits* 1982;**17**(6):969—82.

[4] Grebene AB. *Bipolar and MOS analog integrated circuits*. John Wiley; 1984.

[5] Hastings A. *The art of analog layout*. Prentice Hall; 2001.

[6] Hodges DA, Gray PR, Brodersen RW. Potential of MOS technologies for analog integrated circuits. *IEEE J Solid State Circuits* 1978;**13**(3):285—94.

[7] Jiang R, Tang H, Mayaram K. A simple and accurate method for calculating the low frequency common-mode gain in a MOS differential amplifier with a current-mirror load. *IEEE Trans Educ* 2000;**43**(3):362—4.

[8] Johns DA, Martin K. *Analog integrated circuit design*. John Wiley; 1997.

[9] Lee TH. *The design of CMOS radio-frequency integrated circuits*. Cambridge University Press; 1998.

[10] Lilienfeld J. US Patents 1,745, 175; 1,877, 140 and 1,900,018. Available from the U.S. Patent and Trademark office, www.uspto.gov.

[11] Middlebrook RD, Richer I. Limits on the power-law exponent for field-effect transistor transfer characteristics. *Solid-State Electron* 1963;**6**:542—4.

[12] Pierret RF. *Modular series on solid state devices, field effect devices*, vol. 4. Addison-Wesley; 1983. Singh J. Semiconductor devices basic principles. John Wiley; 2001.

[13] Sze SM. *Semiconductor devices physics and technology*. 2nd ed. John Wiley; 2002.

[14] Winarski TY. Dielectrics in MOS devices, DRAM capacitors, and inter-metal isolation. *IEEE Electr Insul Mag* 2001;**17**(6):34—47.

Large-Signal Switching of Bipolar Transistors and MOSFETs

10

"There is more to life than increasing its speed."

—M. K. Gandhi

IN THIS CHAPTER

▶ In this chapter, we examine in detail large-signal switching of bipolar transistors and metal oxide–semiconductor field-effect transistors (MOSFETs). For bipolar junction transistors (BJTs), a model used to estimate the switching speed, called the "charge control model", is introduced. For MOSFETs, we focus on gate charge test data in order to estimate switching speed.

Introduction

In the previous chapters, we discussed the dynamic behavior of transistors from a small-signal point of view using the hybrid-pi model for bipolar transistors and incremental models for MOSFETs. In other words, the small-signal models developed so far are valid only for small variations of the voltages and currents around an operating point. We used the method of open-circuit time constants to analyze bandwidth limitations in linear amplifiers.

For bipolar transistors, a further modeling technique, called the *charge control model*, permits us to analyze the behavior of transistor circuits during large-signal switching and during transistor saturation. Similar to the method of open-circuit time constants, the charge control model gives results that are not necessarily numerically accurate; rather, the usefulness of the method is the design insight that we derive from the application of relatively simple models.

An analogous method for determining the large-signal switching behavior of MOSFETs, focusing on the total gate charge necessary to turn the MOSFET ON and OFF, is developed as well.

Development of the large-signal switching model for BJTs

In the following development, we consider the charge control model for the NPN transistor. The results are exactly the same for a PNP transistor. Transistor terminal current definitions are given in Figure 10.1.

Intuitive Analog Circuit Design. http://dx.doi.org/10.1016/B978-0-12-405866-8.00010-3

341

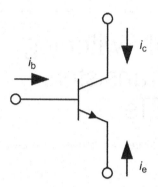

FIGURE 10.1

Terminal definitions for NPN transistor, showing base current i_b, collector current i_c, and emitter current i_e.

From device physics, we derived in an earlier chapter the fact that when a transistor is biased in the forward-active region, the base-to-emitter junction is forward biased and there is extra charge stored in the base region. For an NPN transistor, the *excess minority carrier concentration* is shown in Figure 10.2. In the case of the NPN transistor, the base is made of P material and the excess minority carriers are electrons and have the profile $n(x)$ shown. There is a similar hole profile in the emitter $p(x)$, but for purposes of analyzing switching operation, we need to only consider the base region.

Since transistor collector current is dominated by diffusion in the base; collector current I_c is proportional to the slope of the $n'(x)$ curve. The dotted line shows the carrier concentration profile at a higher collector current. Since the vertical axis has units of charge concentration, the total area under the curve is proportional to the charge stored in the base of the transistor. When the transistor is operated in the normal manner and in the linear region (called the forward-active region), we will call this stored charge q_F, for *forward charge*.

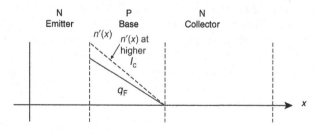

FIGURE 10.2

Excess carrier concentration for NPN transistor in forward-active region. The extra charge in the base ($n'(x)$) causes a net diffusion current of electrons to the right, resulting in a collector current to the left. The total charge in the base when the transistor is forward biased is q_F.

We can summarize this relationship between the collector current and the charge stored in the base as:

$$i_c = \frac{q_F}{\tau_F} \tag{10.1}$$

where τ_F is a device-dependent constant that has units of time. This is the *charge control equation* for the collector current in the forward-active region. We assume that as the base–emitter voltage changes, the concentration $n'(x=0)$ changes instantaneously in response.[1]

Now, we need to consider the base current. Since there is charge stored in the base, we expect that the base current charge control equation has a term related to how fast we add and remove charge q_F from the base. Furthermore, we know that there has to be a DC term (because collector and base current bias levels are related by the large-signal β_F). Using this reasoning, the base current charge control equation is:

$$i_b = \frac{q_F}{\tau_{BF}} + \frac{dq_F}{dt} \tag{10.2}$$

The q_F/τ_{BF} term is the DC base current, and the dq_F/dt term is the charge needed if we change the charge stored in the base. The characteristic time τ_{BF} is related to τ_F in a way that we will explain later.

We also recognize that the emitter current is the negative sum of the collector and base currents, or:

$$i_e = -(i_c + i_b) \tag{10.3}$$

Summarizing, the charge control equations for the forward-active region are:

$$i_b = \frac{q_F}{\tau_{BF}} + \frac{dq_F}{dt}$$
$$i_c = \frac{q_F}{\tau_F} \tag{10.4}$$
$$i_e = -(i_c + i_b)$$

The charge stored in the base (q_F) is the large-signal version of the base diffusion capacitance discussed in Chapter 4. Note that these equations do not account for charge stored in the base–emitter and base–collector depletion capacitances. We will fix this later.

BJT reverse-active region

In the forward-active region, the base–emitter junction is forward biased and the collector–base junction is reverse biased. We could just as easily operate the transistor with the collector–base junction forward biased and the base–emitter junction reverse biased. This mode of operation is called the *reverse-active region*, and analysis is exactly analogous with the forward-active region, with the roles of

---

[1]This is approximately true as long as we do not allow the transistor v_{be} to change too fast.

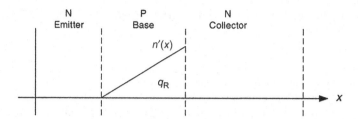

FIGURE 10.3

Excess carrier concentration for NPN transistor in the reverse-active region. Note that in the reverse-active region, the collector–base junction is forward biased and the base–emitter junction is reverse biased. The total charge in the base in this case is q_R'.

the collector and the emitter reversed.[2] The carrier profile is as shown in Figure 10.3. The charge control equations in the reverse-active region are:

$$i_b = \frac{q_R}{\tau_{BR}} + \frac{dq_R}{dt}$$

$$i_e = \frac{q_R}{\tau_R} \tag{10.5}$$

$$i_c = -(i_e + i_b)$$

   Transistors are optimized to run in the forward-active region. When operated in the reverse-active region, as might be expected, transistor parameters such as speed and β_F are degraded.

BJT saturation

Saturation is the case when both base–emitter and base–collector junctions are forward biased. We can consider this mode of operation as a mixture of the forward-active and reverse-active operations, as shown in Figure 10.4. In this case, the total charge in the base is $q_F + q_R$.

   The total charge control equations including forward-active and reverse-active operations are found by summing the previous results:

$$i_b = \frac{q_F}{\tau_{BF}} + \frac{dq_F}{dt} + \frac{q_R}{\tau_{BR}} + \frac{dq_R}{dt}$$

$$i_c = \frac{q_F}{\tau_F} - \frac{dq_R}{dt} - q_R \left[\frac{1}{\tau_R} + \frac{1}{\tau_{BR}} \right] \tag{10.6}$$

$$i_e = -(i_c + i_b)$$

   Since we have shown that the forward-active region is dominated by one time constant (τ_F) and the reverse-active region is dominated by another time constant

---

[2]The charge stored in the base in the reverse-active region is q_R.

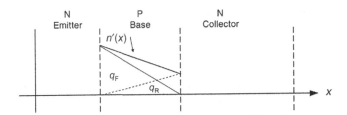

FIGURE 10.4

Excess carrier concentration for NPN transistor in saturation region. In saturation, both junctions are forward biased.

(τ_R), it makes sense that the saturation region will have two time constants, but not necessarily τ_F and τ_R. In fact, we will show that the dynamics in saturation has time constants that are some weighted functions of these individual time constants.

We can solve for the natural frequencies of the growth of charge in the transistor by solving the homogeneous case (and making the Laplace substitution d/dt $\rightarrow$ s):

$$0 = q_F\left(s + \frac{1}{\tau_{BF}}\right) + q_R\left(s + \frac{1}{\tau_{BR}}\right)$$

$$0 = \frac{q_F}{\tau_F} - q_R\left(s + \frac{1}{\tau_R} + \frac{1}{\tau_{BR}}\right)$$

(10.7)

We can solve these simultaneous equations, resulting in:

$$s^2 + \left(\frac{1}{\tau_F} + \frac{1}{\tau_{BF}} + \frac{1}{\tau_R} + \frac{1}{\tau_{BR}}\right)s + \left(\frac{1}{\tau_F\tau_{BR}} + \frac{1}{\tau_R\tau_{BF}} + \frac{1}{\tau_{BR}\tau_{BF}}\right) = 0 \quad (10.8)$$

or, simplifying:[3]

$$s^2 + \left(\frac{\beta_F + 1}{\tau_{BF}} + \frac{\beta_R + 1}{\tau_{BR}}\right)s + \left(\frac{1}{\tau_F\tau_{BR}} + \frac{1}{\tau_R\tau_{BF}} + \frac{1}{\tau_{BR}\tau_{BF}}\right) = 0 \quad (10.9)$$

In general, there are two widely spaced poles,[4] with the "fast" pole being comparable to the ω_T of the transistor:

$$s_{\text{fast}} \approx -\left(\frac{\beta_F + 1}{\tau_{BF}} + \frac{\beta_R + 1}{\tau_{BR}}\right) \approx -\left(\frac{1}{\tau_F} + \frac{1}{\tau_R}\right) \approx -\omega_T \quad (10.10)$$

---

[3]We have not worked it out yet, but we use here the identities $\tau_{BF} = \beta_F\tau_F$ and $\tau_{BR} = \beta_R\tau_R$ where β_F is forward beta and β_R is reverse beta. More on this identify later on.

[4]Remember our widely spaced pole approximation from Chapter 2. Let us assume that you have a 2-pole transfer function of the form: $H(s) = \frac{1}{s^2 + As + B}$

If the poles are on the real axis and are widely spaced, we can approximate the pole locations by:

$s_{\text{fast}} \approx -A$

$s_{\text{slow}} \approx -B/A$

Let us try this method on a transfer function $H(s) = \frac{1}{s^2 + 11s + 10} = \frac{1}{(s+1)(s+10)}$. There is a fast pole at -10 rad/s and a slow dominant pole at -1 rad/s. If we apply the widely spaced pole approximation, we get $s_{\text{fast}} \approx -11$ and $s_{\text{slow}} \approx -10/11$.

The low-frequency pole is given by:

$$s_{\text{slow}} \approx -\left(\frac{\beta_F + \beta_R + 1}{\tau_{BF}(\beta_R + 1) + \tau_{BR}(\beta_F + 1)} \right) \qquad (10.11)$$

The fast time constant, sometimes called the "slosh" mode, corresponds to the time scale with which charge redistributes between q_F and q_R. The slow time constant, or the "fill" mode, corresponds to how fast q_F and q_R rise together. This is analogous to even and odd modes in oscillating mass–spring systems. The slosh mode dies out with a very fast time constant, and therefore the charge growth in the saturation region is dominated by the slow time constant, as:

$$\tau_s = \frac{\tau_{BF}(\beta_R + 1) + \tau_{BR}(\beta_F + 1)}{\beta_F + \beta_R + 1} \qquad (10.12)$$

Hence, we can approximate the charge control equation in the saturation region as:

$$i_b - \frac{I_{C,\text{SAT}}}{\beta_F} = \frac{q_S}{\tau_S} + \frac{dq_S}{dt} \qquad (10.13)$$

The term τ_S is the saturation time constant and $I_{C,\text{SAT}}$ is the collector current in saturation. The term q_S is the saturation charge.

BJT base–emitter and base–collector depletion capacitances

In addition to the stored charge in the base region, there is charge stored in the nonlinear base–emitter and base–collector depletion capacitances. We will term these stored charges q_{ve} and q_{vc}, and this corresponds to the large-signal version of the base–emitter depletion capacitance C_{je} and base–collector depletion capacitance C_{jc}. The extra terms that we need to add to the previous charge control equations are:

$$i_{b,\text{SCL}} = \frac{d}{dt}(q_{ve} + q_{vc})$$
$$\qquad (10.14)$$
$$i_{c,\text{SCL}} = -\frac{d}{dt}q_{vc}$$

The notation "SCL" refers to the fact that these current components of base and collector currents charge and discharge the *space charge layers*, also called the *depletion* layers. We will see how to find these depletion capacitances on a datasheet later on.

Relationship between the charge control and the hybrid—pi parameters in bipolar transistors

We will next work out an important relationship between charge control parameters and hybrid-pi parameters. In Figure 10.5(a) we see a circuit that will help us work out this relationship. The small-signal hybrid-pi model for this circuit is shown in Figure 10.5(b).

If we run the transistor at very high frequencies, the hybrid-pi model finds the base current:

$$i_b|_{\omega \to \infty} \approx (C_\pi + C_\mu) \frac{dv_{be}}{dt} \tag{10.15}$$

For very high frequencies, the charge control model predicts (ignoring space charge capacitances):

$$i_b \approx \frac{dq_F}{dt} \approx \frac{d}{dt} (i_c \tau_F) \approx g_m \tau_F \frac{dv_{be}}{dt} \tag{10.16}$$

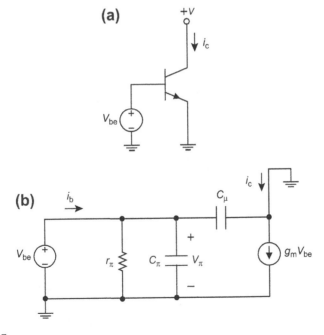

FIGURE 10.5

Circuit for finding relationship between charge control and hybrid-pi parameters.
(a) Transistor circuit. (b) High-frequency hybrid-pi model.

Equating the two equations, we find that $C_\pi + C_\mu \approx g_m \tau_F$, and the result is:

$$\tau_F \approx \frac{(C_\pi + C_\mu)}{g_m} \approx \frac{1}{\omega_T} \tag{10.17}$$

Next, by looking at the charge control equations at very low frequencies, we find:

$$i_b|_{DC} = \frac{q_F}{\tau_{BF}}$$

$$i_c = \frac{q_F}{\tau_F} \tag{10.18}$$

$$\frac{i_c}{i_b|_{DC}} = \frac{\tau_{BF}}{\tau_F} \rightarrow \tau_{BF} = \beta_F \tau_F$$

We now have a relationship between the charge control parameters τ_F and τ_{BF} and the hybrid-pi parameters β_F and ω_T.

Finding depletion capacitances from the datasheet

How do we find the values of depletion capacitances? Well, we are fortunate if manufacturers' data is available. For instance, for the 2N3904 transistor, capacitance data is available for reverse bias voltages in excess of 0.1 V (i.e. junction voltages less than −0.1 V) on both the emitter–base and collector–base junctions, as shown in Figure 10.6.

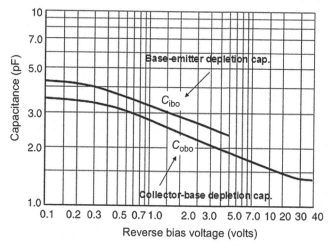

FIGURE 10.6

Manufacturer's datasheet[5] tabulation of depletion capacitance of 2N3904 transistor.

Reprinted with permission from On Semiconductor.

[5]From On Semiconductor, http://www.onsemi.com, reprinted with permission of On Semiconductor.

What do we do if we need charge data for *positive* voltages on the base–emitter junction? An NPN transistor is cutoff for V_{BE} less than 0.4 V or so. The datasheet does not give us capacitance values for positive V_{BE}. Fortunately, there is a little bit of physics that is available to us. The depletion capacitance of a PN junction is given by:

$$C_j = \frac{C_{jo}}{\left[1 - \frac{V_j}{\phi_{BI}}\right]^m} \tag{10.19}$$

C_{jo} is the depletion capacitance at a junction voltage $V_j = 0$; ϕ_{BI} is the *built-in voltage* characteristic of the junction and is given by:

$$\phi_{BI} = \frac{kT}{q} \ln \left[\frac{N_A N_D}{n_i^2}\right] \tag{10.20}$$

where N_A and N_D are the doping levels on the P and N sides and n_i is the intrinsic minority carrier concentration. A typical value for ϕ_{BI} is 0.8 V at normal doping levels. The factor m depends on the type of junction and is 0.5 for an abrupt junction and 0.333 for a linearly graded junction.

A curve fit for the emitter–base depletion capacitance for the 2N3904 transistor is given in Figure 10.7, with $C_{jo} = 4.5$ pF, $\phi_{BI} = 0.8$ V, and $m = 0.333$. Note that the depletion capacitance blows up when the junction voltage approaches the built-in

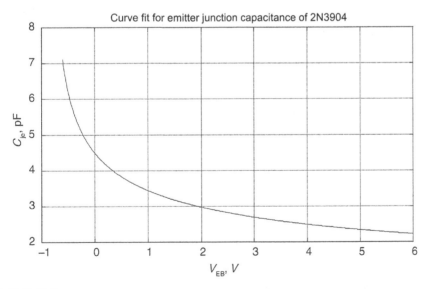

FIGURE 10.7

Curve fit for emitter–base depletion capacitance C_{je} for 2N3904 transistor with $C_{jo} = 4.5$ pF, $\phi_{BI} = 0.8$ V, and $m = 0.333$. Note that in this plot, reverse bias is to the left.

voltage. This curve fit will be sufficient for crude estimates of switching time using the charge control model.

Manufacturers' testing of BJTs

Manufacturers sometimes (but not always) put switching time test results on their datasheets. Shown in Figure 10.8 are test circuits and test results for the 2N3904 NPN transistor.[6] From these test results we can approximate various charge control parameters.

2N3903, 2N3904

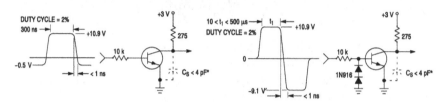

* Total shunt capacitance of test jig and connectors

FIGURE 10.8

Manufacturer's test circuits for the 2N3904 transistor.

Reprinted with permission from On Semiconductor.

Referring to Figure 10.9, the delay time listed ($t_d = 35$ ns) is the delay before the transistor turns on. The risetime (t_r) is the 10–90% risetime of the collector current. Storage time[7] (t_s) is the time it takes to remove saturation charge before coming out of saturation. The falltime t_f is the time it takes collector current to fall from 90% of full value to 10%.

SWITCHING CHARACTERISTICS

Delay Time	(V_{CC} = 3.0 Vdc, V_{BE} = 0.5 Vdc, I_C = 10 mAdc, I_{B1} = 1.0 mAdc)		t_d	–	35	ns
Rise Time			t_r	–	35	ns
Storage Time	(V_{CC} = 3.0 Vdc, I_C = 10 mAdc, I_{B1} = I_{B2} = 1.0 mAdc)	2N3903 2N3904	t_s	– –	175 200	ns
Fall Time			t_f	–	50	ns

FIGURE 10.9

Manufacturer's switching time results found on 2N3904 transistor datasheet.

Reprinted with permission from On Semiconductor.

---

[6]Taken from On Semiconductor 2N3904 datasheet, found at http://www.onsemi.com, reprinted with permission of On Semiconductor.
[7]Note that this storage delay time t_s is *not* the same as the saturation region time constant τ_S. However, We can use the value of t_s supplied by the manufacturer to estimate τ_S.

Charge control model examples

Next, we will see how to use the charge control model by working through a number of examples.

Example 10.1: Transistor inverter with base current drive

In this problem, a transistor is controlled by driving a pulsed base current as in Figure 10.10. Let us analyze the dynamics of the transistor switching, assuming transistor parameters $\tau_F = 0.3$ ns, $\beta_F = 416$, $\tau_R = 240$ ns, and $\beta_R = 0.7$. We will find the switching profile of this transistor circuit under these conditions. Assume that the current pulse transitions high at $t = 0$. Much later, after all transients have died down, the current pulse transitions back to zero.

Crossing the forward-active region

Ignoring space charge capacitances, we will find base charge $q_F(t)$ and collector current $i_c(t)$ when crossing through the forward-active region. Since we are ignoring depletion capacitances, the necessary charge control equation is relatively simple. First, recognize that when we first step the base current, the transistor will enter the forward-active region. There is no turn-on delay time in this example since

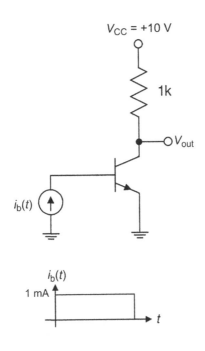

FIGURE 10.10

Transistor with a long base current drive pulse of amplitude 1 mA.

we have assumed that depletion capacitances are negligible. After a while, the transistor will saturate, since there is sufficient voltage drop across the resistor, or:

$$\beta_F I_B R_L > V_{CC} \tag{10.21}$$

When the transistor is in the forward-active region, the charge control equations are:

$$i_b = \frac{q_F}{\tau_{BF}} + \frac{dq_F}{dt}$$

$$i_c \approx \frac{q_F}{\tau_F} \tag{10.22}$$

Since we are driving the base with a stepped current source, the differential equation for the base current is:

$$\frac{dq_F}{dt} + \frac{q_F}{\tau_{BF}} = I_B u_{-1}(t) \tag{10.23}$$

where $\tau_{BF} = \beta_F \tau_F = 125$ ns and $u_{-1}(t)$ is the unit step. This charge control equation has the solution:

$$q_F(t) = I_B \tau_{BF} \left(1 - e^{\frac{-t}{\tau_{BF}}} \right)$$

$$i_c(t) \approx \frac{q_F(t)}{\tau_F} \approx \beta_F I_B \left(1 - e^{\frac{-t}{\tau_{BF}}} \right) = (416 \text{ mA}) \left(1 - e^{\frac{-t}{125 \text{ ns}}} \right) \tag{10.24}$$

The associated waveforms for forward charge $q_F(t)$ and collector current $i_c(t)$ are shown in Figure 10.11.

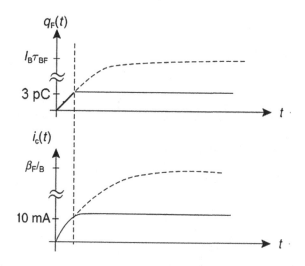

FIGURE 10.11

Growth of forward charge $q_F(t)$ and collector current $i_c(t)$ when crossing the forward-active region in Example 10.1. The final value of collector current is 10 mA, after which the transistor enters saturation and the forward charge q_F and collector current remain constant.

Note that the collector current rises with characteristic time constant $\tau_{BF} = 125$ ns. The final value of collector current is approximately 10 mA (at which point the transistor saturates). This final value of collector current is reached with a risetime $t_r \approx 3$ ns, and at this time the forward charge $q_F = i_c \tau_F = 3$ pC. Of course, the actual time to cross the forward-active region will be somewhat higher due to the base current needed to charge the base–emitter and base–collector depletion capacitances.

Going into saturation

Since $\beta_F I_B > I_{C,SAT}$, the transistor does not remain forward active for all time. Let us find the point at which saturation occurs. Then we will find the final values (in saturation) of the saturation charge[8] q_S and evaluate the time constant τ_S. The charge control equations for base and collector current in saturation are given by:[9]

$$i_b - \frac{I_{C,SAT}}{\beta_F} = \frac{q_S}{\tau_S} + \frac{dq_S}{dt'}$$ (10.25)

$$i_c = I_{C,SAT}$$

The time constant in the saturation region is:

$$\tau_S = \frac{\tau_{BF}(\beta_R + 1) + \tau_{BR}(\beta_F + 1)}{\beta_F + \beta_R + 1}$$

$$= \frac{(125 \text{ ns})(0.7 + 1) + (168 \text{ ns})(416 + 1)}{416 + 0.7 + 1} = 168 \text{ ns}$$ (10.26)

The final value of saturation charge $q_{S\text{final}}$ is found by:

$$q_{S, \text{final}} = \left(I_B - \frac{I_{C,SAT}}{\beta_F}\right)\tau_S = \left((1 \text{ mA}) - \frac{(10 \text{ mA})}{416}\right)(168 \text{ ns}) = 163 \text{ pC}$$ (10.27)

This saturation charge (163 pC) is the charge that needs to be removed from the base before the transistor will leave saturation, when we try to turn it off. The growth of saturation charge $q_S(t')$ is shown in Figure 10.12.

Leaving saturation

Now, after waiting a long time, the base drive is turned off, i.e. $i_B(t) = 0$. Let us figure out how long the transistor remains saturated, with $i_C = I_{C,SAT}$. Since the transistor is saturated, but the base drive value is zero, the charge control equation is now:[10]

$$-\frac{I_{C,SAT}}{\beta_F} = \frac{q_S}{\tau_S} + \frac{dq_S}{dt''}$$ (10.28)

---

[8]We will assume that the base current drive remains on long enough to fully saturate the transistor and to fully fill up the final value of saturation charge.

[9]Note that we have changed the time scale; we assume now that we enter saturation at $t' = 0$. We will change the timescale throughout the problem as we progress.

[10]We have again changed the time scale.

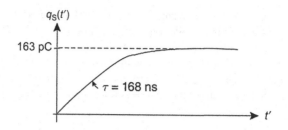

FIGURE 10.12

Growth of saturation charge $q_S(t')$ while going into saturation. During this interval the collector current is constant at 10 mA. (Note the change of time scale).

Note that recombination in the base causes the saturation charge to decrease. This equation needs to be solved with initial value of $q_S(t'')$ being 163 pC. We need to solve this equation and find out how long it takes $q_S(t'')$ to drop to zero.[11] The solution of this equation is:

$$q_S(t'') = 163 \text{ pC} - (167 \text{ pC})\left(1 - e^{\frac{-t''}{\tau_S}}\right) \tag{10.29}$$

This equation shows that $q_S(t'')$ drops to zero at $t'' = 627$ ns, which is when the transistor leaves saturation. Therefore, our saturation delay time is $t_s = 627$ ns (Figure 10.13).

Crossing through the forward-active region again while turning off

The transistor enters the forward-active region again after saturation charge q_S decays to zero. We will next find $q_F(t''')$ and the collector current. We expect the turn-off time of the collector current to be significantly slower than the turn-on

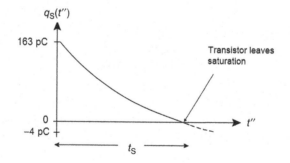

FIGURE 10.13

Decay of saturation charge $q_S(t'')$ while coming out of saturation showing saturation storage delay time t_s. During this delay interval, the collector current is constant at 10 mA.

---

[11]If we let the charge control equation run forever, the equation says that the final value of q_S will be −4 pC. However, we know that the transistor leaves saturation when the saturation charge drops to zero. We will find the time when q_S drops to zero.

time, since we are not actively pulling any base current out of the base. We have to rely on recombination in the base to decrease $q_F(t''')$ and hence collector current. The charge control equation in this case is:

$$\frac{dq_F}{dt'''} + \frac{q_F}{\tau_{BF}} = 0 \tag{10.30}$$

Note that the right-hand side of Eqn (10.30) is zero, since there is no base current. At the beginning of the forward-active region, the collector current is 10 mA and the forward charge $q_F = I_c\tau_F = (10\text{ mA})(0.3\text{ ns}) = 3\text{ pC}$. The solution to this charge control equation is:

$$q_F(t) = q_{Fo}e^{\frac{-t'''}{\tau_{BF}}} = (3\text{ pC})e^{\frac{-t'''}{125\text{ ns}}}$$

$$i_c(t) = \frac{q_F(t''')}{\tau_F} = (10\text{ mA})e^{\frac{-t''}{125\text{ ns}}} \tag{10.31}$$

The decay of forward charge when the transistor turns off is shown in Figure 10.14.

Therefore, once the transistor leaves saturation, the collector current will decay with characteristic time constant of 125 ns; the current will decay to zero in a few time constants (or around 375 ns).

The LTSPICE simulation (Figure 10.15) shows a saturation turn-off delay time of approximately 625 ns and a slowly decaying collector current. This LTSPICE simulation is reasonably consistent with what we calculated using the charge control model. We expect this to be so, since we used a SPICE model with the exact

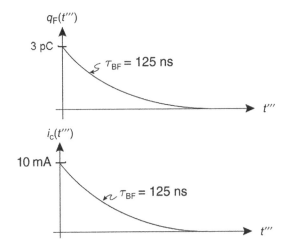

FIGURE 10.14

Decay of forward charge $q_F(t''')$ and resultant collector current $i_c(t''')$ when crossing the forward-active region and turning the transistor off. The collector current decays to zero with characteristic time constant τ_{BF}.

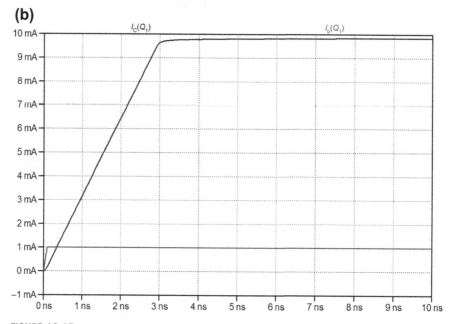

FIGURE 10.15

Simulation of base current-driven transistor switch of Example 10.1. (a) LTSPICE circuit, where we have created a transistor model using the parameters given in this problem. (b) Startup detail for the first 10 ns.

parameters used in the problem. Note that if you built this circuit, the times would be a little slower, due to the depletion capacitances, which we ignored in this analysis. A more detailed look at the LTSPICE simulation results is presented in Table 10.1.

Example 10.2: Effect of depletion capacitances on switching speed

Example 10.1 is a good academic exercise to get us started using the charge control model, but in some of the details, the model we used does not reflect the reality

(C)

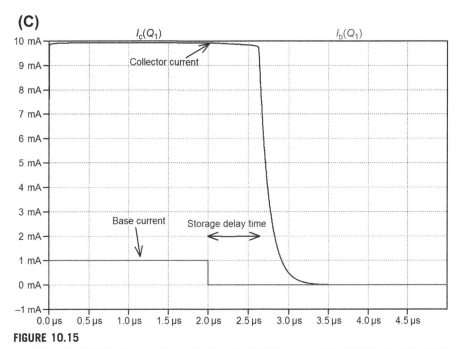

FIGURE 10.15

(*Continued*). (c) Detail of the storage delay time and collector current falltime. In this simulation, the current pulse is turned on at $t = 0$ and turned off at $t = 2$ µs. The storage delay time is about 625 ns. The falltime of the collector current is about 350 ns. (For color version of this figure, the reader is referred to the online version of this book.)

Table 10.1 Comparison of Charge Control Model with LTSPICE Model Results for Example 10.1

	Charge Control Model	LTSPICE Simulation
Risetime of collector current (crossing forward-active region)	3 ns	3 ns
Storage delay time	627 ns	625 ns
Falltime of collector current (crossing forward-active region)	375 ns	350 ns

of real-world transistors. Specifically, we ignored the transistor base–emitter and base–collector depletion capacitances. We will now put in some depletion capacitances and see how the results are affected. We expect things to slow down a bit.

Table 10.2 Charge Control Model of Depletion Capacitance and Corresponding SPICE Transistor Model Parameters

	Charge Control Model	SPICE Model Parameters
Depletion capacitance at zero bias	C_{jo}	CJE for base–emitter junction; CJC for collector–base junction
Grading-dependent exponent	m	MJE for base–emitter junction; MJC for collector–base junction
Junction built-in voltage	ϕ_{bi}	VJE for base–emitter junction; VJC for collector–base junction

Remember from Chapters 3 and 4 that depletion capacitances are junction voltage dependent, as:

$$C_j = \frac{C_{jo}}{\left(1 - \frac{V_j}{\phi_{bi}}\right)^m} \tag{10.32}$$

where $m - 1/2$ for an abrupt junction and $m = 1/3$ for a linearly graded junction, C_{jo} is the capacitance at zero junction voltage, V_j is the junction voltage, and ϕ_{bi} is the built-in voltage. When we create a SPICE transistor model including junction capacitances, there are model parameters that must be used, as shown in Table 10.2.

Let us create an LTSPICE model assuming that we want a few picofarads of capacitance for both transistor junctions. The resultant LTSPICE model is shown in Figure 10.16(a) where we have added the six depletion capacitance LTSPICE model parameters to the transistor model.

A simulation of the startup of the transistor (Figure 10.16(b)) shows that things do indeed slow down; there is a turn-on delay of about 5 ns (when the collector current is zero as the base–emitter junction charges up to a few tenths of a volt). After the base–emitter junction is charged up to ~0.5 V or so, the transistor turns on and the collector current rises. It takes about 20 ns for the collector current to rise to its final value of 10 mA.

A simulation of the saturation storage delay time followed by fall of the collector current is shown in Figure 10.16(c). We see that the saturation delay time is about the same as the previous example without depletion capacitances. The falltime of the collector current increases significantly.

The lesson here is to double-check your device models when you attempt to simulate.

Example 10.3: Transistor inverter with base voltage drive

Consider the transistor inverter shown in Figure 10.17(a). Assume that we use a transistor with the following specifications:

- $f_T = 300$ MHz
- $\beta_F = 200$

(a)

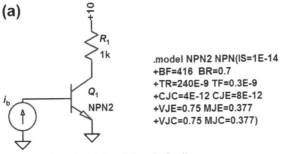

.model NPN2 NPN(IS=1E-14
+BF=416 BR=0.7
+TR=240E-9 TF=0.3E-9
+CJC=4E-12 CJE=8E-12
+VJE=0.75 MJE=0.377
+VJC=0.75 MJC=0.377)

PULSE(0 0.001 0 0.1 ns 0.1 ns 2u 5u 1)
.tran 0 5us 0 0.01n
file: Example 10_1 charge control with junction caps added.asc

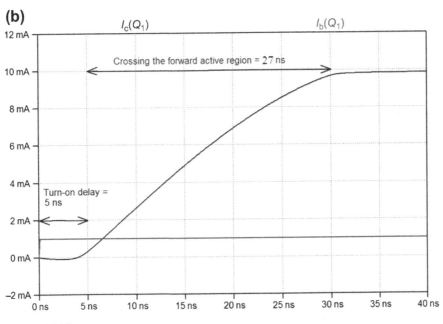

FIGURE 10.16

Simulation, with depletion capacitances added for Example 10.2. (a) LTSPICE circuit. (b)
Startup detail for the first 40 ns showing a turn-on delay time of 5 ns followed by the collector
current crossing the forward-active region, taking about 27 ns.

- $\tau_R = 10$ ns
- $\beta_R = 10$
- For depletion capacitances, use the graphs in Figure 10.18.

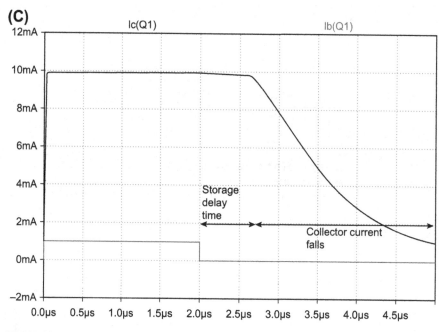

FIGURE 10.16

(*Continued*). (c) Detail of the storage delay time and collector current falltime. The falltime increased significantly when depletion capacitances were added to the model. (For color version of this figure, the reader is referred to the online version of this book.)

We will use the charge control model to find the approximate switching profile. There are six regions of operation:

1. Traversing the cutoff region (t_{d1}). During this time, the nonlinear depletion region capacitances C_{je} and C_{jc} are charged to the value that puts the transistor on the beginning of the forward-active region. We can approximate the voltage when the transistor begins to carry collector current as $V_{BE} \approx 0.4$ V.

2. Traversing the forward-active region (t_r). During this time, V_{BE} increases from 0.4 V to approximately 0.7 V and the collector current rises to its final value.

3. Going into hard saturation. When $V_{CB} \approx 0$, the transistor enters the saturation region. In hard saturation, excess base charge builds up in the base as the base—collector junction becomes more and more forward biased.

After we have waited a long time so that the transistor is fully in saturation, we have the turnoff waveforms:

4. Turn-off (saturation) delay (t_s) when saturation charge is removed. In this period, the collector current remains approximately constant.

5. Traversing the forward-active region (t_f) where the collector current drops back to zero.

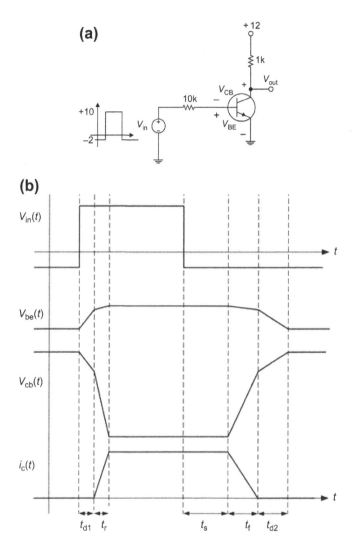

FIGURE 10.17

Transistor inverter with base voltage drive for Example 10.3. (a) Circuit. At time $t=0$, input voltage abruptly transitions from -2 V to $+10$ V. (b) Switching waveforms.

6. Traversing the cutoff region again (t_{d2}). The space charge layer capacitances are discharged as V_{BE} moves to its final value. During this time, the collector current is approximately zero.

Calculations for the various regions are as follows:

Cutoff region: At $t=0$, we assume that the input instantaneously transitions from -2 to $+10$ V. The turn-on delay time t_{d1} is the time that it takes to charge the nonlinear collector–base and base–emitter capacitances so that $V_{BE} \cong 0.4$ V

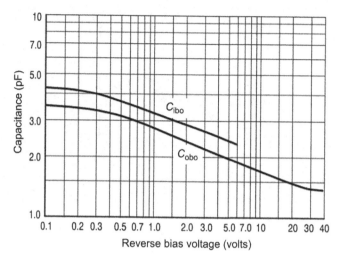

FIGURE 10.18

Graphs for finding junction capacitances C_{je} (i.e. C_{ibo}) and C_{jc} (I.e. C_{obo}) for Example 10.3.

Reprinted with permission by On Semiconductor.

or so, which we will take as the beginning of the forward-active region. During this charging interval, the base current does vary a little bit, since the V_{BE} changes, but for simplicity, we will use the average base current to determine the switching time. Terminal conditions are:

At the beginning of the cutoff region, at $t = 0$:

$V_{BE} = -2$ V
$V_{CB} = +14$ V
$C_{je} = 2.9$ pF @ $V_{BE} = -2$ V
$q_{ve} = C_{je}V_{BE} = (2.9 \text{ pF})(-2 \text{ V}) = -5.8$ pC
$C_{jc} = 1.5$ pF @ $V_{CB} = +14$ V
$q_{vc} = C_{jc}V_{CB} = (1.5 \text{ pF})(14 \text{ V}) = 21$ pC
$i_b(t = 0^+) = (10 \text{ V} - (-2 \text{ V}))/10\text{k} = 1.2$ mA
$i_c(t = 0) = 0$

At the very beginning of the forward-active region, $t = t_{d1}$:

$V_{BE} = 0.4$ V
$V_{CB} = +11.6$
$C_{je} \approx 8$ pF @ $V_{BE} = 0.4$ V (this is an estimate, since the curves on the datasheet do not extend to positive voltages).
$q_{ve} = C_{je}V_{BE} = (8 \text{ pF})(0.4 \text{ V}) = +3.2$ pC
$C_{jc} = 1.6$ pF @ $V_{CB} = +11.6$ V
$q_{vc} = C_{jc}V_{CB} = (1.6 \text{ pF})(11.6 \text{ V}) = 18.6$ pC
$i_b(t = t_{d1}) = (10 \text{ V} - 0.4 \text{ V})/10\text{k} = 0.96$ mA
$i_c(t = t_{d1}) = 0$

The average base current $i_{b,avg} = 1.08$ mA during this charging interval in the cutoff region. The change in charge in the nonlinear base–emitter capacitance is $\Delta q_{ve} = (5.8 \text{ pC} + 3.2 \text{ pC})$, resulting in $\Delta q_{ve} = 9$ pC. A similar calculation for the collector–base junction results in $\Delta q_{vc} = 2.4$ pC. The turn-on delay time is approximated by:

$$t_{d1} \approx \frac{\Delta q_{ve} + \Delta q_{vc}}{i_{b,avg}} = \frac{11.4 \text{ pC}}{1.08 \text{ mA}} \approx 11 \text{ ns} \qquad (10.33)$$

During this interval, the collector current is zero.

Traversing the forward-active region: At time $t = t_{d1}$, we assume that the transistor begins to turn on and the collector current begins to rise, until saturation is reached. The rise time t_r depends on the rate with which we can both charge the nonlinear depletion capacitances and supply the excess base charge that supports the collector current. During the forward-active region, the charge control equations are:

$$i_c \approx \frac{q_F}{\tau_F}$$

$$i_b = \frac{q_F}{\tau_{BF}} + \frac{d}{dt}(q_F + q_{ve} + q_{vc}) \qquad (10.34)$$

At the beginning of saturation, we will assume that $V_{CB} = 0$ and $V_{BE} = 0.7$, requiring $I_{C,SAT} = 11.3$ mA. Remember that the forward transit time is related to the transistor f_T by $\tau_F = 1/(2\pi f_T) = 0.53$ ns. Therefore, the forward charge required is $\Delta q_F = I_{C,SAT}\tau_F = (11.3 \text{ mA})(0.53 \text{ ns}) = 6$ pC. This is the base charge needed to completely turn the transistor on, until it begins to saturate. An additional charge is needed to charge C_{je} from 0.4 to 0.7 V and C_{jc} from 11.6 to 0 V. The calculations are as follows:

At $t = t_{d1}$:

$V_{BE} = 0.4$ V
$V_{CB} = 11.6$ V
$C_{je} \approx 8$ pF @ $V_{BE} = 0.4$ V and $q_{ve} = 3.2$ pC
$C_{jc} = 1.6$ pF @ $V_{CB} = 11.6$ V and $q_{vc} = 18.6$ pC
$i_b(t = t_{d1}) = 0.96$ mA

At $t = t_{d1} + t_r$:

$V_{BE} = 0.7$ V
$V_{CB} = 0$ V
$C_{je} \approx 8$ pF @ $V_{BE} = 0.7$ V and $q_{ve} \approx 5.6$ pC
$C_{jc} = 3.5$ pF @ $V_{CB} \approx 0$ and $q_{vc} \approx 0$
$i_b(t = t_{d1} + t_r) = 0.93$ mA

The average base current $i_{b,avg} = 0.945$ mA during this charging interval, and the change in charge in the capacitors is $\Delta q_{ve} = 2.4$ pC and $\Delta q_{vc} = 18.6$ pC. Therefore, the switching time is (approximately):

$$t_r \approx \frac{q_F + \Delta q_{ve} + \Delta q_{vc}}{i_{b,avg}} = \frac{27 \text{ pC}}{0.945 \text{ mA}} = 28.6 \text{ ns} \qquad (10.35)$$

Going into saturation: At $t = t_{d1} + t_r$, the transistor enters saturation and the transistor V_{BE} is clamped at approximately 0.7 V. The charge control equations in saturation are:

$$i_c = I_{C,SAT}$$

$$i_b - \frac{I_{C,SAT}}{\beta_F} = \frac{dq_S}{dt} + \frac{q_S}{\tau_S} \tag{10.36}$$

The second term in the second charge control equation for base current ($I_{C,SAT}/\beta_F$) is the base current needed to support recombination in the base. In the saturation region, the collector current is constant, and the base saturation charge builds up with a time constant given by:

$$\tau_S = \frac{\tau_{BF}(\beta_R + 1) + \tau_{BR}(\beta_F + 1)}{\beta_F + \beta_R + 1} \tag{10.37}$$

The total saturation charge once the transistor is in full saturation is given by:

$$q_S = \left(i_b - \frac{I_{C,SAT}}{\beta_F} \right) \tau_S \tag{10.38}$$

For this example, $\tau_S = 100$ ns and $q_S = 87.6$ pC. q_S is the saturation charge that must be removed before the transistor can be brought out of saturation.

Saturation storage (turn-off) delay: Let us assume that we have left the input voltage pulse on long enough so that the transistor is fully saturated. During the turn-off delay time (also called storage delay time), the collector current remains constant at $I_{C,SAT}$, while the saturation charge is removed. The saturation charge q_S is removed by two mechanisms: reverse base current and recombination. The charge control equations are the same as in the previous case:

$$i_c = I_{C,SAT}$$

$$i_b - \frac{I_{C,SAT}}{\beta_F} = \frac{dq_S}{dt} + \frac{q_S}{\tau_S} \tag{10.39}$$

During the saturation turn-off delay, the input voltage is -2 V and the V_{BE} is clamped at approximately 0.7 V; hence the base current $i_b = -0.27$ mA. The recombination current is $I_{C,SAT}/\beta_F = 11.3$ mA/200 = 0.056 mA. The base current charge control equation is:

$$-0.326 \text{ mA} = \frac{dq_S}{dt} + \frac{q_S}{\tau_S} \tag{10.40}$$

Solving this charge control equation for the time when q_s drops to zero results in a saturation delay $t_s = 131$ ns, as shown in Figure 10.19.

Forward-active region (current falltime): After leaving saturation, we again enter the forward-active region. In order to turn the transistor off and traverse the forward-active region, we need to remove the base charge q_F as well as discharge the nonlinear depletion capacitances. At the beginning of the current falltime interval:

$V_{BE} = 0.7$ V
$V_{CB} \approx 0$

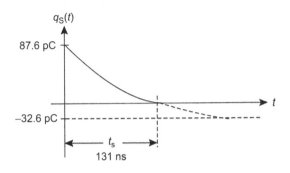

FIGURE 10.19

Example 10.3: Decay of saturation charge $q_S(t)$ while coming out of saturation showing a saturation storage delay time $t_s = 131$ ns. During this storage interval, the collector current is constant and the saturation charge decays from 87.6 pC to zero.

$C_{je} \approx 8$ pF @ $V_{BE} = 0.7$ V
$C_{jc} = 3.5$ pF @ $V_{CB} = 0$
$i_b = 0.27$ mA

At the end of the falltime interval:

$V_{be} = 0.4$ V
$V_{CB} = 11.6$ V
$C_{je} \approx 8$ pF @ $V_{BE} = 0.4$ V
$C_{jc} = 1.6$ pF @ $V_{CB} = 11.6$ V
$i_b = 0.24$ mA

The base charge q_F that we need to remove is the same as in the turn-on case ($q_F = 6$ pC). The change in charge in the depletion regions is also the same as in the turn-on case. We find that the turn-off is slower than the turn-on case, because the base current is smaller (approximately -0.255 mA):

$$t_f \approx \frac{q_F + \Delta q_{ve} + \Delta q_{vc}}{i_{b,avg}} = \frac{27 \text{ pC}}{0.255 \text{ mA}} = 106 \text{ ns} \qquad (10.41)$$

Note that even though the total charge supplied is the same as in the turn-on case, the turn-off time is much longer since the average base current is lower.

Cutoff: During this interval, we discharge the nonlinear capacitances back to their final values. The total gate charge required is the same as the turn-on case, where the charge is 11.44 pC. In this case, the average base current over the interval is approximately 0.12 mA (an average of 0.24 mA at the beginning of cutoff and 0 mA at the end), resulting in a total turnoff delay time of:

$$t_{d2} \approx \frac{\Delta q_{ve} + \Delta q_{vc}}{i_{b,avg}} = \frac{11.4 \text{ pC}}{0.13 \text{ mA}} = 95 \text{ ns} \qquad (10.42)$$

An LTSPICE simulation for the inverter is shown in Figure 10.20 with the input voltage pulse being 2000 ns wide (this pulse is sufficiently wide so that the transistor

(a)

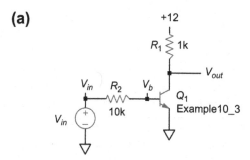

PULSE(−2 10 0 0.1n 0.1n 2000n 4000n)
.model Example10_3 NPN(IS=1E-14 BF=200 BR=10 TR=10E-9 TF=0.53E-9
+CJC=4E-12 CJE=8E-12 VJE=0.75 MJE=0.377 VJC=0.75 MJC=0.377)
.tran 0 4us 0 0.01n
file: Example 10_3 voltage driven transistor.asc

(b)

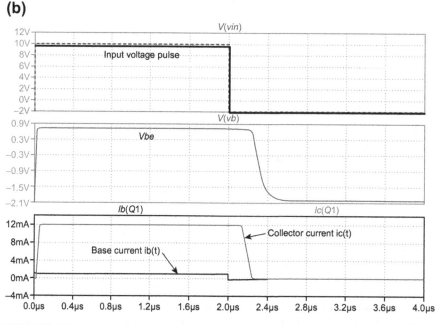

FIGURE 10.20

LTSPICE result for Example 10.3. (a) Circuit. (b) Simulation results. Note that the saturation delay time (occuring starting at time t = 2 μs) is roughly 130 ns. (For color version of this figure, the reader is referred to the online version of this book.)

will go into hard saturation). Table 10.3 shows a comparison of the charge control calculations with the SPICE calculations. This result shows that the charge control model is a useful approximation technique for calculating the switching times. More importantly, the model shows how to speedup the switching process. In this

Table 10.3 Comparison of Charge Control Model Results with LTSPICE for Example 10.3

Item	Charge Control	LTSPICE	% Diff
1. Cutoff	11 ns	20 ns	+82%
2. FAR	28.6 ns	40 ns	+40%
4. Storage	131 ns	130 ns	−1%
5. FAR	106 ns	100 ns	−6%

example, the output risetime and falltime can be significantly reduced by reducing the base resistor.

Now, in order to speedup the switching, we reduce the base resistor from 10 to 1 kΩ (Figure 10.21). As expected, the risetime and falltime of the output is much faster because we are driving the base of the transistor with about 10× the base current. However, the saturation delay time is approximately the same as the previous example. Why?

We can figure out why the saturation delay time remains approximately constant by considering the charge control equation for the transistor in saturation. Note that

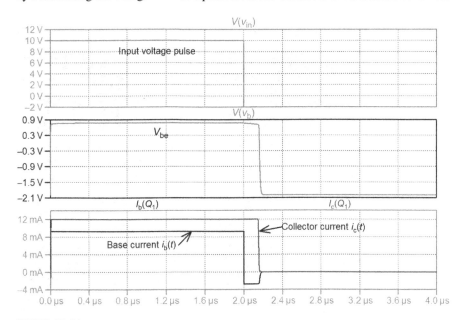

FIGURE 10.21

Waveforms when base resistor is reduced from 10 to 1 kΩ in Example 10.3. The collector current risetime and falltime are much faster and the base current is about 10× higher. Note that the saturation delay time is still roughly 130 ns. (For color version of this figure, the reader is referred to the online version of this book.)

the final value of saturation charge q_s depends on how hard we drive the base current of the transistor. When we reduce the base resistor, we get 10× as much base current, but also 10× as much saturation charge. This results in a saturation delay time that is approximately constant.

Example 10.4: Laboratory experiment: 2N3904 inverter

An inverter built with a 2N3904 was tested, using the circuit of Figure 10.22. The circuit was driven with a 250-kHz pulse train, with the input transitioning

(a)

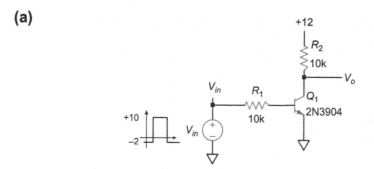

(b)

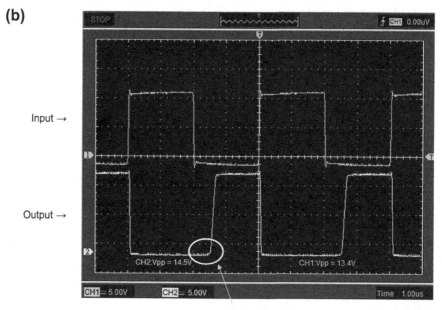

Saturation delay

FIGURE 10.22

2N3904 inverter with base voltage drive for Example 10.4. (a) Circuit. (b) Input and output, vertical scale 5 V/div, horizontal scale 1 μs/div. (For color version of this figure, the reader is referred to the online version of this book.)

between −2 and +10 V. We see that the inverter does provide a fairly fast risetime of around 40 ns, with detail shown in Figure 10.23(a). When coming out of saturation, the storage delay time is about 550 ns (Figure 10.23(b)).

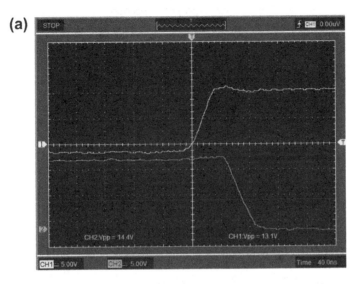

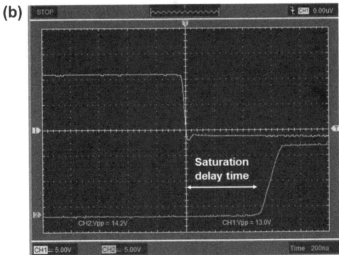

FIGURE 10.23

Detail of output from laboratory experiment of Example 10.4. (a) Startup detail (40 ns/div). Top trace = input. (b) Detail of saturation storage time delay (200 ns/div) showing a delay time of about 550 ns. (For color version of this figure, the reader is referred to the online version of this book.)

Example 10.5: Speedup capacitor

We saw in the previous example that saturation slows down the switching of a bipolar transistor significantly. Before coming out of saturation, the stored saturation charge q_S must be removed before the transistor can enter the forward-active region and shut off. The use of a proper "speedup capacitor" across the base resistor can add a negative impulse of current to quickly pull the saturation charge out of the base when turning off the transistor. If you have the right value of speedup capacitor, ideally the saturation delay time is zero, as shown in the

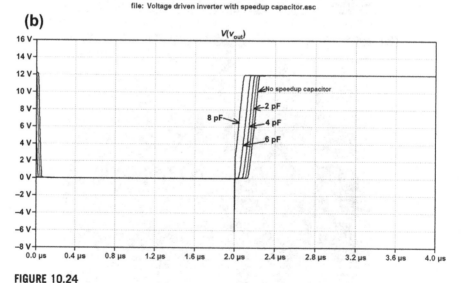

FIGURE 10.24

Transistor inverter with speedup capacitor for Example 10.5. (a) LTSPICE model. (b) Waveforms for various values of speedup capacitor 0–8 pF. At about 8 pF of speedup capacitor, the saturation delay time is near zero. (For color version of this figure, the reader is referred to the online version of this book.)

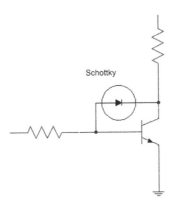

FIGURE 10.25

Clamp to keep transistor switch from saturating.

simulation of Figure 10.24. For this example, the optimal value of the speedup capacitor is around 7 pF.

Another scheme to keep a transistor switch from saturating is the use of a Schottky clamp, as shown in Figure 10.25. The Schottky diode turns on at a lower voltage than the base—collector junction and steals base current, keeping the bipolar transistor from saturating.

Example 10.6: Schottky diode clamp

In Figure 10.26(a), we see a voltage-driven transistor inverter with a Schottky diode clamp. The clamp is designed to keep the transistor from saturating. In Figure 10.26(b), we see the output response with and without the diode clamp. The clamp indeed keeps the transistor from saturating, and hence there is no saturation storage delay time.

Example 10.7: Laboratory experiment: 2N3904 inverter with 47-pF speedup capacitor

In this experiment, we test the same circuit as in Example 10.4, except we add a 47-pF speedup capacitor across the base resistor (Figure 10.27(a)). Looking at the output, we see that the saturation delay time is substantially eliminated (Figure 10.27(b)).

Example 10.8: Nonsaturating current switch

A nonsaturating current switch implemented with 2N2222 signal transistors is shown in Figure 10.28. The bias current in the emitter can be switched from one

(a) Voltage driven inverter with Schottky clamp
In this example, we show the difference in the output waveform
of the voltage-driven saturating inverter, with and without
a Schottky diode clamp. The clamp substantially reduces the saturation
storage delay time by keeping the transistor from saturating

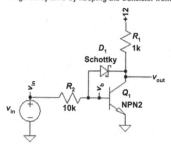

PULSE(-2 10 0 0.1n 0.1n 2000n 4000n)

.model NPN2 NPN(IS=1E-14 BF=200 BR=10 TR=10E-9 TF=0.53E-9
+CJC=4E-12 CJE=8E-12 VJE=0.75 MJE=0.377 VJC=0.75 MJC=0.377)

.model Schottky D(Is=31.7u Rs={Rs} N=1.373
+ Eg=.69 Xti=2 Iave=1 Vpk=20 type=Schottky)

.tran 0 4us 0 0.1n
.step param Rs list 0.051 1e9

file: Voltage driven inverter with diode clamp.asc

(b)

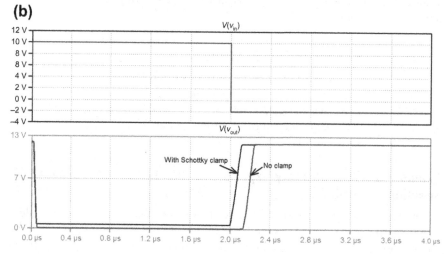

FIGURE 10.26

LTSPICE study of Schottky diode clamp for Example 10.6. (a) Circuit. (b) Output response
with and without clamp. (For color version of this figure, the reader is referred to the online
version of this book.)

(a)

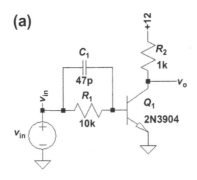

(b)

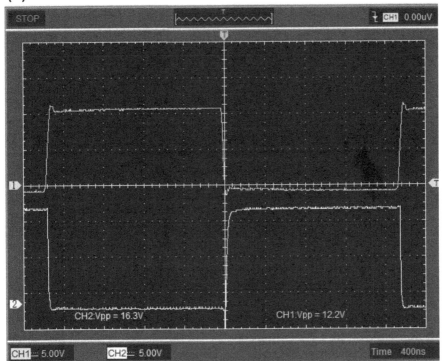

FIGURE 10.27

Speedup capacitor demonstration of Example 10.7. (a) Circuit. (b) Result with 47-pF speedup capacitor. Top trace, input voltage; bottom trace, inverter output voltage showing saturation turn-off delay time largely eliminated. (For color version of this figure, the reader is referred to the online version of this book.)

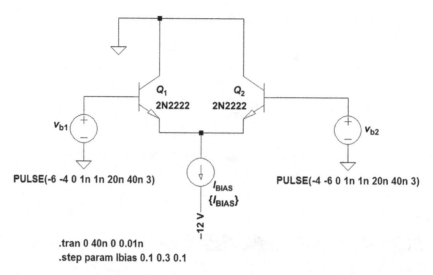

PULSE(-6 -4 0 1n 1n 20n 40n 3)   PULSE(-4 -6 0 1n 1n 20n 40n 3)

.tran 0 40n 0 0.01n
.step param Ibias 0.1 0.3 0.1

FIGURE 10.28

Nonsaturating current switch for Example 10.8. (For color version of this figure, the reader is referred to the online version of this book.)

transistor to the other, without saturating either one. If we ignore depletion capacitances, the relevant charge control equations for the base and collector currents are:

$$i_b = \frac{q_F}{\tau_{BF}} + \frac{dq_F}{dt}$$

$$i_c = \frac{q_F}{\tau_F}$$

(10.43)

In order to turn the transistors ON and OFF fast, we need to supply base current to charge and discharge the base forward charge q_F for each transistor. Therefore, in large part, the switching speed is determined by how much base current we can supply. The base current in this voltage-driven case is set by the voltage swing of the base drivers and the base spreading resistance r_x of the transistors. For the 2N2222 transistor,[12] the value of the base spreading resistance is very roughly $r_x \approx 40\ \Omega$ at high levels (>20 mA) of collector current.

We will assume a 2-V swing for the base drivers $v_{b1}(t)$ and $v_{b2}(t)$, so the maximum base current is $\sim$50 mA. We can solve for the switching time as:

$$\tau_{SW} \approx \frac{q_{Final}}{\langle i_b \rangle} \approx \frac{I_{C,final}\tau_F}{\langle i_b \rangle} \approx \frac{I_{C,final}}{\omega_T \langle i_b \rangle}$$

(10.44)

[12]This is a crude approximation since the value of the base spreading resistance varies with collector current bias level. On the 2N2222 datasheet information is available for base resistance at a collector current of 20 mA; from this information, we can calculate that the base resistance is roughly 40 Ω. At low levels of collector current, we would expect the base resistance of the 2N2222 transistor to be two to three times higher.

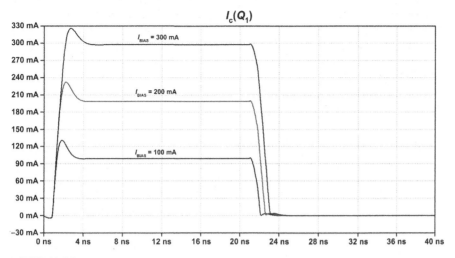

FIGURE 10.29

Current switch simulation for Example 10.8, showing collector current of Q_1 as bias current source I_{BIAS} is varied from 100 mA to 300 mA. (For color version of this figure, the reader is referred to the online version of this book.)

where the notation $\langle i_b \rangle$ denotes the average value of base current during switching. The actual switching time will be longer, due to the fact that the same base current must charge the nonlinear base–collector and base–emitter depletion capacitances. The important result from this is that the switching time will

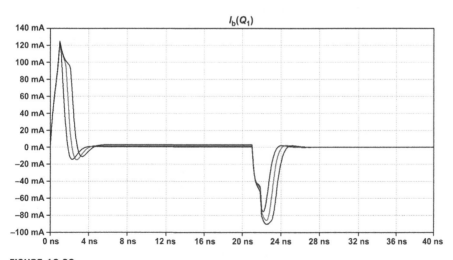

FIGURE 10.30

Current switch simulation for Example 10.8, showing base current of Q_1. (For color version of this figure, the reader is referred to the online version of this book.)

approximately scale with the collector current, if the base current remains constant. Figure 10.29 is a transient LTSPICE simulation for I_{BIAS} levels of 100, 200, and 300 mA. We note that the switching time is a few nanoseconds[13] in each case and this switching time does increase approximately proportionally with the bias current level, as predicted by the charge control model. Figure 10.30 is a transient simulation showing base current. During the transition of the collector current, the base currents are high.

Example 10.9: Emitter switching

A compound MOSTET−BJT switch is shown in Figure 10.31. This type of topology has been used in power switches where high-voltage isolation to the load is needed. The MOSFET provides the high-speed current switching, and the bipolar transistor is a common-base buffer that provides voltage isolation. Note that in the emitter switching case, there is no problem with transistor saturation. Hence, we expect this type of switch to be fast. We will estimate the switching speed of this type of topology assuming that the MOSFET provides a step of current to the emitter of the transistor.

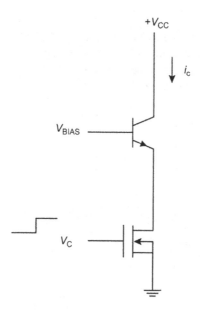

FIGURE 10.31

Emitter switching of Example 10.9, with the low-side MOSFET providing the fast switching current and the transistor providing the high-voltage buffering.

[13]Note that a nanosecond is a very short period. Light travels about a foot in a nanosecond.

Referring to Figure 10.31 and ignoring depletion capacitances, the relevant charge control equations are:

$$i_b = \frac{q_F}{\tau_{BF}} + \frac{dq_F}{dt}$$

$$i_c = \frac{q_F}{\tau_F} \qquad (10.45)$$

$$i_e = (i_b + i_c) = -\frac{q_F}{\tau_{BF}} - \frac{dq_F}{dt} - \frac{q_F}{\tau_F}$$

Rearranging the above emitter charge control equation results in:

$$-\frac{q_F}{\tau_{BF}} - \frac{dq_F}{dt} - \frac{q_F}{\tau_F} = -I_E u_{-1}(t) \qquad (10.46)$$

where $u_{-1}(t)$ is the unit step. We can make a further approximation for the emitter charge control equation by noting that $\tau_{BF} \gg \tau_F$ and hence $q_F/\tau_{BF} \ll q_F/\tau_F$, resulting in:

$$\frac{dq_F}{dt} + \frac{q_F}{\tau_F} \approx I_E u_{-1}(t) \qquad (10.47)$$

The solution to this charge control equation is an exponential rise in transistor forward charge q_F with characteristic time constant τ_F.

$$q_F(t) \approx I_E \tau_F \left(1 - e^{\frac{-t}{\tau_F}}\right)$$

$$i_c = \frac{q_F(t)}{\tau_F} \approx I_E \left(1 - e^{\frac{-t}{\tau_F}}\right) \qquad (10.48)$$

This solution shows that the collector current rises with characteristic time τ_F, which is comparable to $1/\omega_T$ of the transistor. As shown in the LTSPICE plot in Figure 10.32, emitter switching is fast.

Large-signal switching of MOSFETs

Next, we will do a quick investigation of the large-signal switching speed of MOSFETs. MOSFETs are voltage-controlled devices; if you charge up the gate to a sufficiently high voltage, the MOSFET will turn ON. Estimating the switching speed of a MOSFET circuit is generally easier than a transistor circuit. Turning ON a MOSFET involves charging the nonlinear gate–source and gate–drain capacitances. Manufacturers of power MOSFETs usually test their devices and give you "gate charge" information, from which you can figure out how to drive your MOSFET to achieve a desired switching speed.

A MOSFET switching time test circuit is shown in Figure 10.33(a). Let us figure out how this circuit works, referring to the switching waveforms in Figure 10.33(b).

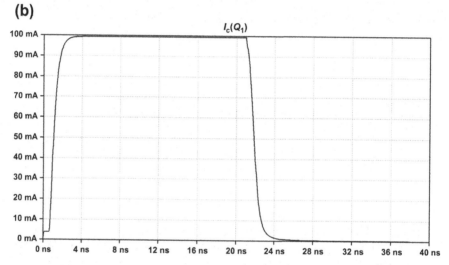

FIGURE 10.32

Emitter switch—LTSPICE simulation result of Example 10.9, using 2N2222 transistor. (a) Circuit. (b) Collector current in response to a 20-ns, 100-mA current pulse at the emitter. (For color version of this figure, the reader is referred to the online version of this book.)

When the input voltage pulse is low, the MOSFET is OFF and the current I_o keeps the clamping diode ON. After the input voltage transitions high (to +10 V or so), the MOSFET gate–source and gate–drain capacitors charge as the MOSFET gate voltage rises. The MOSFET stays OFF until the gate–source voltage rises to the MOSFET threshold voltage V_T. Charging up to the threshold voltage occurs in time period "1".

Once the gate–source voltage is at the threshold voltage and higher, the drain current $i_D(t)$ begins increasing. The relationship between gate–source voltage V_{GS} and drain current I_D is set by the forward transconductance of the MOSFET, which is a nonlinear function of drain current. In time interval "2", the drain current rises to the value of the bias current source I_o, while the gate–source voltage rises above the threshold voltage.

(a)   **(b)**

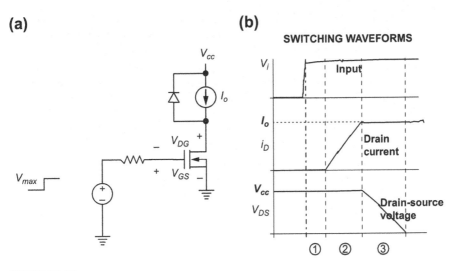

SWITCHING WAVEFORMS

FIGURE 10.33

Gate charge test circuit. (a) Circuit that is typically used. (b) Waveforms showing time intervals "1", "2", and "3".

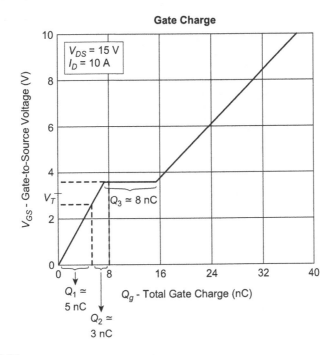

FIGURE 10.34

Gate charge test data for the Si4410DY N-channel power MOSFET, with markups by the author.

Reprinted with permission from Vishay.

Once the drain current reaches the value of the current source I_0, the diode can now turn OFF, and the drain of the MOSFET is no longer clamped at the power supply voltage. In time interval "3" the MOSFET drain voltage drops, and the gate—drain capacitor is charged by the input voltage source. At the end of time interval "3", the MOSFET is fully ON, with drain—source voltage equal to the drain current (I_0) multiplied by the ON resistance of the MOSFET ($R_{DS,ON}$). The ON resistance of the MOSFET is a parameter that you can look up on the datasheet.

Next, we will figure out how to use manufacturers' test data of "Gate Charge" to figure out switching speed. We will look at a circuit using the Vishay Si4410DY, a surface-mount power MOSFET. Parameters of interest the threshold voltage, which is about 2.5 V. Next, we will look at the "Gate Charge" test data for this MOSFET, shown in Figure 10.34, with markups by the author.

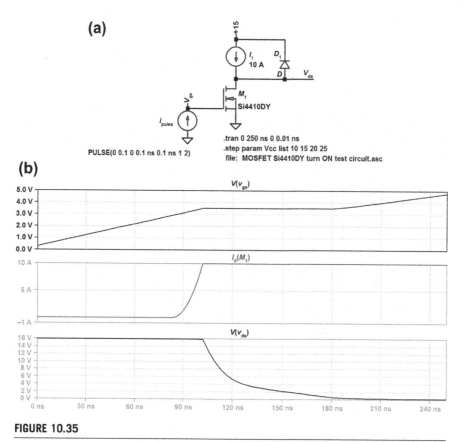

FIGURE 10.35

LTSPICE circuit for illustration of MOSFET switching speed. (a) Circuit. (b) Waveforms of interest, including gate—source voltage, drain current, and drain—source voltage. (For color version of this figure, the reader is referred to the online version of this book.)

- $Q_1 =$ charge needed to bring V_{GS} from 0 V to the threshold voltage, which is about 2.5 V. This charge is about 5 nC.
- $Q_2 =$ charge needed to bring V_{GS} from the threshold voltage, to about 3.5 V, which is enough V_{GS} to support the 10 A drain current. $Q_2 \sim 3$ nC.
- $Q_3 =$ charge needed to charge the drain–gate capacitance. During this interval, the drain voltage drops from 15 V to almost zero. $Q_3 \sim 8$ nC.

In Figure 10.35(a), we see an LTSPICE simulation of the switching time test circuit. In this case, the gate is driven by a 0 to 100 mA current pulse at time $t = 0$. A ballpark prediction of the various time intervals for switching is:

- $t_1 =$ time to charge V_{GS} up to the threshold voltage V_T: $Q_1/i_{pulse} = 5$ nC/0.1 A $= 50$ ns
- $t_2 =$ time for the drain current to rise from zero to 10 A $= Q_2/i_{pulse} = 30$ ns
- $t_3 =$ time for the drain–source voltage to drop from +15 V to near zero $= Q_3/i_{pulse} = 80$ ns

These estimates are not so bad, especially considering that we do not exactly know the LTSPICE parameters for the MOSFET (Figure 10.35(b)). In order to turn the MOSFET OFF, we need to remove all the charge we added to the gate turning it ON.

Example 10.10: A more practical MOSFET gate driver

A more practical MOSFET driving circuit uses a pulsed voltage source to drive the gate. (You can buy ICs specifically designed to drive MOSFETs; they are called "gate drivers"). You want a gate driver that can provide lots of current quickly to charge and discharge the MOSFET gate capacitances. In Figure 10.36(a), we have a model of a voltage-driven MOSFET, driven by a 100-ns wide 0- to 10-V pulse, which rises and falls in 10 ns. Waveforms showing the turn-on and turn-off transitions are shown in Figure 10.36(b). We note that the instantaneous power dissipation in the MOSFET is quite high (over 10 W) during the switching interval.

Chapter 10 problems

Problem 10.1

A pulsed transistor circuit is shown in Figure 10.37(a). The transistor is driven by a pulse $v_i(t)$ (Figure 10.37(b)), which transitions from -5 to $+5$ V at time $t = 0$. At time $t = 10$ μs, the input pulse $v_i(t)$ transitions from -5 to $+5$ V. Assume that when the transistor is saturated $V_{BE} = 0.7$ V. Transistor parameters are as follows: $\tau_f = 0.5$ ns, $\beta_F = 50$, $\tau_r = 50$ ns, and $\beta_R = 5$. The depletion capacitances can be found from the following:

$$C_j = \frac{C_{jo}}{\sqrt{1 - \frac{V_j}{\phi_{bi}}}}$$

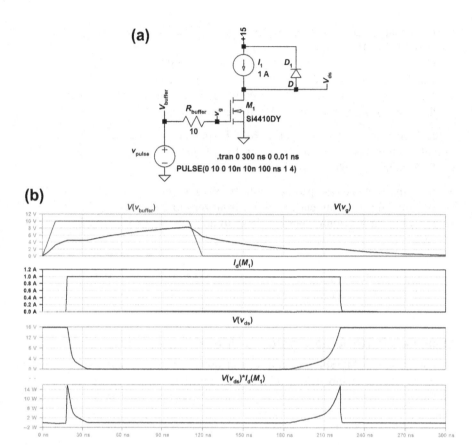

FIGURE 10.36

Voltage-driven MOSFET circuit of Example 10.10. (a) Circuit. (b) Waveforms of interest, including the drive voltage, gate–source voltage, drain current, drain–source voltage, and instantaneous MOSFET power dissipation. (For color version of this figure, the reader is referred to the online version of this book.)

For each junction $\phi_{bi} = 0.9$ V. For the base–emitter junction, $C_{jo} = 35$ pF. For the base–collector junction, $C_{jo} = 10$ pF. Plots of C_{je} and C_{jc} are shown in Figure 10.38.

a. At $t = 0+$, the input pulse $v_i(t)$ transitions high. Approximately how long does the transistor remain in cutoff before entering the forward-active region? In doing this part of the problem, identify all terminal conditions that help you determine the switching time.

b. After leaving the cutoff region (near $V_{BE} = 0.7$ V), the transistor enters the forward-active region. Determine the time to cross the forward-active region $t_{far,ON}$. Sketch $v_{be}(t)$ and collector current $i_c(t)$ in the interval $0 < t < 500$ ns, indicating cutoff time and collector current risetime

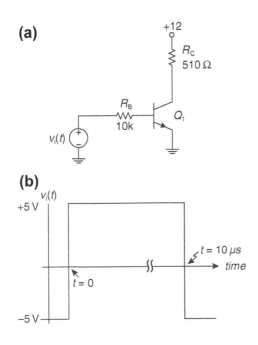

FIGURE 10.37

Transistor circuit for Problem 10.1. (a) Circuit. (b) Base drive pulse.

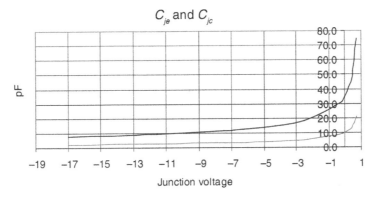

FIGURE 10.38

Depletion capacitances for Problem 10.1.

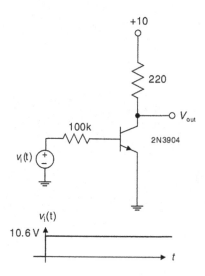

FIGURE 10.39

Transistor switch for Problem 10.2.

c. At time $t = 10$ μs, the input pulse transitions from $+5$ to -5 V. Find and sketch the collector current for $t > 10$ μs.

Problem 10.2

The amplifier in Figure 10.39 is driven by a 10.6-V voltage step at $t = 0$.

a. Using the charge control model, sketch $i_c(t)$ and $v_{out}(t)$ for the transistor amplifier shown below. Include the effect of space charge capacitances. Make reasonable assumptions, including using average values for base current, capacitances, etc. For the 2N3904, assume that $\beta_F = 100$.

b. After you finish the hand calculations using the charge control model, simulate your circuit using SPICE. Compare your results and explain any discrepancies. Note: for the input voltage, use a SPICE pulse generator with a risetime much faster ($>10\times$) than the dynamics you expect. Make the duration of the transient simulation run long enough so that you see the final value of collector current and output voltage.

Problem 10.3

In this problem, a 2N3904 transistor is controlled by a voltage drive as shown in Figure 10.40. Analyze the dynamics of the transistor. Assume transistor parameters: $\beta_F = 100$. Ignore space charge capacitances.

a. Assume that the transistor remains in the forward-active region for all time. Determine the time constants, final values, etc., and sketch $q_F(t)$.

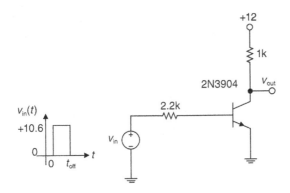

FIGURE 10.40

Transistor switch for Problem 10.3.

b. How long (t_a) does it take to traverse the active region on turn ON?

c. Since $\beta_{Fi_B} > I_{C(SAT)}$, the transistor will not remain forward active for all time. Indicate on your graphs the point at which saturation occurs. Find q_{B0}, the value of q_F that puts the transistor on the edge of saturation. Find the value of $I_{C,SAT}$ and evaluate the time constant τ_S.

d. Now, after waiting a long time, the transistor is fully saturated. After reaching full saturation, what is the value of saturation charge?

e. After the transistor has reached full saturation, the input pulse goes low. Obviously, the transistor does not remain saturated, but enters the forward-active region when q_S is zero. Determine the storage delay time, the time during which the device remains saturated after the input voltage pulse drops to zero. Sketch the saturation charge q_S.

f. Compare turn-on and turn-off times, and explain the difference.

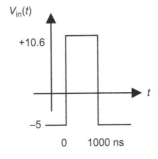

FIGURE 10.41

New drive waveform for Problem 10.3.

g. Simulate your circuit using LTSPICE and compare your charge control result with the simulation results. Attempt to explain any major discrepancies.

h. It is noted that the storage delay time decreases if the input pulse duration is reduced. Explain.

Next, the transistor is driven with voltage pulse shown in Figure 10.41. Simulate your circuit using LTSPICE and explain any major differences in the switching waveforms as compared to part (7).

Problem 10.4

A transistor is in a switching circuit as shown in Figure 10.42(a). At time $t = 0$ the base drive current source transitions instantaneously from 0 mA to 10 mA. The current source remains energized for 1000 ns, after which it turns OFF. The transistor has $\omega_T = 10^9$ rad/s, $\beta_F = 100$, $\beta_R = 5$, $\tau_R = 50$ ns, and $V_{CE,SAT} = 0$. Assume that the

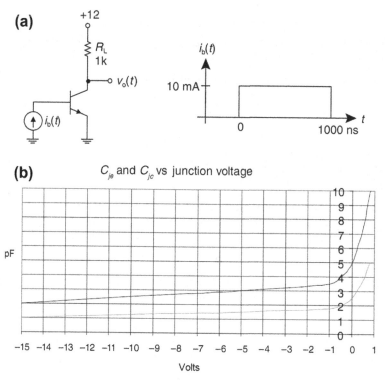

FIGURE 10.42

(a) Circuit for Problem 10.4. (b) Depletion capacitances vs. junction voltage. Top trace, C_{je}; bottom trace, C_{jc}.

transistor enters the forward-active region at $V_{BE} = 0.5$ V. Depletion capacitance values are shown in Figure 10.42(b).

a. At $t = 0+$, the current source transitions HIGH. Find the transistor turn-on delay t_d.

b. After crossing the cutoff region, the transistor enters the forward-active region. Calculate the approximate risetime of collector current t_r.

c. After crossing the forward-active region, the transistor enters saturation. Calculate the final value of saturation charge q_S.

d. After the current source turns off, the transistor remains saturated for a time. Calculate the storage delay time t_s.

e. After the transistor leaves saturation, it again crosses the forward-active region when the collector current drops to zero. Calculate the approximate current falltime t_f. (In this section, estimate the falltime as the time it takes the collector current to fall from 100% of initial value to 10% of initial value.)

f. Sketch the collector current and transistor V_{CE}, labeling all axes.

Problem 10.5

For this transistor in Figure 10.43, use: $C_{je} = 10$ pF, $C_{jc} = 2$ pF, $f_T = 300$ MHz, $\beta_F = 100$, $\beta_R = 5$, and $\tau_R = 50$ ns.

a. The circuit is driven by a 0- to 5-V step at 1000 ns. What is the initial collector current at $t = 0$?

b. Does the transistor ever saturate, and why or why not?

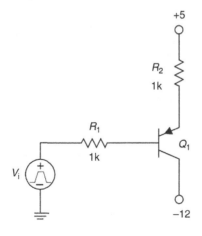

FIGURE 10.43

Circuit for Problem 10.5.

c. At $t = 1000$ ns, the input voltage steps from 0 to 5 V. Calculate the delay time before the collector current begins changing and also calculate the transition time of collector current approximately as the transistor transitions through the forward-active region. State your assumptions and back them up.

d. Sketch the input waveform and the transistor collector current.

Further reading

[1] Barna A. Analytic approximations for propagation delays in current-mode switching circuits including collector-base capacitances. *IEEE J Solid-State Circuits* 1981;**16**(5):597−9.

[2] Bashkow TR. Effect of nonlinear collector capacitance on collector current rise time. *IRE Trans Electron Devices* 1956:167−72.

[3] Casaravilla G, Silveira F. Emitter drive: a technique to drive a bipolar power transistor switching at 100 kHz. In: *Proceedings of the 1990 IEEE colloquium in South America* August 31−September 15, 1990. p. 188−92.

[4] Chuang CT, Chin K. High-speed low-power direct-coupled complementary push-pull ECL circuit. *IEEE J Solid-State Circuits* 1994;**29**(7):836−9.

[5] Easley JW. The effect of collector capacity on the transient response of junction transistors. *IRE Trans Electron Devices* 1957:6−14.

[6] Ebers JJ, Moll JL. Large-signal behavior of junction transistors. *Proc IRE* 1954:1761.

[7] Ebers JJ, Moll JL. Large-signal behavior of junction transistors. *Proc IRE* 1954:1761−72.

[8] Ghannam MY, Mertens RP, van Overstraeten RJ. An analytical model for the determination of the transient response of CML and ECL gates. *IEEE Trans Electron Devices* 1990;**37**(1):191−201.

[9] Gray PE, Searle CL. *Electronic principles physics, models and circuits*. John Wiley; 1969.

[10] Jensen RW. Charge control transistor model for the IBM electronic circuit analysis program. *IEEE Trans Circuit Theory* 1966;**CT-13**(4):428−37.

[11] Karadzinov LV, Arsov GL, Dzekov TA, Jeffries DJ. Charge-control piecewise-linear bipolar junction transistor model. In: *Proceedings of the IEEE international symposium on industrial electronics, ISIE '96'* June 17−20, 1996. p. 561−6.

[12] Konstadinidis GK, Berger HH. Optimization of buffer stages in bipolar VLSI systems. *IEEE J Solid-State Circuits* 1992;**27**(7):1002−13.

[13] Kuno HJ. Rise and fall time calculations of junction transistors. *IEEE Trans Electron Devices* 1964:151−5.

[14] Moll JL. Large-signal transient response of junction transistors. *Proc IRE* 1954:1773−84.

[15] Muller RS, Kamins TI. *Device electronics for integrated circuits*. 2nd ed. John Wiley; 1986.

[16] Musumeci S, Pagano R, Raciti A, Porto C, Ronsisvalle C, Scollo R. Characterization, parameter identification and modeling of a new monolithic emitter- switching bipolar transistor. In: *Proceedings of the 39th industry applications conference* October 3−7, 2004. p. 1924−31.

[17] Neudeck GW. *The PN junction diode* In *Modular series on solid state devices*, vol. II. Addison-Wesley; 1983.

PN2222, PN2222A

PN2222A is a Preferred Device

General Purpose Transistors

NPN Silicon

ON Semiconductor®

http://onsemi.com

Features

• Pb−Free Packages are Available*

MAXIMUM RATINGS

Rating		Symbol	Value	Unit
Collector-Emitter Voltage	PN2222 PN2222A	V_{CEO}	30 40	Vdc
Collector-Base Voltage	PN2222 PN2222A	V_{CBO}	60 75	Vdc
Emitter-Base Voltage	PN2222 PN2222A	V_{EBO}	5.0 6.0	Vdc
Collector Current − Continuous		I_C	600	mAdc
Total Device Dissipation @ T_A = 25°C Derate above 25°C		P_D	625 5.0	mW mW/°C
Total Device Dissipation @ T_C = 25°C Derate above 25°C		P_D	1.5 12	W mW/°C
Operating and Storage Junction Temperature Range		T_J, T_{stg}	−55 to +150	°C

THERMAL CHARACTERISTICS

Characteristic	Symbol	Max	Unit
Thermal Resistance, Junction-to-Ambient	$R_{\theta JA}$	200	°C/W
Thermal Resistance, Junction-to-Case	$R_{\theta JC}$	83.3	°C/W

Stresses exceeding Maximum Ratings may damage the device. Maximum Ratings are stress ratings only. Functional operation above the Recommended Operating Conditions is not implied. Extended exposure to stresses above the Recommended Operating Conditions may affect device reliability.

*For additional information on our Pb−Free strategy and soldering details, please download the ON Semiconductor Soldering and Mounting Techniques Reference Manual, SOLDERRM/D.

COLLECTOR
3

2
BASE

1
EMITTER

MARKING DIAGRAM

TO−92
CASE 29−11

PN 222x AYWW •

PN222	= Device Code
x	= A or 2
A	= Assembly Location
Y	= Year
WW	= Work Week
•	= Pb−Free Package

(Note: Microdot may be in either location)

ORDERING INFORMATION

See detailed ordering and shipping information in the package dimensions section on page 6 of this data sheet.

Preferred devices are recommended choices for future use and best overall value.

March, 2006 − Rev. 2

Publication Order Number:
PN2222/D

[18] Robinson FVP, Williams BW. Optimizing snubbers for high-current emitter-switched transistors. In: *Proceedings of the third international conference on power electronics and variable-speed drives* July 13—15, 1988. p. 177—80.

[19] Sharaf KM, Elmasry MI. An accurate analytical propagation delay model for high-speed CML bipolar circuits. *IEEE J Solid-State Circuits* 1994;**29**(1):31—45.

[20] Stork JMC. Bipolar transistor scaling for minimum switching delay and energy dissipation. In: *Technical Digest of International Electron Devices Meeting* December 11—14, 1988. p. 550—3.

[21] Thompson MT, Schlecht MF. High power laser diode driver based on power converter technology. *IEEE Trans Power Electron* 1997;**12**(1):46—52.

[22] Winkel JTE. Extended charge-control model for bipolar transistors. *IEEE Trans Electron Devices* 1973;**ED-20**(4):389—94.

2N2222 NPN transistor datasheet excerpts[14]

---

[14]Reprinted with permission of On Semiconductor.

PN2222, PN2222A

ELECTRICAL CHARACTERISTICS (T_A = 25°C unless otherwise noted)

Characteristic		Symbol	Min	Max	Unit
OFF CHARACTERISTICS					
Collector–Emitter Breakdown Voltage (I_C = 10 mAdc, I_B = 0)	PN2222 PN2222A	$V_{(BR)CEO}$	30 40	– –	Vdc
Collector–Base Breakdown Voltage (I_C = 10 μAdc, I_E = 0)	PN2222 PN2222A	$V_{(BR)CBO}$	60 75	– –	Vdc
Emitter–Base Breakdown Voltage (I_E = 10 μAdc, I_C = 0)	PN2222 PN2222A	$V_{(BR)EBO}$	5.0 6.0	– –	Vdc
Collector Cutoff Current (V_{CE} = 60 Vdc, $V_{EB(off)}$ = 3.0 Vdc)	PN2222A	I_{CEX}	–	10	nAdc
Collector Cutoff Current (V_{CB} = 50 Vdc, I_E = 0) (V_{CB} = 60 Vdc, I_E = 0) (V_{CB} = 50 Vdc, I_E = 0, T_A = 125°C) (V_{CB} = 50 Vdc, I_E = 0, T_A = 125°C)	 PN2222 PN2222A PN2222 PN2222A	I_{CBO}	– – – –	0.01 0.01 10 10	μAdc
Emitter Cutoff Current (V_{EB} = 3.0 Vdc, I_C = 0)	PN2222A	I_{EBO}	–	100	nAdc
Base Cutoff Current (V_{CE} = 60 Vdc, $V_{EB(off)}$ = 3.0 Vdc)	PN2222A	I_{BL}	–	20	nAdc
ON CHARACTERISTICS					
DC Current Gain (I_C = 0.1 mAdc, V_{CE} = 10 Vdc) (I_C = 1.0 mAdc, V_{CE} = 10 Vdc) (I_C = 10 mAdc, V_{CE} = 10 Vdc) (I_C = 10 mAdc, V_{CE} = 10 Vdc, T_A = –55°C) (I_C = 150 mAdc, V_{CE} = 10 Vdc) (Note 1) (I_C = 150 mAdc, V_{CE} = 1.0 Vdc) (Note 1) (I_C = 500 mAdc, V_{CE} = 10 Vdc) (Note 1)	 PN2222A only PN2222 PN2222A	h_{FE}	35 50 75 35 100 50 30 40	– – – – 300 – – –	–
Collector–Emitter Saturation Voltage (Note 1) (I_C = 150 mAdc, I_B = 15 mAdc) (I_C = 500 mAdc, I_B = 50 mAdc)	 PN2222 PN2222A PN2222 PN2222A	$V_{CE(sat)}$	– – – –	0.4 0.3 1.6 1.0	Vdc
Base–Emitter Saturation Voltage (Note 1) (I_C = 150 mAdc, I_B = 15 mAdc) (I_C = 500 mAdc, I_B = 50 mAdc)	 PN2222 PN2222A PN2222 PN2222A	$V_{BE(sat)}$	– 0.6 – –	1.3 1.2 2.6 2.0	Vdc
SMALL–SIGNAL CHARACTERISTICS					
Current–Gain – Bandwidth Product (Note 2) (I_C = 20 mAdc, V_{CE} = 20 Vdc, f = 100 MHz)	PN2222 PN2222A	f_T	250 300	– –	MHz
Output Capacitance (V_{CB} = 10 Vdc, I_E = 0, f = 1.0 MHz)		C_{obo}	–	8.0	pF
Input Capacitance (V_{EB} = 0.5 Vdc, I_C = 0, f = 1.0 MHz)	PN2222 PN2222A	C_{ibo}	– –	30 25	pF
Input Impedance (I_C = 1.0 mAdc, V_{CE} = 10 Vdc, f = 1.0 kHz) (I_C = 10 mAdc, V_{CE} = 10 Vdc, f = 1.0 kHz)	PN2222A PN2222A	h_{ie}	2.0 0.25	8.0 1.25	kΩ
Voltage Feedback Ratio (I_C = 1.0 mAdc, V_{CE} = 10 Vdc, f = 1.0 kHz) (I_C = 10 mAdc, V_{CE} = 10 Vdc, f = 1.0 kHz)	PN2222A PN2222A	h_{re}	– –	8.0 4.0	X 10^{-4}
Small–Signal Current Gain (I_C = 1.0 mAdc, V_{CE} = 10 Vdc, f = 1.0 kHz) (I_C = 10 mAdc, V_{CE} = 10 Vdc, f = 1.0 kHz)	PN2222A PN2222A	h_{fe}	50 75	300 375	–

1. Pulse Test: Pulse Width ≤ 300 μs, Duty Cycle ≤ 2.0%.
2. f_T is defined as the frequency at which $|h_{fe}|$ extrapolates to unity.

PN2222, PN2222A

ELECTRICAL CHARACTERISTICS (T_A = 25°C unless otherwise noted) (Continued)

Characteristic		Symbol	Min	Max	Unit
SMALL–SIGNAL CHARACTERISTICS					
Output Admittance (I_C = 1.0 mAdc, V_{CE} = 10 Vdc, f = 1.0 kHz) (I_C = 10 mAdc, V_{CE} = 10 Vdc, f = 1.0 kHz)	 PN2222A PN2222A	h_{oe}	 5.0 25	 35 200	µMhos
Collector Base Time Constant (I_E = 20 mAdc, V_{CB} = 20 Vdc, f = 31.8 MHz)	 PN2222A	$rb'C_c$	–	150	ps
Noise Figure (I_C = 100 µAdc, V_{CE} = 10 Vdc, R_S = 1.0 kΩ, f = 1.0 kHz)	 PN2222A	NF	–	4.0	dB
SWITCHING CHARACTERISTICS (PN2222A only)					
Delay Time	(V_{CC} = 30 Vdc, $V_{BE(off)}$ = –0.5 Vdc,	t_d	–	10	ns
Rise Time	I_C = 150 mAdc, I_{B1} = 15 mAdc) (Figure 1)	t_r	–	25	ns
Storage Time	(V_{CC} = 30 Vdc, I_C = 150 mAdc,	t_s	–	225	ns
Fall Time	I_{B1} = I_{B2} = 15 mAdc) (Figure 2)	t_f	–	60	ns

SWITCHING TIME EQUIVALENT TEST CIRCUITS

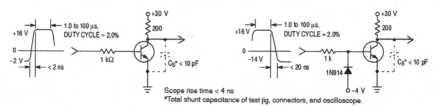

Scope rise time < 4 ns
*Total shunt capacitance of test jig, connectors, and oscilloscope.

Figure 1. Turn–On Time   **Figure 2. Turn–Off Time**

PN2222, PN2222A

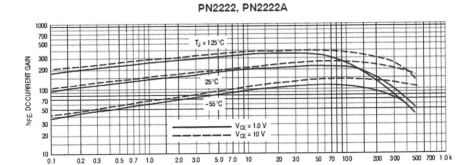

Figure 3. DC Current Gain

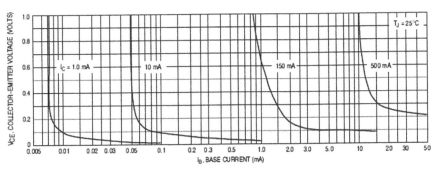

Figure 4. Collector Saturation Region

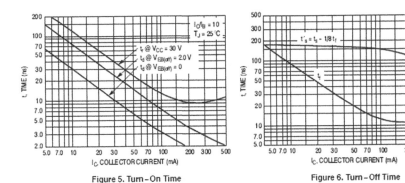

Figure 5. Turn–On Time

Figure 6. Turn–Off Time

PN2222, PN2222A

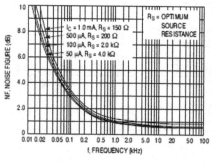

Figure 7. Frequency Effects

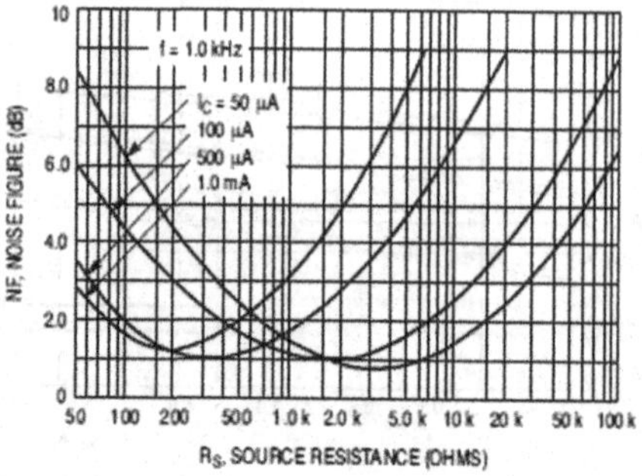

Figure 8. Source Resistance Effects

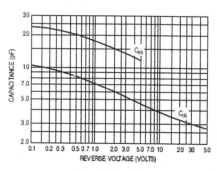

Figure 9. Capacitances

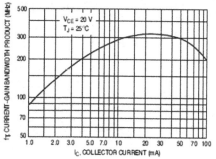

Figure 10. Current–Gain Bandwidth Product

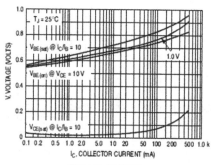

Figure 11. "On" Voltages

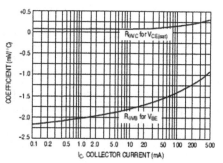

Figure 12. Temperature Coefficients

Si4410DY N-channel MOSFET datasheet excerpts[15]

Si4410DY

Vishay Siliconix

N-Channel 30-V (D-S) MOSFET

PRODUCT SUMMARY		
V$_{DS}$ (V)	r$_{DS(on)}$ (Ω)	I$_D$ (A)
30	0.0135 @ V$_{GS}$ = 10 V	10
	0.020 @ V$_{GS}$ = 4.5 V	8

FEATURES
- TrenchFET® Power MOSFET

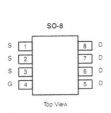

SO-8

Top View

ABSOLUTE MAXIMUM RATINGS (T$_A$ = 25°C UNLESS OTHERWISE NOTED)				
Parameter		Symbol	Limit	Unit
Drain-Source Voltage		V$_{DS}$	30	V
Gate-Source Voltage		V$_{GS}$	± 20	
Continuous Drain Current (T$_J$ = 150°C)[a]	T$_A$ = 25°C	I$_D$	10	A
	T$_A$ = 70°C		8	
Pulsed Drain Current		i$_{DM}$	50	
Continuous Source Current (Diode Conduction)[a]		I$_S$	2.3	
Maximum Power Dissipation[a]	T$_A$ = 25°C	P$_D$	2.5	W
	T$_A$ = 70°C		1.6	
Operating Junction and Storage Temperature Range		T$_J$, T$_{stg}$	-55 to 150	°C

THERMAL RESISTANCE RATINGS			
Parameter	Symbol	Limit	Unit
Maximum Junction-to-Ambient[a]	R$_{thJA}$	50	°C/W
Maximum Junction-to-Foot (Drain)	R$_{thJF}$	22	

Notes
a.   Surface Mounted on FR4 Board, t ≤ 10 sec.

Document Number: 71726
S-40838—Rev. L, 03-May-04

www.vishay.com
1

[15]Information courtesy of Vishay Intertechnology, Inc.

Si4410DY
Vishay Siliconix

SPECIFICATIONS (T$_J$ = 25°C UNLESS OTHERWISE NOTED)

Parameter	Symbol	Test Condition	Min	Typ	Max	Unit
Static						
Gate Threshold Voltage	V$_{GS(th)}$	V$_{DS}$ = V$_{GS}$, I$_D$ = 250 μA	1.0		3.0	V
Gate-Body Leakage	I$_{GSS}$	V$_{DS}$ = 0 V, V$_{GS}$ = ±20 V			±100	nA
Zero Gate Voltage Drain Current	I$_{DSS}$	V$_{DS}$ = 30 V, V$_{GS}$ = 0 V			1	μA
		V$_{DS}$ = 30 V, V$_{GS}$ = 0 V, T$_J$ = 55°C			25	
On-State Drain Current[a]	I$_{D(on)}$	V$_{DS}$ ≥ 5 V, V$_{GS}$ = 10 V	20			A
Drain-Source On-State Resistance[a]	r$_{DS(on)}$	V$_{GS}$ = 10 V, I$_D$ = 10 A		0.011	0.0135	Ω
		V$_{GS}$ = 4.5 V, I$_D$ = 5 A		0.015	0.020	
Forward Transconductance[a]	g$_{fs}$	V$_{DS}$ = 15 V, I$_D$ = 10 A		38		S
Diode Forward Voltage[a]	V$_{SD}$	I$_S$ = 2.3 A, V$_{GS}$ = 0 V		0.7	1.1	V
Dynamic[b]						
Gate Charge	Q$_g$	V$_{DS}$ = 15 V, V$_{GS}$ = 5 V, I$_D$ = 10 A		20	34	nC
Total Gate Charge	Q$_{gt}$			37	60	
Gate-Source Charge	Q$_{gs}$	V$_{DS}$ = 15 V, V$_{GS}$ = 10 V, I$_D$ = 10 A		7		
Gate-Drain Charge	Q$_{gd}$	V$_{DS}$ = 15 V, V$_{GS}$ = 10 V, I$_D$ = 10 A		7.0		
Gate Resistance	R$_g$		0.5	1.5	2.6	Ω
Turn-On Delay Time	t$_{d(on)}$			19	30	ns
Rise Time	t$_r$	V$_{DD}$ = 25 V, R$_L$ = 25 Ω		9	20	
Turn-Off Delay Time	t$_{d(off)}$	I$_D$ = 1 A, V$_{GEN}$ = 10 V, R$_G$ = 6 Ω		70	100	
Fall Time	t$_f$			20	80	
Source-Drain Reverse Recovery Time	t$_{rr}$	I$_F$ = 2.3 A, di/dt = 100 A/μs		40	80	

Si4410DY

Vishay Siliconix

TYPICAL CHARACTERISTICS (25°C UNLESS NOTED)

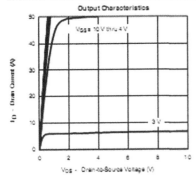

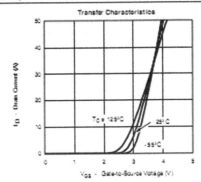

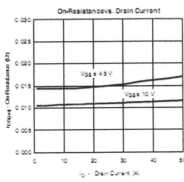

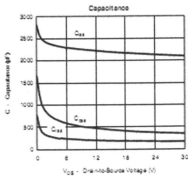

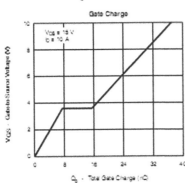

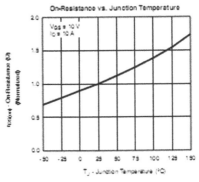

Si4410DY
Vishay Siliconix

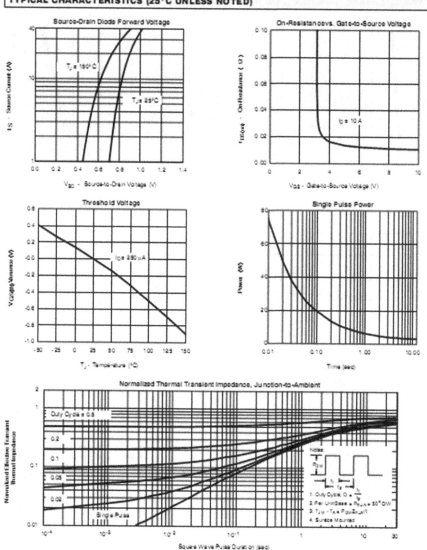

Document Number: 71726
S-40838—Rev. L, 03-May-04

Review of Feedback Systems

"A governor is a part of a machine by means of which the velocity of the machine is kept nearly uniform, notwithstanding variations in the driving-power or the resistance."

— **James C. Maxwell, "On Governors," 1867**

IN THIS CHAPTER

▶ The arena of "classical control systems" covers single-input, single-output (SISO) linear time-invariant, continuous time systems. This includes the design of classical servomechanisms and most operational amplifier circuits. This chapter offers introductory material and numerous examples of the design of feedback control systems.

Introduction and some early history of feedback control

A feedback system is one that compares its output to a desired input and takes corrective action to force the output to follow the input. Arguably, the beginnings of automatic feedback control[1] can be traced back to the work of James Watt in" the 1700s. Watt did a lot of work on steam engines, and he adapted[2] a centrifugal *governor* to automatically control the speed of a steam engine. The governor was composed of two rotating metal balls that would fly out due to centrifugal force. The amount of "fly-out" was then used to regulate the speed of the steam engine by adjusting a throttle. This was an example of proportional control.

The steam engines of Watt's day worked well with the governor, but as steam engines became larger and better engineered, it was found that there could be

---

[1]Others may argue that the origins of feedback control trace back to the water clocks and float regulators of the ancients. See, e.g., Otto Mayr's *The Origins of Feedback Control*, The MIT Press, 1970.

[2]The centrifugal governor was invented by Thomas Mead c. 1787, for which he received British Patent #1628.

Intuitive Analog Circuit Design. http://dx.doi.org/10.1016/B978-0-12-405866-8.00011-5

stability problems in the engine speed. One of the problems was *hunting,* or an engine speed that would surge and decrease, apparently hunting for a stable operating point. This phenomenon was not well understood until the latter part of the nineteenth century, when James Maxwell[3] (yes, the same Maxwell famous for all those equations) developed the mathematics of the stability of the Watt governor using differential equations.

Invention of the negative feedback amplifier

We now jump forward to the twentieth century. In the early days of telephone, practical difficulties were encountered in building a transcontinental telephone line. The first transcontinental telephone system, built in 1914, used #8 copper wire weighing about 1000 pounds per mile. Loss due to the resistance of the wire was approximately 60 dB.[4] Several vacuum tube amplifiers were used to boost the amplitude. These amplifiers have limited bandwidth and significant nonlinear distortion. The effects of cascading amplifiers (Figure 11.1) resulted in intolerable amounts of signal distortion.

Harold Black graduated from Worcester Polytechnic Institute in 1921 and joined Bell Laboratory. At this time, a major task facing AT&T was the improvement of the telephone system and the problem of distortion in cascaded amplifiers. In 1927, Black[5] was considering the problem of distortion in amplifiers and came up with the idea of the negative feedback amplifier.

> *"Then came the morning of Tuesday, August 2, 1927, when the concept of the negative feedback amplifier came to me in a flash while I was crossing the Hudson River on the Lackawanna Ferry, on the way to work. For more than 50 years I have pondered how and why the idea came, and I can't say any more today than I could that morning. All I know is that after several years of hard work on the problem, I suddenly realized that if I fed the amplifier output back to the input, in reverse phase, and kept the device from oscillating (singing, as we called it then), I would have exactly what I wanted: a means of canceling out the*

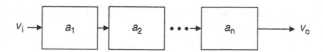

FIGURE 11.1

Amplifier cascade.

[3]James C. Maxwell, "On Governors", *Proceedings of the Royal Society,* 1867, pp. 270–283.
[4]William McC. Siebert, *Circuits, Signals and Systems,* The MIT Press, 1986.
[5]Harold Black, "Inventing the Negative Feedback Amplifier," *IEEE Spectrum,* December 1977, pp. 55–60. See also Harold Black's US Patent #2,102, 671, "Wave Translation System", filed April 22, 1932, and issued December 21, 1937, and Black's early paper "Stabilized Feed-Back Amplifiers," *Bell System Technical Journal,* 1934.

distortion in the output. I opened my morning newspaper and on a page of The New York Times I sketched a simple diagram of a negative feedback amplifier plus the equations for the amplification with feedback. I signed the sketch, and 20 minutes later, when I reached the laboratory at 463 West Street, it was witnessed, understood, and signed by the late Earl C. Bleassing.

I envisioned this circuit as leading to extremely linear amplifiers (40 to 50 dB of negative feedback), but an important question is: How did I know I could avoid self-oscillations over very wide frequency bands when many people doubted such circuits would be stable? My confidence stemmed from work that I had done two years earlier on certain novel oscillator circuits and three years earlier in designing the terminal circuits, including the filters, and developing the mathematics for a carrier telephone system for short toll circuits."

The block diagram form as envisioned by Black, and the description given in Black's patent are shown in Figure 11.2. This is a single-input, single output (SISO) control loop.

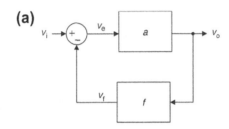

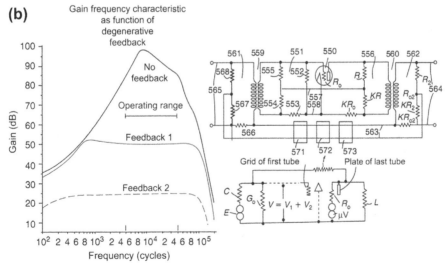

FIGURE 11.2

Classical single-input, single-output (SISO) control loop, as envisioned by Black. (a) Block diagram form. (b) Excerpt from Black's US patent #2,102,671 issued in 1937.

A typical closed-loop negative feedback system as is commonly implemented is shown in Figure 11.3. The "plant" in this diagram might represent, for instance, the power stage in an audio amplifier. A properly designed control system can maintain the output at a desired level in the face of external disturbances and uncertainties in the model of the plant. The goal of the feedback system is to force the output to track the input, perhaps with some gain and frequency-response shaping.

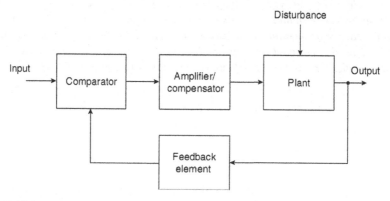

FIGURE 11.3

Typical feedback system showing functionality of individual blocks.

In this configuration, the output signal is fed back to the input, where it is compared with the desired input. The difference between the two signals is amplified and applied to the plant input.

In order to design a successful feedback system, several issues must be resolved.

- First, how do you generate the model of the plant, given that many systems do not have well-defined transfer functions?
- Once you have the model of the plant, how do you close the loop, resulting in a stable system with a desired gain and bandwidth, and an acceptable amount of overshoot?

Control system basics

A classical feedback loop, as envisioned by Black, is shown in Figure 11.4. Note that there is an external disturbance in this system, the voltage v_d.

In this system, a is the forward path gain and f is the feedback gain. The forward gain a and feedback factor f may depend on frequency (and, hence, the plant should be denoted as $a(s)$), but for notational simplicity we will drop the Laplace variable s.

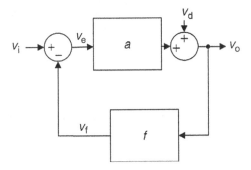

FIGURE 11.4

Classical SISO control loop, with input voltage v_i, output voltage v_o, and external disturbance v_d.

Initially, let us set the disturbance v_d to zero. The "error" term v_e is the difference between the input and the fed-back portion of the output. We can solve for the transfer function with the result:

$$v_o = av_e$$
$$v_e = v_i - v_f \tag{11.1}$$
$$v_f = fv_o$$

The closed-loop gain is ($A \equiv$ closed-loop gain):

$$A = \frac{v_o}{v_i} = \frac{a}{1 + af} \tag{11.2}$$

or

$$A = \frac{\text{Forward gain}}{1 - \text{Loop transmission}} \tag{11.3}$$

Note what happens in the limit of $af \gg 1$:

$$A \approx \frac{1}{f} \tag{11.4}$$

This is the key to designing a successful feedback system; if you can guarantee that $af \gg 1$ for the frequencies that you are interested in, *then your closed-loop gain will not be dependent on the details of the plant gain a(s)*. This is very useful, since in some cases the feedback function f can be implemented with a simple resistive divider, which can be cheap and accurate.

Loop transmission and disturbance rejection

The term in the denominator of the gain equation is $1 + af$, where the term $-af$ is called the loop transmission (L.T.). This term is the gain going around the whole feedback loop; you can find the loop transmission by doing a thought experiment:

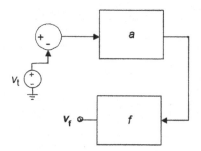

FIGURE 11.5

Block diagram manipulation to find the loop transmission.

cut the feedback loop in one place as shown in Figure 11.5, inject a test signal $v_t = 1$ V, and find out what returns where you cut. The gain around the loop is the loop transmission; in this case, the loop transmission is just $-af$.

Now, let us find the closed-loop gain from the disturbance input to the output, using the block diagram in Figure 11.4:

$$\frac{v_o}{v_d} = \frac{1}{1 + af} = \frac{1}{1 - \text{L.T.}} \tag{11.5}$$

Note that if the loop transmission is large at frequencies of interest, then the output due to the disturbance will be small. The term $1 + af$ is called the *desensitivity* of the system. Let us figure out the fractional change in closed-loop gain A due to a change in forward path gain a.

$$A = \frac{a}{1 + af}$$

$$\frac{dA}{da} = \frac{(1 + af) - af}{(1 + af)^2} = \left(\frac{1}{1 + af}\right)\left(\frac{1}{1 + af}\right) = \frac{A}{a}\left(\frac{1}{1 + af}\right) \tag{11.6}$$

$$\therefore \frac{dA}{A} = \frac{da}{a}\left(\frac{1}{1 + af}\right)$$

This result means that if $af \gg 1$, then the fractional change in closed-loop gain (dA/A) is much smaller than the fractional change in forward path gain (da/a).

Example 11.1: Distortion rejection

In this example, we will see how negative feedback can be used to reduce distortion of an amplifier. In Figure 11.6(a), we see a "push–pull" amplifier, which is often used as the output stage of op-amps and power audio amplifiers. If we apply a sine wave to the input of the push–pull amplifier, the output will follow th input (in the "emitter follower" sense) when the magnitude of the input voltage is greater than about 0.6 V. So, for a 2 V pp input sine wave, the output voltage

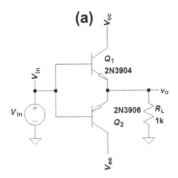

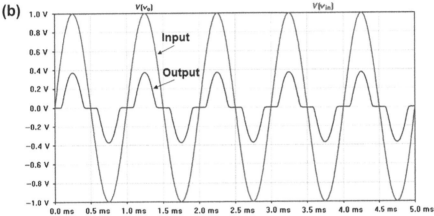

FIGURE 11.6

Open-loop push—pull amplifier for Example 11.1. (a) Circuit. (b) Response for 2 V pp sine wave at 1 kHz. (For color version of this figure, the reader is referred to the online version of this book.)

(Figure 11.6(b)) shows significant distortion. The dead zone in the output voltage is called "crossover distortion".

Using negative feedback, we can substantially eliminate the crossover distortion by placing the push—pull amplifier inside a negative feedback loop, as shown in Figure 11.7(a). This amplifier is configured as a unity-gain buffer, and negative feedback forces the output to follow the input. Note that the input and output voltages track each other very well (Figure 11.7(b)), and the distortion is greatly compensated when $1 + a(s)f(s) \gg 1$.

Approximate closed-loop gain of a feedback loop

We can make a couple of approximations in the limit of large and small loop transmission For large loop transmission ($a(s)f(s) \gg 1$), as we have shown before, the

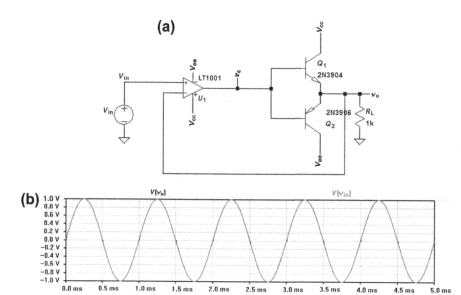

FIGURE 11.7

Push–pull amplifier inside a feedback loop for Example 11.1. (a) Circuit. (b) Response for 2 V pp sine wave at 1 kHz. (For color version of this figure, the reader is referred to the online version of this book.)

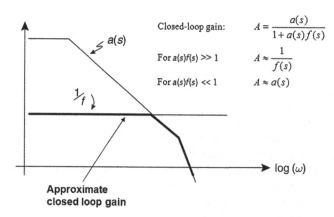

FIGURE 11.8

Plot for estimating closed-loop transfer function (A) graphically. The curve $a(s)$ depicts the frequency dependence of the forward path gain. The line $1/f$ is the inverse of the feedback gain, shown here for resistive feedback. The thick line indicates our estimate for closed-loop transfer function A. For $a(s)f \gg 1$, the closed-loop gain is approximately $1/f$. For $a(s)f \ll 1$, the closed-loop gain is approximately $a(s)$.

closed loop gain $A \approx 1/f$. For small loop transmission $(a(s)f(s) << 1)$, the closed-loop gain is approximately $a(s)$. If we plot $a(s)$ and $1/f$ on the same set of axes, we can find an approximation for the closed-loop gain as the lower of the two curves, as shown in Figure 11.8.

Pole locations, damping and relative stability

So far, we have not discussed the issue regarding the stability of closed-loop systems. There are many definitions of stability in the literature, but we will consider bounded input, bounded output (BIBO) stability. In other words, we will consider the stability problem given that we will only excite our system with bounded inputs. The system is BIBO stable if *bounded inputs* generate *bounded outputs,* a condition that is met if all poles are in the left-half plane (Figure 11.9).

Next we will consider some systems and their impulse responses and frequency responses. In Figure 11.10 we see the pole location, impulse response, and frequency response for a single-pole system, with the pole in the left-half plane. The transfer function of this system is:

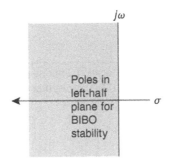

FIGURE 11.9

Closed-loop pole locations in the left-half plane for BIBO stability.

$$H(s) = \frac{1}{\tau s + 1}$$ (11.7)

$$\tau = 1 \text{ s}$$

This system is BIBO stable, as shown by the impulse response, which is a decaying exponential with time constant τ.

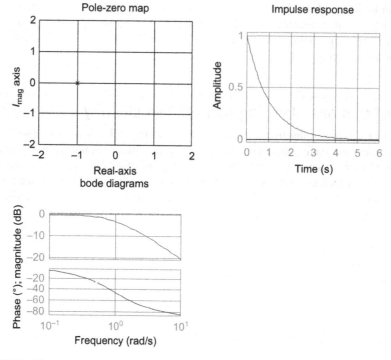

FIGURE 11.10

Single left-half-plane pole showing pole location, impulse response, and frequency response of the system. This system is BIBO stable. (For color version of this figure, the reader is referred to the online version of this book.)

In Figure 11.11 we see the pole location, impulse response, and frequency response for a single-pole system, with the pole in the right-half plane. The transfer function of this system is:

$$H(s) = \frac{1}{\tau s - 1}$$

$$\tau = 1 \text{ s}$$

(11.8)

The pole has a positive real part, and the system is BIBO *unstable*, as shown by the impulse response, which is a growing exponential with time constant τ.

Next, we will look at second-order systems. A second-order system with two complex poles can be characterized by:

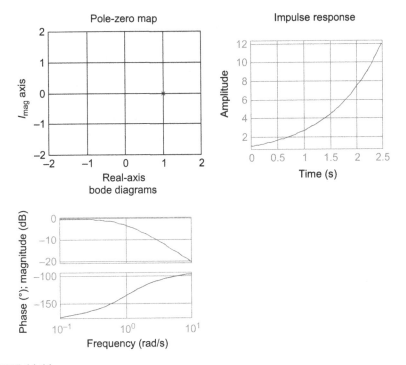

FIGURE 11.11

Single right-half-plane pole showing pole location, impulse response, and frequency response of the system. This system is BIBO unstable. (For color version of this figure, the reader is referred to the online version of this book.)

$$H(s) = \frac{1}{\frac{s^2}{\omega_n^2} + \frac{2\zeta s}{\omega_n} + 1} \qquad (11.9)$$

In Figure 11.12, we see a system with natural frequency $\omega_n = 1$ rad/s, damping ratio $\zeta = 0.1$, and two poles in the left-half plane. The impulse response is a decaying exponential. The system is BIBO stable, but the relative stability of the system depends on how high the damping ratio is.

In Figure 11.13 we see a system with natural frequency $\omega_n = 1$ rad/s and damping ratio $\zeta = 0$, with two poles in on the $j\omega$-axis. The impulse response is a sinusoid at 1 rad/s.

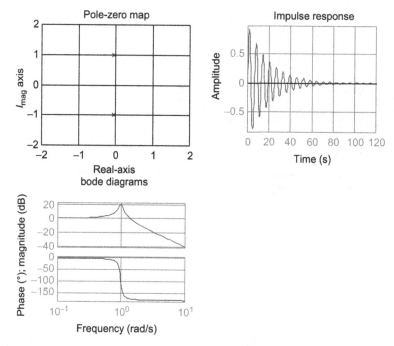

FIGURE 11.12

Complex pair of poles in the left-half plane with natural frequency 1 rad/s and damping ratio 0.1 showing pole locations, impulse response, and frequency response of the system. This system is BIBO stable but relatively less stable as the damping ratio decreases. (For color version of this figure, the reader is referred to the online version of this book.)

In Figure 11.14, we see a system with natural frequency $\omega_n = 1$ rad/s and damping ratio $\zeta = -0.1$, with two poles in the right-half plane. The system is BIBO unstable, and the impulse response is a growing sinusoid.

The effects of feedback on relative stability

When you apply feedback around a system that has poles and zeros, the closed-loop poles move. Consider the feedback system with an integrator and unity feedback (Figure 11.15). The input/output transfer function is:

$$\frac{v_o}{v_i} = \frac{\frac{A}{s}}{\frac{A}{s} + 1} = \frac{A}{s + A} = \frac{1}{\frac{s}{A} + 1} \tag{11.10}$$

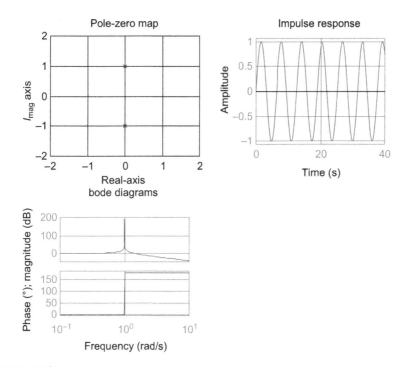

FIGURE 11.13

Complex pair of poles on the $j\omega$-axis with natural frequency 1 rad/s showing pole locations, impulse response, and frequency response of the system. This system is BIBO stable, but the impulse response oscillates at the natural frequency. (For color version of this figure, the reader is referred to the online version of this book.)

Note that as the integrator constant A increases, the closed-loop bandwidth increases as well, with the closed-loop pole staying on the real axis at $s = -A$. As long as A is positive, this system is BIBO stable for any value of A.

The second-order system (Figure 11.16(a)) is also easy to work out, with closed-loop transfer function:

$$\frac{v_o}{v_i} = \frac{\frac{K}{(\tau_a s + 1)(\tau_b s + 1)}}{1 + \frac{K}{(\tau_a s + 1)(\tau_b s + 1)}} = \frac{K}{K + (\tau_a s + 1)(\tau_b s + 1)}$$

$$= \left(\frac{K}{K+1}\right) \frac{1}{\left(\frac{\tau_a \tau_b}{1+K}\right) s^2 + \left(\frac{\tau_a \tau_b}{1+K}\right) s + 1} \tag{11.11}$$

The pole locations are plotted in Figure 11.16(b), with the locus of closed-loop poles shown for K increasing. Note the fundamental tradeoff between high DC open-loop gain (which means a small closed-loop DC error) and loop stability. For K approaching infinity, the closed-loop poles are very underdamped.

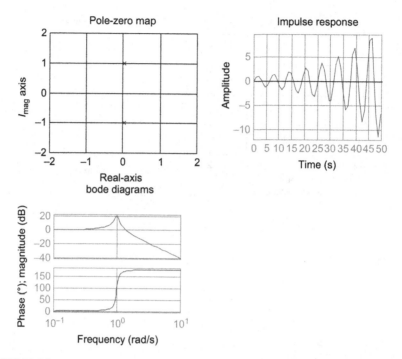

FIGURE 11.14

Complex pair of poles in the right-half s-plane showing pole locations, impulse response, and frequency response of the system. This system is BIBO unstable, with the impulse response being a growing sinusoid. (For color version of this figure, the reader is referred to the online version of this book.)

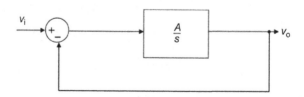

FIGURE 11.15

First-order system composed of an integrator inside a negative feedback loop. The closed-loop pole location varies as we vary the integrator constant A.

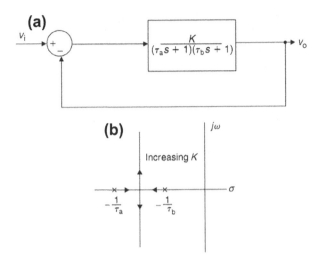

FIGURE 11.16

Second-order system inside a negative feedback loop. (a) Block diagram. (b) Location of closed-loop roots (also known as "root locus") as K increases.

Routh stability criterion (a.k.a. the "Routh test")

The Routh test is a mathematical test that can be used to determine how many roots of the characteristic equation lie in the right-half plane. When we use the Routh test, we do not calculate the location of the roots—rather we determine whether there are any roots at all in the right-half plane, without explicitly determining where they are.

The procedure for using the Routh test is as follows:

- Write the characteristic polynomial:

$$1 - \text{L.T.} = a_0 s^n + a_1 s^{n-1} + \cdots + a_n \tag{11.12}$$

Note that since we have written the characteristic polynomial $(1 - \text{L.T.})$, we now are interested in finding whether there are *zeros* of $(1 - \text{L.T.})$ in the right-half plane. Zeros of $(1 - \text{L.T.})$ in the right-half plane correspond to closed-loop poles in the right-half plane. Furthermore, we assume that $a_n \neq 0$ for the analysis to proceed.
- Next, we see if any of the coefficients are zero or have a different sign from the others. A necessary (but not sufficient) condition for stability is that there are no nonzero coefficients in the characteristic equation and that all coefficients have the same sign.
- If all coefficients have the same sign, we next form a matrix with rows and columns in the following pattern, which is shown for n even.[6] The table is

---

[6]For n odd, a_n terminates the second row.

filled horizontally and vertically until zeros are obtained in the rows. The third row and following rows are calculated from the previous two rows.

$$
\begin{array}{cccccc}
a_0 & a_2 & a_4 & \cdots & \cdots & \cdots \\
a_1 & a_3 & a_5 & \cdots & \cdots & \cdots \\
b_1 & b_2 & b_3 & \cdots & \cdots & \cdots \\
c_1 & c_2 & c_3 & \cdots & \cdots & \cdots \\
\cdots & \cdots & \cdots & \cdots & \cdots & \cdots \\
0 & 0 & 0 & 0 & 0 & 0
\end{array}
$$

$$
\begin{aligned}
b_1 &= \frac{a_1 a_2 - a_0 a_3}{a_1} \\
b_2 &= \frac{a_1 a_4 - a_0 a_5}{a_1} \\
b_3 &= \frac{a_1 a_6 - a_0 a_7}{a_1} \\
c_1 &= \frac{b_1 a_3 - a_1 b_2}{b_1} \\
c_2 &= \frac{b_1 a_5 - a_1 b_3}{b_1}
\end{aligned}
\tag{11.13}
$$

- The number of poles in the right-half plane is equal to the number of sign changes in the first column of the Routh matrix.

Example 11.2: Using the Routh test

Let us apply the Routh test to the transfer function:

$$
H(s) = \frac{1}{(s+1)(s+2)(s+3)(s-2)} = \frac{1}{s^4 + 4s^3 - s^2 - 16s - 12}
\tag{11.14}
$$

In this case, we already know that there is one right-half-plane pole at $s = +2$ rad/s, but we will use the Routh test to verify this. The Routh matrix is:

$$
\begin{array}{cccc}
1 & -1 & -12 & 0 \\
4 & -16 & 0 & 0 \\
\left(\dfrac{-\begin{vmatrix} 1 & -1 \\ 4 & -16 \end{vmatrix}}{-1} = -12 \right) & \left(\dfrac{-\begin{vmatrix} 1 & -12 \\ 4 & 0 \end{vmatrix}}{-1} = 48 \right) & \left(\dfrac{-\begin{vmatrix} 1 & 0 \\ 4 & 0 \end{vmatrix}}{-1} = 0 \right) & 0 \\
\left(\dfrac{-\begin{vmatrix} 4 & -16 \\ -12 & 48 \end{vmatrix}}{-12} = 0 \right) & \left(\dfrac{-\begin{vmatrix} 4 & 0 \\ -12 & 0 \end{vmatrix}}{-12} = 0 \right) & 0 & 0
\end{array}
$$

$$
\tag{11.15}
$$

We see that there is one sign change in the first column, with the elements of the matrix changing from $+4$ to -12. Hence, there is one right-half-plane pole, as expected.

Let us next apply the Routh test to a system with three poles inside a unity-feedback loop (Figure 11.17). We will use the Routh test to determine the values of K that result in stable operation of this feedback loop. The closed-loop transfer function for this system is:

$$\frac{v_o(s)}{v_i(s)} = \frac{\frac{K}{(s+1)^3}}{1 + \frac{K}{(s+1)^3}} = \left(\frac{K}{1+K}\right)\left(\frac{1}{\frac{s^3}{K+1} + \frac{3s^2}{K+1} + \frac{3s}{1+K} + 1}\right) \tag{11.16}$$

The denominator polynomial is:

$$D(s) = a_0 s^3 + a_1 s^2 + a_2 s + a_3 = \left(\frac{1}{K+1}\right)s^3 + \left(\frac{3}{K+1}\right)s^2 + \left(\frac{3}{K+1}\right)s + 1 \tag{11.17}$$

The Routh matrix is:

$$\begin{array}{cc}
\left(\dfrac{1}{K+1}\right) & \left(\dfrac{3}{K+1}\right) \\[3mm]
\left(\dfrac{3}{K+1}\right) & 1 \\[3mm]
\dfrac{\left(\dfrac{3}{K+1}\right)^2 - \left(\dfrac{1}{K+1}\right)}{\left(\dfrac{1}{K+1}\right)} = \dfrac{8-K}{3(K+1)} & 0 \\[5mm]
1 & 0 \\[2mm]
0 & 0
\end{array} \tag{11.18}$$

Note that if $K > 8$, there are two sign changes in the first column. Therefore, for $K = 8$, we expect two poles on the $j\omega$-axis, and for $K > 8$, the system is unstable with two poles in the right-half plane. For $K < 8$, the system is stable with all three poles in the left-half plane.

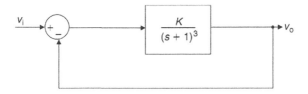

FIGURE 11.17

Three poles inside unity-feedback loop, with stability determined by the Routh test. (Example 11.2)

The phase margin and gain margin tests

The previous analyses tell us what the bandwidth and DC gain of a closed-loop system is, but do not consider the question of whether a system will oscillate or have large amounts of overshoot. By using a simple Bode plot technique and a method known as the *phase margin method,* the relative stability of a feedback system can be determined. Phase margin is a very useful measure of the stability of a feedback system. The method for finding phase margin for a negative feedback system is as follows (Figure 11.18):

- Plot the negative of the magnitude and angle of the L.T. In a negative feedback loop, the negative of the L.T. is $a(s)f(s)$.
- Find the frequency where the magnitude of $a(j\omega)f(j\omega)$ drops to $+1$. This is the *crossover frequency* ω_c.
- The difference between the angle at the cross-over frequency and $-180°$ is the *phase margin* ϕ_m.
- The *gain margin* (G.M.) is defined as the change in open-loop gain required to make the system unstable. Systems with greater G.M.s can withstand greater changes in system parameters before becoming unstable in closed loop.

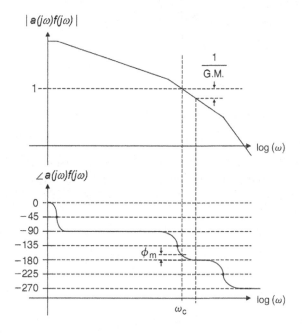

FIGURE 11.18

Plot of the negative of the L.T. $a(j\omega)f(j\omega)$ magnitude and phase, showing crossover frequency ω_c, phase margin (ϕ_m) and gain margin (G.M.)

- The *phase margin* is defined as the negative change in open-loop phase shift required to make a closed-loop system unstable.
- In general, a well-designed feedback loop has a phase margin of at least 45° and a G.M. > 3 or so. Sometimes in real-world systems you need to live with less gain margin, or phase margin, with associated overshoot in the transient response.

Relationship between damping ratio and phase margin

The damping ratio and phase margin in feedback systems are directly related. For a second-order system, a low phase margin in general implies a low damping ratio. For a standard second-order system with damping ratio less than about 0.6, the relationship is approximately:

$$\varsigma \approx \frac{\phi_m}{100} \tag{11.19}$$

Therefore, a damping ratio of 0.6 corresponds to a phase margin of 60°. The actual relationship over the range of damping ratios $0 < \zeta < 2$ is shown in Figure 11.19.

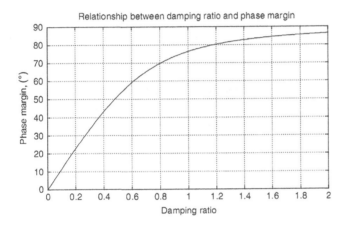

FIGURE 11.19

Relationship between phase margin and damping ratio.

Phase margin, step response, and frequency response

Once we know the phase margin, we can estimate other important items of interest in the closed-loop response of our system. For instance, we may want to know how much the step response overshoots, as shown in Figure 11.20a. If we know the phase margin, we can estimate the damping ratio by $\zeta \cong \phi_m/100$ and use it to find the peak overshoot:

$$P_0 = 1 + e^{-\frac{\pi\zeta}{\sqrt{1-\zeta^2}}} \tag{11.20}$$

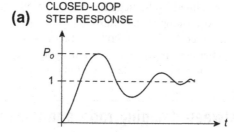

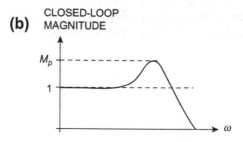

FIGURE 11.20

Generic closed-loop system response. (a) Step response, showing the peak overshoot (P_o) in the step response. (b) Frequency response, showing peaking in the magnitude (M_p).

We may also want to know the magnitude of the peaking in the frequency response (Figure 11.20(b)). A semiempirical relationship derived from typical closed-loop transfer functions results in:

$$M_p \approx \frac{1}{\sin(\phi_m)} \tag{11.21}$$

These relationships show that as phase margin increases, the system becomes relatively more stable, with less step response peak overshoot (P_o) and less frequency response peaking (M_p).

Example 11.3: Unity-gain amplifier

Next we will consider a numerical example, assuming an op-amp with open-loop transfer function

$$a(s) = \frac{10^5}{(0.1s + 1)(10^{-5}s + 1)} \tag{11.22}$$

The op-amp is connected as a unity-gain amplifier (Figure 11.21(a)), with resultant block diagram in Figure 11.21(b). We will apply the phase margin test and estimate the step response peak overshoot and frequency response peak. The negative of the L.T. for this system is:

$$-\text{L.T.} = \frac{10^5}{(0.1s + 1)(10^{-5} + 1)} \tag{11.23}$$

(a)   **(b)**

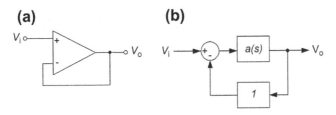

FIGURE 11.21

Unity-gain amplifier for Example 11.3. (a) Circuit. (b) Block diagram. (For color version of this figure, the reader is referred to the online version of this book.)

When we plot the magnitude of $-a(s)f(s)$ we see a cross-over frequency of 3.08×10^5 rad/s and a phase margin of 18° (Figure 11.22). Next, let us estimate damping ratio, step response peak overshoot, and frequency response peaking.

$$\zeta \approx \frac{\phi_m}{100} \rightarrow \zeta \approx 0.18$$

$$M_p \approx \frac{1}{\sin(\phi_m)} = 3.23(10 \text{ dB}) \tag{11.24}$$

$$P_o = 1 + e^{\frac{-\pi\zeta}{\sqrt{1-\zeta^2}}} \rightarrow P_o \approx 1.58$$

$a(s)f(s)$

$G_m = \text{Inf},\ P_m = 17.966°.\ (\text{at } 3.0842e + 005 \text{ rad/s})$

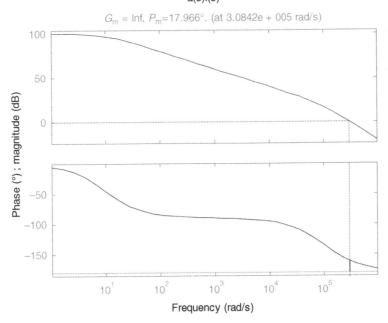

FIGURE 11.22

Negative of the L.T. $a(s)f(s)$ for the circuit in Example 11.3, showing a cross-over frequency $\omega_c = 3.08 \times 10^5$ rad/s and a phase margin $\phi_m = 18°$. (For color version of this figure, the reader is referred to the online version of this book.)

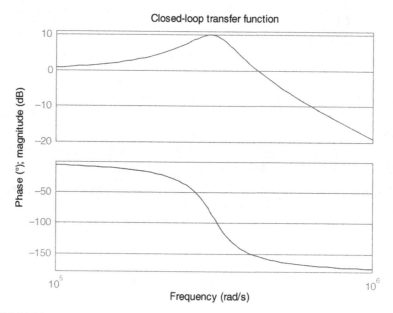

FIGURE 11.23

Closed-loop frequency response for the unity gain amplifier in Example 11.3, showing frequency response peaking $M_p \sim 10$ dB. (For color version of this figure, the reader is referred to the online version of this book.)

When we compare these estimates to the closed-loop frequency response (Figure 11.23) and closed-loop unit step response (Figure 11.24), we see that the phase margin test does a good job of estimating these parameters.

Example 11.4: Gain of $-k$ amplifier

Next we will show how to generate the block diagram for an inverting op-amp circuit with a gain of $-k$ (Figure 11.25(a)). From superposition, we know that the voltage at the negative input of the op-amp ($v-$) is a function of both the output voltage v_o and the input voltage v_i, as:

$$v- = v_i\left(\frac{kR}{R+kR}\right) + v_o\left(\frac{R}{R+kR}\right) = v_i\left(\frac{k}{1+k}\right) + v_o\left(\frac{1}{1+k}\right) \qquad (11.25)$$

We also know how the output relates to the negative input voltage:

$$v_o = -a(s)v- \qquad (11.26)$$

We see the resultant block diagram in Figure 11.25(b). The resultant closed-loop transfer function is:

$$\frac{v_o}{v_i} = \left(\frac{k}{1+k}\right)\left(\frac{-a(s)}{1+\frac{a(s)}{1+k}}\right) \approx -k \quad \text{if} \quad \frac{a(s)}{1+k} \gg 1 \qquad (11.27)$$

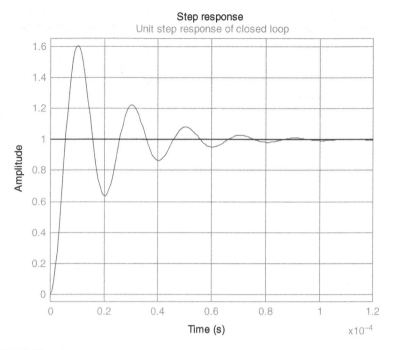

FIGURE 11.24

Closed-loop unit step response for the unity gain amplifier in Example 11.3, showing step response peaking $P_o \sim 1.6$.

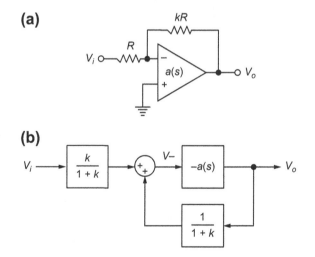

FIGURE 11.25

Op-amp circuit with a closed-loop gain of $-k$ (Example 11.4). (a) Circuit. (b) Block diagram for the circuit of Example 11.3.

Loop compensation techniques—lead and lag networks

Several networks are available to compensate feedback networks. These networks can be added in series to the plant to modify the closed-loop transfer function, or placed in other locations in the feedback system. Next, a quick look at "lead" and "lag" networks is presented.

The lag network (Figure 11.26(a)) is often used to reduce the gain of the L.T. so that cross-over occurs at a benign frequency. The transfer function of the lag network is:

$$H(s) = \frac{v_o(s)}{v_i(s)} = \frac{R_2Cs + 1}{(R_1 + R_2)Cs + 1} = \frac{\tau s + 1}{\alpha \tau s + 1}$$

$$\alpha = \frac{R_1 + R_2}{R_2} \tag{11.28}$$

$$\tau = R_2C$$

The Bode plot of the lag network (Figure 11.26(b)) shows that the network produces magnitude reduction at frequencies between the pole and the zero. When

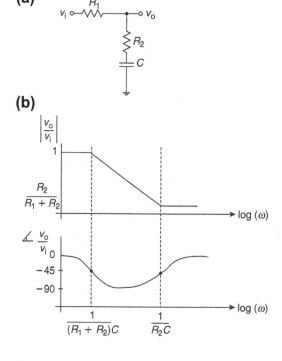

(a)

(b)

FIGURE 11.26

Lag network. (a) Circuit. (b) Bode plot of magnitude and phase angle of the frequency response of the lag network.

using a lag network, you will typically place the lag zero well below the cross-over frequency of the loop. This ensures that the lag network does not provide too much negative phase shift at cross-over.

The lead network (Figure 11.27(a)) is used to provide positive phase shift in the vicinity of the cross-over frequency. The transfer function of the lead network is:

$$H(s) = \frac{v_o(s)}{v_i(s)} = \left(\frac{R_2}{R_1 + R_2}\right)\left(\frac{R_1 Cs + 1}{\dfrac{R_1 R_2}{R_1 + R_2}Cs + 1}\right) = \frac{1}{\alpha}\frac{\alpha\tau s + 1}{\tau s + 1}$$

(11.29)

$$\alpha = \frac{R_1 + R_2}{R_2}$$

$$\tau = (R_1 \parallel R_2)C$$

(a)

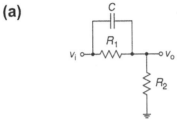

(b)

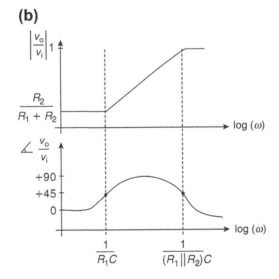

FIGURE 11.27

Lead network. (a) Circuit. (b) Bode plot of magnitude and phase angle of the frequency response of the lead network.

The Bode plot of the lead (Figure 11.27(b)) shows that the lead provided $+45°$ of positive phase shift at the zero frequency, while at the zero, there is only $+3$ dB of gain increase. When using a lead network, one generally places the lead zero near the cross-over frequency of the loop to take advantage of the positive phase shift provided by the lead. The lead pole is then above cross-over.

Parenthetical comment on some interesting feedback loops

The inquisitive student may wonder whether a system that has a loop transmission magnitude greater than unity where the loop transmission angle is $-180°$ can be stable or not. In using the G.M./phase margin test, we look at the frequency at which the magnitude drops to unity and do not concern ourselves with other frequencies. By example, we will show next that a system that has a loop transmission magnitude greater than unity where the loop transmission angle is $-180°$ *can* be conditionally stable. It is understood that this is not necessarily an intuitive result, but we will run with it anyway. Consider the system of Figure 11.28(a), which is a unity-feedback system with two zeros and three poles in the forward path. The negative of the loop transmission for this system is:

$$-\text{L.T.} = a(s)f(s) = \frac{100(s+1)^2}{s^2} \tag{11.30}$$

A plot of the negative of the loop transmission is shown in Figure 11.28(b). Note that the loop transmission magnitude is greater than unity at the frequency when the angle of the negative of the loop transmission is $-180°$. In this case, the angle is less than $-180°$ up to 1 rad/s.

We will next use the Routh test to determine the stability of this system. The closed-loop transfer function for this system is:

$$\frac{v_o(s)}{v_i(s)} = \frac{\frac{100(s+1)^2}{s^3}}{1 + \frac{100(s+1)^2}{s^3}} = \frac{100(s^2 + 2s + 1)}{s^3 + 100s^2 + 200s + 100} \tag{11.31}$$

We can use the Routh criteria or we can factor the denominator of this transfer function to determine stability. The denominator polynomial is:

$$D(s) = s^3 + 100s^2 + 200s + 100 \tag{11.32}$$

The Routh matrix is as follows:

$$
\begin{array}{cc}
1 & 200 \\
100 & 100 \\
\dfrac{(100)(200) - (1)(100)}{100} = 199 & 0 \\
\dfrac{(199)(100) - (100)(0)}{199} = 100 & 0 \\
0 & 0
\end{array}
\tag{11.33}
$$

(a)

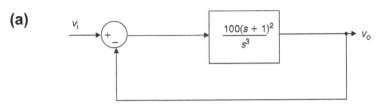

$$\frac{100(s + 1)^2}{s^3}$$

(b)

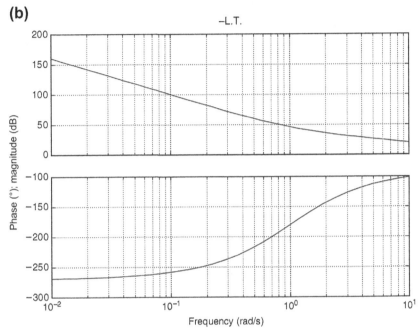

FIGURE 11.28

Unity-feedback system that has a L.T. magnitude greater than unity where the L.T. angle is −180°. (a) System. (b) Bode plot of −L.T. = $a(s)f(s)$ showing that the magnitude is greater than 1 when the angle is more negative than −180°.

The Routh test shows that this system is BIBO stable since there are no sign changes in the first column of the Routh matrix. Numerical analysis shows that the closed-loop poles and zeros are at following frequencies:

Zeros: Two zeros at −1 rad/s
Poles: Poles at −97.97 rad/s, −1.12 rad/s, and −0.92 rad/s

Therefore, all poles are in the left-half plane and the system is BIBO stable. The closed-loop step response (Figure 11.29) confirms that the system is stable. Note the

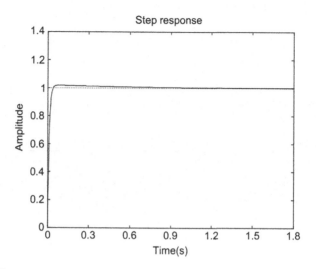

FIGURE 11.29

Step response of unity-feedback system that has a L.T. magnitude greater than unity when the L.T. angle is $-180°$.

long decaying "tail" of the step response, while it settles to unity gain. This long tail is characteristic of systems with closely spaced poles and zeros.[7]

Example 11.5: Another unity-gain amplifier

Consider an operational amplifier with a DC gain of 10^5, a low-frequency pole at 10 rad/s, and a high-frequency pole at 10^6 rad/s. This transfer function $a(s)$ is representative of many commercially available operational amplifiers and is expressed as:

$$a(s) = \frac{10^5}{(0.1s + 1)(10^{-6}s + 1)} \tag{11.34}$$

What is the bandwidth and risetime when this op-amp is configured as a gain of $+1$ amplifier?

Shown in Figure 11.30 is the Bode plot of the forward path gain $a(s)$. Note that the DC gain is 10^5 (100 dB) and the phase starts at $0°$ and falls asymptotically to $-180°$ at frequencies much higher than 10^6 rad/s (above the second pole).

The gain of $+1$ op-amp circuit is shown in Figure 11.31(a) where the op-amp has unity feedback. The block diagram of this circuit is shown in Figure 11.31(b).

For this circuit, the negative of the loop transmission is:

$$-\text{L.T.} = a(s)f(s) = \frac{10^5}{(0.1s + 1)(10^{-6}s + 1)} \tag{11.35}$$

---

[7]Also known as a pole/zero "doublet".

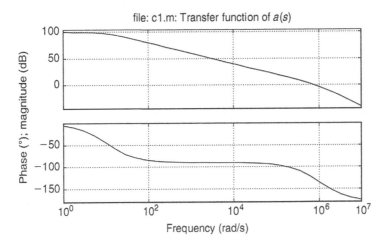

FIGURE 11.30

Open-loop transfer function for forward path gain $a(s)$ for Examples 11.5 and 11.6.

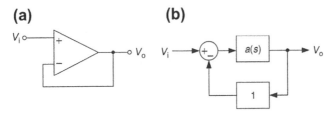

FIGURE 11.31

Gain of $+1$ op-amp circuit. (a) Circuit. (b) Block diagram.

which is the same as the open-loop transfer function of the op-amp $a(s)$ since we have unity feedback. Using MATLAB, the bandwidth and phase margin are calculated, as shown in Figure 11.32. The results show a phase margin of 52° and a cross-over frequency of 786,150 rad/s (125 kHz). From this analysis, we expect some overshoot in the step response (since the phase margin results in a damping ratio of $\sim$0.5 resulting in an estimated peak overshoot $P_o \sim 1.15$), some overshoot in the frequency response ($M_p \sim 1.27$), and a 10–90% risetime[8] of approximately $0.35/125,000 = 2.8$ μs (Figure 11.33).

Example 11.6: Gain of $+10$ amplifier

What is the bandwidth and risetime when the same op-amp from the previous example is configured as a gain of $+10$ amplifier? The gain of $+10$ op-amp circuit

---

[8]In general, an estimate of 10–90% risetime is $0.35/f_c$, where f_c is the cross-over frequency in hertz.

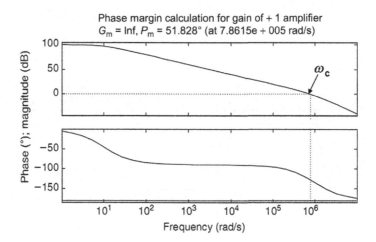

FIGURE 11.32

Phase margin and cross-over frequency calculation for gain of +1 amplifier of Example 11.5 circuit, showing a cross-over frequency $\omega_c = 7.86 \times 10^5$ rad/s and a phase margin $\phi_m = 51.8°$.

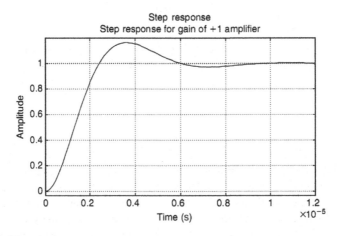

FIGURE 11.33

Step response for gain of +1 amplifier of Example 11.5. Estimated peak overshoot $P_o \sim 1.15$ based on the phase margin of 52°, confirmed by simulation.

is shown in Figure 11.34(a), and the block diagram is shown in Figure 11.34(b). Note that the $9R/R$ voltage divider gives a feedback factor $f = 0.1$.

For this circuit, the negative of the loop transmission is:

$$-\text{L.T.} = a(s)f(s) = \left(\frac{10^5}{(0.1s + 1)(10^{-6}s + 1)} \right)\left(\frac{1}{10} \right) = \frac{10^4}{(0.1s + 1)(10^{-6}s + 1)}$$

$$(11.36)$$

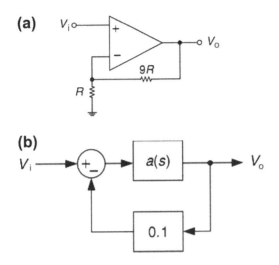

FIGURE 11.34

Gain of +10 op-amp circuit for Example 11.6. (a) Circuit. (b) Block diagram.

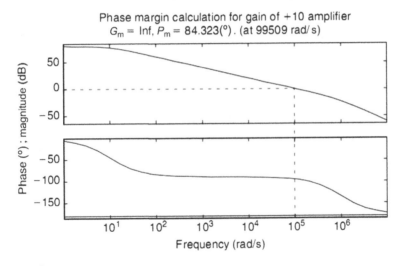

FIGURE 11.35

Phase margin and crossover frequency calculation for gain of +10 op-amp circuit of Example 11.6, showing a crossover frequency $\omega_c = 99,509$ rad/s and a phase margin $\phi_m = 84°$.

When we plot the negative of the loop transmission magnitude and phase (Figure 11.35), the results show a phase margin of 84° and a crossover frequency of 99,509 rad/s (15.8 kHz). From this analysis, we expect that once we close the loop, there will be no overshoot in the step response and no peaking in the frequency response, since the phase margin is close to 90°. The frequency response

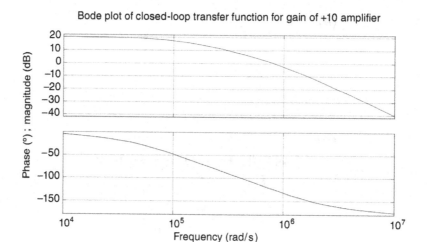

FIGURE 11.36

Bode plot of closed-loop transfer function for gain of +10 (20 dB) amplifier of Example 11.6. The low-frequency magnitude is +10 (20 dB) and the closed-loop bandwidth is about 10^5 rad/s.

is well behaved as shown in Figure 11.36. We expect a 10–90% risetime of approximately $0.35/15,800 = 22$ μs with no overshoot, and this is confirmed in the step response (Figure 11.37).

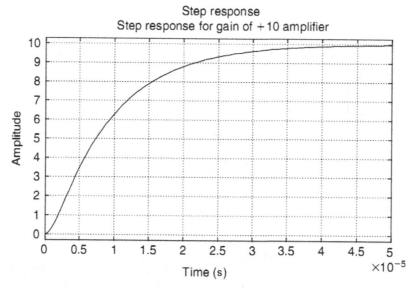

FIGURE 11.37

Step response for gain of +10 amplifier of Example 11.6.

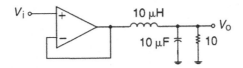

FIGURE 11.38

Ideal op-amp driving a reactive load for Example 11.7.

Example 11.7: Integral control of a reactive load

Consider an ideal op-amp driving a reactive load as shown in Figure 11.38. Assume that the ideal op-amp has infinite bandwidth and can source and sink infinite current. Given this, the transfer function is:

$$H(s) = \frac{v_o(s)}{v_i(s)} = \frac{1}{LCs^2 + \frac{L}{R}s + 1} = \frac{1}{10^{-10}s^2 + 10^{-6}s + 1} \qquad (11.37)$$

This second-order reactive load $H(s)$ has the following:

- Poles: $-5 \times 10^3 \pm 9.99 \times 10^4 j$
- Natural frequency: 10^5 rad/s
- Damping ratio: 0.05
- $Z_o = 1\,\Omega$
- Q: 10

When plotting this transfer function $H(s)$, we see the underdamped response of Figure 11.39.

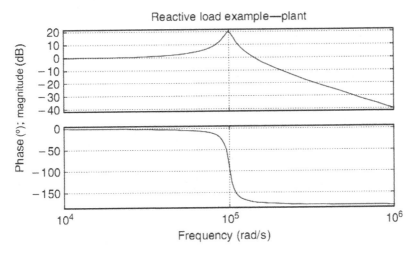

FIGURE 11.39

Plot of frequency response of $H(s)$ for example 11.7.

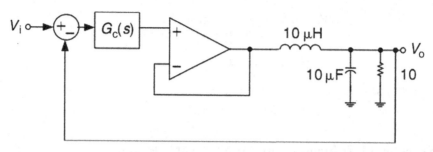

FIGURE 11.40

Closed-loop controller for Example 11.7.

In this example, we will design a closed-loop controller to regulate the output voltage (Figure 11.40). We will assume that the forward path compensator $G_c(s)$ includes an integrator, so that there will be zero DC error[9] in V_0.

As a first attempt, let us try a compensator transfer function of the form:

$$G_c(s) = \frac{4 \times 10^3}{s} \tag{11.38}$$

The resultant loop transmission is:

$$-\text{L.T.} = \left(\frac{1}{10^{-10}s^2 + 10^{-6}s + 1}\right)\left(\frac{4 \times 10^3}{s}\right) \tag{11.39}$$

Plotting this loop transmission magnitude and phase yields some interesting results, as shown in Figure 11.41.

Results are:

- Cross-over frequency: 4006 rad/s (637 Hz)
- Phase margin: 90°
- Given this, we expect a well-controlled step response, with risetime ~0.5 ms.

The resultant step response of the closed-loop system is shown in Figure 11.42.

Things look OK except for the oscillatory behavior on the rising edge. What is going on here?

The Bode plot of the loop transmission phase and magnitude tells the story. Although the phase margin is fine, the gain margin is not so great (only about 8 dB), due to the underdamped pole pair. In order to help this oscillation problem, we will add a pole above the cross-over frequency to damp out the underdamped pair. Let us try adding a pole at 5×10^4 rad/s to damp the complex poles at 10^5 rad/s. This results in a compensator transfer function of:

$$G_c(s) = \left(\frac{4 \times 10^3}{s}\right)\left(\frac{1}{2 \times 10^{-5}s + 1}\right) \tag{11.40}$$

---

[9]Having an integrator in $G_c(s)$ ensures that the DC error of the loop is zero. If there were an error, the integrator would forever integrate to infinity. Hence, there must be zero DC error.

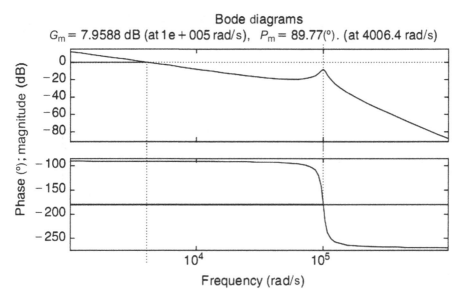

FIGURE 11.41

Plot of magnitude and phase of the L.T., Attempt #1 (Example 11.7), showing a cross-over frequency of 4006 rad/s, a phase margin of 89.8°, and a gain margin of 8 dB.

The results of a MATLAB crossover frequency and phase margin test using the revised controller are shown in Figure 11.43. Note that, by adding the low-pass filter, we have significantly improved the gain margin of this circuit. Results show:

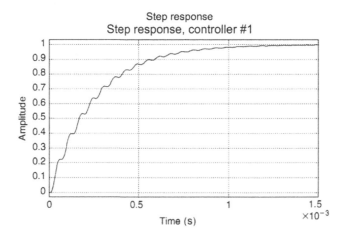

FIGURE 11.42

Step response, Attempt #1 (Example 11.7), showing high-frequency oscillation on the rising waveform due to low gain margin.

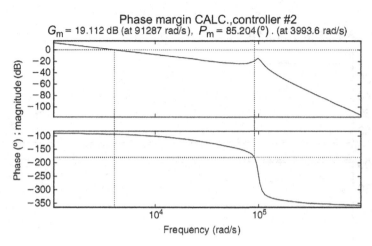

FIGURE 11.43

Plot of the L.T. magnitude and phase, Attempt #2 (Example 11.7), showing cross-over frequency of 3993 rad/s, phase margin of 85°, and gain margin G.M. increased to 19 dB.

- Crossover frequency: 3993 rad/s (635 Hz)
- Phase margin: 85°
- Better-behaved step response, as shown in Figure 11.44.

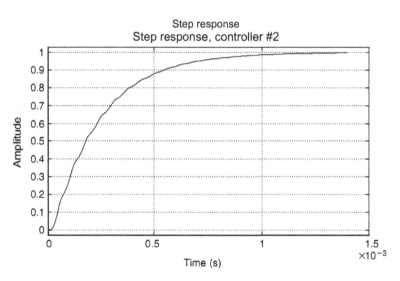

FIGURE 11.44

Step response, Attempt #2 (Example 11.7), with greatly reduced oscillations due to improved gain margin.

Example 11.8: Photodiode amplifier

A photodiode amplifier is shown in Figure 11.45(a). The photodiode puts out a current proportional to the light hitting it, and the *transimpedance* connection of the op-amp converts this photodiode current into an output voltage. The ideal input-to-output transfer function is:

$$\frac{v_o}{i_p} = -R_f \tag{11.41}$$

When modeling this circuit, the photodiode is modeled as a current source in parallel with a parasitic capacitance C_p, as shown in Figure 11.45(b).

The block diagram for this system is shown in Figure 11.46. Note that the input is the photodiode current and the output is the op-amp output voltage.

The closed-loop transfer function of this system is:

$$\frac{v_o}{i_p} = -\frac{R_f}{R_f C_p s + 1}\left(\frac{a(s)}{1 + \frac{a(s)}{R_f C_p s + 1}}\right) \tag{11.42}$$

(a)

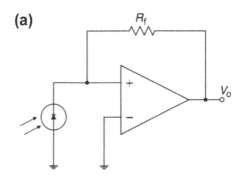

(b)

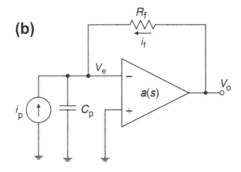

FIGURE 11.45

Photodiode amplifier for Example 11.8. (a) Circuit, with op-amp configured as a transimpedance amplifier. (b) Model, showing equivalent model of photodiode that includes a current source i_p and parasitic capacitance C_p.

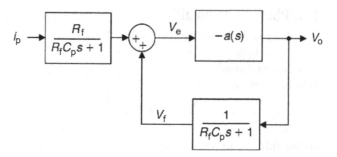

FIGURE 11.46

Block diagram of photodiode amplifier for Example 11.8. The block $a(s)$ is the open-loop gain of the operational amplifier.

Note that if the loop transmission is much larger than 1, the approximate transfer function is $-R_f$. The loop transmission of this system, easily found by inspection, is:

$$-\text{L.T.} = \frac{a(s)}{R_f C_p s + 1} \tag{11.43}$$

In general, an operational amplifier will have a dominant pole and a second pole near the cross-over frequency, resulting in:

$$-\text{L.T.} = \frac{a_o}{s(\tau s + 1)(R_f C_p s + 1)} \tag{11.44}$$

This means that there are three poles (at least) in the L.T. If we attempt to close a feedback loop with a bandwidth greater than $1/R_f C_p$, there are potential problems with stability.

A PSPICE model of an actual system, using the CLC426 op-amp, was created (Figure 11.47), including dominant pole, second pole, and output resistance. Parasitics in this circuit are as follows:

- L_{cable}: inductance of the cable connecting photodiode to printed-circuit board. Approximate inductance is 10 nH/cm of length. A value of 50 nH was used for all simulations.
- L_{pr}: inductance in series with feedback resistance; approximately 10 nH.
- L_{pc}: inductance in series with feedback capacitor; approximately 5 nH.
- C_p. capacitance of photodiode; approximately 60 pF.
- C_{in}: input capacitance of op-amp; approximately 5 pF.

The $R_f C_p$ combination results in a pole inside the feedback loop that results in potential instability. Plotted in Figure 11.48 is frequency response of output voltage vs. photodiode current as the input (v_o/i_p), of the original circuit with $R_f = 15$ kΩ, showing that the response is very underdamped. Other unmodeled poles can result in oscillation.

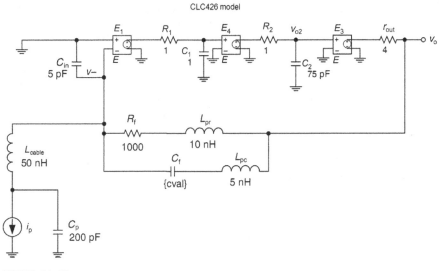

FIGURE 11.47

Photodiode amplifier model for Example 11.8. The op-amp is a CLC426 op-amp and a PSPICE model created by the author.

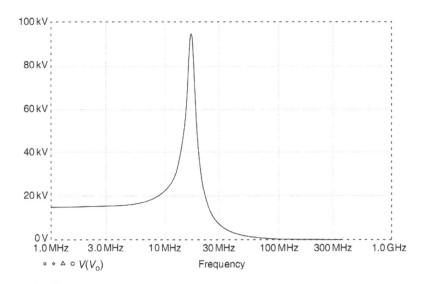

FIGURE 11.48

PSPICE results of frequency response of the photodiode amplifier of Example 11.8, showing potential instability near 15 MHz.

By adding a capacitance across the feedback resistor, a *lead* transfer function is created as in Figure 11.49. The added lead zero can create positive phase shift near the cross-over frequency, hence improving stability. The feedback factor becomes:

$$f(s) = \frac{\frac{1}{C_p s}}{\frac{1}{C_p s} + \frac{R_f}{R_f C_f s + 1}} = \frac{1}{1 + \frac{R_f C_p s}{R_f C_f s + 1}} = \frac{R_f C_f s + 1}{R_f (C_f + C_p) s + 1} \qquad (11.45)$$

Simulations were modified for $R_f = 1\ \text{k}\Omega$ and C_f adjustable from 2 pF to 10 pF. Results show (Figure 11.50) that it may be possible to achieve ~50 MHz bandwidth by proper adjustment of C_f. This, of course, depends on the accuracy of these simulations. Measurements should be made on the prototype to verify/refute these models.

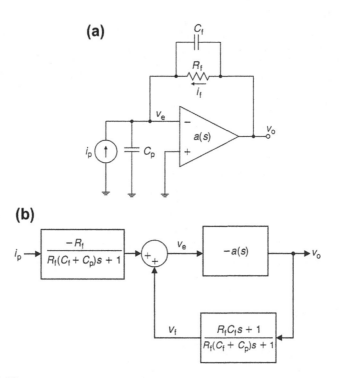

FIGURE 11.49

Photodiode amplifier of Example 11.8 with added feedback capacitor C_f creating a lead compensator. (a) Circuit. (b) Block diagram of lead-compensated circuit.

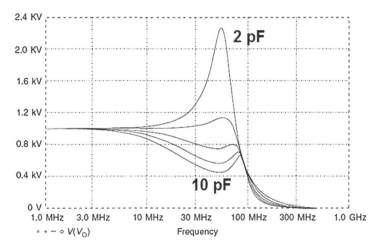

FIGURE 11.50

Frequency response results for photodiode amplifier of Example 11.8 with lead compensation capacitor $C_f = 2, 4, 6, 8,$ and 10 pF, with $R_f = 1$ kΩ and $C_p = 60$ pF. It looks like a feedback capacitor of about 4 pF is optimal.

Example 11.9: The bad effects of a time delay inside a feedback loop

Introduction of a pure time delay inside a feedback loop can cause stability problems. A pure time delay has a magnitude of $+1$ over all frequencies and a negative phase shift that is proportional to frequency. The reader should understand by now that extra negative phase shift inside a feedback loop can result in instability.

Let us consider the model of a microphone, amplifier, and speaker (Figure 11.51(a)) with the microphone being at a distance d from the speaker. Due to the finite speed of sound (about 1126 ft/s in dry air at room temperature), there is a time delay T in the feedback loop composed of the output of the speaker to the input of the microphone.

A time delay in the Laplace domain has transfer function e^{-sT} where "s" is the Laplace variable and T is the time delay. We will also assume that the magnitude of the fed-back sound decreases as distance d squared and that the amplifier–speaker combination has a sound level gain of K. The resultant block diagram is shown in Figure 11.51(b).

The closed-loop gain for this system is:

$$\frac{\text{Output}}{\text{Input}} = \frac{K}{1 - \frac{Ke^{-sT}}{d^2}} \tag{11.46}$$

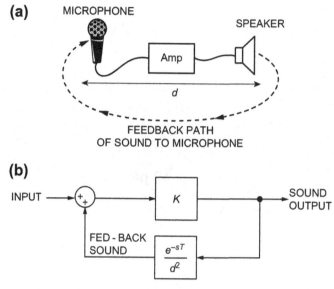

FIGURE 11.51

Speaker–microphone feedback for Example 11.9. (a) System with microphone, amplifier, and speaker output with a feedback path from the speaker to microphone. (b) Block diagram, assuming the sound level decreases as distance squared and that the amplifier and speaker have a gain of K.

where K is a constant corresponding to where the volume knob is set on the amplifier, T is the time delay, and d is the distance from the speaker to the microphone. The poles of this transfer function are found when:

$$\frac{Ke^{-sT}}{d^2} = 1 \tag{11.47}$$

Solving for "s", we find:

$$s = -\left(\frac{1}{T}\right)\ln\left(\frac{d^2}{K}\right) \tag{11.48}$$

In order for "s" to always be negative (meaning all poles are in the left-half plane and there is no screeching oscillation), we need

$$\frac{d^2}{K} > 1 \quad \text{for stability} \tag{11.49}$$

This corresponds to our intuition because we know that to reduce the screeching feedback we either need to turn down the volume (decrease K) or increase the distance between the microphone and speaker (increase d).

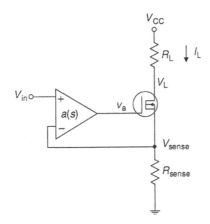

FIGURE 11.52

MOSFET current source for Example 11.10. The voltage V_{sense} senses MOSFET drain current.

Example 11.10: MOSFET current source

Figure 11.52 shows a metal oxide–semiconductor field-effect transistor (MOSFET) current source, with an op-amp used in a negative feedback configuration to maintain control of the MOSFET drain current. If the op-amp is ideal and if the MOSFET is operating in the linear region, the input–output transfer function is:

$$\frac{I_L}{V_{in}} \approx \frac{1}{R_{sense}} \tag{11.50}$$

This result is contingent on the feedback control system being stable (i.e. not oscillating). A small-signal model is shown in Figure 11.53. Following are the parameters:

τ_h: time constant of op-amp high-frequency pole
r_{out}: output resistance of op-amp
C_{gs}: MOSFET gate–source capacitance
C_{gd}: MOSFET gate–drain capacitance
g_m: MOSFET transconductance
R_L: load resistance

The transfer function of the MOSFET source follower is estimated using the method of open-circuit time constants. This method assumes that a single pole

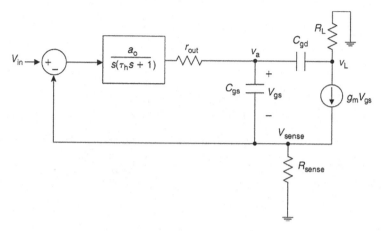

FIGURE 11.53

MOSFET current source of Example 11.10—small-signal model.

dominates the transfer function; the resultant transfer function from the output of the op-amp to v_{sense} is:

$$\frac{v_{sense}}{v_a} \approx \frac{A_o}{\tau s + 1}$$

$$A_o = \frac{g_m R_{sense}}{1 + g_m R_{sense}}$$

$$\tau = \left[\frac{r_{out} + R_{sense}}{1 + g_m R_{sense}}\right] C_{gs} + [r_{out} + R_L + G_M r_{out} R_L] C_{gd} \tag{11.51}$$

$$G_M = \frac{g_m}{1 + g_m R_{sense}}$$

The system was modeled assuming a TLO84 op-amp (with gain–bandwidth product of 4 MHz) and an IRF7403 MOSFET with $g_m = 10$ A/V, $C_{gs} = 1040$ pF, and $C_{gd} = 160$ pF and a load resistance $R_L = 1\,\Omega$. We know that the transconductance of a MOSFET varies with current level, but we will try to approximate it as being constant around some nominal operating point. In Figure 11.54 we see the results showing a cross-over frequency of 6.7 Mrad/s and a phase margin of 28°. Due to the low phase margin we expect a peak overshoot $P_o \sim 1.4$. This is confirmed in the simulation result shown in Figure 11.55.

By adding lag compensation (Figure 11.56(a)) the system can be stabilized to provide more phase margin. Figure 11.56(b) shows the block diagram of the system. The lag compensation adds a zero in the L.T. at $-R_i C$ and a lag pole at $-(R_f + R_i)C$.

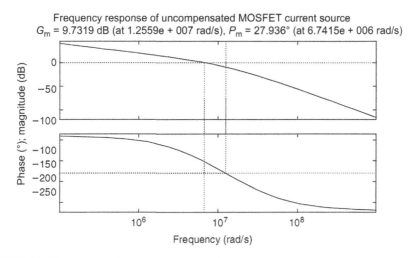

FIGURE 11.54

MOSFET current source of Example 11.10. Bode plot of L.T. showing crossover frequency of 6.7 Mrad/s, G.M. of 9.7 dB, and a phase margin of 28°.

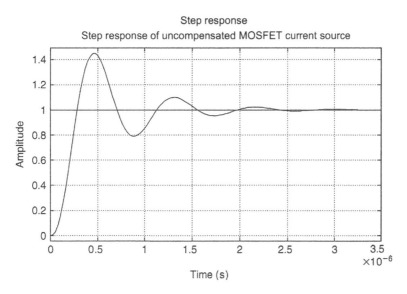

FIGURE 11.55

MOSFET current source of Example 11.10. Step response of uncompensated system, showing significant peak overshoot due to low phase margin. (This value of peak overshoot ($P_o \sim 1.14$) is consistent with a phase margin of 28°

(a)

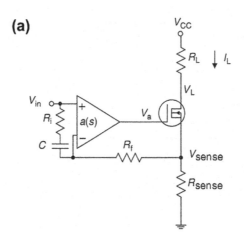

(b)

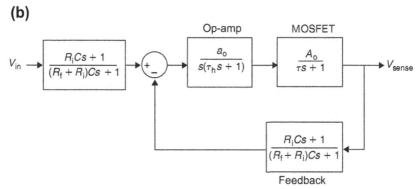

FIGURE 11.56

MOSFET current source of Example 11.10 with lag compensation (a) Circuit with lag components R_i, C, and R_f added. (b) Block diagram.

Step response results for the resultant system with the following parameters are shown inFigure 11.57.

- $R_i = 47 \text{ k}\Omega$
- $R_f = 470 \text{ k}\Omega$
- $C = 1000 \text{ pF}$

Note that the system has a better-behaved step response, with no overshoot, but that the 10–90% risetime has significantly increased as compared to the uncompensated case.

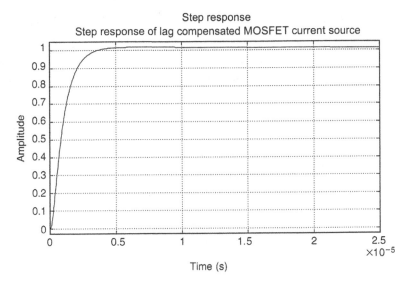

Step response

FIGURE 11.57

MOSFET current source of Example 11.10 with lag compensation, step response.

Example 11.11: Electrodynamic Maglev example

Magnetic levitation (Maglev[10]) systems using superconducting magnets have low damping. Furthermore, it has been demonstrated that these electrodynamic suspensions may have slightly negative damping under certain operating conditions (with poles in the right-half plane). Therefore, a control system is needed to prevent under-damped or unstable vertical oscillations.

In order to suspend the magnet statically, the downward gravitational pull is canceled by an upward magnetic force. A deviation from the equilibrium position results in a restoring force, similar to a mass and a spring. The magnetic levitating force acting on the magnet is given by:

$$f_z = -k_m z = -C i_M^2 z \tag{11.52}$$

where k_m is the equivalent spring constant, z is the vertical distance with reference to the magnet *null position,* i_M is magnet current, and C is a constant that accounts for magnet and coil geometry and relative velocity between the magnet and levitating coils.

---

[10]"Maglev" is a term used generically for a number of systems utilizing magnetic suspensions to levitate a vehicle. Currently (2012) there are revenue-producing Maglev systems operating in China and Japan, with others proposed.

Assuming that there are incremental changes in forces, magnet vertical position, and magnet currents, a linearized model can be generated relating incremental changes in magnet vertical position to changes in incremental magnet current. Vertical force, vertical position, and magnet current are given as the sum of a DC component and an incremental component:

$$f_z = F_z + \tilde{f}_z$$
$$z = Z_o + \tilde{z} \tag{11.53}$$
$$i_M = I_M + \tilde{i}_m$$

Putting this into the force equation results in:

$$f_z \approx -CI_M^2 Z_o - CI_M^2 \tilde{z} - 2CI_M Z_o \tilde{i}_m \tag{11.54}$$

where second-order and higher terms have been neglected. With M being mass and g being acceleration due to gravity, at equilibrium, there is a resultant magnetic force that balances the force of gravity:

$$F_Z = Mg = -CI_M^2 Z_o \tag{11.55}$$

Newton's law applied to the magnet results in:

$$M\frac{d^2\tilde{z}}{dt^2} = f_M - Mg = -CI_M^2 \tilde{z} - 2CI_M Z_o \tilde{i} \tag{11.56}$$

resulting in:

$$\frac{M}{CI_M^2}\frac{d^2\tilde{z}}{dt^2} + \tilde{z} = \frac{2Mg}{k_m I_M}\tilde{i}_m \tag{11.57}$$

Using the spring constant k and converting the equation to the frequency domain results in:

$$\left(\frac{M}{k_m}s^2 + 1\right)z(s) = \frac{2Mg}{k_m I_M}i_m(s) \tag{11.58}$$

resulting in the transfer function between magnet position and magnet control current:

$$\frac{z(s)}{i_m(s)} = \frac{2Mg}{k_m I_M\left(\frac{M}{k_m}s^2 + 1\right)} \tag{11.59}$$

This result shows that this suspension has two $j\omega$-axis poles (Figure 11.58), as in a simple, lossless mass–spring system:

$$s_{p1,2} = \pm j\sqrt{\frac{k_m}{M}} \tag{11.60}$$

Representative numbers for a Maglev suspension magnet section are as follows:

$M = 10{,}000$ kg
$k_m = 10^5$ N/cm $= 10^7$ N/m
$I_M = 10^4$ A

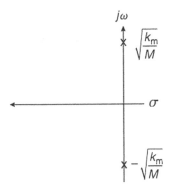

FIGURE 11.58

Pole plot of plant for Maglev example (Example 11.11), showing poles on the $j\omega$-axis.

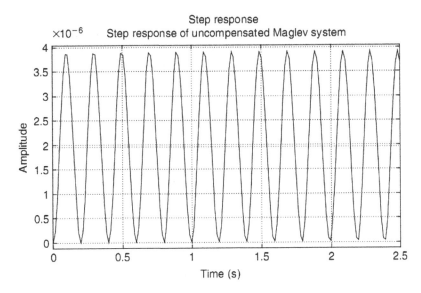

FIGURE 11.59

Step response of uncompensated Maglev suspension of Example 11.11.

This results in a resonant frequency of $\omega_o = 31.6$ rad/s (or $f_o = 5$ Hz) and a plant transfer function:

$$\frac{z(s)}{i_m(s)} = \frac{1.96 \times 10^{-6}}{(10^{-3}s^2 + 1)} = \frac{A}{\frac{s^2}{\omega_o^2} + 1} \tag{11.61}$$

The poles are on the $j\omega$-axis, corresponding to the underdamped suspension (Figure 11.59). In order to improve ride quality for the passengers, the suspension poles must be moved into the left-half plane by selecting appropriate control.

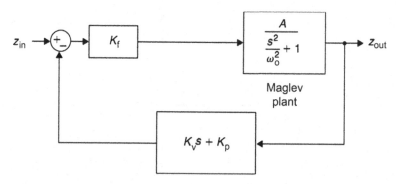

FIGURE 11.60

Control system block diagram for Maglev example of Example 11.11, showing forward path gain (K_f), velocity feedback (K_v), and position feedback (K_p).

The system is compensated using velocity and position feedback, as shown in Figure 11.60. Velocity feedback is equivalent to adding damping to the system.

A control system with $K_v = 10^5$ and $K_p = 10^4$ results in a system with closed-loop transfer function:

$$H(s) = \frac{1.64 \times 10^{-6}}{8.36 \times 10^{-4}s^2 + 1.64 \times 10^{-2}s + 1} \tag{11.62}$$

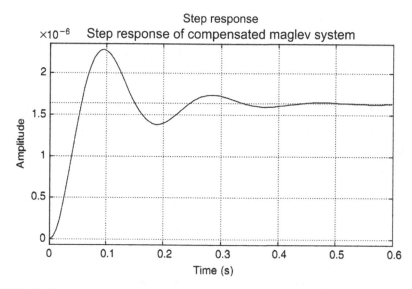

FIGURE 11.61

Step response of compensated Maglev system of Example 11.11.

The closed-loop poles have a damping ratio of $\zeta = 0.28$ and pole locations $-9.8 \pm j(33.2)$ rad/s. Therefore, we expect some oscillation near 33 rad/s (5.2 Hz). The resultant control system results in a much more well-behaved step response (Figure 11.61).

Example 11.12: Electromagnetic levitation and stability

A different kind of levitation uses electromagnets and magnetic attraction. A classical control system experiment is shown in Figure 11.62, where a steel ball is magnetically levitated below an electromagnet. The coil is driven by a current i_M and the ball is suspended at a distance z_B below the pole face of the electromagnet. The magnetic field created by the coil current creates an upward magnetic force f_{MAG} acting on the ball, which can be expressed as:

$$f_{MAG} = \frac{Ki_M^2}{z_B^2} \tag{11.63}$$

where K is some constant. It is clear that while we could ideally set the coil current to exactly offset the weight of the ball, the position of the ball will not be stable. If the ball jiggles up a little bit (toward the magnet) while keeping the coil current constant, the extra force accelerates the ball toward the magnet. Likewise, if the ball lowers a little bit, the ball would accelerate downward. Next, let us prove this instability using mathematics.

We will linearize the force, current, and ball position about an operating point, using:

$$f_{MAG} = F_{MAG} + f_{mag}$$
$$i_M = I_M + i_m \tag{11.64}$$
$$z_B = Z_B + z_b$$

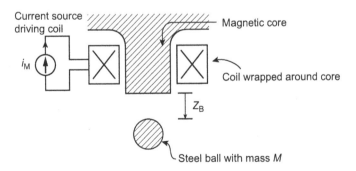

FIGURE 11.62

Electromagnetic levitation system of Example 11.12. The coil is driven by current I and the steel ball is suspended at a distance Z_B below the magnet pole face.

where F_{MAG} denotes the nominal magnetic force on the ball and f_{mag} is the perturbed force. We can write the magnetic force equations as:

$$F_{MAG} + f_z = K\frac{(I_M + i_m)^2}{(Z_B + z_b)^2} = K\frac{I_M^2 + 2I_M i_m + i_m^2}{Z_B^2 + 2Z_B z_b + z_b^2} \approx K\frac{I_M^2 + 2I_M i_m}{Z_B^2 + 2Z_B z_b} \quad (11.65)$$

where we have thrown out the very small i_m^2 and z_b^2 terms. Next, we group the denominator as:

$$F_{MAG} + f_{mag} \approx K\frac{I_M^2 + 2I_M i_m}{Z_B^2\left[1 + \frac{2z_b}{Z_B}\right]} \quad (11.66)$$

With small perturbations, $z_b << Z_B$ and we can make use of the identity $1/(1+x) \cong 1 - x$ for $x << 1$ to further clear out the numerator:

$$F_{MAG} + f_{mag} \approx K\frac{\left[I_M^2 + 2I_M i_m\right]\left[1 - \frac{2z_b}{Z_B}\right]}{Z_B^2} \approx K\frac{\left[I_M^2 + 2I_M i_m - \frac{2z_b}{Z_B}I_M^2\right]}{Z_B^2} \quad (11.67)$$

The F_{MAG} term on the left side is exactly balanced by the $\frac{KI_M^2}{Z_B^2}$ on the right-hand side, these being the DC terms that offset the gravitational (Mg) force acting on the ball due to gravity. We then are left with the small-signal part of the equations:

$$f_{mag} \approx \frac{2KI_M i_m}{Z_B^2} - \frac{2KI_M^2 z_b}{Z_B^3} \quad (11.68)$$

In order to figure out the stability of this suspension, we next make use of Newton's law of motion, (i.e. $f = Ma$), with:

$$f_{mag} \approx \frac{2KI_M i_m}{Z_B^2} - \frac{2KI_M^2 z_b}{Z_B^3} = -M\frac{d^2 z_b}{dt^2} \quad (11.69)$$

Hence, the equation of motion of the ball is:

$$\frac{d^2}{dt^2}z_b - \left(\frac{2KI_M^2}{MZ_B^3}\right)z_b = -\left(\frac{2KI_M}{MZ_B^2}\right)i_m \quad (11.70)$$

Next, we will define a constant $A^2 = 2KI_M^2/MZ_B^3$ and Laplace transform the equation of motion, resulting in:

$$z_b(s)\left[s^2 - A^2\right] = -A^2\left(\frac{Z_B}{I_M}\right)i_m(s) \quad (11.71)$$

Finally, the transfer function relating ball position $z_b(s)$ to coil current $i_m(s)$ is:

$$\frac{z_b(s)}{i_m(s)} = \frac{-A^2\left(\frac{Z_B}{I_M}\right)}{(s+A)(s-A)} = \frac{-\left(\frac{Z_B}{I_M}\right)}{\left(\frac{s}{A}+1\right)\left(\frac{s}{A}-1\right)} \quad (11.72)$$

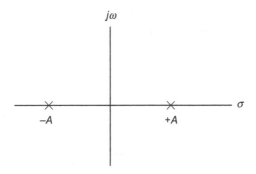

FIGURE 11.63

Pole plot for electromagnetic levitation system for Example 11.12, showing one pole in the left-hand plane and a pole in the right-half-plane at $+A$ rad/s, indicating instability.

We note that there are two real-axis poles (Figure 11.63) with one right-half-plane pole at $+A$ rad/s. This system, without feedback, is unstable. Interestingly enough, this was proved as early as 1842 by Samuel Earnshaw.

Fortunately, through the use of feedback we can stabilize this suspension. Figure 11.64 shows a feedback loop where we are measuring ball position $z_b(s)$ (using some kind of optical or magnetic sensor) and we are applying position and velocity feedback ($K_p + sK_v$) on the position error signal. We have defined another constant $C = -Z_B/I_M$ in order to simplify the block diagram a bit.

The closed-loop transfer function relating the ball position $z_b(s)$ to a set point $z_{set}(s)$ is:

$$\frac{z_b(s)}{z_{set}(s)} = \frac{[K_p + sK_v]\frac{C}{(\frac{s}{A}+1)(\frac{s}{A}-1)}}{1 + [K_p + sK_v]\frac{C}{(\frac{s}{A}+1)(\frac{s}{A}-1)}} = \frac{(K_p + sK_v)}{\frac{s^2}{A^2 C} + K_v s + (K_p - 1/C)} \quad (11.73)$$

This system is stable with all roots in the left-half plane if we set the position constant $K_p > 1/C$. Once we set K_p to achieve stability, we can adjust the damping ratio by varying the velocity constant K_v.

This type of attractive magnetic suspension is used in high-speed ground transportation. A deployed Maglev system using electromagnetic levitation is shown in Figure 11.65. This system is in operation in Shanghai, China, and uses closed-loop control to control the levitation airgap.

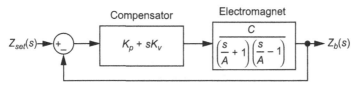

FIGURE 11.64

Stabilizing the magnetic suspension of Example 11.12 with the use of position and velocity feedback compensation. The feedback loop compares the actual ball position z_B with the desired position z_{set}.

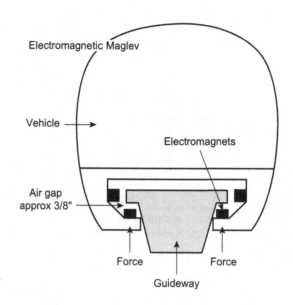

FIGURE 11.65

Depiction of Electromagnetic Maglev Suspension, deployed in Shanghai, China.

Reprinted from F. Zhao and R. Thornton, "Automatic Design of a Maglev Controller in State Space,"
M.I.T. Artificial Intelligence Laboratory, memo 1303, December, 1991, reprinted with permission from author.

Example 11.13: Laboratory experiment—TL084 unity-gain buffer driving a capacitive load

We will now investigate the effects of capacitive loading on the stability of an op-amp circuit. Shown in Figure 11.66(a) is a unity-gain buffer built with a TL084 operational amplifier (a 3-MHz gain–bandwidth product, JFET-input op-amp). When we apply a 500-kHz, 250-mV pp square wave, the resultant output is shown in Figure 11.66(b).

Next, we will load the output with a 470-pF capacitor, as shown in Figure 11.67(a). The resultant output (Figure 11.67(b)) shows significant overshoot and ringing. We will next analyze what is going on.

The block diagram of the unity-gain buffer with no loading is shown in Figure 11.68(a). The block $a(s)$ is the open-loop gain of the operational amplifier; in this case, there is enough phase margin in the L.T. to result in a well-behaved step response.

In Figure 11.68(b), we see the block diagram of the buffer with capacitive load C_L. The op-amp has an output resistance r_o, and this output resistance interacts with the load capacitor C_L to create a pole, which is inside the feedback loop. (For typical op-amps, the output resistance is greater than 100 Ω or so, depending on the particulars of the design). This pole degrades the phase margin, resulting in the ringing shown before.

(a)

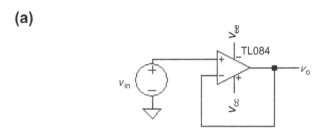

(b)

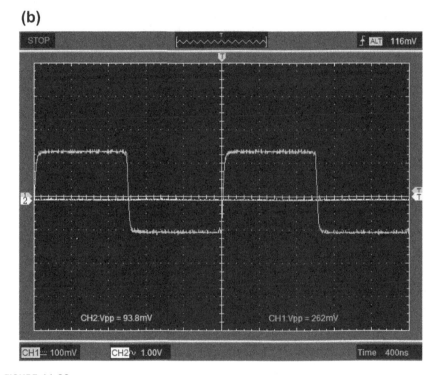

FIGURE 11.66

Unity-gain buffer built with a TL084 op-amp as described in Example 11.13. (a) Circuit. (b) Response to a 250-mV pp square wave at 500 kHz showing a risetime $\tau_R \sim 50$ ns. (For color version of this figure, the reader is referred to the online version of this book.)

(a)

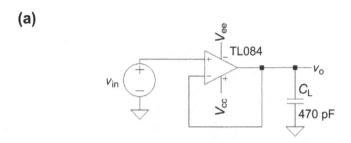

(b)

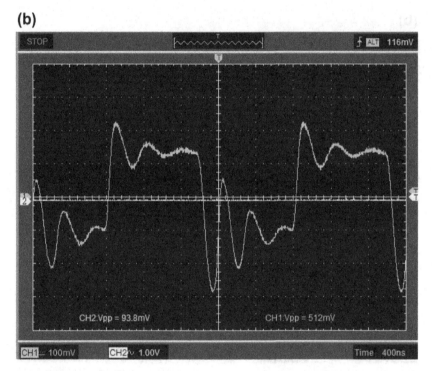

FIGURE 11.67

Unity-gain buffer built with a TL084 op-amp, loaded with a 470-pF load capacitor (Example 11.13). (a) Circuit. (b) Response to a 250 mV pp square wave at 500 kHz showing the output on the verge of instability. (For color version of this figure, the reader is referred to the online version of this book.)

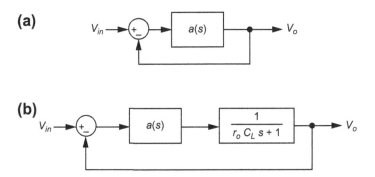

FIGURE 11.68

Unity-gain buffer block diagrams, with $a(s)$ being the op-amp open-loop gain (Example 11.13). (a) No capacitive loading. (b) Capacitive loading adds another pole inside the feedback loop, degrading phase margin.

We can improve phase margin and reduce ringing by isolating the capacitive load from the op-amp, as shown in Figure 11.69(a), where we have added an external resistor R_b of 100 Ω. The resultant step response (Figure 11.69(b)) is much better behaved. We have sacrificed some speed as compared to the no-capacitive-load case.

A block diagram (Figure 11.70) helps to explain this improvement in stability. Adding the external resistor R_b creates a zero inside the feedback loop, with the zero at frequency $\omega_z = -1/R_B C_L$. This zero helps cancel the negative phase shift from the pole due to the load capacitor. The pole outside the feedback loop at $\omega_p = -1/R_B C_L$ is a low-pass filter relating output voltage v_o to fed-back voltage v_f.

Example 11.14: Op-amp driving inductor load

You can configure an op-amp as a current source, as shown in Figure 11.71(a), where an inductor load L_{LOAD} ideally has a constant current $i_L = v_i/R_{SENSE}$ running through it. The op-amp has open-loop transfer function $a(s)$ and output resistance r_o. We will next investigate the practical stability issues associated with this circuit, using the block diagram of Figure 11.71(b).

The block $a(s)$ of the op-amp open-loop transfer function typically has a high gain, one dominant pole, and a second pole at a much higher frequency, with other poles at even higher frequency. For this example, we will assume $a(s)$ has the form of a typical op-amp:

$$a(s) = \frac{2.3 \times 10^5}{(0.01s + 1)(2.4 \times 10^{-8}s + 1)} \tag{11.74}$$

The corresponding open-loop magnitude and phase for $a(s)$ is shown in Figure 11.72. Furthermore, for this example, we will assume that the op-amp output resistance $r_o = 300$ Ω and the sense resistor $R_{SENSE} = 10$ Ω.

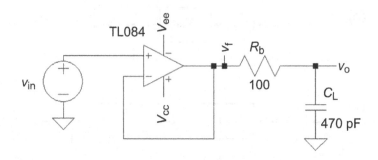

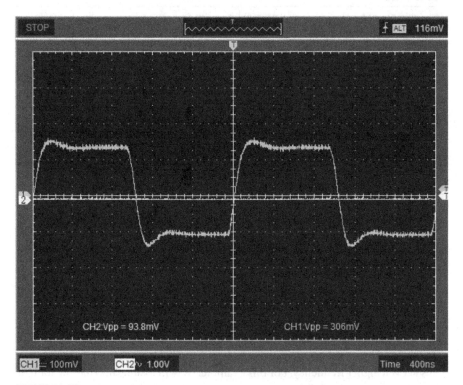

FIGURE 11.69

Unity-gain buffer with resistor R_b added to isolate the op-amp from the capacitive load (Example 11.13). (a) Circuit. (b) Resultant step response, with risetime about 200 ns. (For color version of this figure, the reader is referred to the online version of this book.)

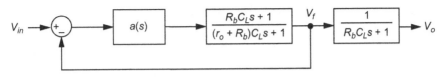

FIGURE 11.70

Block diagram for the circuit in Figure 11.69.

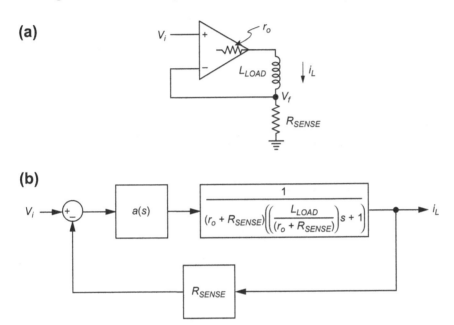

FIGURE 11.71

Circuit for example 11.14. (a) Op-amp driving an inductive load in a current source configuration. (b) Block diagram.

The negative of the loop transmission for this system is given by:

$$-\text{L.T.} = \left[\frac{2.3 \times 10^5}{(0.01s + 1)(2.4 \times 10^{-8}s + 1)}\right]\left[\frac{R_{\text{SENSE}}}{(r_o + R_{\text{SENSE}})\left(\left(\frac{L_{\text{LOAD}}}{r_o + R_{\text{SENSE}}}\right)s + 1\right)}\right]$$

$$(11.75)$$

We note that there are three poles in the feedback loop (two from the op-amp, and one due to the inductive load), and the potential for instability depends on the value of the load inductor L that is inside the loop. This potential instability is illustrated in Figure 11.73, where the closed-loop transfer function i_L/v_i is plotted for load inductor values ranging from 10 µH to 100 mH. We see in this example that the frequency response is relatively flat for inductances up to 100 µH. For higher

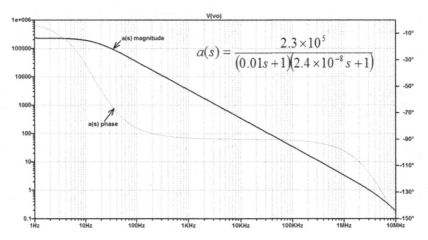

FIGURE 11.72

Op-amp open-loop magnitude and phase for Example 11.14.

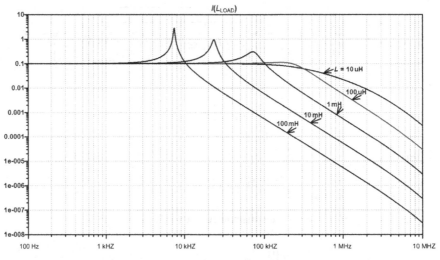

FIGURE 11.73

Closed-loop frequency response for Example 11.14, with $R_{SENSE} = 10\,\Omega$ and load inductor L varying from 10 μH to 100 mH. (For color version of this figure, the reader is referred to the online version of this book.)

inductance values, the phase margin of the system is degraded with resultant peaking in the frequency response. At higher inductance values, we might expect the step response to be relatively unstable as well; this is shown in Figure 11.74.

Note: MATLAB scripts for selected control system examples from this chapter are given in the Appendix.

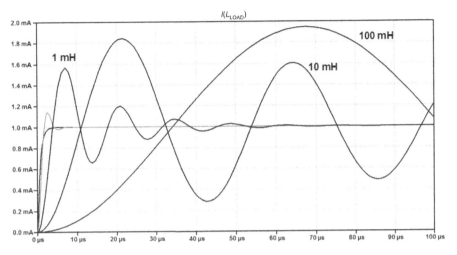

FIGURE 11.74

Closed-loop step response for Example 11.14, with the input voltage stepping 10 mV, with load inductor L varying from 10 µH to 100 mH. (For color version of this figure, the reader is referred to the online version of this book.)

Chapter 11 problems

Problem 11.1

Consider the negative feedback amplifier in Figure 11.75. Assume that the input voltage $V_i = 1$ V and the disturbance input $v_d = 1$ V. Feedback factor $f = 0.1$.

a. For $a = 100$, find output voltage v_o and error voltage v_e.

b. For $a = 100,000$ find v_o and v_e.

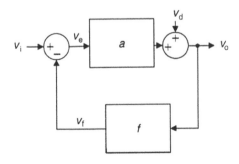

FIGURE 11.75

Negative feedback amplifier for Problem 11.1.

Problem 11.2

Now, consider the circuit in Figure 11.76 where $a(s) = K/s$, an integrator with $K = 1000$.

There is unity feedback. Find the transfer function $v_o(s)/v_i(s)$. Plot the pole-zero plot and Bode (magnitude and phase) plot of this transfer function.

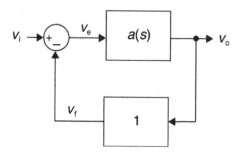

FIGURE 11.76

Negative feedback amplifier for Problem 11.2.

Problem 11.3

This problem considers a nonlinear element inside a negative feedback loop (Figure 11.77). The incremental gain for the nonlinear element is:

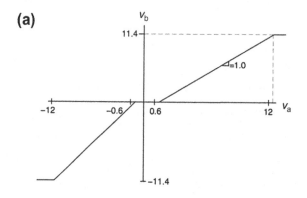

FIGURE 11.77

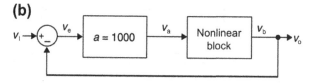

Negative feedback amplifier for Problem 11.3. (a) Nonlinear block. (b) Feedback loop.

$v_b/v_a = 0$ for $|v_a| < 0.6$ V

$v_b/v_a = 1$ for 0.6 V $< |v_a| < 12$ V

$v_b/v_a = 0$ for $|v_a| > 12$ V

The gain profile of this nonlinear element is typical of a transistor power output stage with cross-over distortion.

For this closed-loop amplifier, plot v_o for v_i ranging from -12 V to $+12$ V.

Problem 11.4

For the feedback system of Figure 11.78:

a. Find and plot the ideal input–output transfer function $v_o/v_i(s)$.

b. Find $-$L.T.

c. Plot the Bode plot of the L.T. magnitude and phase.

d. Find the cross-over frequency and phase margin.

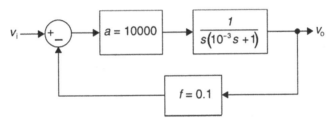

FIGURE 11.78

Negative feedback amplifier for Problem 11.4.

Problem 11.5

The video amplifier shown in Figure 11.79 is a high-bandwidth, positive-gain stage designed to amplify a video signal v_i. The 50-Ω resistive divider models the cabling from the video generator. Assume that the op-amp open-loop transfer function $a(s)$ is:

$$a(s) = \frac{10^5}{(10^{-3}s + 1)(10^{-8}s + 1)}$$

You may assume that all other op-amp parameters are ideal (i.e. infinite slew rate, zero input current, zero output resistance, etc.).

a. Draw a block diagram of this system, with the input to your block diagram being signal v_i and the output being v_o.

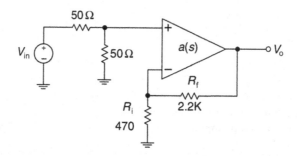

FIGURE 11.79

Video amplifier for Problem 11.5.

b. Find the ideal closed-loop gain of this configuration. (Ideal means that in this part of the problem, you do not consider any bandwidth limitations.)

c. Find the negative of the −L.T. Plot −L.T. magnitude and phase and estimate cross-over frequency, G.M., and phase margin.

d. Now, assume a unit step at the input v_i. Plot $v_o(t)$, assuming that the op-amp does not slew rate limit.

Problem 11.6

Using the Routh test, find the number of poles in the right-half plane of the transfer function:

$$H(s) = \frac{40}{s^5 + 10s^4 - 5s^3 - 50s^2 + 4s + 40}$$

Problem 11.7

We showed in this chapter that it is impossible to stably levitate an unconstrained magnetic ball with an electromagnet in the absence of feedback. Instead of using an electromagnet, could you stably levitate a steel ball with a permanent magnet?

Further reading

[1] Abramovitch D. Phase-locked loops: a control centric tutorial. In: *Proceedings of the 2002 American control conference* May 8–10, 2002. p. 1–15.

[2] Abramovitch D. The outrigger: a prehistoric feedback mechanism. In: *Proceedings of the 42nd IEEE conference on decision and control* December 9–12, 2003. p. 2000–9.

[3] Abramovitch D, Franklin G. A brief history of disk drive control. *IEEE Contr Syst Mag* 2002;**22**(3):28–42.

[4]   Bennett S. Development of the PID controller. *IEEE Contr Syst Mag* 1993;**13**(6):58−62, 64−65.

[5]   Bernstein DS. Feedback control: an invisible thread in the history of technology. *IEEE Contr Syst Mag* April 2002;**22**(2):53−68.

[6]   Black HS. Stabilized feed-back amplifiers. *Electr Eng* 1934;**53**(1):114−20. Reprinted in Proc IEEE February 1999;87(2):379−85.

[7]   Black HS. Inventing the negative feedback amplifier. *IEEE Spectrum* 1977:55−60.

[8]   Black HS. United States Patent #2,102,671, Wave translation system, issued December 21, 1937. Available from: www.uspto.gov.

[9]   Calleja H. An approach to amplifier frequency compensation. *IEEE Trans Educ* 2003;**46**(1):43−9.

[10] Denny M. Watt steam governor stability. *Eur J Phys* 2002;**23**:339−51.

[11] Desoer C. In memoriam: Harold Stephen Black (1898−1983). *IEEE Trans Autom Control* 1984;**29**(8):673−4.

[12] Earnshaw S. On the nature of the molecular forces which regulate the constitution of the luminiferous ether. *Trans Cambridge Philos Soc* 1842;**7**:97−112.

[13] Fasol KH. A short history of hydropower control. *IEEE Contr Syst Mag* 2002;**22**(4):68−76.

[14] Headrick MV. Origin and evolution of the anchor clock escapement. *IEEE Contr Syst Mag* 2002;**22**(2):41−52.

[15] Herwald S. Recollections of the early development of servomechanisms and control systems. *IEEE Contr Syst Mag* 1984;**4**(4):29−32.

[16] Jury E. On the history and progress of sampled-data systems. *IEEE Contr Syst Mag* 1987;**7**(1):16−21.

[17] Kline R. Harold Black and the negative-feedback amplifier. *IEEE Contr Syst Mag* 1993;**13**(4):82−5.

[18] Lee TH. *The design of CMOS radio-frequency integrated circuits*. Cambridge University Press; 1998.

[19] Lepschy AM, Mian GA, Viaro U. Feedback control in ancient water and mechanical clocks. *IEEE Trans Educ* 1992;**35**(1):3−10.

[20] Lewis FL. *Applied optimal control and estimation*. Prentice-Hall; 1992.

[21] Kent L. *Internal and external op-amp compensation: a control-centric tutorial*. ACC; 2004.

[22] Lundberg KH, Roberge JK. Classical dual-inverted-pendulum control. In: *Proceeding of the IEEE CDC 2003* December 9−12, 2003. p. 4399−404. Maui, Hawaii.

[23] Mancini R. The saga of Harry Black. *EDN Mag* March 15, 2001:34.

[24] Maxwell JC. On governors. *Proc R Soc* 1867:270−83.

[25] Mayr O. *The origins of feedback control*. The MIT Press; 1970.

[26] Michel AN. Stability: the common thread in the evolution of feedback control. *IEEE Contr Syst Mag* 1996;**16**(3):50−60.

[27] Pidhayny D. The origins of feedback control. *IEEE Trans Autom Control* 1972;**17**(2):283−4.

[28] Roberge JK. *Operational amplifiers: theory and practice*. John Wiley; 1975.

[29] Siebert WMcC. *Circuits, signals and systems*. The MIT Press; 1986.

[30] Tilbury D, Luntz J, Messner W. Controls education on the WWW: tutorials for MATLAB and Simulink. In: *Proceedings American control conference, Philadelphia PA* June 1998. p. 1304−8.

Basic Operational Amplifier Topologies and a Case Study

"The operational amplifier is responsible for a dramatic and continuing revolution in our approach to analog system design. The availability of high performance, inexpensive devices influences the entire spectrum of circuits and systems, ranging from simple, mass-produced circuits to highly sophisticated equipment designed for complex data collection or processing operations."

—James K. Roberge, "Operational Amplifiers" 1975

IN THIS CHAPTER

▶ The basic operational amplifier (op-amp) is discussed from a topological point of view. A step-by-step design case study illustrates the basic building blocks in a monolithic op-amp. At the end of this chapter, we will consider some of the real-world limitations of op-amps.

Basic operational amplifier operation

The ideal op-amp (Figure 12.1) has the following characteristics:

- *Differential inputs.* The output is an amplified version of the difference between the + and − terminals.
- *Infinite gain.* The gain is infinite.
- *Infinite bandwidth.* There are no bandwidth limitations.
- *Infinite slew rate.* There is no limit to the rate with which the output can change. In other words, there is no limit to dV_{out}/dt.
- *Zero input current.* The input current to both inputs is zero.
- *Zero output impedance.* The output impedance is zero.
- *Zero power dissipation.* The ideal op-amp does not draw or dissipate any power.
- *Infinite power supply rejection.* The output is not dependent on variations in power supply voltage.
- *Infinite common-mode signal rejection.* The output does not depend on the value of the common-mode signal.

Intuitive Analog Circuit Design. http://dx.doi.org/10.1016/B978-0-12-405866-8.00012-7

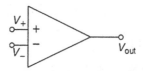

FIGURE 12.1

An ideal operational amplifier showing differential inputs V+ and V−. The ideal op-amp has zero input current and infinite gain that amplifies the difference between V+ and V−.

The ideal op-amp is, of course, nonexistent, but device manufacturers have done a better and better job over the years by designing devices that approach the ideal. For instance, it is common to find devices with direct current (DC) gains that are much higher than 10^6 and/or gain-bandwidth products of >100 MHz.[1]

The usual method for doing the first-cut analysis of closed-loop op-amp circuits is to assume a "virtual ground". This term is a bit of a misnomer, because the input terminals in general do not need to be at ground potential. However, in an op-amp, operating with a negative feedback, the difference between the two voltage inputs is ideally 0 V. If the op-amp positive (+) terminal is at ground, the negative (−) terminal will be at approximately ground. If, in a different configuration the + terminal is at +6 V, the − terminal will also be at approximately +6 V.

A basic two-stage op-amp is shown in Figure 12.2. This is a two-stage op-amp because it consists of two gain stages: the input differential gain stage followed by a second common-emitter gain stage. The input differential amplifier stage (Q_1 and Q_2) has a high differential-mode gain and a low common-mode gain. The second gain stage (Q_3) provides additional gain and also provides a DC level-shifting function. Compensation capacitor C_c provides a low-frequency pole and by the process of pole splitting[2] causes the next-highest frequency pole to move to a higher frequency. Pole splitting has important ramifications in the overall amplifier stability.

Emitter follower Q_4 buffers the high-gain node from the output. The output stage (Q_5 and Q_6) is a class AB *push–pull* stage. This output stage can source or sink current. When sourcing current, Q_5 is ON and when sinking current Q_6 is ON. The diodes at the bases of Q_5 and Q_6 provide two functions: First, they reduce *crossover distortion* in the output stage. Second, by proper sizing of D_1

---

[1]Compare this to the baseline specifications of the 741 op-amp (a device from the 1960s still for sale), with a nominal DC gain of 200,000 and gain–bandwidth product of 1 MHz. A Linear Technology LT1226 (a 1990s-era device) has a gain–bandwidth product of 1000 MHz.
[2]We discussed pole splitting in detail in Chapter 7.

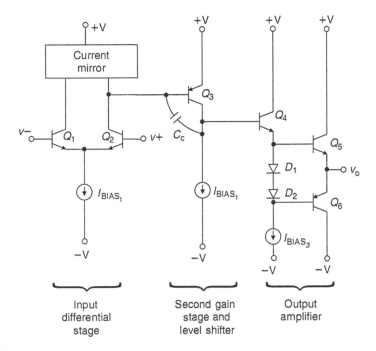

FIGURE 12.2

A basic two-stage op-amp with an input differential gain stage (Q_1–Q_2) followed by a second gain and level-shifting stage (Q_3–Q_4 with a pole-splitting capacitor C_c) and an output buffer (Q_4–Q_6 and D_1–D_2).

and D_2[3] relative to the sizes of Q_5 and Q_6, we can set a modest bias current in the output stage transistors, which lowers the incremental output resistance of the output amplifier.

A push–pull output stage without biasing diodes is shown in the LTSPICE circuit of Figure 12.3. The output waveforms show a significant DC offset and *crossover distortion* or a dead zone of approximately ±0.6 V around $v_o = 0$.

This push–pull circuit can be modified to improve voltage offset as in Figure 12.4. The diodes between the bases of the transistors set up an approximate 1.2V bias that begins to turn on the output transistors. Resistors R_1 and R_2 bias the diode string.

The small resistor R_3 is provided to reduce the chances of thermal runaway. Here is how thermal runaway would work in this circuit: let's suppose that the output transistors Q_1 and Q_2 are carrying a significant amount of current. The output

---

[3]In some real-world op-amps, the voltage drop provided by these two diodes is alternately provided by an alternate circuit topology. See, for example, the 741 op-amp, which uses a "V_{BE} multiplier" to provide this function.

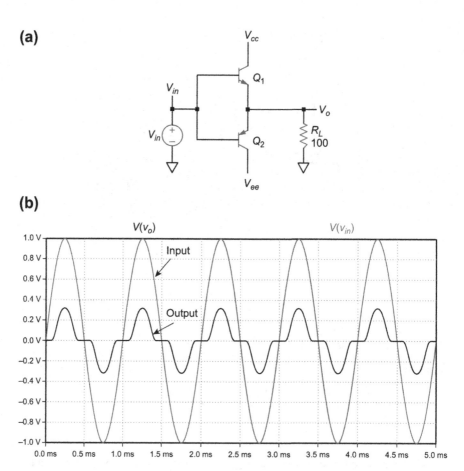

FIGURE 12.3

Push–pull output stage. (a) Circuit with one NPN and one PNP transistor. (b) Output response for 2 V pp sine wave input exhibiting crossover distortion. (For color version of this figure, the reader is referred to the online version of this book.)

transistors will heat up. Since the transistors are now at a higher temperature, they require less base–emitter voltage to sustain the same output current.[4] Therefore, the current in the output stage increases. The transistors heat up more, and the circuit undergoes "thermal runaway", which destroys one or both of the output transistors. The inclusion of the small resistor R_3 ensures that if the collector currents begin to increase significantly the voltage drop across R_3 will become significant, hence stealing away base-emitter voltage from the transistors.

---

[4]Remember that the temperature coefficient of V_{BE} is approximately -2.2 mV/°C.

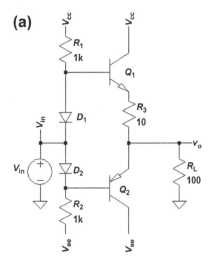

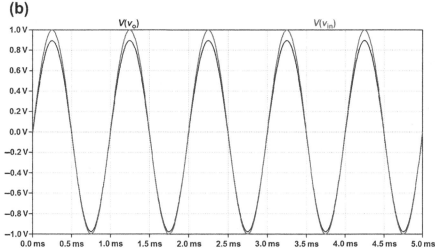

FIGURE 12.4

Push–pull stage with base biasing diodes. (a) Basic circuit. (b) Output response for 2 V pp sine wave input. (For color version of this figure, the reader is referred to the online version of this book.)

Let's next study the power dissipation in the basic push–pull amplifier (Figure 12.5(a)), assuming ideal components (i.e., no crossover distortion) and a symmetric power supply voltages. The NPN transistor is on for positive load current I_L. The power dissipation in the NPN transistor is as follows:

$$P_D = V_{CE}I_L = (V_S - V_o)I_L = V_S I_L - I_L^2 R_L \qquad (12.1)$$

(a)

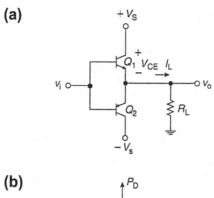

(b)

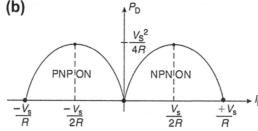

FIGURE 12.5

Circuit for studying power dissipation in a push–pull amplifier. (a) Basic circuit. (b) Transistor power dissipation P_D vs. load current I_L for NPN and PNP transistors.

Therefore, the shape of the power dissipation vs. load current curve has the parabolic shapes in Figure 12.5(b). The maximum power dissipation is found by taking the derivative of the power dissipation with respect to load current.

$$\frac{dP_D}{dI_L} = V_S - 2I_L R_L \tag{12.2}$$

The shape of the power vs. I_L curve is the same for the PNP transistor. The maximum power dissipation occurs at load current $I_L = V_S/(2R_L)$, and at an output voltage that is half the supply voltage.[5]

Example 12.1: Case study: design, analysis and simulation of a discrete op-amp

Let us start off by designing an op-amp using discrete components. This resulting design will not be state of the art, but rather the design process illustrates the various building blocks that exist in most monolithic integrated circuit op-amps. The design insight afforded by this exercise will help us to understand the

---

[5]One might ask why the power dissipation is zero at maximum collector current. In this simplified model, we assume that the saturation voltage of the transistor is zero; hence, a finite collector current multiplied by zero V_{CE} results in zero transistor power dissipation.

limitations of real-world integrated circuit op-amps. The limitations of op-amps are discussed in more detail later on in this chapter.

Differential input stage

In an op-amp, some form of differential input stage is needed. One possible implementation is the current mirror-loaded differential input stage (Figure 12.6). The components of this stage are as follows:

- Differential input transistors Q_1 and Q_2.
- Current mirror transistors Q_4 and Q_5. The mirror has emitter degeneration resistors R_4 and R_5.
- Bias current source Q_3.

Let us do a DC sweep of v_{in+} while grounding v_{in-} (the base of Q_2). This will give us an idea as to the voltage offset of the input stage. Ideally, we would like the output v_1 to be in the middle of the (approximately) linear gain region $v_{in+} = 0$ V. In practicality, however, this voltage offset will not be zero due to device mismatches

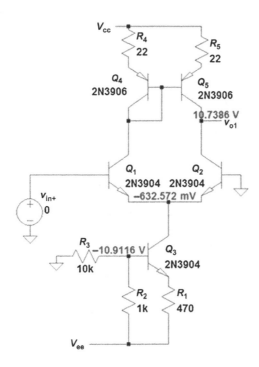

FIGURE 12.6

Differential input stage for Example 12.1 with input transistors Q_1 and Q_2, and current mirror Q_4 and Q_5. We take the output voltage to be v_{o1}, at the connection of the collectors of Q_2 and Q_5. (For color version of this figure, the reader is referred to the online version of this book.)

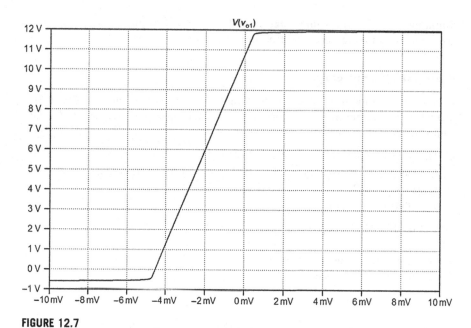

FIGURE 12.7

DC sweep of the differential input stage of Example 12.1.

and errors in the Q_4/Q_5 current mirror. An LTSPICE simulation shows this effect (Figure 12.7); the output at v_{o1} transitions from approximately ground to approximately $+12$ V while v_{in+} is varied from approximately -5 to $+1$ mV. Note that the gain of this stage (gain being v_{o1}/v_{in+}) is significant.

Let us put a DC bias of -2 mV at v_{in+}, and also put a 2-mV peak–peak 1-kHz sine wave at the input (Figure 12.8(a)). Note that the DC bias level of v_{o1} sits at about 6 V (Figure 12.8(b)), which means that Q_2 and Q_5 are both ON and in the forward-active region. Therefore, we expect active gain with the transistors biased under these conditions. The gain from v_{in+} to v_1 is approximately 575.

Continuing on with the design, we now recognize that further transistors are needed for buffering and level-shifting functions. The output at v_1 only has positive voltages; in a practical amplifier, we transition both positive and negative voltages at the output. Therefore, we need to level shift the output voltage from the differential stage. One possible way to do this is with a current source-loaded common-emitter amplifier, as shown in Figure 12.9. The 10-pF compensating capacitor C_c does the pole-splitting function, and the Q_6/Q_7 combination provides an additional voltage gain.

Emitter follower buffering and output push–pull stage

To this input stage, we now need to add an output push–pull stage (Figure 12.10(a)). The output transistors Q_9/Q_{10} can sink and source current. Emitter follower Q_8 is

(a)

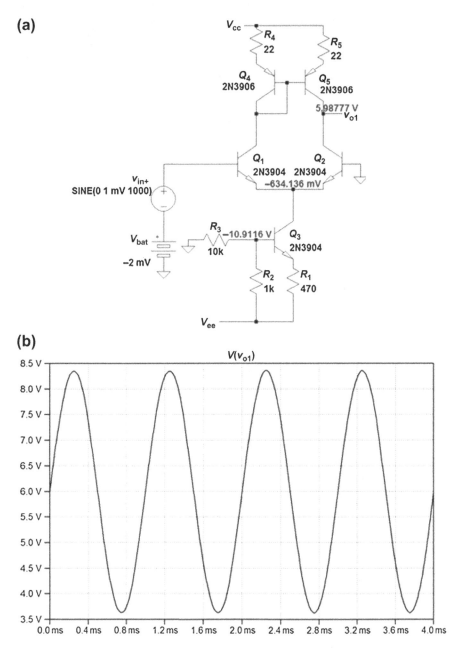

(b)

FIGURE 12.8

Results of AC sine wave sweep for Example 12.1. (a) Circuit. (b) LTSPICE result for a 2 mV-pp variation in v_{in+}. (For color version of this figure, the reader is referred to the online version of this book.)

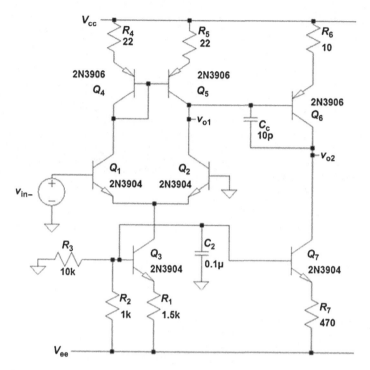

FIGURE 12.9

Design with second gain stage (Q_6/Q_7) for Example 12.1. (For color version of this figure, the reader is referred to the online version of this book.)

added so that the output transistor does not significantly load down the high-gain node at the collector of Q_6. After applying a DC bias voltage of -0.5 mV at v_{in-}, we can find the small-signal open-loop transfer function $a(s)$ of the amplifier, in Figure 12.10(b). The DC gain is about 100,000, and there is a low-frequency pole at about 1 kHz and a second pole at about 20 MHz.

We next configure the op-amp for various positive gains, depending on the value of the feedback resistor (Figure 12.11(a)). We see the closed-loop transfer function for various values of R_f (Figure 12.11(b)). There is some instability evidenced at low gains (as seen in the peaking of the gain curve), so we really should tweak this design further. (This is left as an exercise for the reader.)

A brief review of LM741 op-amp schematic

We will now quickly review the topology of the LM741 op-amp, manufactured by many companies since the 1960s (Figure 12.12). This is a two-stage op-amp with pole splitting.

(a)

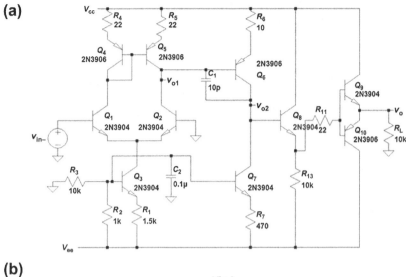

(b)

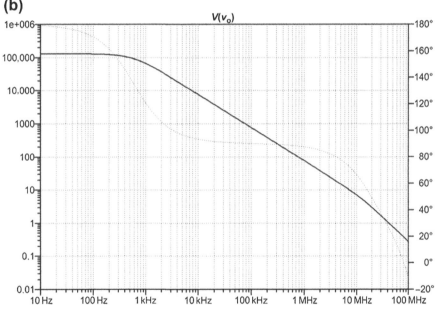

FIGURE 12.10

Final open-loop amplifier for Example 12.1. (a) Circuit. (b) Open-loop transfer function a(s) showing a DC gain of about 100,000, a low-frequency pole at about 1 kHz, and a high-frequency pole at about 20 MHz. (For color version of this figure, the reader is referred to the online version of this book.)

(a)

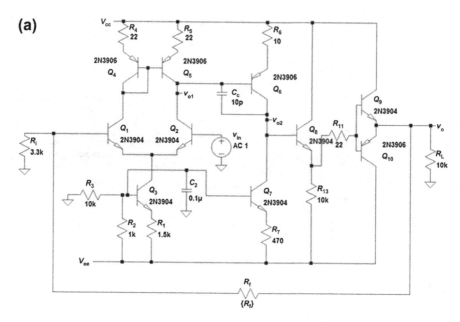

(b)

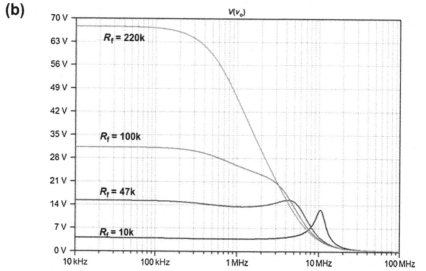

FIGURE 12.11

Amplifier in a positive gain configuration (Example 12.1). (a) Circuit. (b) Closed-loop transfer function for $R_f = 10$, 47, 100, and 220 kΩ. (For color version of this figure, the reader is referred to the online version of this book.)

It is to be noted that this design is shown for illustrative purposes only. There are many design improvements that could be made, and these include better biasing, more intelligent topology selection, and the like. So, please, refrain from launching diatribes about how it is a crummy design!

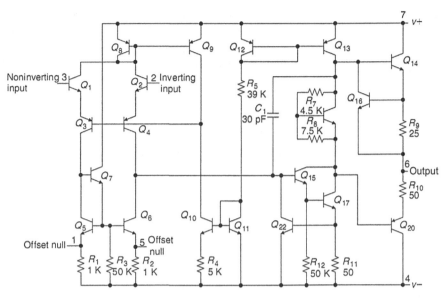

FIGURE 12.12

Schematic of the LM741 op-amp, from the National Semiconductor Corporation.

Reprinted with permission of the National Semiconductor Corporation.

Input differential gain stage: Q and Q_2 are the differential input transistors. Q_5 and Q_6 are the current mirror loads for the differential stage. Q_7 is a beta helper transistor that improves the gain ratio of the mirror. Q_8 is the bias current source.

Bias current mirrors: Q_8 and Q_9; Q_{10} and Q_{11}; Q_{12} and Q_{13}. These mirrors set the bias current levels throughout the op-amp.

Second gain stage: Comprises a common-emitter amplifier with 50-Ω emitter degeneration Q_{17}. Transistor Q_{15} provides buffering between the output of the first gain stage and the input of the second gain stage. The 30-pF capacitor C_1 provides a pole-splitting function.

Output push—pull amplifier. Transistors Q_{14} and Q_{20} form the heart of the push—pull. A V_{BE} multiplier[6] between the bases of Q_{14} and Q_{20} partially biases the push—pull ON. Transistor Q_{15} provides current limiting for positive output currents.

Some real-world limitations of op-amps
Voltage offset

In Chapter 5, we analyzed the differential amplifier, and assumed ideal (i.e. perfectly matched) devices. However, in the real world, there is a mismatch among devices,

---

[6]This "V_{BE} multiplier" circuit provides an approximately 1.6 V_{BE} drop.

and in the op-amp front end, this mismatch manifests itself as a voltage offset. The voltage offset is the voltage that must be applied differentially to force the output of the op-amp to zero. In commercially available op-amps, the voltage offset is typically fractions of a millivolt, up to a few millivolts.

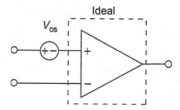

FIGURE 12.13

An operational amplifier showing voltage source modeling the voltage offset V_{os}.

We can model voltage offset by adding a voltage generator of value V_{os} to an ideal op-amp, as shown in Figure 12.13. The voltage offset can be especially troublesome in high-gain configurations, such as in Figure 12.14. Note that the offset generator is in series with the input signal v_i and that the output voltage of this configuration is

$$v_o = 1000v_i + 1000V_{os} \tag{12.3}$$

For small input signals, the voltage offset can swamp out the input signal.

A datasheet excerpt from National Semiconductor shows the input offset voltage for the LM741 op-amp (Figure 12.15). Note that different grades of op-amps have slightly different specifications, but for the 741 the typical offset is ~1 mV while the maximum offset is a few millivolts.

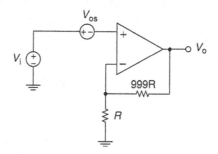

FIGURE 12.14

Gain of +1000 amplifier showing the effects of voltage offset V_{os}.

Parameter	Conditions	LM741A			LM741			LM741C			Units
		Min	Typ	Max	Min	Typ	Max	Min	Typ	Max	
Input Offset Voltage	$T_A = 25°C$										
	$R_S \leq 10\,k\Omega$					1.0	5.0		2.0	6.0	mV
	$R_S \leq 50\,\Omega$		0.8	3.0							mV
	$T_{AMIN} \leq T_A \leq T_{AMAX}$										
	$R_S \leq 50\,\Omega$			4.0							mV
	$R_S \leq 10\,k\Omega$						6.0			7.5	mV

FIGURE 12.15

The datasheet excerpt from the LM741 operational amplifier showing voltage offset (from the National Semiconductor Corporation).

Reprinted with permission of the National Semiconductor Corporation.

Voltage offset drift with temperature

Another important design issue is voltage offset drift with temperature. Referring to Figure 12.16, we see a differential input stage that is typical of that in op-amps. We note that the voltage offset is

$$V_{os} = V_{BE1} - V_{BE2} \tag{12.4}$$

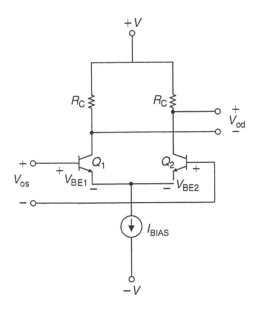

FIGURE 12.16

Circuit for finding the offset voltage drift with temperature in a bipolar op-amp.

This voltage is the differential voltage that occurs when the output differential voltage is zero. For a transistor under forward bias, there is an exponential relationship between collector current and base–emitter voltage, which for transistor Q_1 is

$$I_{C1} = I_S\left(e^{\frac{qV_{BE1}}{kT}} - 1\right) \approx I_S e^{\frac{qV_{BE1}}{kT}} \tag{12.5}$$

where I_S is the reverse saturation current of the transistor. We now find the equation for the base–emitter voltage of bipolar transistor Q_1:

$$V_{BE1} = \frac{kT}{q}\ln\left(\frac{I_{C1}}{I_S}\right) \tag{12.6}$$

The temperature coefficient of a single V_{BE} is found by

$$\frac{dV_{BE1}}{dT} = \frac{k}{q}\ln\left(\frac{I_{C1}}{I_S}\right) - \frac{kT}{qI_S}\frac{dI_S}{dT} = \frac{V_{BE1}}{T} - \frac{kT}{qI_S}\frac{dI_S}{dT} \tag{12.7}$$

The second term in this expression represents the leakage current temperature coefficient. We can now find the total voltage offset drift of the differential amplifier as

$$\frac{dV_{os}}{dT} \approx \frac{dV_{BE1}}{dT} = \left(\frac{V_{BE1} - V_{BE2}}{T}\right) - \frac{kT}{qI_{S1}} + \frac{kT}{qI_{S2}} \tag{12.8}$$

In matched transistors, $I_{S1} = I_{S2}$ and the component of voltage offset drift due to leakage current drift cancels, leaving us with

$$\frac{dV_{os}}{dT} \approx \frac{V_{BE1} - V_{BE2}}{T} \approx \frac{V_{os}}{T} \tag{12.9}$$

Using this expression, let us predict the voltage offset drift for a 741 op-amp with a maximum 4-mV offset at room temperature:

$$\frac{dV_{os}}{dT} \approx \frac{4\ \text{mV}}{300\ \text{K}} \approx 13\ \frac{\mu V}{K} \tag{12.10}$$

This value is consistent with the datasheet value from the 741 datasheet (Figure 12.17), which lists 15 μV/°C.

Average Input Offset Voltage Drift			15							μV/°C

FIGURE 12.17

Datasheet excerpt from the LM741 op-amp showing voltage offset drift with temperature (from the National Semiconductor Corporation).

Reprinted with permission of the National Semiconductor Corporation.

Input bias and input offset current

The input differential amplifier of an op-amp requires base current (in the case of a bipolar-input op-amp). This input current is specified on an op-amp datasheet as *input bias* current, as shown in Figure 12.18. The bias currents are small but finite. Datasheets also specify *input offset* current, which is the guaranteed maximum difference between I_{B+} and I_{B-}.

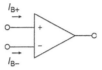

FIGURE 12.18

An operational amplifier showing small but finite input bias currents I_{B+} and I_{B-}.

Differential input resistance

There is a differential-mode input resistance that can be modeled as a large-valued resistance across the op-amp input terminals (Figure 12.19).

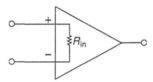

FIGURE 12.19

An operational amplifier showing differential input resistance.

Slew rate

We have seen before that a small capacitor is needed to tailor the frequency response of the op-amp open-loop characteristic. By pole splitting, we create a dominant low-frequency pole. Another effect of this compensation capacitance is slew-rate limiting. There is a finite amount of current available from the bias current I_{BIAS} supplying the input stage to drive this feedback capacitance. Hence, the output voltage time rate of change is limited by the capacitance C_c and the I_{BIAS} current available to charge this capacitance, as

$$\frac{dv}{dt} = \frac{I_{BIAS}}{C_c} \qquad (12.11)$$

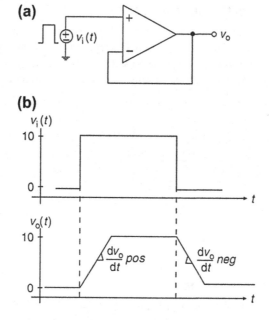

FIGURE 12.20

A circuit that shows the effects of slew-rate limiting. (a) Voltage follower. (b) Input and output waveforms.

Slew-rate limiting manifests itself when we attempt to switch large amplitude signals. Consider the voltage follower of Figure 12.20, where we are switching a 0–10-V signal. We note that the output shows slew-rate limiting on the positive-going and negative-going edges. Note that the slew rates in the two directions are not necessarily the same.

Output resistance and capacitive loading

All op-amps have a finite output resistance. As we will see, this finite output resistance (Figure 12.21) can have a significant effect on closed-loop stability when driving capacitive loads.

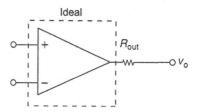

FIGURE 12.21

Operational amplifier showing output resistance.

Noise

Lots more on noise in op-amps can be found in Chapter 16.

Example 12.2: Op-amp driving capacitive load

Let us consider a typical op-amp open-loop transfer function:

$$a(s) \approx \frac{10^5}{(0.01s + 1)(10^{-7}s + 1)} \tag{12.12}$$

The open-loop transfer function $a(s)$ of this op-amp is plotted in Figure 12.22, where we see the low-frequency pole at 100 rad/s and high-frequency pole at 10^7 rad/s. This type of transfer function is typical of many commercially available op-amps.

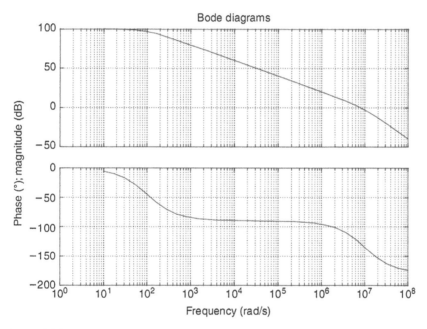

FIGURE 12.22

Plot of $a(s)$ for a typical op-amp for Example 12.2.

We will now see what happens if we use this op-amp to drive a capacitive load in a follower configuration (Figure 12.23(a)). A model of the op-amp showing the op-amp output resistance is shown in Figure 12.23(b). In this example, we will assume that the output resistance of the op-amp is 100 Ω.

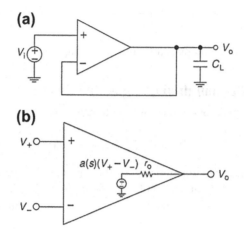

FIGURE 12.23

Op-amp follower driving capacitive load for Example 12.2. (a) Circuit. (b) Model showing op-amp output resistance r_o.

The block diagram of this capacitively loaded amplifier is shown in Figure 12.24.

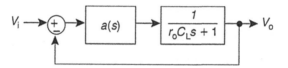

FIGURE 12.24

Capacitively loaded op-amp follower block diagram for Example 12.2.

When we plot the magnitude and the phase of the loop transmission (Figure 12.25), we note that there is a poor phase margin of $22°$. The resultant step response of the closed-loop system (Figure 12.26) shows a significant overshoot, as expected, with this low value of the phase margin. The low phase margin is due to the low-pass filter inside the feedback loop, due to the op-amp output resistance interacting with the load capacitor.

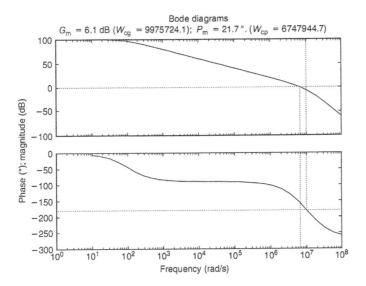

FIGURE 12.25

The plot of loop transmission gain and phase for the follower of Example 12.2, showing the phase margin of 21.7° when loaded with 1000 pF.

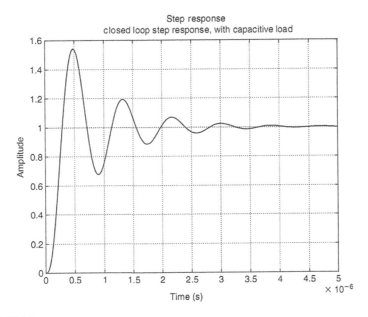

FIGURE 12.26

Unit step response of a capacitively loaded follower of Example 12.2.

Chapter 12 Problems

Problem 12.1

Referring to the µA741A op-amp datasheet, find the maximum frequency (in hertz) that you can drive a 20 V pp sine wave without slew-rate limiting, using this op-amp. Use typical numbers from the datasheet to determine this number.

Problem 12.2

The 741A circuit in Figure 12.27 drives a resistive load. The input is a 20 V pp sine wave at 100 Hz. Again, using typical numbers from the datasheet, plot the output voltage as a function of time.

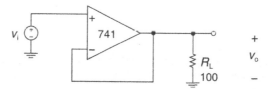

FIGURE 12.27

Op-amp driving resistive load.

Problem 12.3

In the 741A circuit of Figure 12.28, find the output voltage. Include the effects of op-amp input bias current and the voltage offset of the op-amp. Use typical values from the datasheet.

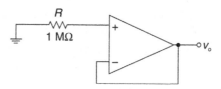

FIGURE 12.28

Circuit for Problem 12.3 illustrating effects of input bias current and voltage offset of op-amps.

Problem 12.4

This problem revisits this issue of an op-amp follower driving a capacitive load. One way to mitigate the destabilizing effects of driving a capacitive load is to isolate the capacitive load C_L from the op-amp follower circuit using an external resistor R_{EXT}, as in Figure 12.29. Draw the block diagram including signals v_i, v_o, and v_f, and comment on how this scheme can be used to stabilize this circuit by improving its phase margin.

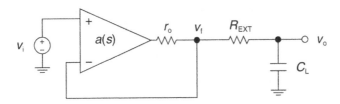

FIGURE 12.29

Circuit for Problem 12.4, showing a mitigating strategy to improve the stability of op-amp follower driving capacitive load C_L. The op-amp has output resistance r_o intrinsic to the device and has transfer function $a(s)$.

Further reading

A wealth of information is provided in the following references. Some of the author's favorites are the references by Bob Widlar, one of the original architects of operational amplifiers. The Solomon and Gray and Meyer references are also excellent overviews of op-amp technology.

[1] Allen PE. Slew induced distortion in operational amplifiers. *IEEE J Solid-State Circuits* February 1977;**12**(1):39−44.

[2] Bowers DF, Wurcer SA. Recent developments in bipolar operational amplifiers. In: *Proceedings of the 1999 bipolar/BiCMOS circuits and technology meeting*. September 26−28, 1999. p. 38−45.

[3] Brown JL. Differential amplifiers that reject common-mode currents. *IEEE J Solid-State Circuits* December 1971;**6**(6):385−91.

[4] Chuang CT. Analysis of the settling behavior of an operational amplifier. *IEEE J Solid-State Circuits* February 1982;**17**(1):74−80.

[5] Comer DT, Comer DJ. A new amplifier circuit with both practical and tutorial value. *IEEE Trans Educ* February 2000;**43**(1):25.

[6] Erdi G. Common-mode rejection of monolithic operational amplifiers. *IEEE J Solid-State Circuits* December 1970;**5**(6):365−7.

[7] Gray P, Meyer R. Recent advances in monolithic operational amplifier design. *IEEE Trans Circuits Syst* May 1974;**21**(3):317−27.

[8] Hearn WE. Fast slewing monolithic operational amplifier. *IEEE J Solid-State Circuits* February 1971;**6**(1):20−4.

[9] Huijsing JH, Tol F. Monolithic operational amplifier design with improved HF behaviour. *IEEE J Solid-State Circuits* April 1976;**11**(2):323−8.

[10] Roberge JK. *Operational amplifiers*. John Wiley; 1975.

[11] Ruediger VG, Hosticka BJ. The response of 741 op amps to very short pulses. *IEEE J Solid-State Circuits* October 1980;**15**(5):908−10.

[12] Solomon JE. "The monolithic op amp: a tutorial study. *IEEE J Solid-State Circuits* December 1974;**9**(6):314−32.

[13] Soloman JE. A tribute to Bob Widlar. *IEEE J Solid-State Circuits* August 1991;**26**(8):1087−9.

[14] Soundararajan K, Ramakrishna K. Characteristics of nonideal operational amplifiers. *IEEE Trans Circuits Syst* January 1974;**21**(1):69−75.

[15] Treleaven D, Trofimenkoff F. Modeling operational amplifiers for computer-aided circuit analysis. *IEEE Trans Circuits Syst* 1971;**18**(1):205−7.

[16] Widlar RJ. A new breed of linear ICs runs at 1-volt levels. *Electronics* March 29, 1979:115−9.

[17] Widlar RJ. DC error reduction in bipolar opamps, 1980 IEEE Solid State Circuits Conference, vol. 23, February 1980. p. 204−5.

[18] Widlar RJ. Design techniques for monolithic operational amplifiers. *IEEE J Solid-State Circuits* August 1969;**4**(4):184−91.

[19] Widlar RJ. Low voltage techniques [for micropower operational amplifiers]. *IEEE J Solid-State Circuits* December 1978;**13**(6):838−46.

[20] Widlar RJ. Some circuit design techniques for linear integrated circuits. *IEEE Trans Circuit Theory* December 1965;**CT-12**(4):586−90.

[21] Widlar RJ, Dobkin R, Yamatake M. New op amp ideas, National Semiconductor Application Note 211, December 1978.

[22] Widlar RJ, Yamatake M. A 150W opamp, 1985 IEEE Digest of Technical Papers, Solid State Circuits Conference, vol. 27, February 1985. p. 140−1.

[23] Wooley BA, Pederson DO. A computer-aided evaluation of the 741 amplifier. *IEEE J Solid-State Circuits* December 1971;**6**(6):357−66.

[24] Yang HC, Allsot DJ. Considerations for fast settling operational amplifiers. *IEEE Trans Circuits Syst* March 1990;**37**(3):326−34.

Review of Current Feedback Operational Amplifiers

"A wide-band direct-coupled transistor amplifier exhibits greatly improved settling time characteristics as the result of circuitry permitting the use of current feedback rather than voltage-feedback in order to reduce the sensitivity of settling time and bandwidth to feedback elements without thereby affecting the manner in which feedback is applied externally by the user, reducing the sensitivity of settling time to the effects of temperature, eliminating saturation and turn-off problems within the amplifier that are related to bias control, to large input signals, and to high-frequency input signals or those having fast rise times, and minimizing the sensitivity of settling time to power supply voltages."

—US Patent #4,502,020, 1985

IN THIS CHAPTER

▶ Current-feedback operational amplifiers are devices in which the main feedback path from output to input carries a current, and not a voltage, as in voltage-feedback op-amps. Current-feedback op-amps do not suffer from the gain–bandwidth product paradigm as do voltage-feedback op-amps, and can achieve very high bandwidths. However, they do have other limitations.

Conventional voltage-feedback op-amp and the constant "gain–bandwidth product" paradigm

A conventional voltage-feedback op-amp is shown in Figure 13.1. Let us assume that the op-amp response is dominated by a single pole at a very low

Intuitive Analog Circuit Design. http://dx.doi.org/10.1016/B978-0-12-405866-8.00013-9

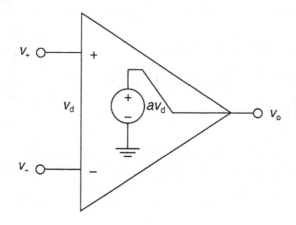

FIGURE 13.1

A conventional voltage-feedback op-amp. The output voltage is $v_o = av_d = a(v_+ - v_-)$.

frequency. We can express the op-amp open-loop transfer function under this assumption as follows:[1]

$$a(s) \approx \frac{a_o}{\tau s + 1} \approx \frac{K}{s} \tag{13.1}$$

where a_o is the low frequency gain of the amplifier and $1/\tau$ is the pole frequency of the op-amp dominant pole. Note that we are approximating the op-amp as behaving as an integrator for frequencies of interest. This simplifies the mathematics in the following discussion considerably.

Let us put this amplifier into a positive-gain configuration of Figure 13.2(a). The ideal low-frequency closed-loop gain (A_v) of this amplifier is as follows:

$$A_v = 1 + \frac{R_2}{R_1} = \frac{R_1 + R_2}{R_1} \tag{13.2}$$

Using the block diagram of Figure 13.2(b), let us figure out the overall transfer function of this amplifier:

$$\frac{v_o}{v_i} = \frac{\frac{K}{s}}{1 + \left(\frac{K}{s}\right)\left(\frac{R_1}{R_1 + R_2}\right)} = \frac{K}{s + \frac{K}{A_v}} = \frac{A_v}{\frac{A_v}{K}s + 1} \tag{13.3}$$

---

[1]Since we are interested in the details of the high-frequency breakpoint, we will make the simplifying assumption that the low-frequency pole is at or near DC. This simplifies the math significantly. In a generic op-amp there is a dominant pole at a very low frequency and higher-frequency poles that can affect closed-loop stability. The dominant pole is at approximately g_m/C_c, where g_m is the transconductance of the input stage and C_c is the compensating capacitor value. For typical op-amps, this dominant pole is at a few hertz.

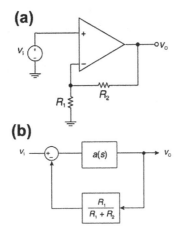

FIGURE 13.2

Conventional voltage-feedback op-amp. (a) Configured as a positive-gain amplifier.
(b) Block diagram.

Note that this amplifier has a low-frequency gain of A_V (as expected) and a bandwidth of (K/A_V). So, if the closed-loop gain goes up, the closed-loop bandwidth goes down. This is the familiar gain—bandwidth product for a voltage-feedback amplifier, which is illustrated in Figure 13.3. As we will see, the current-feedback op-amp is not subject to this same gain—bandwidth product limitation.

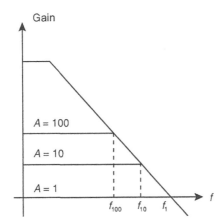

FIGURE 13.3

Illustration of a gain—bandwidth product for a conventional voltage-feedback amplifier.
As the closed-loop gain is increased, the closed-loop bandwidth decreases.

Slew-rate limitations in a conventional voltage-feedback op-amp

Another limitation of conventional op-amps is slew-rate limiting. In a conventional op-amp, a compensating capacitor is used to tailor the open-loop frequency response so that it is dominated by a single pole. In a conventional internally compensated op-amp (such as the LM741 in Figure 13.4), there is a compensating capacitor internal to the device. The maximum rate at which we can charge and discharge this capacitor (and hence the output voltage) depends on the bias current for the previous differential stage (provided originally by Q_8). The specification for slew rate for this op-amp is $\sim 0.5\ \mu s$, consistent with an input bias current of $\sim 15\ \mu A$.[3]

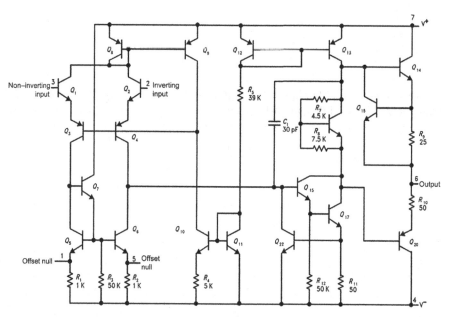

FIGURE 13.4

An LM741 simplified schematic.[2] Note the 30-pF compensation capacitor C_1.

[2]This particular datasheet is from the National Semiconductor; other vendors make this op-amp as well. Reprinted with permission of the National Semiconductor.
[3]The slew-rate calculation is simple enough: $dV/dt = I/C$ where I is the input stage bias current and C is the compensation capacitor value. A detailed look at the inner workings of the LM741 can be found in Gray, Hurst, Lewis, and Meyer, *Analysis and Design of Analog Integrated Circuits*, 4th edition, Chapter 6.8. It is shown in this reference that $I \sim 19\ \mu A$ which results in a slew rate of ~ 0.6 V/μs for a 30-pF compensating capacitor.

The current-feedback op-amp

The basic current-feedback op-amp (Figure 13.5) differs from the conventional voltage-feedback op-amp in several respects. The negative input ($v-$) has a very low input impedance (by design), since this negative node is the place where we will be feeding back current. A unity-gain buffer (Q_1–Q_3 and Q_2–Q_4) forces the $v-$ input to follow the $v+$ input.[4]

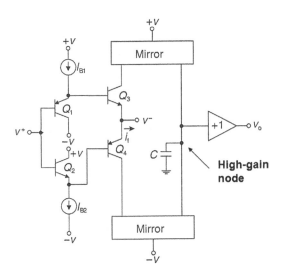

FIGURE 13.5

One topology of the current-feedback op-amp.

Since we use a negative *current* feedback to the inverting $v-$ input,[5] we want low output impedance at this $v-$ node, so that the input impedance will not significantly affect the amount of current that is fed back. The input amplifier Q_3/Q_4 and associated current mirrors convert the feedback current i_f flowing out of the negative input to a voltage at the high-gain node at the output of the current mirrors. This voltage at the high-gain node is further buffered by a unity-gain buffer to the output.

Note that in Figure 13.5 no voltage amplification occurs inside the Q_1–Q_4 amplifier. The input followers Q_1–Q_2–Q_3–Q_4 have a voltage gain of +1. The current mirrors provide a current gain of +1. At the high-gain node, the mirror

---

[4]This type of four-transistor follower is sometimes called a "diamond buffer" or a "diamond follower". See, for example, Burr-Brown (later, Texas Instruments) application note AB-181 by K. Lehmann, "Diamond Transistor OPA660", and W. Lillis and A. Wang, "Complementary current mirror for correcting input offset voltage of diamond follower, especially as input stage for wideband amplifier", US Patent # 4,893,091, issued January 9, 1990.

[5]The concept of current feedback is not new; designs for tube amplifiers using cathode feedback date back to the 1920s.

current is converted to a voltage, which is buffered to the output. Therefore, there is no "Miller effect" in the current-feedback amplifier.

To analyze approximately the gain and bandwidth of the current-feedback op-amp, let us consider a simplified model that captures the first-order effects. Figure 13.6 shows this current-feedback op-amp reduced to a simplified form. The input resistance R_{in} is the parallel combination of the output resistances of Q_3 and Q_4:[6]

$$R_{in} \approx \frac{1}{g_{m3}} \left\| \frac{1}{g_{m4}} \right.$$ (13.4)

For a Q_3/Q_4 bias current of 250 mA, this input resistance is approximately 52 Ω.[7]

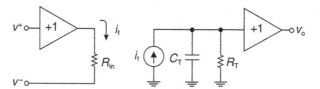

FIGURE 13.6

A simplified equivalent diagram of a current-feedback op-amp. The feedback current i_f is mirrored to the high-gain node.

To first order, this amplifier has a single pole due to the $R_T C_T$ time constant of the high-gain node. The resistance R_T at the high-gain node is a combination of the output resistances of the high-side and low-side current mirrors. Assuming that we have cascode or Wilson current mirrors (to increase the output resistance), the resistance at the high-gain node is as follows:

$$R_T \approx \frac{r_{\mu,npn}}{2} \left\| \frac{r_{\mu,npn}}{2} \right.$$ (13.5)

Assuming a 250-μA mirror bias current and base-width modulation factors $\eta_{npn} = \eta_{pnp} = 10^{-3}$, and with $\beta_{Fnpn} = \beta_{Fpnp} = 100$, the high-gain node resistance $R_T = 2.6$ MΩ.[8] The capacitor C_T is set by device parameter and layout parasitics.

If we are interested in DC accuracy, the DC voltage gain of this amplifier is of importance. The open-loop voltage gain is as follows:

$$a_{v,ol} \approx \frac{R_T}{R_{in}} \approx \frac{2.6 \times 10^6}{52} \approx 50,000$$ (13.6)

---

[6]This simplified calculation ignores the base-spreading resistance r_x.

[7]Remember that the input resistance looking into the emitter of a transistor in the forward active region is approximately $1/g_m$, if the base resistance r_x and the source resistance seen from the base terminal are neglected. The input resistance, in close form, is $r_{in,emitter} \approx \dfrac{R_s + r_x + r_\pi}{h_{fe} + 1} = \dfrac{1}{g_m}$ if $R_s, r_x \ll r_\pi$

[8]These numbers are consistent with the Analog Devices AD844, which lists a nominal input resistance at the inverting input of 50 Ω, and a high-gain node resistance of 3 MΩ.

A current-feedback amplifier connected in a positive-gain configuration is shown in Figure 13.7. A block diagram of this system is shown in Figure 13.8.

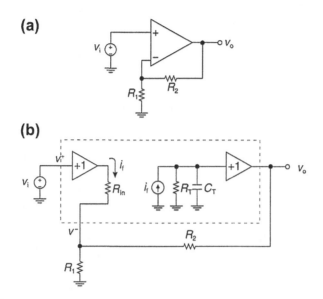

FIGURE 13.7

A current-feedback op-amp configured for positive closed-loop gain. (a) Circuit. (b) Model.

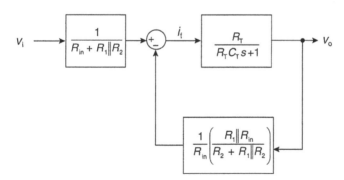

FIGURE 13.8

The block diagram of a current-feedback op-amp configured for positive closed-loop gain.

The overall gain for this amplifier is as follows:

$$\frac{v_o}{v_i} = \left(\frac{1}{R_{in} + R_1 \| R_2}\right) \left(\frac{R_T}{\dfrac{R_T C_T s + 1}{1 + \left(\frac{R_T}{R_T C_T s + 1}\right)\left(\frac{1}{R_{in}}\right)\left(\frac{R_1 \| R_{in}}{R_2 + R_1 \| R_{in}}\right)}}\right) \tag{13.7}$$

Although this looks like a mess, we can simplify this by recognizing that the input resistance to the unity-gain input buffer is very small ($R_{in} \ll R_1$ and R_2) and that the resistance at the high-gain node is very large (or $R_T \gg R_1$ and R_2). Using these assumptions, we arrive at the gain of this amplifier as follows:

$$
\frac{v_o}{v_i} \approx \left(\frac{R_1 + R_2}{R_1 R_2}\right)\left(\frac{R_T}{R_T C_T s + 1 + \frac{R_T}{R_2}}\right) \approx \left(\frac{R_1 + R_2}{R_1}\right)
$$
$$
\times \left(\frac{1}{R_2 C_T s + 1 + \frac{R_2}{R_T}}\right) \approx G\left(\frac{1}{R_2 C_T s + 1}\right)
$$

(13.8)

Note that the ideal closed-loop gain G is the same as with the voltage-feedback amplifier. Also, note the absence of the gain–bandwidth product in this case. The overall bandwidth does depend on the value of the feedback capacitor R_2, but we can independently set the closed-loop gain (by varying R_1) without affecting the bandwidth.[9]

Next, we will comment on stability. We note that the closed-loop bandwidth as shown before is roughly $1/(R_2 C_T)$, ignoring high-frequency poles, as shown in Figure 13.9. We also note that there are higher-frequency poles unaccounted for in this simplified model. Therefore, manufacturers will specify a minimum feedback resistance[10] requirement to ensure stability.

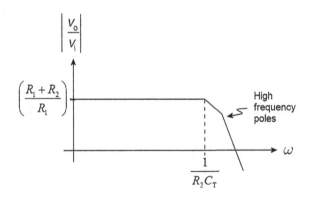

FIGURE 13.9

The Bode plot illustrating the effect of feedback resistor R_2 on closed-loop gain and bandwidth.

[9]Note that this works to first order only. Second-order effects do come into play to limit the bandwidth as well.
[10]Note that if you try to push up the bandwidth by reducing R_2, eventually you will "run into" the high-frequency poles, creating a potential stability problem.

Absence of slew-rate limit in current-feedback op-amps

In the current-feedback op-amp, we do not have the same slew-rate limit as in a conventional voltage-feedback op-amp. Note that in the current-feedback op-amp, the current injected into the negative ($v-$) input to the input stage gets mirrored and eventually charges and discharges the capacitance C_T at the high-gain node. Therefore, for a large voltage swing at the output of the amplifier, there is more current fed back to the $v-$ terminal. Following this line of reasoning, there is no intrinsic slew-rate limit in a current-feedback op-amp.[11]

Example 13.1: An admittedly very crude current-feedback op-amp discrete design

Let us look at an illustrative design of a discrete current-feedback op-amp circuit and simulate with 2N3904 and 2N3906 transistors. Note that this is by no means an optimized design; the resultant design is likely to have problems with voltage offset, transient response, and the like. Rather, this simple example illustrates many of the concepts that we have gone over in the previous analytic discussion.

Shown in Figure 13.10 is the design, configured initially for positive gain. In this configuration, the closed-loop gain (at low frequencies) is approximately

$$G = 1 + \frac{R_f}{R_i} = \frac{R_1 + R_2}{R_1} \tag{13.9}$$

Going through the design, we see the following functional blocks:

- Q_1-Q_4 is the input buffer, providing high input impedance at $v+$ (the bases of Q_1 and Q_2) and a low input impedance at $v-$ (the emitters of Q_3 and Q_4). Q_1 and Q_2 are biased by current sources, and the collector currents of Q_3 and Q_4 are approximately:

$$I_{C3} \approx I_{C4} \approx \sqrt{I_{C1}I_{C2}} \tag{13.10}$$

We will see in a later section in this book (on the translinear principle) how this calculation is done.

- Q_5/Q_7 and Q_6/Q_8 are current mirrors that mirror the collector currents of Q_3 and Q_4 to the high-gain node. These mirrors act as a current buffer, mirroring the feedback current to the high-gain node. The high-gain node is the collector connection of Q_7 and Q_8. The output resistance at this node is the parallel

---

[11]There are, of course, second-order limitations that affect the ultimate maximum slew rate of current feedback op-amps. There is some parasitic capacitance at the high-gain node. However, a survey of some current feedback op-amps shows that the slew rate is indeed high. See, for example, Analog Devices's AD844 (2000 V/µs); National Semiconductor's LM6181 (2000 V/µs); Texas Instruments' TMS3110 (1300 V/µs); Linear Technology LT1227 (1100 V/µs). There are lots more.

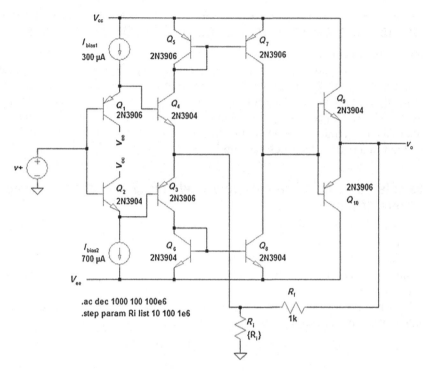

FIGURE 13.10

Discrete-design current-feedback op-amp configured for positive gain (Example 13.1). (For color version of this figure, the reader is referred to the online version of this book.)

combination of the output resistances of current mirror transistors Q_7 and Q_8; the total resistance at this node to ground is this current mirror resistance in parallel with the input resistance of the follower stage Q_9/Q_{10}.

- Q_9/Q_{10} is a unity-gain push–pull output stage, which provides a low impedance at the output v_o node.
- R_f is the feedback resistor, and R_i is the input resistor at the inverting input $v-$. Note that our previous analysis of current-feedback op-amps showed that the main bandwidth limitation is our selection of feedback resistor R_f.

The LTSPICE bias point values (not shown in Fig 13.10) show that the voltage offset of this amplifier is about -10 mV.

Results showing the incremental gain for this amplifier, for values $R_i = 1$ MΩ, 100, and 10 Ω are shown in Figure 13.11. Note that the closed-loop bandwidth is roughly constant over a wide range of closed-loop gains.

In Table 13.1 we see a summary of the LTSPICE results compared to results from an ideal amplifier. The resultant amplifier bandwidth and gain–bandwidth product

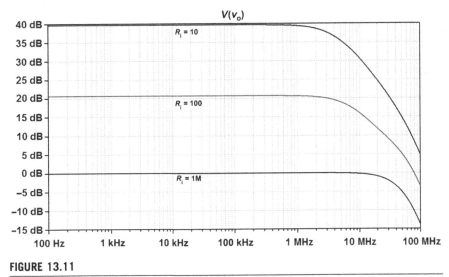

FIGURE 13.11

Magnitude response of a discrete current-feedback op-amp of Example 13.1. (For color version of this figure, the reader is referred to the online version of this book.)

Table 13.1 LTSPICE results for a current-feedback op-amp design example of Example 13.1

R_i	Ideal Gain	Ideal Gain (dB)	LTSPICE Gain	LTSPICE Gain (dB)	LTSPICE Small-Signal Bandwidth (MHz)	Gain-Bandwidth Product (MHz)
1 MΩ	1.001	0.009	0.9995	0	36.7	36.7
100 Ω	11.0	20.8	10.8	20.7	7.1	147
10 Ω	101	40.1	96.3	39.7	3.8	366

are given. Note that there are some minor differences in calculated gains vs. LTSPICE gains, due in part to loading effects not accounted for in the simplified analyses.

The step response for the gain-of-101 configuration is shown in Figure 13.12. The input step was set to 1 mV. The rise time of the output is approximately 100 ns, consistent with our small-signal bandwidth of 3.8 MHz.[12] Note the poor

---

[12] The input step needs to be small enough so that the operating point of the transistors does not significantly change and the amplifier does not slew-rate limit, hence the use of the small 1-mV input step. Remember our calculation for the rise time of a first-order system, where it was found that rise time = 0.35/bandwidth, where bandwidth is in hertz. This op-amp has more than one pole (more than a dozen in this example). However, the amplifier small-signal AC response and small-signal step response is indeed dominated by a single pole, due to loading at the high-gain node.

voltage offset of this amplifier, as illustrated by the node voltage map of Figure 13.13, where the amplifier is configured for a gain of $+101$.[13]

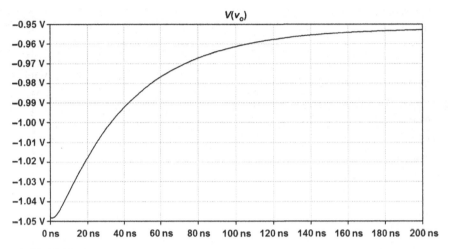

FIGURE 13.12

LTSPICE results for a 1-mV input step response of a discrete current-feedback op-amp, configured as a gain of $+101$. Note the risetime of about 100 nanoseconds; note also that the DC offset in the output is a result of the -10 mV voltage offset (due to the crude design).

Manufacturer's datasheet information for a current-feedback amplifier

Let us have a look at the excerpts of a manufacturer's datasheet. A section of the front page of the datasheet of the National LM6181 is shown in Figure 13.14. Note the high slew rate (2000 V/μs). This op-amp may typically be used in video amplifier applications where high bandwidth, high slew rate, and low differential phase and gain are needed.

In Figure 13.15, we see the schematic of the LM6181 current-feedback op-amp, with some internal circuits highlighted. The bulk of the circuitry is either voltage followers or current mirrors. There is a four-transistor input buffer providing the positive and negative op-amp inputs. The feedback current is mirrored through a cascade current mirror. Cascoding provides higher current source output impedance at the high-gain node, and hence more open-loop voltage gain. The voltage at the high-gain node is buffered to the output by another four-transistor unity-gain buffer. Note that the output includes short-circuit protection; if the voltage across the output resistors R exceeds approximately 0.6 V, the output transistors are shut down.

---

[13]This is why the output starts at -680 mV (instead of zero). In a real-world CF op-amp, extra biasing circuitry is added to reduce this voltage offset significantly.

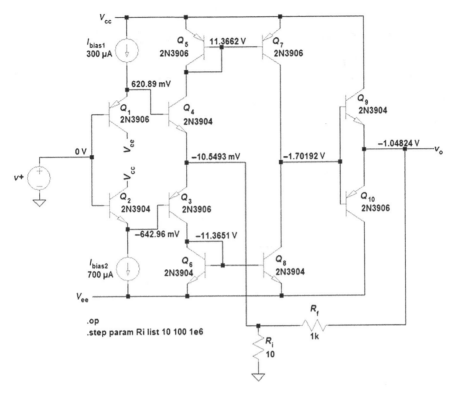

FIGURE 13.13

Node voltages, with a current-feedback op-amp configured for a gain of +101. Note that the output voltage is about −1 V, corresponding to a voltage offset referred to the input of about −10 mV. (For color version of this figure, the reader is referred to the online version of this book.)

The closed-loop frequency response curves (Figure 13.16) show that the bandwidth is roughly constant over a wide gain setting from −1 to −10.

A more detailed model and some comments on current-feedback op-amp limitations

A more detailed model of the current-feedback op-amp, modeling some further effects that may affect bandwidth, gain, and stability is shown in Figure 13.17. For instance, C_{in-} models the parasitic capacitance at the inverting input node. This capacitance is due to internal transistor capacitances and any external parasitic capacitances due to bond wires and integrated circuit leads. The resistance r_o models the small (but finite) output resistance of the output unity-gain buffer. After using the

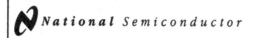

National Semiconductor

May 1998

LM6181
100 mA, 100 MHz Current Feedback Amplifier

General Description

The LM6181 current-feedback amplifier offers an unparalleled combination of bandwidth, slew-rate, and output current. The amplifier can directly drive up to 100 pF capacitive loads without oscillating and a 10V signal into a 50Ω or 75Ω back-terminated coax cable system over the full industrial temperature range. This represents a radical enhancement in output drive capability for an 8-pin DIP high-speed amplifier making it ideal for video applications.

Built on National's advanced high-speed VIP™ II (Vertically Integrated PNP) process, the LM6181 employs current-feedback providing bandwidth that does not vary dramatically with gain; 100 MHz at A_V = −1, 60 MHz at A_V = −10. With a slew rate of 2000V/µs, 2nd harmonic distortion of −50 dBc at 10 MHz and settling time of 50 ns (0.1%) the LM6181 dynamic performance makes it ideal for data acquisition, high speed ATE, and precision pulse amplifier applications.

Features

(Typical unless otherwise noted)
- Slew rate: 2000 V/µs
- Settling time (0.1%): 50 ns
- Characterized for supply ranges: ±5V and ±15V
- Low differential gain and phase error: 0.05%, 0.04°
- High output drive: ±10V into 100Ω
- Guaranteed bandwidth and slew rate
- Improved performance over EL2020, OP160, AD844, LT1223 and HA5004

Applications

- Coax cable driver
- Video amplifier
- Flash ADC buffer
- High frequency filter
- Scanner and Imaging systems

FIGURE 13.14

National Semiconductor LM6181 current-feedback op-amp.

Reprinted with permission of the National Semiconductor Corporation.

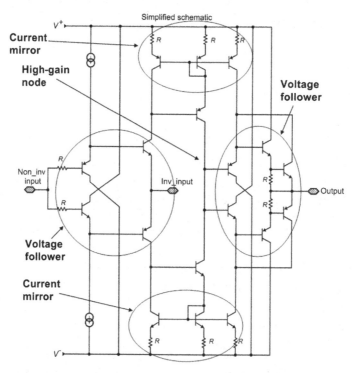

FIGURE 13.15

The simplified schematic of an LM6181 current-feedback op-amp, with various circuits highlighted.

From the LM6181 datasheet, reprinted with permission from the National Semiconductor.

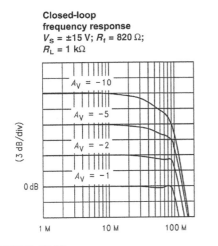

Closed-loop
frequency response
$V_S = \pm15$ V; $R_f = 820\ \Omega$;
$R_L = 1$ kΩ

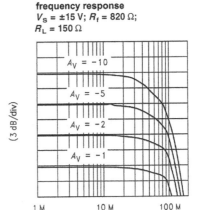

Closed-loop
frequency response
$V_S = \pm15$ V; $R_f = 820\ \Omega$;
$R_L = 150\ \Omega$

FIGURE 13.16

National Semiconductor LM6181 current-feedback op-amp small-signal bandwidth.

Reprinted with permission of the National Semiconductor Corporation.

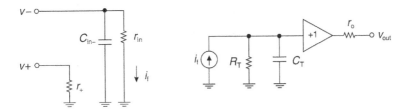

FIGURE 13.17

A more detailed model of the current-feedback op-amp. Parasitics that have been added to the previous model include: r_+ (input resistance at the positive input terminal); C_{in-} (input capacitance at the negative input terminal), and r_o (output resistance).

simplified equivalent circuit model (Figure 13.6) for initial analysis, it often behooves the designer to take into account these additional circuit components that will affect circuit operation as well.

There are other design issues to take into account when using a current-feedback op-amp, as compared to a voltage-feedback op-amp.

DC performance. Operation of the current-feedback op-amp with acceptable DC performance (i.e. voltage offset, common-mode rejection ratio (CMRR), etc.) relies on matching between NPN and PNP transistors in the front end. Further degradation in the voltage offset is caused by the Early voltage of the transistors in the input buffer.

Low impedance at inverting input. Many current-feedback amplifiers, by their nature, rely on a low impedance at the inverting input.

Minimum value of feedback resistor. Due to stability issues, a minimum value of a feedback resistor must be used in closed-loop configurations to ensure stability.

Chapter 13 problems

Problem 13.1

Using a conventional op-amp model, draw a circuit using the 741 op-amp providing a gain of $+100$. Find the bandwidth of this amplifier circuit.

Problem 13.2

Using the AD844 datasheet, extract current-feedback op-amp parameters and fill in the equivalent circuit model of Figure 13.6.

Problem 13.3

Using open-circuit time constants, estimate the bandwidth of the four-transistor input circuit (Q_1–Q_4) of Figure 13.5. Assume that Q_3 and Q_4 are each biased at a collector current I_o and that each transistor has a current gain–bandwidth product of $\omega_T = 6.3 \times 10^9$ rad/sec. Assume that each transistor has $C_\mu = 0.5$ pF, and assume that $I_{B1} = I_{B2} = I_o = 1$ mA. Make other assumptions as needed and state them.

Problem 13.4

Using the circuit of Figure 13.5, find the collector current in transistors Q_3 and Q_4, assuming that the input transistors Q_1 and Q_2 are biased as shown. Assume that $i_f = 0$ and that the input $v+$ is grounded for purposes of this calculation. Further, assume that the reverse saturation current is $I_{S,NPN}$ for Q_2 and Q_3, and is $I_{S,PNP}$ for Q_1 and Q_4.

Further reading

[1] Bowers DF. The so-called current-feedback operational amplifier-technological breakthrough or engineering curiosity?. In: *IEEE International Symposium on Circuits and Systems, ISCAS '93* May 3–6, 1993. p. 1054–7.

[2] Bowers DF, Wurcer SA. Recent developments in bipolar operational amplifiers. In: *Proceedings of the 1999 Bipolar/BiCMOS Circuits and Technology Meeting* September 26–28, 1999. p. 38–45.

[3] Franco S. Current feedback amplifiers benefit high-speed designs. *EDN* January 5, 1989:161–72. Analytical foundations of current-feedback amplifiers, 1993. In: *IEEE international symposium on circuits and systems (ISCAS '93, vol. 2)*, May 3–6, 1993. p. 1050–3.

[4] Harvey B. Current feedback opamp limitations: a state-of-the-art review, 1993. In: *IEEE International Symposium on Circuits and Systems, (ISCAS '93)* May 3–6, 1993:1066–9.

[5]   Harvey B. Dual-amplifier designs increase the accuracy of current-feedback amps. *EDN* 1993:129−34.

[6]   Harvey B. Take advantage of current-feedback amps for high-frequency gain. *EDN* 1993:215−22.

[7]   Intersil, Inc. Current feedback amplifier theory and applications, Application Note # AN9420.1, April 1995.

[8]   Intersil, Inc. An intuitive approach to understanding current feedback amplifiers, Application Note # AN9787, February 1998.

[9]   Intersil, Inc. Converting from voltage-feedback to current-feedback amplifiers, Application Note # AN9663.1, April 1999.

[10] Intersil, Inc. A designer's guide for the HA-5033 video buffer, Application Note AN548.1, November 1996.

[11] Koullias IA. A wideband low-offset current-feedback op amp design. In: *Proceedings of the 1989 bipolar circuits and technology meeting, Minneapolis* September 18−19, 1989. p. 120−3.

[12] Lidgey FJ, Hayatleh K. Current-feedback operational amplifiers and applications. *Electron Commun Eng J* 1997;**9**(4):176−82.

[13] Lillis W, Wang A. Complementary current mirror for correcting input offset voltage of diamond follower, especially as input stage for wide-band amplifier. U.S. Patent # 4,893, 091, Filed October 11, 1988; issued January 9, 1990.

[14] Mancini R. Converting from voltage-feedback to current-feedback amplifiers. *Electron Des (Analog Applications Issue)* June 26, 1995:37−46.

[15] National Semiconductor. Current feedback op amp applications circuit guide, Application Note # OA-07, May 1988.

[16] National Semiconductor. Frequent faux pas in applying wideband current feedback amplifiers, Application Note # OA-15, August 1990.

[17] National Semiconductor. Current-feedback myths debunked, Application Note # OA-20, July 1992.

[18] National Semiconductor. Current feedback amplifiers, Application Note # OA-31, November 1992.

[19] National Semiconductor. Current feedback loop gain analysis and performance enhancement, Application Note # OA-13, January 1993.

[20] National Semiconductor. Stability analysis of current feedback amplifiers, Application Note # OA-25, May 1995.

[21] National Semiconductor. Current vs. voltage feedback amplifiers, Application Note # OA-30, January 1998.

[22] Palumbo G. Current feedback amplifier: stability and compensation. In: *Proceedings of the 40th Midwest symposium on circuits and systems* August 3-6, 1997. p. 249−52.

[23] Palumbo G, Pennisi S. Current-feedback amplifiers versus voltage operational amplifiers. *IEEE Trans Circuits Syst Fundam Theory Appl* 2001;**48**(5):617−23.

[24] Sauerwald M. Current feedback and voltage feedback. The choice amplifiers: which one to use, when and why, Northcon/94 conference record, October 11−13, 1994, p. 285−9.

[25] Smith D. Current-feedback amps enhance active-filter speed and performance. *EDN* 1990:167−72.

[26] Smith, D. High speed operational amplifier architectures. *Proceedings of the 1993 IEEE bipolar circuits and technology meeting*, pp. 141−148.

[27] Smith D, Koen M, Witulski A. Evolution of high-speed operational amplifier architectures. *IEEE J Solid-State Circuits* 1994;**29**(10):1166−79.

[28] Tammam AA, Hayatleh K, Hart B, Lidgey FJ. High performance current-feedback op-amps. In: *Proceedings of the 2004 international symposium on circuits systems, (ISCAS '04)* May 23−26, 2004. p. 825−8.

[29] Texas Instruments, Inc. Voltage feedback vs. current feedback op amps applications report, TI Appl Note, November 1998.

[30] Toumazou C, Payne A, Lidgey J. Current-feedback versus voltage feedback amplifiers: history, insight and relationships. In: *1993 IEEE international symposium on circuits and systems, (ISCAS '93)* May 3−6, 1993. p. 1046−9.

[31] Wilson B. Current-mode amplifiers. In: *IEEE international symposium on circuits and systems* 1989. p. 1576−9.

[32] Wong J. Current-feedback op amps extend high frequency performance. *EDN* 1989:211−6.

Appendix: LM6181 current-feedback op-amp

 *National Semiconductor*

May 1998

LM6181
100 mA, 100 MHz Current Feedback Amplifier

General Description

The LM6181 current-feedback amplifier offers an unparalleled combination of bandwidth, slew-rate, and output current. The amplifier can directly drive up to 100 pF capacitive loads without oscillating and a 10V signal into a 50Ω or 75Ω back-terminated coax cable system over the full industrial temperature range. This represents a radical enhancement in output drive capability for an 8-pin DIP high-speed amplifier making it ideal for video applications.

Built on National's advanced high-speed VIP™ II (Vertically Integrated PNP) process, the LM6181 employs current-feedback providing bandwidth that does not vary dramatically with gain; 100 MHz at $A_V = -1$, 60 MHz at $A_V = -10$. With a slew rate of 2000V/μs, 2nd harmonic distortion of –50 dBc at 10 MHz and settling time of 50 ns (0.1%) the LM6181 dynamic performance makes it ideal for data acquisition, high speed ATE, and precision pulse amplifier applications.

Features

(Typical unless otherwise noted)
- Slew rate: 2000 V/μs
- Settling time (0.1%): 50 ns
- Characterized for supply ranges: ±5V and ±15V
- Low differential gain and phase error: 0.05%, 0.04°
- High output drive: ±10V into 100Ω
- Guaranteed bandwidth and slew rate
- Improved performance over EL2020, OP160, AD844, LT1223 and HA5004

Applications

- Coax cable driver
- Video amplifier
- Flash ADC buffer
- High frequency filter
- Scanner and Imaging systems

Typical Application

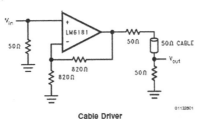

Cable Driver

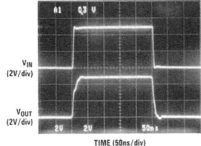

V_{IN} (2V/div)

V_{OUT} (2V/div)

TIME (50ns/div)

VIP™ is a registered trademark of National Semiconductor Corporation.

LM6181 100 mA, 100 MHz Current Feedback Amplifier

Reprinted with permission from National Semiconductor

LM6181

Absolute Maximum Ratings (Note 1)

If Military/Aerospace specified devices are required, please contact the National Semiconductor Sales Office/ Distributors for availability and specifications.

Supply Voltage	±18V
Differential Input Voltage	±6V
Input Voltage	±Supply Voltage
Inverting Input Current	15 mA
Soldering Information	
Dual-In-Line Package (N)	
Soldering (10 sec)	260°C
Small Outline Package (M)	
Vapor Phase (60 seconds)	215°C
Infrared (15 seconds)	220°C
Output Short Circuit	(Note 7)

Storage Temperature Range	−65°C ≤ T$_J$ ≤ +150°C
Maximum Junction Temperature	150°C
ESD Rating (Note 2)	±3000V

Operating Ratings

Supply Voltage Range	7V to 32V
Junction Temperature Range (Note 3)	
LM6181AM	−55°C ≤ T$_J$ ≤ +125°C
LM6181AI, LM6181I	−40°C ≤ T$_J$ ≤ +85°C
Thermal Resistance (θ$_{JA}$, θ$_{JC}$)	
8-pin DIP (N)	102°C/W, 42°C/W
8-pin SO (M-8)	153°C/W, 42°C/W
16-pin SO (M)	70°C/W, 38°C/W

±15V DC Electrical Characteristics

The following specifications apply for Supply Voltage = ±15V, R$_F$ = 820Ω, and R$_L$ = 1 kΩ unless otherwise noted. **Boldface** limits apply at the temperature extremes; all other limits T$_J$ = 25°C.

Symbol	Parameter	Conditions	LM6181AM Typical (Note 4)	LM6181AM Limit (Note 5)	LM6181AI Typical (Note 4)	LM6181AI Limit (Note 5)	LM6181I Typical (Note 4)	LM6181I Limit (Note 5)	Units
V$_{OS}$	Input Offset Voltage		2.0	3.0 **4.0**	2.0	3.0 **3.5**	3.5	5.0 **5.5**	mV max
TC V$_{OS}$	Input Offset Voltage Drift		5.0		5.0		5.0		µV/°C
I$_B$	Inverting Input Bias Current		2.0	5.0 **12.0**	2.0	5.0 **12.0**	5.0	10 **17.0**	µA max
	Non-Inverting Input Bias Current		0.5	1.5 **3.0**	0.5	1.5 **3.0**	2.0	3.0 **5.0**	
TC I$_B$	Inverting Input Bias Current Drift		30		30		30		nA/°C
	Non-Inverting Input Bias Current Drift		10		10		10		
I$_B$ PSR	Inverting Input Bias Current Power Supply Rejection	V$_S$ = ±4.5V, ±16V	0.3	0.5 **3.0**	0.3	0.5 **3.0**	0.3	0.75 **4.5**	µA/V max
	Non-Inverting Input Bias Current Power Supply Rejection	V$_S$ = ±4.5V, ±16V	0.05	0.5 **1.5**	0.05	0.5 **1.5**	0.05	0.5 **3.0**	
I$_B$ CMR	Inverting Input Bias Current Common Mode Rejection	−10V ≤ V$_{CM}$ ≤ +10V	0.3	0.5 **0.75**	0.3	0.5 **0.75**	0.3	0.75 **1.0**	
	Non-Inverting Input Bias Current Common Mode Rejection	−10V ≤ V$_{CM}$ ≤ +10V	0.1	0.5 **0.5**	0.1	0.5 **0.5**	0.1	0.5 **0.5**	
CMRR	Common Mode Rejection Ratio	−10V ≤ V$_{CM}$ ≤ +10V	60	50 **50**	60	50 **50**	60	50 **50**	dB min
PSRR	Power Supply Rejection Ratio	V$_S$ = ±4.5V, ±16V	80	70 **70**	80	70 **70**	80	70 **65**	dB min
R$_O$	Output Resistance	A$_V$ = −1, f = 300 kHz	0.2		0.2		0.2		Ω

±15V DC Electrical Characteristics (Continued)

The following specifications apply for Supply Voltage = ±15V, R_F = 820Ω, and R_L = 1 kΩ unless otherwise noted. **Boldface** limits apply at the temperature extremes; all other limits T_J = 25°C.

Symbol	Parameter	Conditions	LM6181AM		LM6181AI		LM6181I		Units
			Typical (Note 4)	Limit (Note 5)	Typical (Note 4)	Limit (Note 5)	Typical (Note 4)	Limit (Note 5)	
R_{IN}	Non-Inverting Input Resistance		10		10		10		MΩ min
V_O	Output Voltage Swing	R_L = 1 kΩ	12	11 **11**	12	11 **11**	12	11 **11**	V min
		R_L = 100Ω	11	10 **7.5**	11	10 **8.0**	11	10 **8.0**	
I_{SC}	Output Short Circuit Current		130	100 **75**	130	100 **85**	130	100 **85**	mA min
Z_T	Transimpedance	R_L = 1 kΩ	1.8	1.0 **0.5**	1.8	1.0 **0.5**	1.8	0.8 **0.4**	MΩ
		R_L = 100Ω	1.4	0.8 **0.4**	1.4	0.8 **0.4**	1.4	0.7 **0.35**	min
I_S	Supply Current	No Load, V_O = 0V	7.5	10 **10**	7.5	10 **10**	7.5	10 **10**	mA max
V_{CM}	Input Common Mode Voltage Range		V^+ − 1.7V V^- + 1.7V		V^+ − 1.7V V^- + 1.7V		V^+ − 1.7V V^- + 1.7V		V

±15V AC Electrical Characteristics

The following specifications apply for Supply Voltage = ±15V, R_F = 820Ω, R_L = 1 kΩ unless otherwise noted. **Boldface** limits apply at the temperature extremes; all other limits T_J = 25°C.

Symbol	Parameter	Conditions	LM6181AM		LM6181AI		LM6181I		Units
			Typical (Note 4)	Limit (Note 5)	Typical (Note 4)	Limit (Note 5)	Typical (Note 4)	Limit (Note 5)	
BW	Closed Loop Bandwidth −3 dB	A_V = +2	100		100		100		MHz min
		A_V = +10	80		80		80		
		A_V = −1	100	80	100	80	100	80	
		A_V = −10	60		60		60		
PBW	Power Bandwidth	A_V = −1, V_O = 5 V_{PP}	60		60		60		
SR	Slew Rate	Overdriven	2000		2000		2000		V/µs min
		A_V = −1, V_O = ±10V, R_L = 150Ω (Note 6)	1400	1000	1400	1000	1400	1000	
t_s	Settling Time (0.1%)	A_V = −1, V_O = ±5V R_L = 150Ω	50		50		50		ns
t_r, t_f	Rise and Fall Time	V_O = 1 V_{PP}	5		5		5		
t_p	Propagation Delay Time	V_O = 1 V_{PP}	6		6		6		
$i_{n(+)}$	Non-Inverting Input Noise Current Density	f = 1 kHz	3		3		3		pA/$\sqrt{Hz}$
$i_{n(-)}$	Inverting Input Noise Current Density	f = 1 kHz	16		16		16		pA/$\sqrt{Hz}$
e_n	Input Noise Voltage Density	f = 1 kHz	4		4		4		nV/$\sqrt{Hz}$

LM6181

±15V AC Electrical Characteristics (Continued)

The following specifications apply for Supply Voltage = ±15V, R_F = 820Ω, R_L = 1 kΩ unless otherwise noted. **Boldface** limits apply at the temperature extremes; all other limits T_J = 25°C.

Symbol	Parameter	Conditions	LM6181AM Typical (Note 4)	LM6181AM Limit (Note 5)	LM6181AI Typical (Note 4)	LM6181AI Limit (Note 5)	LM6181I Typical (Note 4)	LM6181I Limit (Note 5)	Units
	Second Harmonic Distortion	2 V_{PP}, 10 MHz	−50		−50		−50		dBc
	Third Harmonic Distortion	2 V_{PP}, 10 MHz	−55		−55		−50		
	Differential Gain	R_L = 150Ω A_V = +2 NTSC	0.05		0.05		0.05		%
	Differential Phase	R_L = 150Ω A_V = +2 NTSC	0.04		0.04		0.04		Deg

LM6181

±5V DC Electrical Characteristics

The following specifications apply for Supply Voltage = ±5V, R_F = 820Ω, and R_L = 1 kΩ unless otherwise noted. **Boldface** limits apply at the temperature extremes; all other limits T_J = 25°C.

Symbol	Parameter	Conditions	LM6181AM		LM6181AI		LM6181I		Units
			Typical (Note 4)	Limit (Note 5)	Typical (Note 4)	Limit (Note 5)	Typical (Note 4)	Limit (Note 5)	
V_{OS}	Input Offset Voltage		1.0	2.0 **3.0**	1.0	2.0 **2.5**	1.0	3.0 **3.5**	mV max
TC V_{OS}	Input Offset Voltage Drift		2.5		2.5		2.5		µV/°C
I_B	Inverting Input Bias Current		5.0	10 **22**	5.0	10 **22**	5.0	17.5 **27.0**	µA max
	Non-Inverting Input Bias Current		0.25	1.5 **1.5**	0.25	1.5 **1.5**	0.25	3.0 **5.0**	
TC I_B	Inverting Input Bias Current Drift		50		50		50		nA/°C
	Non-Inverting Input Bias Current Drift		3.0		3.0		3.0		
I_B PSR	Inverting Input Bias Current Power Supply Rejection	V_S = ±4.0V, ±6.0V	0.3	0.5 **0.5**	0.3	0.5 **0.5**	0.3	1.0 **1.0**	µA/V max
	Non-Inverting Input Bias Current Power Supply Rejection	V_S = ±4.0V, ±6.0V	0.05	0.5 **0.5**	0.05	0.5 **0.5**	0.05	0.5 **0.5**	
I_B CMR	Inverting Input Bias Current Common Mode Rejection	$-2.5V \leq V_{CM} \leq +2.5V$	0.3	0.5 **1.0**	0.3	0.5 **1.0**	0.3	1.0 **1.5**	
	Non-Inverting Input Bias Current Common Mode Rejection	$-2.5V \leq V_{CM} \leq +2.5V$	0.12	0.5 **1.0**	0.12	0.5 **0.5**	0.12	0.5 **0.5**	
CMRR	Common Mode Rejection Ratio	$-2.5V \leq V_{CM} \leq +2.5V$	57	50 **47**	57	50 **47**	57	50 **47**	dB min
PSRR	Power Supply Rejection Ratio	V_S = ±4.0V, ±6.0V	80	70 **70**	80	70 **70**	80	64 **64**	
R_O	Output Resistance	A_V = −1, f = 300 kHz	0.25		0.25		0.25		Ω
R_{IN}	Non-Inverting Input Resistance		8		8		8		MΩ min
V_O	Output Voltage Swing	R_L = 1 kΩ	2.6	2.25 **2.2**	2.6	2.25 **2.25**	2.6	2.25 **2.25**	V min
		R_L = 100Ω	2.2	2.0 **2.0**	2.2	2.0 **2.0**	2.2	2.0 **2.0**	
I_{SC}	Output Short Circuit Current		100	75 **70**	100	75 **70**	100	75 **70**	mA min
Z_T	Transimpedance	R_L = 1 kΩ	1.4	0.75 **0.35**	1.4	0.75 **0.4**	1.0	0.6 **0.3**	MΩ min
		R_L = 100Ω	1.0	0.5 **0.25**	1.0	0.5 **0.25**	1.0	0.4 **0.2**	
I_S	Supply Current	No Load, V_O = 0V	6.5	8.5 **8.5**	6.5	8.5 **8.5**	6.5	8.5 **8.5**	mA max
V_{CM}	Input Common Mode Voltage Range		V⁺ − 1.7V V⁻ + 1.7V		V⁺ − 1.7V V⁻ + 1.7V		V⁺ − 1.7V V⁻ + 1.7V		V

www.national.com

LM6181

±5V AC Electrical Characteristics

The following specifications apply for Supply Voltage = ±5V, R_F = 820Ω, and R_L = 1 kΩ unless otherwise noted. **Boldface** limits apply at the temperature extremes; all other limits T_J = 25°C.

Symbol	Parameter	Conditions	LM6181AM		LM6181AI		LM6181I		Units
			Typical (Note 4)	Limit (Note 5)	Typical (Note 4)	Limit (Note 5)	Typical (Note 4)	Limit (Note 5)	
BW	Closed Loop Bandwidth –3 dB	A_V = +2	50		50		50		MHz min
		A_V = +10	40		40		40		
		A_V = –1	55	35	55	35	55	35	
		A_V = –10	35		35		35		
PBW	Power Bandwidth	A_V = –1, V_O = 4 V_{PP}	40		40		40		
SR	Slew Rate	A_V = –1, V_O = ±2V, R_L = 150Ω (Note 6)	500	375	500	375	500	375	V/µs min
t_s	Settling Time (0.1%)	A_V = –1, V_O = ±2V R_L = 150Ω	50		50		50		ns
t_r, t_f	Rise and Fall Time	V_O = 1 V_{PP}	8.5		8.5		8.5		
t_p	Propagation Delay Time	V_O = 1 V_{PP}	8		8		8		
$i_{n(+)}$	Non-Inverting Input Noise Current Density	f = 1 kHz	3		3		3		pA/$\sqrt{Hz}$
$i_{n(-)}$	Inverting Input Noise Current Density	f = 1 kHz	16		16		16		pA/$\sqrt{Hz}$
e_n	Input Noise Voltage Density	f = 1 kHz	4		4		4		nV/$\sqrt{Hz}$
	Second Harmonic Distortion	2 V_{PP}, 10 MHz	–45		–45		–45		dBc
	Third Harmonic Distortion	2 V_{PP}, 10 MHz	–55		–55		–55		
	Differential Gain	R_L = 150Ω A_V = +2 NTSC	0.063		0.063		0.063		%
	Differential Phase	R_L = 150Ω A_V = +2 NTSC	0.16		0.16		0.16		Deg

Note 1: Absolute Maximum Ratings indicate limits beyond which damage to the device may occur. Operating ratings indicate conditions the device is intended to be functional, but device parameter specifications may not be guaranteed under these conditions. For guaranteed specifications and test conditions, see the Electrical Characteristics.

Note 2: Human body model 100 pF and 1.5 kΩ.

Note 3: The typical junction-to-ambient thermal resistance of the molded plastic DIP(N) package soldered directly into a PC board is 102°C/W. The junction-to-ambient thermal resistance of the S.O. surface mount (M) package mounted flush to the PC board is 70°C/W when pins 1, 4, 8, 9 and 16 are soldered to a total 2 in² 1 oz. copper trace. The 16-pin S.O. (M) package must have pin 4 and at least one of pins 1, 8, 9, or 16 connected to V⁻ for proper operation. The typical junction-to-ambient thermal resistance of the S.O. (M-8) package soldered directly into a PC board is 153°C/W.

Note 4: Typical values represent the most likely parametric norm.

Note 5: All limits guaranteed at room temperature (standard type face) or at operating temperature extremes (bold face type).

Note 6: Measured from +25% to +75% of output waveform.

Note 7: Continuous short circuit operation at elevated ambient temperature can result in exceeding the maximum allowed junction temperature of 150°C. Output currents in excess of ±130 mA over a long term basis may adversely affect reliability.

Note 8: For guaranteed Military Temperature Range parameters see RETS6181X.

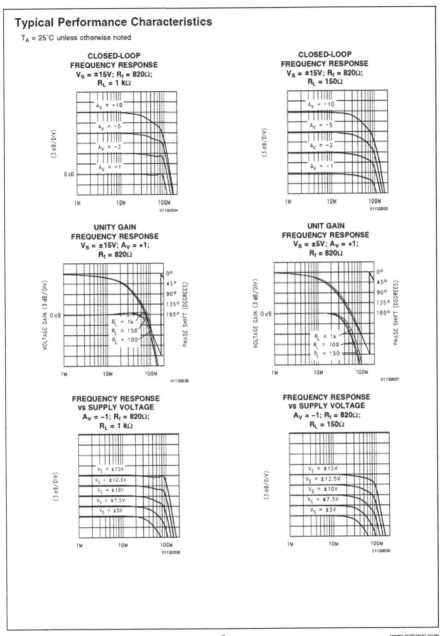

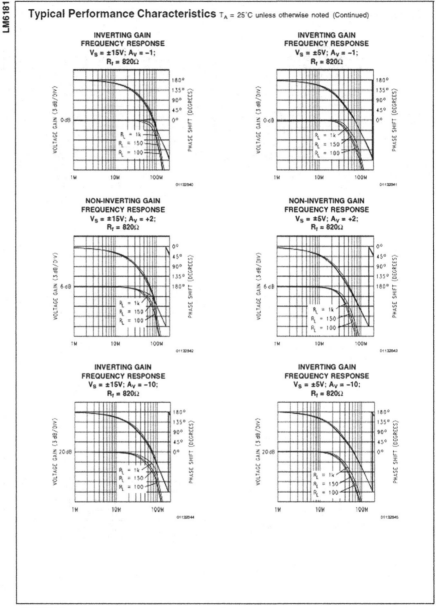

Typical Performance Characteristics $T_A = 25°C$ unless otherwise noted (Continued)

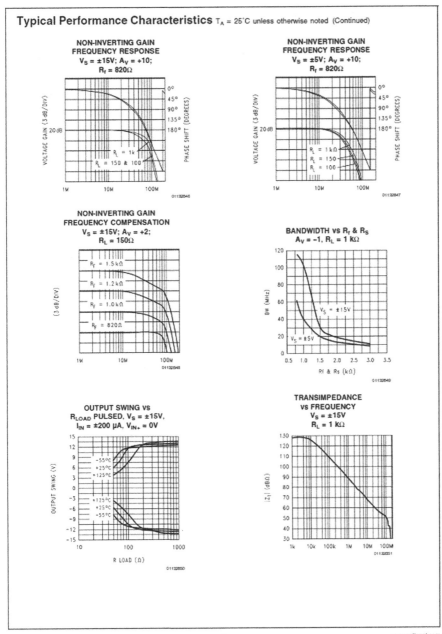

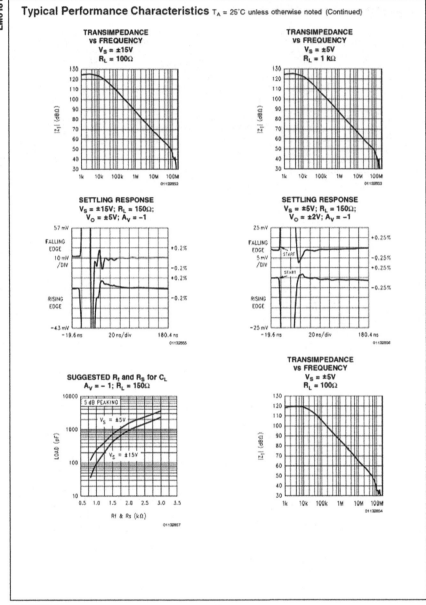

Typical Performance Characteristics T_A = 25°C unless otherwise noted (Continued)

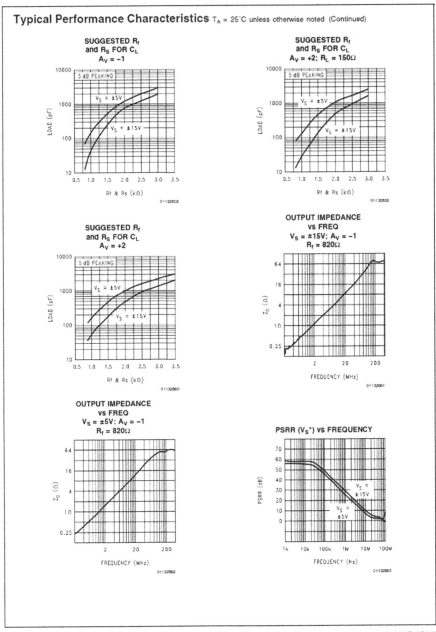

Typical Performance Characteristics $T_A = 25°C$ unless otherwise noted (Continued)

LM6181

Typical Performance Characteristics $T_A = 25°C$ unless otherwise noted (Continued)

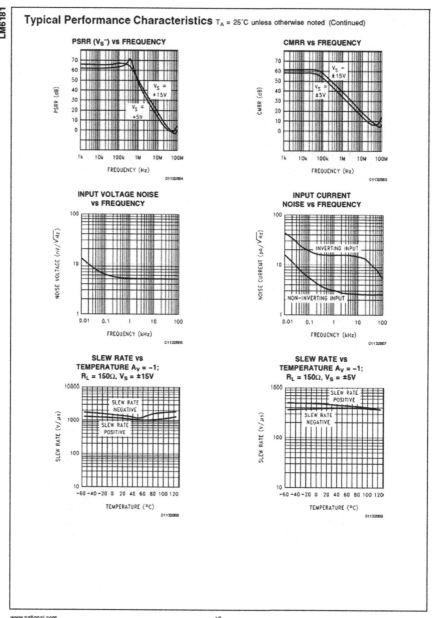

Typical Performance Characteristics $T_A = 25°C$ unless otherwise noted (Continued)

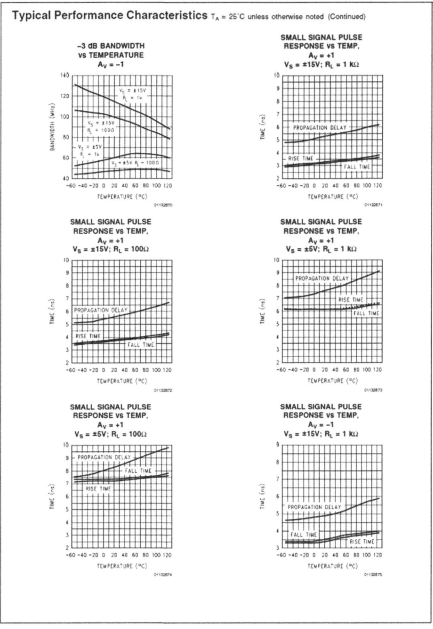

Typical Performance Characteristics T_A = 25°C unless otherwise noted (Continued)

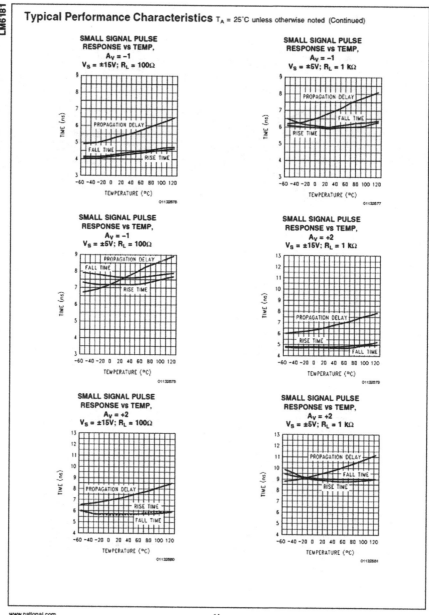

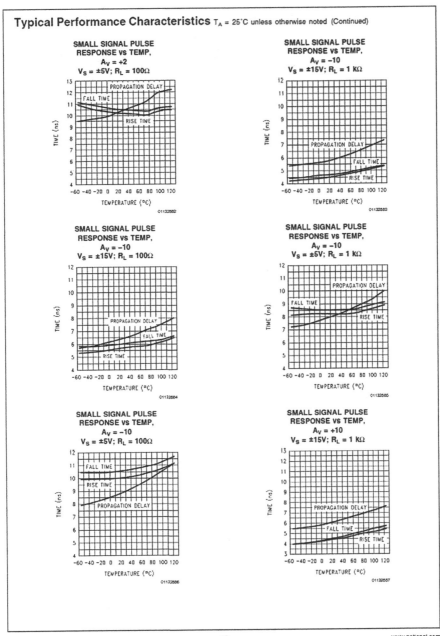

LM6181

Typical Performance Characteristics $T_A = 25°C$ unless otherwise noted (Continued)

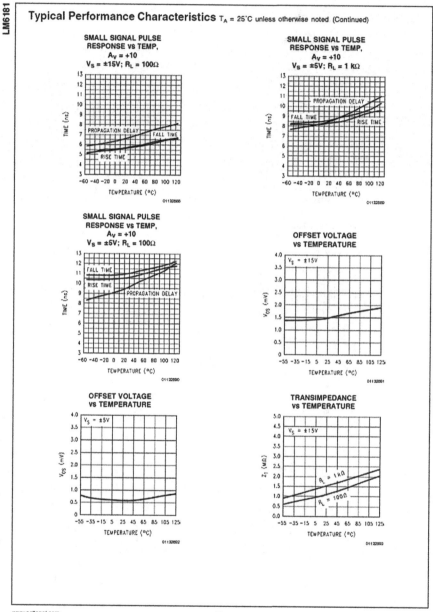

Typical Performance Characteristics $T_A = 25°C$ unless otherwise noted (Continued)

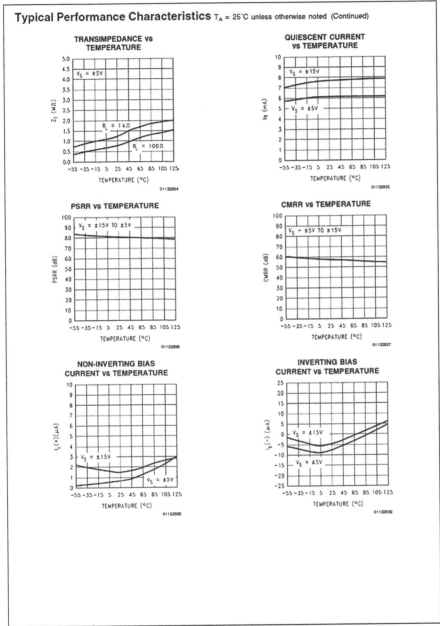

LM6181

Typical Performance Characteristics T_A = 25°C unless otherwise noted (Continued)

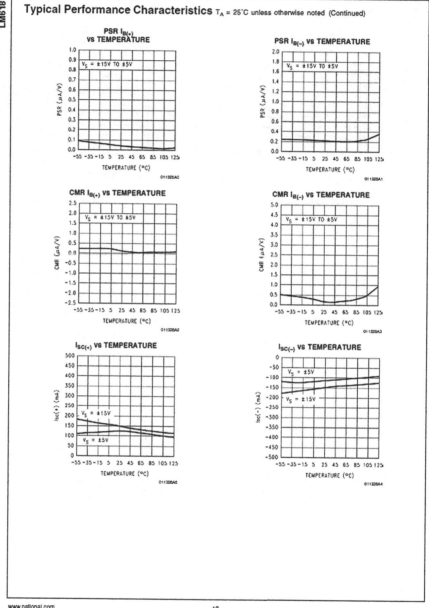

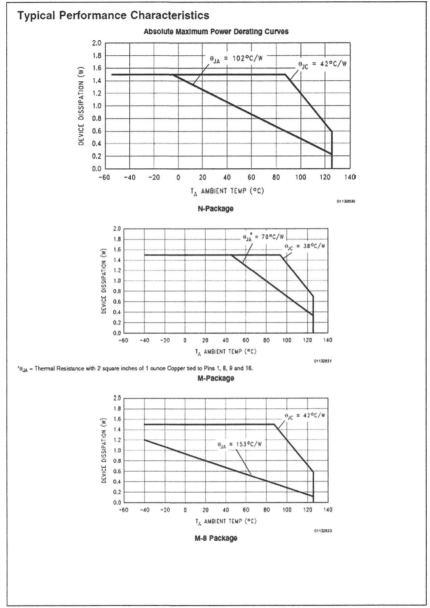

Typical Performance Characteristics

Absolute Maximum Power Derating Curves

N-Package

M-Package

$^*\theta_{JA}$ – Thermal Resistance with 2 square inches of 1 ounce Copper tied to Pins 1, 8, 9 and 16.

M-8 Package

Typical Performance Characteristics (Continued)

Simplified Schematic

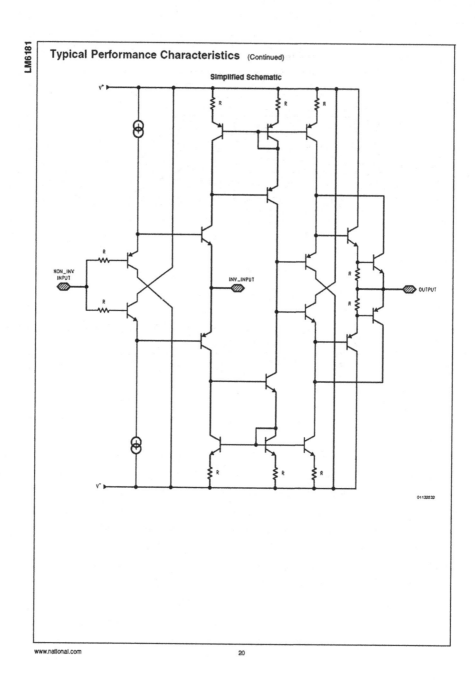

01122832

Typical Applications

CURRENT FEEDBACK TOPOLOGY

For a conventional voltage feedback amplifier the resulting small-signal bandwidth is inversely proportional to the desired gain to a first order approximation based on the gain-bandwidth concept. In contrast, the current feedback amplifier topology, such as the LM6181, transcends this limitation to offer a signal bandwidth that is relatively independent of the closed-loop gain. *Figure 1a* and *Figure 1b* illustrate that for closed loop gains of –1 and –5 the resulting pulse fidelity suggests quite similar bandwidths for both configurations.

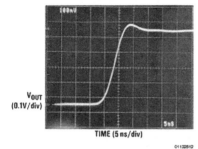

V_{OUT}
(0.1V/div)

TIME (5 ns/div)

1a

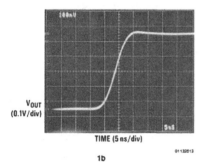

V_{OUT}
(0.1V/div)

TIME (5 ns/div)

1b

FIGURE 1. 1a, 1b: Variation of Closed Loop Gain from –1 to –5 Yields Similar Responses

The closed-loop bandwidth of the LM6181 depends on the feedback resistance, R_f. Therefore, R_S and not R_f, must be varied to adjust for the desired closed-loop gain as in *Figure 2*.

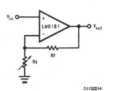

FIGURE 2. R_S is Adjusted to Obtain the Desired Closed Loop Gain, A_{VCL}

POWER SUPPLY BYPASSING AND LAYOUT CONSIDERATIONS

A fundamental requirement for high-speed amplifier design is adequate bypassing of the power supply. It is critical to maintain a wideband low-impedance to ground at the amplifiers supply pins to insure the fidelity of high speed amplifier transient signals. 10 µF tantalum and 0.1 µF ceramic bypass capacitors are recommended for each supply pin. The bypass capacitors should be placed as close to the amplifier pins as possible (0.5" or less).

FEEDBACK RESISTOR SELECTION: R_f

Selecting the feedback resistor, R_f, is a dominant factor in compensating the LM6181. For general applications the LM6181 will maintain specified performance with an 820Ω feedback resistor. Although this value will provide good results for most applications, it may be advantageous to adjust this value slightly. Consider, for instance, the effect on pulse responses with two different configurations where both the closed-loop gains are 2 and the feedback resistors are 820Ω and 1640Ω, respectively. *Figure 3a* and *Figure 3b* illustrate the effect of increasing R_f while maintaining the same closed-loop gain — the amplifier bandwidth decreases. Accordingly, larger feedback resistors can be used to slow down the LM6181 (see –3 dB bandwidth vs R_f typical curves) and reduce overshoot in the time domain response. Conversely, smaller feedback resistance values than 820Ω can be used to compensate for the reduction of bandwidth at high closed loop gains, due to 2nd order effects. For example *Figure 4* illustrates reducing R_f to 500Ω to establish the desired small signal response in an amplifier configured for a closed loop gain of 25.

LM6181

Typical Applications (Continued)

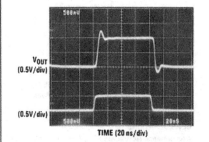

V_{OUT} (0.5V/div)

(0.5V/div)

TIME (20 ns/div)

3a: $R_f = 820\Omega$

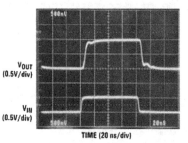

V_{OUT} (0.5V/div)

V_{IN} (0.5V/div)

TIME (20 ns/div)

3b: $R_f = 1640\Omega$

FIGURE 3. Increasing Compensation with Increasing R_f

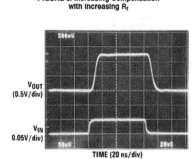

V_{OUT} (0.5V/div)

V_{IN} 0.05V/div)

TIME (20 ns/div)

FIGURE 4. Reducing R_f for Large Closed Loop Gains, $R_f = 500\Omega$

SLEW RATE CONSIDERATIONS

The slew rate characteristics of current feedback amplifiers are different than traditional voltage feedback amplifiers. In voltage feedback amplifiers slew rate limiting or non-linear amplifier behavior is dominated by the finite availability of the 1st stage tail current charging the compensation capacitor. The slew rate of current feedback amplifiers, in contrast, is not constant. Transient current at the inverting input determines slew rate for both inverting and non-inverting gains. The non-inverting configuration slew rate is also determined by input stage limitations. Accordingly, variations of slew rates occur for different circuit topologies.

DRIVING CAPACITIVE LOADS

The LM6181 can drive significantly larger capacitive loads than many current feedback amplifiers. Although the LM6181 can directly drive as much as 100 pF without oscillating, the resulting response will be a function of the feedback resistor value. *Figure 5* illustrates the small-signal pulse response of the LM6181 while driving a 50 pF load. Ringing persists for approximately 70 ns. To achieve pulse responses with less ringing either the feedback resistor can be increased (see typical curves Suggested R_f and R_s for C_L), or resistive isolation can be used (10Ω–51Ω typically works well). Either technique, however, results in lowering the system bandwidth.

Figure 6 illustrates the improvement obtained with using a 47Ω isolation resistor.

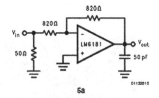

5a

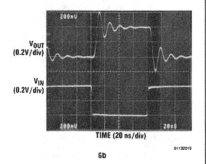

V_{OUT} (0.2V/div)

V_{IN} (0.2V/div)

TIME (20 ns/div)

5b

FIGURE 5. $A_V = -1$, LM6181 Can Directly Drive 50 pF of Load Capacitance with 70 ns of Ringing Resulting in Pulse Response

LM6181

Typical Applications (Continued)

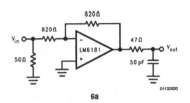

6a

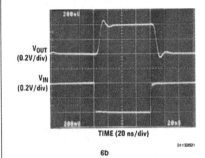

TIME (20 ns/div)

6b

FIGURE 6. Resistive Isolation of C_L Provides Higher Fidelity Pulse Response. R_f and R_S Could Be Increased to Maintain $A_V = -1$ and Improve Pulse Response Characteristics.

CAPACITIVE FEEDBACK

For voltage feedback amplifiers it is quite common to place a small lead compensation capacitor in parallel with feedback resistance, R_f. This compensation serves to reduce the amplifier's peaking in the frequency domain which equivalently tames the transient response. To limit the bandwidth of current feedback amplifiers, do not use a capacitor across R_f. The dynamic impedance of capacitors in the feedback loop reduces the amplifier's stability. Instead, reduced peaking in the frequency response, and bandwidth limiting can be accomplished by adding an RC circuit, as illustrated in *Figure 7b*.

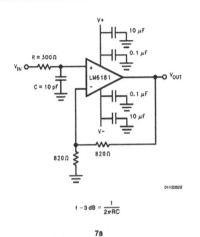

$$f_{-3\,dB} = \frac{1}{2\pi RC}$$

7a

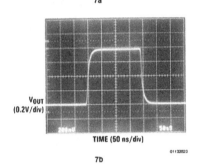

TIME (50 ns/div)

7b

FIGURE 7. RC Limits Amplifier Bandwidth to 50 MHz, Eliminating Peaking in the Resulting Pulse Response

Typical Performance Characteristics

OVERDRIVE RECOVERY

When the output or input voltage range of a high speed amplifier is exceeded, the amplifier must recover from an overdrive condition. The typical recovery times for open-loop, closed-loop, and input common-mode voltage range overdrive conditions are illustrated in *Figures 9, 11, 11, 12* respectively.

The open-loop circuit of *Figure 8* generates an overdrive response by allowing the ±0.5V input to exceed the linear input range of the amplifier. Typical positive and negative overdrive recovery times shown in *Figure 9* are 5 ns and 25 ns, respectively.

Typical Performance Characteristics (Continued)

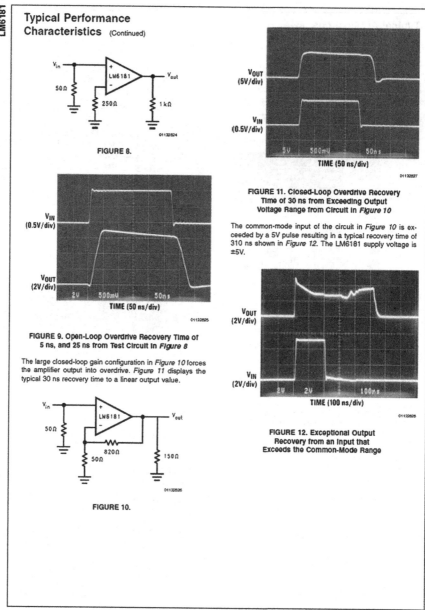

FIGURE 8.

FIGURE 9. Open-Loop Overdrive Recovery Time of 5 ns, and 25 ns from Test Circuit in *Figure 8*

The large closed-loop gain configuration in *Figure 10* forces the amplifier output into overdrive. *Figure 11* displays the typical 30 ns recovery time to a linear output value.

FIGURE 10.

FIGURE 11. Closed-Loop Overdrive Recovery Time of 30 ns from Exceeding Output Voltage Range from Circuit in *Figure 10*

The common-mode input of the circuit in *Figure 10* is exceeded by a 5V pulse resulting in a typical recovery time of 310 ns shown in *Figure 12*. The LM6181 supply voltage is ±5V.

FIGURE 12. Exceptional Output Recovery from an Input that Exceeds the Common-Mode Range

Analog Low-Pass Filters

14

"This chapter will show that modern filter design methods can be used by engineers, and that it is not a restricted field for specialists."

—Anatol I. Zverev, "Handbook of Filter Synthesis" 1967

IN THIS CHAPTER

▶ The basics of analog low-pass filtering are discussed. These techniques are useful for designing analog filters and prototypes used for digital filters or for other types of analog filters (high-pass, band-pass, etc.). This chapter is by no means an all-inclusive tutorial on analog filter design; rather it is introductory in nature, and the reader is referred to other texts for more details, if necessary.

Introduction

The low-pass filter (LPF) is a ubiquitous component in many different kinds of signal-processing systems. Channel separation, A/D antialiasing, and general signal processing are applications of LPFs, just to name a few. Even if you are a digital filter designer, it behooves you to know something about analog filter design, since many digital filters begin as analog prototypes, and are then transformed to the digital domain. Also, digital signal-processing (DSP) systems generally have an analog front end that includes an analog LPF for antialiasing purposes (Figure 14.1).

Another reason to familiarize ourselves with LPFs is that often LPFs are used as prototypes for high-pass or band-pass filters. You first design the LPF to get the attenuation shape you like, and then you transform the LPF into another filter type. We touch on high-pass, band-pass, and band-stop filters at the end of this chapter.

As with any kind of design, the "devil is in the details". Your specification will lead you to choices in filter topology and filter order, depending on the attenuation you need, the ripple that you can live with, and also the group delay variation that

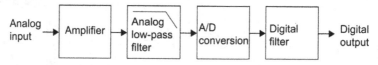

FIGURE 14.1

A typical digital signal-processing chain.

you can live with. In the following sections, we discuss the design issues associated with LPF design. The results can be extended without much trouble to band-pass and high-pass filters as well.

Review of LPF basics

The magnitude response of the ideal LPF is shown in Figure 14.2(a). The gain of this filter is perfectly flat in the passband (for frequencies less than the filter cutoff[1] frequency ω_c), and the response drops to zero for frequencies higher than the cutoff frequency.

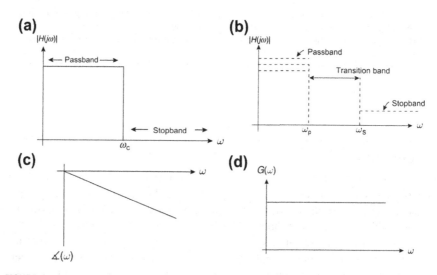

FIGURE 14.2

Response of an low-pass filter. (a) Ideal low-pass filter magnitude $|H(j\omega)|$. (b) Real-world magnitude response $|H(j\omega)|$ showing a possible ripple in the passband and the stopband, and a finite transition width between the passband and the stopband. (c) Ideal LPF phase response $\angle(\omega)$ showing a negative phase shift that increases linearly with frequency. (d) Ideal low-pass filter group delay response G (ω), which is constant.

---

[1]In general, the cutoff frequency of a filter is specified to be the frequency at which the gain through the filter has dropped to $0.707 \times$ of the DC value, or -3 dB.

The magnitude response of a real-world LPF is shown in Figure 14.2(b). The nonideal effects include the following:

- Possible ripple in the filter passband for frequencies $\omega < \omega_p$.
- Possible maximum attenuation floor in the stopband for frequencies $\omega > \omega_s$.
- A finite transition width between the passband and the stopband ($\omega_p < \omega < \omega_p$).

The magnitude response, however, only tells half the story. In addition, we must be concerned with the phase response of filters. As we will see in the following sections, the phase response (and by association the group delay[2] response) affects the transient response of filters. An ideal filter has a linear phase shift with frequency, and hence constant group delay as in Figure 14.2(c) and (d). The following section discusses in detail several different LPF types that to varying degrees approximate the ideal magnitude and phase of a LPF.

Butterworth filter

The Butterworth is a class of filters that provides a maximally flat response in the passband. The pole locations for an Nth-order Butterworth filter are equally spaced around a circle with radius equal to the filter cutoff frequency. A Butterworth filter with a -3-dB cutoff frequency $\omega_{3\,dB} = 1$ rad/s has poles at the following locations (Figure 14.3(a)):

$$-\sin\frac{(2m-1)\pi}{2N} + j\cos\frac{(2m-1)\pi}{2N} \quad m = 1,2...N \quad (14.1)$$

where N is the "filter order", or the number of poles. The frequency response is given by (Figure 14.3(b))

$$|H(j\omega)| = \frac{1}{\sqrt{1+\omega^{2N}}} \quad (14.2)$$

The transfer functions for Butterworth filters of varying orders are shown in Table 14.1. Note that the transfer functions have been broken up into first-order and second-order factors.[3] Breaking up the transfer function in this fashion will help us to implement our filter, since first- and second-order sections are easily synthesized with op-amps. We also show the transfer function for a filter cutoff frequency of 1 rad/s.

---

[2]The group delay of a filter is given by the negative derivative with respect to the frequency of the phase response, or $G(\omega) = -\frac{d\angle(\omega)}{d\omega}$. The group delay is a measure of how much a given frequency component is delayed when passing through the filter. For low pulse distortion, you want all Fourier components to be delayed by the same amount of time, and hence you want a constant group delay, and hence a linear phase response.

[3]This is done to help facilitate the implementation of the transfer function using op-amps. For instance, first-order sections can be implemented with simple RC filters. Second-order transfer functions can be implemented using any number of op-amp circuits, including the Sallen–Key filter. This will be discussed in detail later on in this chapter.

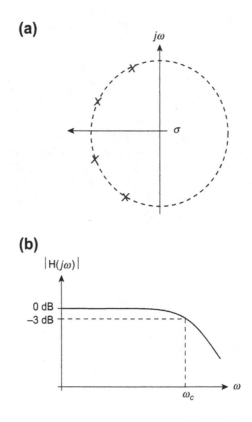

FIGURE 14.3

Butterworth filter. (a) Pole locations for an $N=4$ Butterworth low-pass filter with a cutoff frequency equal to the radius of the circle and poles spaced equally about the circle. (b) Magnitude response.

The magnitude response and step response for Butterworth filters are shown in Figures 14.4 and 14.5, respectively. We note the following:

- As the filter order increases, the sharpness of the attenuation characteristic in the transition band increases. For instance, at 10 rad/s, an eighth-order filter has an attenuation of roughly 160 dB, whereas lower-order filters have less attenuation.
- With regard to step response, the overshoot and delay through the filter increases as the filter order increases.

The preceding allows us to make some general statements regarding the relationship between filter frequency response, group delay, and transient response. A filter with a sharper cutoff will in general have more group delay variations and more overshoot and/or ringing in the step response. This is the reason why group delay

Table 14.1 Transfer Function for Butterworth Filters Broken Up into First-Order and Second-Order Factors. Transfer Functions are Shown for Varying Filter Order N, with a filter cutoff frequency $\omega_{3\,dB} = 1$ rad/s

N	Transfer Function
2	$\dfrac{1}{s^2 + 1.414s + 1}$
3	$\dfrac{1}{(s + 1)(s^2 + s + 1)}$
4	$\dfrac{1}{(s^2 + 0.7654s + 1)(s^2 + 1.8478s + 1)}$
5	$\dfrac{1}{(s + 1)(s^2 + 0.6180s + 1)(s^2 + 1.6180s + 1)}$
6	$\dfrac{1}{(s^2 + 0.5176s + 1)(s^2 + 1.4142s + 1)(s^2 + 1.9318s + 1)}$
7	$\dfrac{1}{(s + 1)(s^2 + 0.4450s + 1)(s^2 + 1.2480s + 1)(s^2 + 1.8019s + 1)}$
8	$\dfrac{1}{(s^2 + 0.3902s + 1)(s^2 + 1.1111s + 1)(s^2 + 1.6629s + 1)(s^2 + 1.9616s + 1)}$

variation is an important design parameter in filter design: more group delay variation results in more pulse distortion of a waveform passing through a filter. Conversely, filters with near-constant group delay pass pulses without significant distortion.

Example 14.1: Determining Butterworth filter order

Let us determine the filter order N needed to achieve a Butterworth filter with a cutoff frequency $f_c = 1$ MHz and 50-dB attenuation at 2.7 MHz. We could estimate the filter order by using the magnitude response plots of Figure 14.4, but for the Butterworth, we can use the magnitude equation directly. First, note that an attenuation of -50 dB is equivalent to a gain magnitude of 3.16×10^{-3}. We can use this in the attenuation equation as follows:

$$3.16 \times 10^{-3} = \frac{1}{\sqrt{1 + \left(\frac{f}{f_c}\right)^{2N}}} = \frac{1}{\sqrt{1 + \left(\frac{2.7}{1}\right)^{2N}}} \rightarrow N = 5.8 \tag{14.3}$$

So, to meet the magnitude specification, we need an $N = 6$ Butterworth filter. An implementation of this filter as a doubly terminated ladder is shown in Figure 14.6(a). (We will show later how to derive the component values.) Also, in Figure 14.6(a), we note the use of a gain of 2.

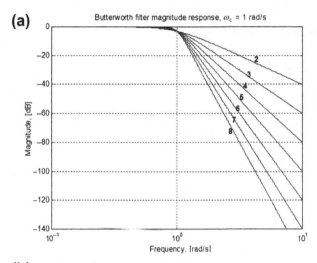

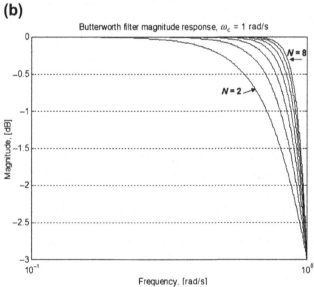

FIGURE 14.4

Butterworth filter magnitude response for filters with a cutoff frequency $\omega_c = 1$ rad/s, filter orders $N = 2$–8. (a) Overall response from 0.1 to 10 rad/s. (b) Passband detail.

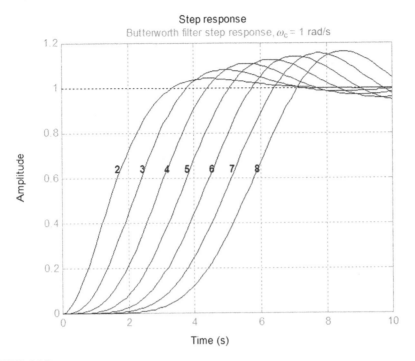

Step response
Butterworth filter step response, $\omega_c = 1$ rad/s

FIGURE 14.5

Butterworth filter step response for filters with a cutoff frequency of 1 rad/s, filter orders $N = 2-8$.

Chebyshev type-1 filter

We can improve the stopband attenuation significantly by allowing some finite ripple in the passband. The Chebyshev type-1 filter has some ripple in the passband, and a sharper cutoff in the transition band than does the Butterworth filter. As we will see, the price one pays for using the Chebyshev design compared to using the Butterworth is more group delay variation and more overshoot in the transient response.

We can form a Chebyshev filter by reducing the real part of the Butterworth pole locations by a factor k_c, with resultant pole locations of an Nth-order Chebyshev given by

$$-(k_c)\sin\frac{(2m-1)\pi}{2N} + j\cos\frac{(2m-1)\pi}{2N} \quad m = 1, 2...N \quad (14.4)$$

In Figure 14.7, we see the pole map of an $N = 5$ Chebyshev filter with a cutoff frequency $\omega_c = 1$ rad/s and 0.25-dB passband ripple. The poles are all inside the unit circle on an ellipse, closer to the $j\omega$ axis than Butterworth filter poles.

(a)

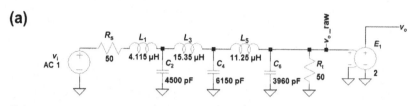

(b)

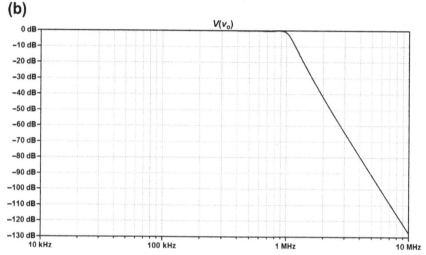

FIGURE 14.6

Ladder filter implementation for an $N=6$ Butterworth filter with a cutoff frequency $f_c = 5\,\text{MHz}$ for Example 14.1. (a) Circuit, with a gain of 2 added to compensate for the DC attenuation of the source and termination resistor. (b) Magnitude response. (For color version of this figure, the reader is referred to the online version of this book.)

We can relate the factor k_c to the passband ripple in decibels (R_{db}) as follows:

$$\varepsilon = \sqrt{10^{\frac{R_{db}}{10}} - 1}$$

$$A = \left(\frac{1}{N}\right)\sinh^{-1}\left(\frac{1}{\varepsilon}\right) \tag{14.5}$$

$$k_c = \tanh(A)$$

These pole locations yield a magnitude transfer function that is sharper than that of the Butterworth filter. Shown in Figure 14.8(a) is the Bode plot of an $N=5$, 0.25-dB ripple Chebyshev filter. Looking at the passband details of this filter, we plot the frequency response from 0.1 to 1 rad/s in Figure 14.8(b), we see that the filter order ($N=5$) equals the number of up and down ripples in the passband.

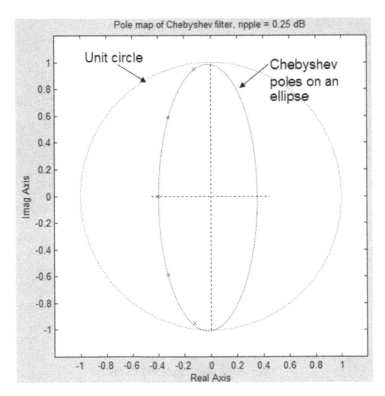

FIGURE 14.7

Pole map of an $N = 5$ Chebyshev filter with 0.25-dB ripple and a cutoff frequency $\omega_c = 1$ rad/s. (For color version of this figure, the reader is referred to the online version of this book.)

In Figures 14.9–14.11, we see the magnitude response for $N = 2$–8 type-1 Chebyshev filters, with 0.1-, 0.25-, and 0.5-dB passband ripple, each filter with a cutoff frequency of 1 rad/s.

Example 14.2: Comparison of $N = 5$ Butterworth and $N = 5$, 0.5 dB ripple Chebyshev filters

Let us compare an $N = 5$ Butterworth LPF and an $N = 5$ Chebyshev type-1 filter with a 0.5-dB passband ripple and a cutoff frequency 1 rad/s. For the Butterworth filter, the pole locations are

$$-0.309 \pm 0.9511j$$
$$-1.0 \tag{14.6}$$
$$-0.809 \pm 0.5878j$$

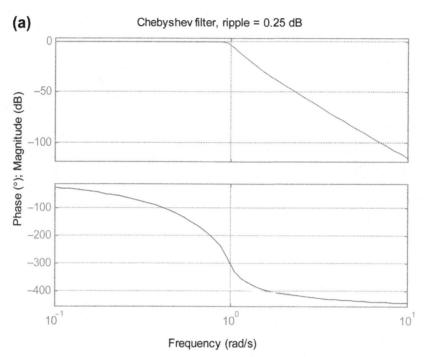

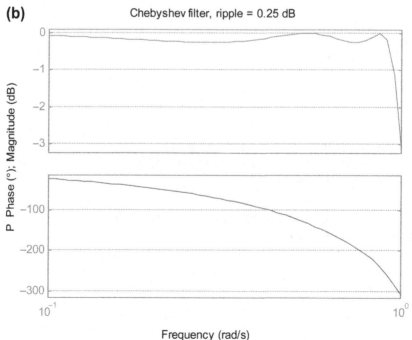

FIGURE 14.8

Chebyshev filter frequency response, order $N = 5$, passband ripple 0.25 dB and cutoff frequency 1 rad/s. (a) Overall response. (b) Details of the passband. (For color version of this figure, the reader is referred to the online version of this book.)

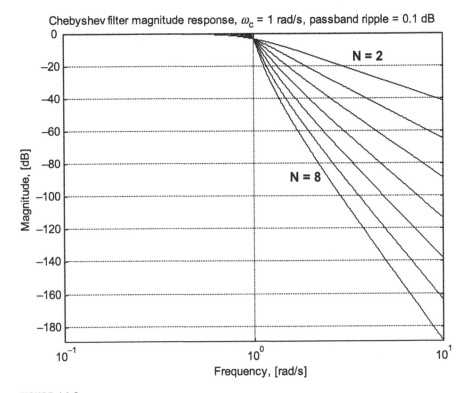

Chebyshev filter magnitude response, $\omega_c = 1$ rad/s, passband ripple = 0.1 dB

FIGURE 14.9

Chebyshev type-1 filter magnitude response, order $N = 2$–8, cutoff frequency $\omega_c = 1$ rad/s, and passband ripple 0.1 dB.

For the Chebyshev filter, to find the pole locations, we first find the factor k_c as follows:

$$\varepsilon = \sqrt{10^{\frac{R_{db}}{10}} - 1} = \sqrt{10^{\frac{0.5}{10}} - 1} = 0.3493$$

$$A = \left(\frac{1}{N}\right)\sinh^{-1}\left(\frac{1}{\varepsilon}\right) = \left(\frac{1}{5}\right)\sinh^{-1}\left(\frac{1}{0.3493}\right) = 0.3548 \qquad (14.7)$$

$$k_c = \tan h(A) = \tan h(0.3548) = 0.3406$$

We multiply the real part of the Butterworth poles by the factor k_c to find the Chebyshev poles as

$$\begin{aligned} &-0.1053 \pm 0.9511j \\ &-0.3406 \\ &-0.2756 \pm 0.5878j \end{aligned} \qquad (14.8)$$

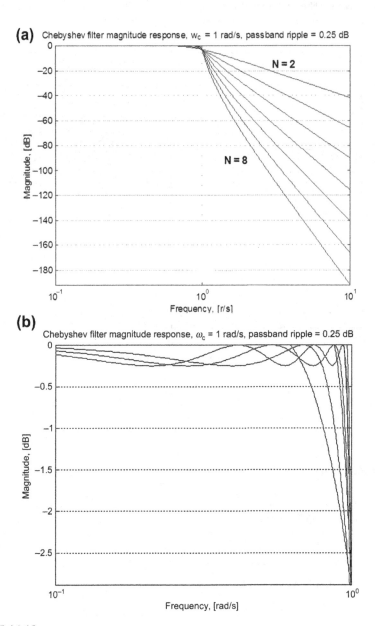

(a) Chebyshev filter magnitude response, w_c = 1 rad/s, passband ripple = 0.25 dB

(b) Chebyshev filter magnitude response, ω_c = 1 rad/s, passband ripple = 0.25 dB

FIGURE 14.10

Chebyshev type-1 low-pass filter magnitude, order $N = 2$–8, cutoff frequency $\omega_c = 1$ rad/s, and passband ripple 0.25 dB. (a) Overall. (b) Passband detail.

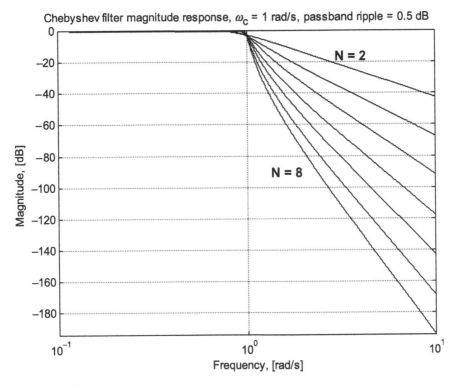

FIGURE 14.11

Chebyshev type-1 low-pass filter magnitude response, order $N = 2$–8, cutoff frequency $\omega_c = 1$ rad/s, and passband ripple 0.5 dB.

In Figure 14.12, we see a pole map of the Butterworth and Chebyshev filters, and note that the Chebyshev poles are closer to the $j\omega$ axis than to the Butterworth poles.

In Figure 14.13(a), we see that the Chebyshev filter has a sharper attenuation in the stopband. In Figure 14.13(b), we see that the Butterworth filter is flatter in the passband.

Group delay of LPFs

The group delay is a measure of the time delay of various frequency components going through a filter. In general, more constant group delay results in less signal distortion when passing a signal through a filter.

The group delay of an analog filter is found by taking the negative derivative of the phase with respect to frequency, or

$$G(\omega) = -\frac{d\angle(\omega)}{d\omega} \tag{14.9}$$

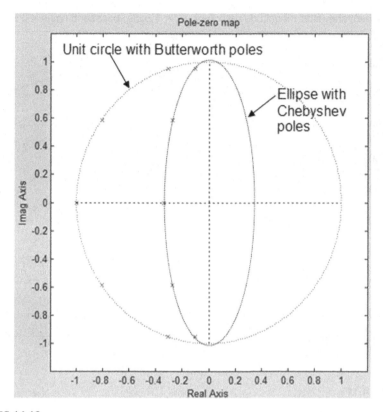

FIGURE 14.12

Pole map of an $N = 5$ Butterworth and $N = 5$, 0.5-dB ripple Chebyshev filter (Example 14.2). (For color version of this figure, the reader is referred to the online version of this book.)

A generic tradeoff in LPFs is that a sharper cutoff in the stopband results in more variation in the group delay. For comparison purposes, we now see

Figure 14.14: Group delay of a Butterworth filter with -3-dB cutoff frequency $\omega_c = 1$ rad/s;

Figure 14.15: Group delay of the Chebyshev type-1 filter with $\omega_c = 1$ rad/s and 0.1-dB ripple;

Figure 14.16: Group delay of the Chebyshev type-1 filter with $\omega_c = 1$ rad/s and 0.25-dB ripple;

Figure 14.17: Group delay of the Chebyshev type-1 filter with $\omega_c = 1$ rad/s and 0.5-dB ripple.

One area where tight group delay specifications are needed is in analog video processing. In the old-style broadcast (analog) television, color information was

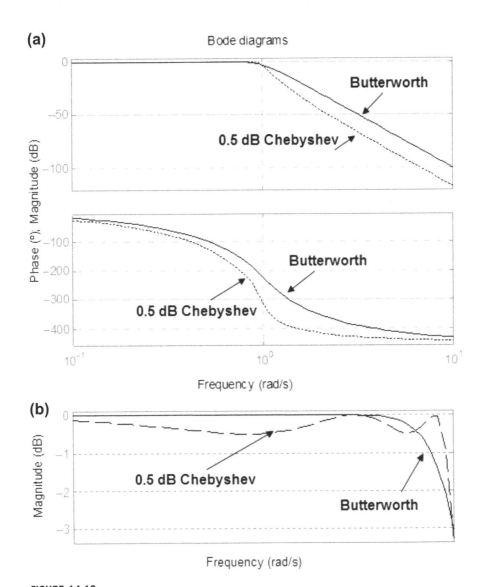

FIGURE 14.13

Bode plots for $N = 5$ Butterworth and $N = 5$, 0.5-dB ripple Chebyshev filter of Example 14.2. The Chebyshev is represented by the dotted line curves. (a) Overall magnitude and phase for frequencies 0.1 to 10 radians/sec. (b) Passband detail for frequencies 0.1 to 1 radians/sec.

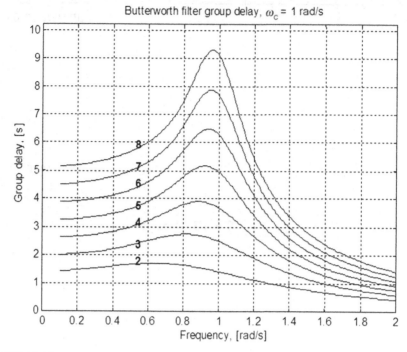

FIGURE 14.14

Butterworth filter group delay response for filters with a cutoff frequency of 1 rad/s, filter orders $N = 2$–8.

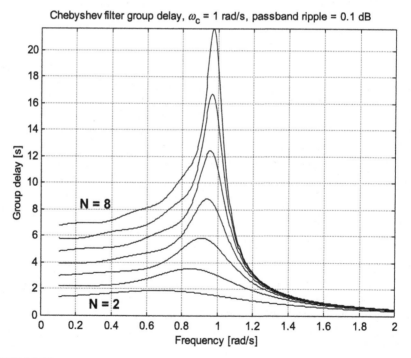

FIGURE 14.15

Chebyshev type-1 filter group delay response for filters with a cutoff frequency 1 rad/s, filter orders $N = 2$–8, and 0.1-dB ripple in the passband.

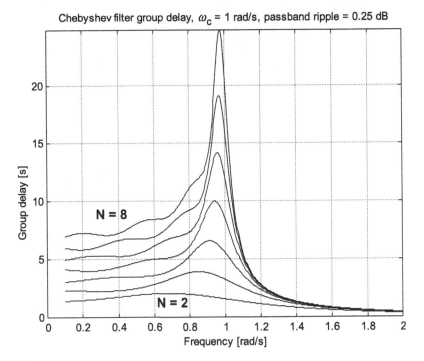

FIGURE 14.16

Chebyshev type-1 filter group delay response for filters with a cutoff frequency 1 rad/s, filter orders $N = 2-8$, and 0.25-dB ripple in the passband.

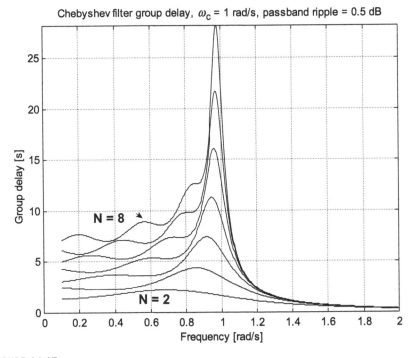

FIGURE 14.17

Chebyshev type-1 filter group delay response for filters with a cutoff frequency 1 rad/s, filter orders $N = 2-8$, and 0.5-dB ripple in the passband.

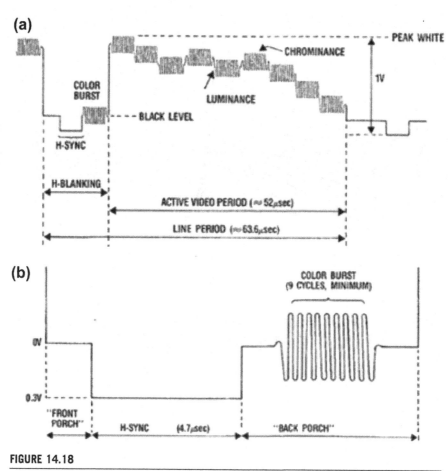

FIGURE 14.18

Details of a broadcast (NTSC) video signal. (a) Details of a total horizontal line of an NTSC signal. (b) Details near the horizontal sync portion.

From: Marc Thompson, "Designing Video Circuits, part 1," ESD Magazine, October 1988.

encoded in the phase difference between the 3.58-MHz "chrominance" information and the 3.58 "color burst" signal, as shown in Figure 14.18. When you run a video signal through an analog filter, you need to be careful to have a filter with a relatively tight group delay specification; otherwise, the phase relationship between the chrominance and burst signal will be distorted and your color information displayed on the TV screen will be misrepresented.

Bessel filter

The Bessel filter (sometimes called the "Thomson" filter) is optimized to provide a constant group delay in the filter passband, while sacrificing sharpness in the

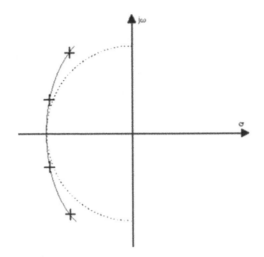

FIGURE 14.19

Pole locations for a Bessel filter, $N = 4$ and cutoff frequency ω_c equal to the radius of the circle.

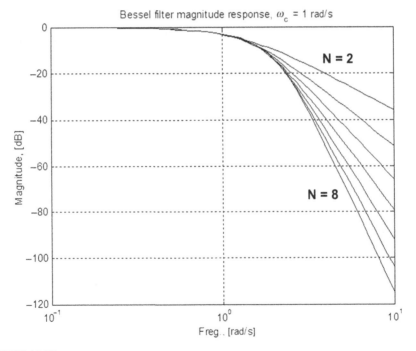

FIGURE 14.20

Bessel filter magnitude response, order $N = 2$–8, shown for filters with -3-dB cutoff frequency of 1 rad/s.

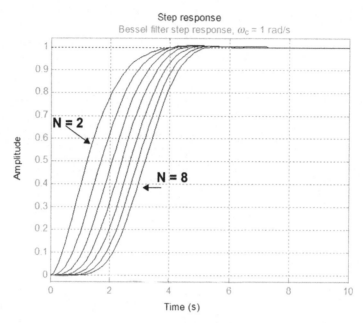

FIGURE 14.21

Bessel filter step response, order $N=2-8$, shown for filters with a −3-dB cutoff frequency of 1 rad/s.

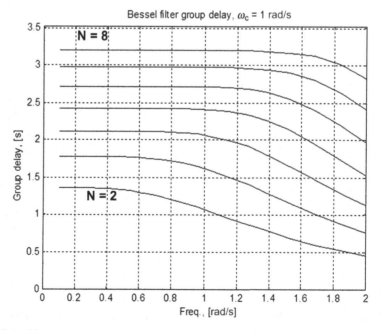

FIGURE 14.22

Bessel filter group delay response, order $N=2-8$, shown for filters with a −3-dB cutoff frequency of 1 rad/s. Note that as the filter order increases, the flatness of the group delay response in the passband improves.

Table 14.2 Pole Locations for Bessel Filters of Varying Filter Orders N, with Filter Cutoff Frequency $\omega_c = 1$ rad/s

N	Real Part ($-\sigma$)	Imaginary Part ($\pm j\omega$)
2	1.1030	0.6368
3	1.0509	1.0025
	1.3270	
4	1.3596	0.4071
	0.9877	1.2476
5	1.3851	0.7201
	0.9606	1.4756
	1.5069	0.3213
	1.5735	
6	1.3836	0.9727
	0.9318	1.6640
7	1.6130	0.5896
	1.3797	1.1923
	0.9104	1.8375
	1.6853	
8	1.7627	0.2737
	0.8955	2.0044
	1.3780	1.3926
	1.6419	0.8253

magnitude response. Bessel filters are sometimes used in applications where a constant group delay is critical, such as in analog video signal processing. The pole locations for the Bessel filter with a cutoff frequency 1 rad/s are outside the unit circle. Pole locations for an $N = 4$ Bessel filter are shown in Figure 14.19.

In Figure 14.20, we see the magnitude response of the Bessel filter for various filter orders. The response is not as sharp as that for the Butterworth filter, with a gradual roll-off in both the passband and stopband. There is little or no overshoot in the step response (Figure 14.21). The group delay exhibits a very flat response in the passband (Figure 14.22).

The pole locations for Bessel filters are shown in Table 14.2.

The resultant transfer functions of the Bessel filters are shown in Table 14.3.

Comparison of Butterworth, Chebyshev, and Bessel filters

A comparison of the magnitude responses for four different types of $N = 5$ filters is shown in Figure 14.23. Note that, as expected, the Chebyshev filters have the sharpest cutoff characteristics, whereas the Bessel response is more gradual in the stopband.

Table 14.3 Transfer Function Broken Up into First-Order and Second-Order Quadratic Factors for Bessel Filters of Varying Filter Orders N, with Filter Cutoff Frequency $\omega_c = 1$ rad/s

N	Transfer Function
2	$\dfrac{1.6221}{s^2 + 2.206s + 1.6221}$
3	$\dfrac{2.7992}{(s + 1.3270)(s^2 + 2.1018s + 2.1094)}$
4	$\dfrac{5.1002}{(s^2 + 2.7192s + 2.0142)(s^2 + 1.9754s + 2.5321)}$
5	$\dfrac{11.3845}{(s + 1.5069)(s^2 + 2.7702s + 2.4370)(s^2 + 1.9212s + 3.1001)}$
6	$\dfrac{26.8328}{(s^2 + 3.1470s + 2.5791)(s^2 + 2.7672s + 2.8605)(s^2 + 1.8636s + 3.6371)}$
7	$\dfrac{69.5099}{(s + 1.6853)(s^2 + 3.2262s + 2.9497)(s^2 + 2.7594s + 3.3251)(s^2 + 1.8208s + 4.2052)}$
8	$\dfrac{98.7746}{(s^2 + 3.5254s + 3.1820)(s^2 + 1.7910s + 4.1895)(s^2 + 2.7560s + 3.8382)(s^2 + 3.2838s + 3.3770)}$

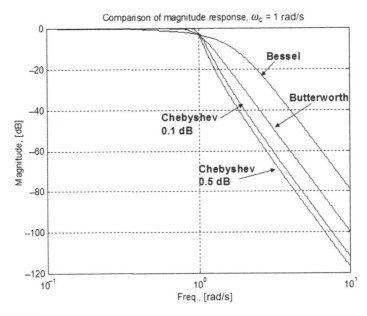

FIGURE 14.23

Comparison of the magnitude response for the $N = 5$ Butterworth, $N = 5$ Chebyshev with 0.1-, and 0.5-dB ripple and $N = 5$ Bessel, each with a cutoff frequency of 1 rad/s.

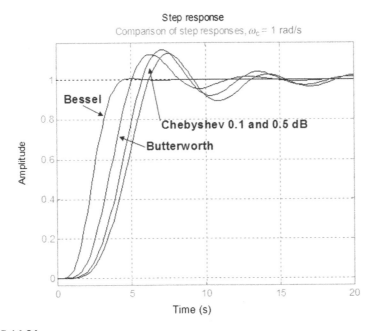

FIGURE 14.24

Comparison of the step response for the $N = 5$ Butterworth, Chebyshev (0.1- and 0.5-dB ripple) and Bessel filter, each with a cutoff frequency of 1 rad/s. Note that there is minimal overshoot in the Bessel response, while the Butterworth and Chebyshev responses are comparable.

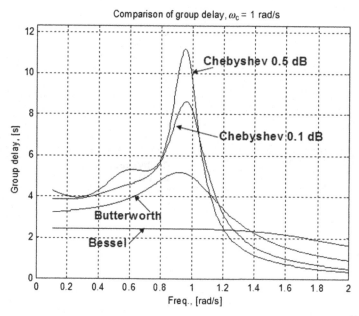

FIGURE 14.25

Comparison of the group delay response for the $N = 5$ Butterworth, Chebyshev (0.1- and 0.5-dB ripple) and Bessel filter, each with a cutoff frequency of 1 rad/s.

A comparison of step responses for the four different filters is shown in Figure 14.24. Note that the Bessel has a minimal overshoot, whereas the overshoots of the Butterworth and Chebyshev filters are comparable. A Chebyshev filter with more passband ripple would have a higher overshoot.

In Figure 14.25, we compare the group delay responses of the four filters. We note that the Bessel filter has a flat group delay response in the passband, whereas the Butterworth has a moderate variation in the group delay in the passband. The Chebyshev exhibits a pronounced group delay peaking in the passband.

Chebyshev type-2 filter

The Chebyshev type-2 filter is maximally flat in the passband, and has an equal-amplitude ripple in the stopband. In the Chebyshev type-2 filter, you specify the frequency at which the stopband begins, and the maximum ripple amplitude. In Figure 14.26, we see the magnitude responses of $N = 2-8$, Chebyshev type-2 LPFs, with a stopband beginning at 1 rad/s and a stopband minimum attenuation of -50 dB.

Elliptic filter

The elliptic, or "Cauer" filter has a ripple in both the passband and the stopband, as shown in Figure 14.27(a). For the generic elliptic filter, we specify the passband and

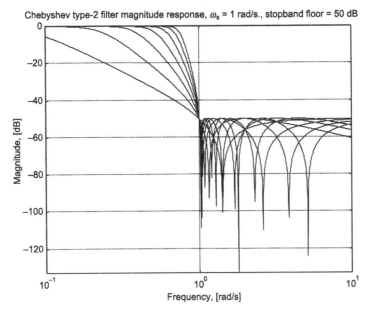

FIGURE 14.26

Magnitude plot for the Chebyshev type-2 filter with a 50-dB minimum attenuation in the stopband, with the stopband beginning at 1 rad/s.

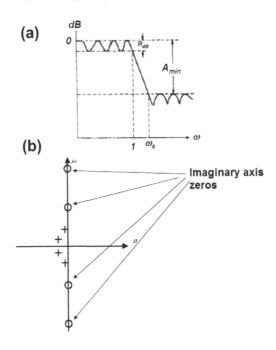

FIGURE 14.27

Generic elliptic low-pass filter magnitude response. (a) R_{db} ripple in the passband, minimum attenuation A_{min}, and transition band from 1 to ω_s rad/s. (b) Pole-zero plot showing imaginary-axis zeros.

Figure 14.27(a) adopted from Zverev.

(a)

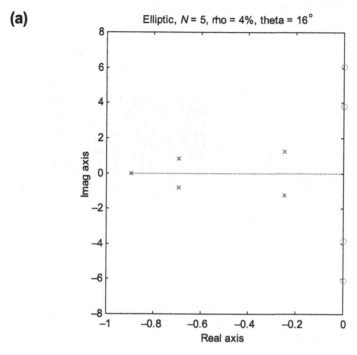

(b)

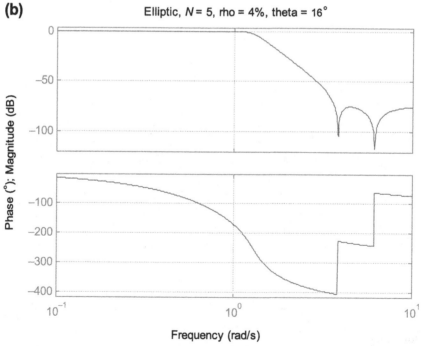

FIGURE 14.28

Elliptic low-pass filter example. (a) Pole-zero plot. (b) Magnitude. (For color version of this figure, the reader is referred to the online version of this book.)

stopband ripple, as well as the minimum attenuation. In Figure 14.27(b), we see the pole-zero plot of a generic elliptic filter with four $j\omega$-axis zeros.

This type of filter has the sharpest transition between the passband and the stopband, at the expense of more variation in the group delay than for the other filters. The elliptic filter in Chapter 2 has the transfer function

$$H(s) = \frac{0.001858\ s^4 + 0.09591\ s^2 + 1}{0.6104\ s^5 + 1.696\ s^4 + 3.128\ s^3 + 3.592\ s^2 + 2.651\ s + 1} \qquad (14.10)$$

The pole-zero map and Bode plot of the magnitude and phase of this filter are shown in Figure 14.28. We note that the use of transmission zeros results in frequencies at which the magnitude of the output response tends to zero. This $N = 5$ elliptic filter has two complex zero pairs on the $j\omega$ axis.

Comparison of filter responses

A qualitative comparison of filter responses is given in Table 14.4.

Table 14.4 Transfer Function Broken Up into First-Order and Second-Order Quadratic Factors for Bessel Filters of Varying Filter Orders N, with Filter Cutoff Frequency $\omega_c = 1$ rad/s

Filter Type	Passband	Stopband	Transition between Passband and Stopband	Group Delay
Butterworth	Maximally flat	Monotonically decreasing	Moderately fast	Moderate variation
Bessel	No ripple	Monotonically decreasing, slowest decrease of all filters	Slow	Maximally flat
Chebyshev type-1	Ripple in passband	Monotonically decreasing	Fast	Moderate-poor, depending on allowable decibel ripple
Chebyshev type-2	No ripple in passband	Ripple in stopband	Fast	Moderate-poor, depending on allowable decibel ripple
Elliptic	Ripple in passband	Ripple in stopband	Fastest	Worst group delay

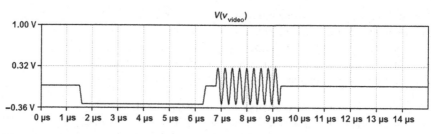

FIGURE 14.29

Video horizontal sync interval for Example 14.3.

Example 14.3: Video signal filtering

In this next example, we will see what happens when we low-pass filter a video signal (Figure 14.29). This signal is the sync and burst portion of an NTSC video signal. We will filter this video signal with a 5-MHz Butterworth, Bessel, and Chebyshev 0.1-dB filters. The implementation of these filters with LC ladders is shown in Figure 14.30(a) where we have intentionally added gain blocks so that the DC gain of each filter is +1. The magnitude response and group delay of each filter is shown in Figure 14.30(b).

The resultant transient response of the video signal after going through each of the filters is shown in Figure 14.31 for comparison. We note that the Bessel filter has the least delay and signal distortion, whereas the delay and distortion going through the Butterworth and Chebyshev filters are more pronounced.

Filter implementation

So, you have done all this hard work to figure out the filter type and order, but how do you build it in practice? There are several methods and we will discuss a few of them in the next section.

RLC ladder

For high-frequency filters, one option is to build a passive ladder using resistors, inductors, and capacitors. The topology for an Nth-order ladder filter that is suitable for implementing the Butterworth, Chebyshev, and Bessel filters is shown in Figure 14.32(a). The filter is made with alternating inductors and capacitors, with source and termination resistors. We note that, for an even-order filter, the filter terminates with a resistor and capacitor and for an odd-order filter the filter terminates with an inductor and resistor. Ladders for $N = 4$ and $N = 5$ filters are shown in Figure 14.32(b) and (c), respectively.

Values have been extensively tabulated for filters with a cutoff frequency 1 rad/s. The termination resistor for this normalized filter is 1.0 Ω, and the filter can have

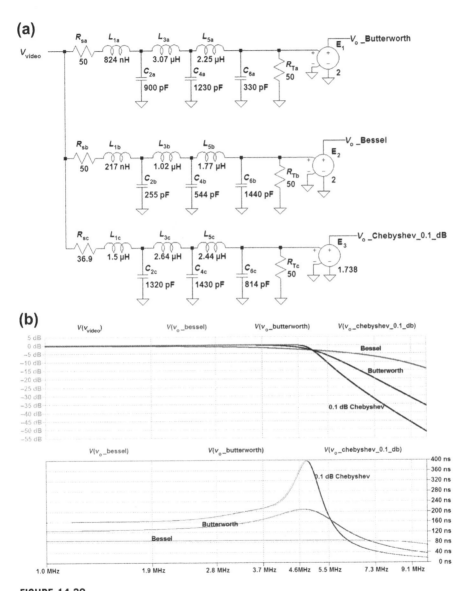

FIGURE 14.30

Video filters for Example 14.3. (a) $N = 6$ filters implemented as ladders. Note that gains have been added so that the DC gain of each filter is unity. (b) Magnitude and group delay response of filters. (For color version of this figure, the reader is referred to the online version of this book.)

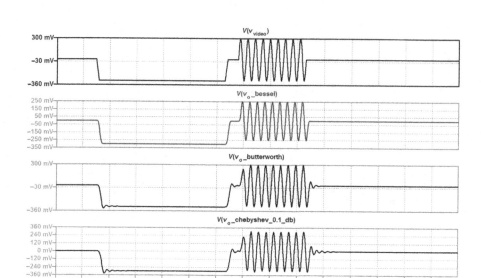

FIGURE 14.31

Response of three $N=6$, 5-MHz filters to the video signal of Example 14.3. (For color version of this figure, the reader is referred to the online version of this book.)

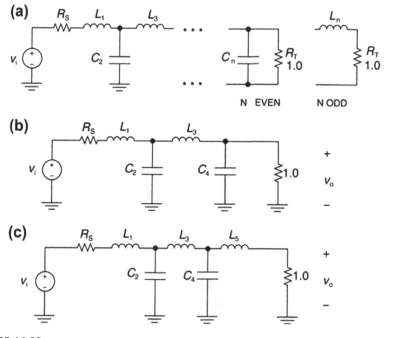

FIGURE 14.32

Ladder filter of order N, suitable for implementing Butterworth, Chebyshev, and Bessel filters. (a) Generic voltage-driven filter. (b) $N=4$ ladder. (c) $N=5$ ladder.

Table 14.5 Butterworth Inductor[4] and Capacitor Values of Varying Filter Orders N, with a Filter Cutoff Frequency $\omega_c = 1$ rad/s, and Termination Resistance $R_T = 1.0\,\Omega$. Note that for This Case, All Values of the Source Resistance are $R_s = 1.0\,\Omega$

N	R_s	L_1	C_2	L_3	C_4	L_5	C_6	L_7	C_8
2	1.0	1.4142	1.4142						
3	1.0	1.0000	2.0000	1.0000					
4	1.0	0.7654	1.8478	1.8478	0.7654				
5	1.0	0.6180	1.6180	2.0000	1.6180	0.6180			
6	1.0	0.5176	1.4142	1.9319	1.9319	1.4142	0.5176		
7	1.0	0.4450	1.2470	1.8019	2.0000	1.8019	1.2470	0.4450	
8	1.0	0.3902	1.1111	1.6629	1.9616	1.9616	1.6629	1.1111	0.3902

Table 14.6 Bessel Inductor and Capacitor Values of Varying Filters Order N, with a Filter Cutoff Frequency $\omega_c = 1$ rad/s, and Termination Resistance $R_T = 1.0\,\Omega$. Note that for This Case, All the Values of the Source Resistance are $R_s = 1.0\,\Omega$

N	R_s	L_1	C_2	L_3	C_4	L_5	C_6	C_7	C_8
2	1.0	0.5755	2.1478						
3	1.0	0.3374	0.9705	2.2034					
4	1.0	0.2334	0.6725	1.0815	2.2404				
5	1.0	0.1743	0.5072	0.8040	1.1110	2.2582			
6	1.0	0.1365	0.4002	0.6392	0.8538	1.1126	2.2645		
7	1.0	0.1106	0.3259	0.5249	0.7020	0.8690	1.1052	2.2659	
8	1.0	0.0919	0.2719	0.4409	0.5936	0.7303	0.8695	1.0956	2.2656

Table 14.7 A 0.1-dB Chebyshev Filter Inductor and Capacitor Values, with a Filter Cutoff Frequency $\omega_c = 1$ rad/s, and Termination Resistance $R_T = 1.0\,\Omega$. Note that the Values of the Source Resistance Varies as Filter Order N Varies

N	R_s	L_1	C_2	L_3	C_4	L_5	C_6	C_7	C_8
2	0.738	1.2087	1.6382						
3	1.0	1.4328	1.5937	1.4328					
4	0.738	0.9924	2.1476	1.5845	1.3451				
5	1.0	1.3013	1.5559	2.2411	1.5559	1.3013			
6	0.738	0.9419	2.0797	1.6581	2.2473	1.5344	1.2767		
7	1.0	1.2615	1.5196	2.2392	1.6804	2.2392	1.5196	1.2615	
8	0.738	0.9234	2.0454	1.6453	2.2826	1.6841	2.2300	1.5091	1.2515

Table 14.8 A 0.25-dB Chebyshev Filter Inductor and Capacitor Values, with a Filter Cutoff Frequency $\omega_c = 1$ rad/s, and Termination Resistance $R_T = 1.0\ \Omega$

N	R_s	L_1	C_2	L_3	C_4	L_5	C_6	L_7	C_8
2	0.5	0.6552	2.7632						
3	1.0	1.6325	1.4360	1.6325					
4	0.5	0.6747	3.6860	1.0247	1.8806				
5	1.0	1.5046	1.4436	2.4050	1.4436	1.5046			
6	0.5	0.6867	3.2074	0.9308	3.8102	1.2163	1.7088		
7	1.0	1.5120	1.4169	2.4535	1.5350	2.4535	1.4169	1.5120	

Table 14.9 A 0.5-dB Chebyshev Filter Inductor and Capacitor Values, with a Filter Cutoff Frequency $\omega_c = 1$ rad/s, and Termination Resistance $R_T = 1.0\ \Omega$

N	R_s	L_1	C_2	L_3	C_4	L_5	C_6	L_7	C_8
2	0.504	0.9827	1.9497						
3	1.0	1.8636	1.2804	1.8636					
4	0.504	0.9202	2.5864	1.3036	1.8258				
5	1.0	1.8068	1.3025	2.6912	1.3025	1.8068			
6	0.504	0.9053	2.5774	1.3675	2.7133	1.2991	1.7961		
7	1.0	1.7896	1.2961	2.7177	1.3848	2.7177	1.2961	1.7896	
8	0.504	0.8998	2.5670	1.3697	2.7585	1.3903	2.7175	1.2938	1.7852

any input resistance R_s that you desire. Tabulated ladder element values for the Butterworth, Bessel, and Chebyshev filters of various orders are shown in Table 14.5–14.9. (Note that for the Chebyshev there are other combination values of source resistance and LC ladder elements that work. Here, I show only one value. Other values can be found in the work of Zverev).

Example 14.4: Design example: fifth-order 1-MHz Chebyshev LPF with a 0.5-dB passband ripple

We will use the filter charts to design a fifth-order 1-MHz low-pass Chebyshev filter with a 0.5-dB ripple in the passband. From the filter chart (Table 14.9), we find the corresponding values for a filter with a cutoff frequency of 1 rad/s as

$R_s = R_T = 1\ \Omega$
$L_1 = 1.8068$
$C_2 = 1.3025$
$L_3 = 2.6914$

$C_4 = 1.3025$
$L_5 = 1.8068$

We next need to select more reasonable values for source and termination resistors (instead of the 1.0-Ω normalized values). For this design example, we will choose $R_s = R_T = 50\ \Omega$. We now make use of an unnormalization process to transform the filter with a cutoff of 1 rad/s to our desired -3-dB cutoff frequency of 1 MHz. The unnormalization process is as follows:

$$C = \frac{C_n}{2\pi f_c R}$$

$$L = \frac{L_n R}{2\pi f_c} \tag{14.11}$$

where C_n and L_n are the normalized values found from the filter charts, f_c is the desired new -3-dB cutoff frequency, and R is the termination resistor value used in the new filter. Applying this process to our filter results in

$$L_1' = \frac{L_1 R}{2\pi f_c} = \frac{(1.8068)(50)}{(2\pi)(10^6)} = 14.378\ \mu H$$

$$C_2' = \frac{C_2}{2\pi f_c R} = \frac{(1.3025)}{(2\pi)(10^6)(50)} = 4146\ pF$$

$$L_3' = \frac{L_3 R}{2\pi f_c} = \frac{(2.6914)(50)}{(2\pi)(10^6)} = 21.417\ \mu H \tag{14.12}$$

$$C_4' = \frac{C_4}{2\pi f_c R} = \frac{(1.3025)}{(2\pi)(10^6)(50)} = 4146\ pF$$

$$L_5' = \frac{L_4 R}{2\pi f_c} = \frac{(1.8068)(50)}{(2\pi)(10^6)} = 14.378\ \mu H$$

The resultant circuit and frequency response for the 1-MHz filter are shown in Figure 14.33.

Example 14.5: Elliptic filter ladder

The detailed design of elliptic filters is beyond the scope of this book, but a few comments are in order. Elliptic filters, also called "brick wall" filters, have very sharp filter cutoff characteristics. Again, this is done at the expense of a very nonlinear group delay. One flavor of elliptic filters has a zero ripple in the passband but a finite ripple in the stopband. This is accomplished by having zeroes in the transfer

---

[4] All inductor and capacitor value charts adopted from A. Zverev, *Handbook of Filter Synthesis*, John Wiley, 1967. For other source resistances R_s, the reader is invited to read this reference.

(a)

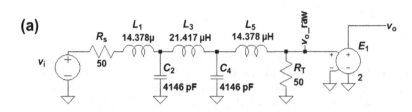

(b)

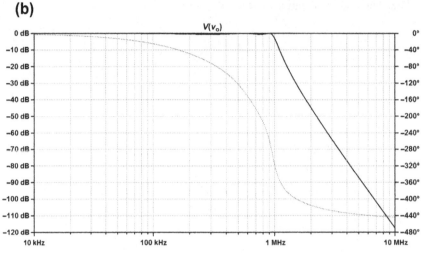

(c)

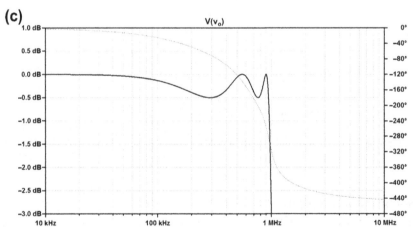

FIGURE 14.33

A 1-MHz, 0.5 dB Chebyshev low-pass filter implemented as a ladder for Example 14.4. (a) LTSPICE circuit. (b) Frequency response. (c) Passband detail. Note that the −6-dB attenuation due to the ladder resistive divider has been compensated for by adding a gain of 2. (For color version of this figure, the reader is referred to the online version of this book.)

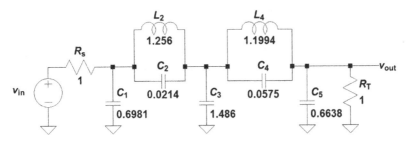

FIGURE 14.34

Elliptic low-pass filter prototype (1 rad/s cutoff with 0.01-dB passband ripple) for Example 14.5. (For color version of this figure, the reader is referred to the online version of this book.)

function. A 1-rad/s-elliptic LPF with a 0.01-dB passband ripple is shown in Figure 14.34. Note that the filter has parallel LC sections that generate zeros in the transfer function. Also note that the 1 rad/s number applies to the frequency where the ripple first exceeds the passband ripple specification, not the −3-dB point of the filter.

We can now unnormalize this filter as before to generate a filter with a frequency where the passband ripple is first exceeded by 5 MHz (Figure 14.35(a)). The frequency response (Figure 14.35(b)) shows a very fast roll-off in the transition band, and has a minimum attenuation floor. There are two frequencies where the response of the elliptic filter drops to zero, corresponding to the complex zero pairs on the $j\omega$-axis. In Figure 14.35(c), we see the ripple detail in the passband.

All-pass filters

An all-pass filter is a filter that has a magnitude response of unity, but which provides a phase shift. You can use all-pass filters to tailor group delay responses in your signal-processing chain. You may find that you will need to cascade your filter with an all-pass filter to meet the group delay specification. A first-order all-pass circuit is shown in Figure 14.36(a). Note that this all-pass provides a DC gain of −1. If you want, you can cascade an inverting op-amp stage with the all-pass to take care of this phase inversion.

The transfer function, angle, and group delay for a first-order all-pass filter are as follows:

$$H(s) = \frac{RCs - 1}{RCs + 1} = \frac{s - \frac{1}{RC}}{s + \frac{1}{RC}} = \frac{s - a}{s + a}$$

$$\angle H(s) = -2\tan^{-1}\frac{\omega}{a} \tag{14.13}$$

$$D(j\omega) = \frac{2a}{a^2 + \omega^2}$$

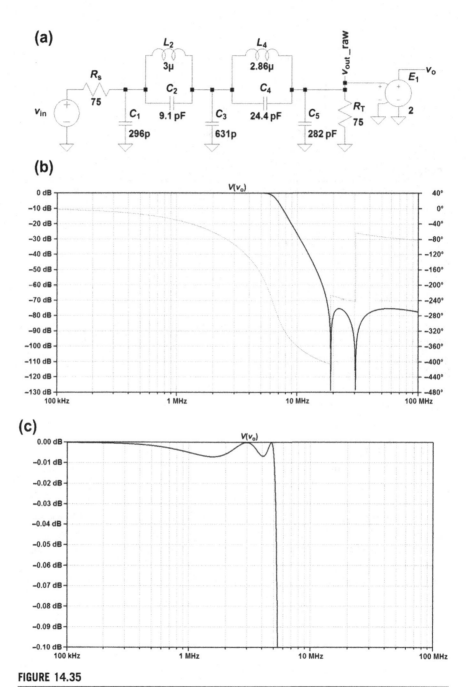

FIGURE 14.35

Unnormalized elliptic low-pass filter prototype (5-MHz passband for Example 14.5). (a) LTSPICE Circuit. (b) Magnitude response. (c) Passband details showing a passband width of 5 MHz for an allowable passband ripple of 0.01 dB. (For color version of this figure, the reader is referred to the online version of this book.)

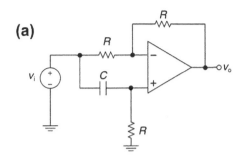

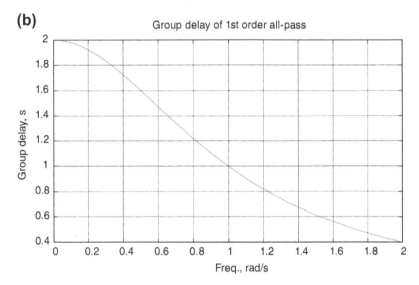

FIGURE 14.36

First-order all-pass filter. (a) Circuit. (b) Group delay.

The group delay characteristics for $RC = 1$ s is shown in Figure 14.36(b). Note that the DC group delay is twice the value of RC.

Example 14.6: Design case study: 1-MHz LPF

In this case study, we will design and simulate an analog LPF meeting the following specifications (Table 14.10):

This design could be implemented with either a ladder filter with passive elements (resistors, inductors, and capacitor) or a cascade of Sallen–Key active second-order sections. If you attempt to use a Bessel filter, the filter order will be quite high. If you look at elliptic filters, the group delay specification may be very difficult to meet.

Table 14.10 Design Case Study Specification for Example 14.6	
Filter type	Low-pass
Nominal −3-dB bandwidth	1 MHz
Passband gain	0-dB nominal, within 0.25 dB of nominal up to 750 kHz
Attenuation	>50 dB at 2.5 MHz
Group delay response	Group delay variation from the DC value <1000 ns up to 1 MHz

In this design, $\omega_s/\omega_c = 2.5$ and minimum attenuation at ω_s is −50 dB. A Bessel filter was not implemented due to the high filter order required. To meet the specification for a −50-dB gain at a normalized frequency of 2.5, a Bessel filter of order $N > 10$ would be needed. A sixth-order Butterworth filter would *almost* make the specification, since

$$20 \log_{10}\left(\frac{1}{\sqrt{1 + 2.5^{12}}}\right) = -47.8 \text{ dB} \tag{14.14}$$

So, if we use a Butterworth, we will need at least $N = 7$. The advantage of the Butterworth is that there is no magnitude ripple in the passband.

Another alternative will be to try a Chebyshev design with $N = 7$ or less and some passband ripple. From our previous work on the Chebyshev filter, it looks like an $N = 6$ Chebyshev with 0.1-dB ripple in the passband will meet the specification. An $N = 5$ Chebyshev with a 0.25-dB passband ripple barely misses the attenuation specification. So, let us go with the $N = 6$, 0.1-dB Chebyshev design (Figure 14.37).

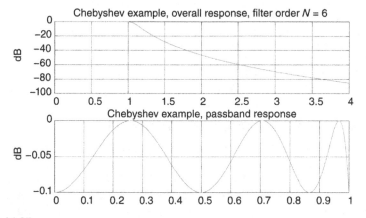

FIGURE 14.37

Chebyshev design, $N = 6$, 0.1-dB passband ripple for Example 14.6.

Using a ladder filter topology for an $N = 6$ Chebyshev with 0.1-dB ripple, normalized component values for $\omega_c = 1$ rad/s and $R_T = 1\ \Omega$ are

$R_s = 0.738$
$L_1 = 0.9419$
$C_2 = 2.0797$
$L_3 = 1.6581$
$C_4 = 2.2473$
$L_5 = 1.5344$
$C_6 = 1.2767$

Remember, we scale the normalized filter by

$$L = \frac{L_{norm}R}{\omega_c}$$

$$C = \frac{C_{norm}}{\omega_c R}$$

(14.15)

(a)

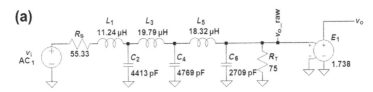

(b)

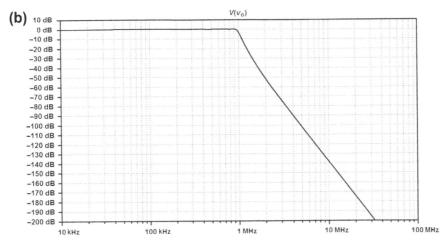

FIGURE 14.38

Chebyshev ladder filter design for Example 14.6. (a) Circuit, with gain of 1.738 added so that DC filter gain = 0 dB. (b) Magnitude response. (For color version of this figure, the reader is referred to the online version of this book.)

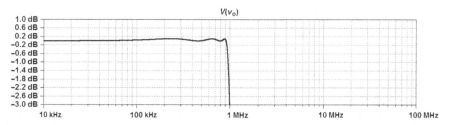

FIGURE 14.39

Chebyshev ladder filter design frequency response, passband details. Vertical scale in decibels (Example 14.6).

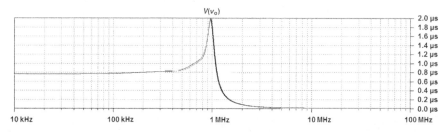

FIGURE 14.40

Chebyshev ladder filter design group delay for Example 14.6. There is approximately 1200 ns of group delay variation in the passband.

From this, we calculate the unnormalized component values for a filter with −3-dB point of 1 MHz (Figure 14.38(a)), and use a termination resistor value of 75 Ω. The overall magnitude of the frequency response of the Chebyshev ladder filter is shown in Figure 14.38. We have chosen a termination resistor value $R_T = 75$ Ω, and the source resistor is $0.738R_T = 55.33$ Ω. The stopband details show that the magnitude is attenuated >60 dB at 2.5 MHz, as expected.

The ripple in the passband is shown in Figure 14.39.

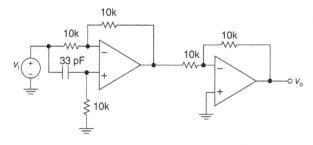

FIGURE 14.41

Group delay equalizer for the Chebyshev filter example for Example 14.6.

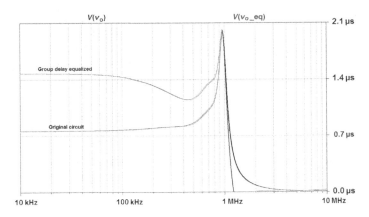

FIGURE 14.42

Chebyshev ladder filter design group delay equalized, compared to the original circuit for Example 14.6. (For color version of this figure, the reader is referred to the online version of this book.)

LTSPICE results show that the variation in the group delay in the DC to 1-MHz range is approximately 1200 ns, which violates the group delay specification (Figure 14.40).

From the above, we see that the delay variation in the DC to 1-MHz range is >1000 ns. So, let us cascade our Chebyshev design with a first-order all-pass network in an attempt to fill in the group delay hole below 1 MHz (Figure 14.41).

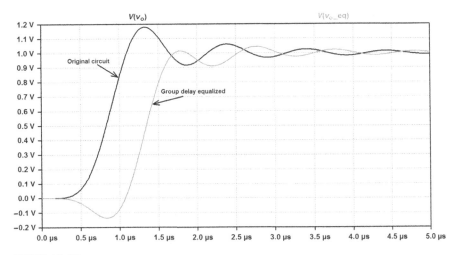

FIGURE 14.43

Chebyshev ladder filter design group delay equalized, step response vs. step response of the original Chebyshev filter (Example 14.6). (For color version of this figure, the reader is referred to the online version of this book.)

This type of circuit is sometimes called a group delay "equalizer". We will assume that we have ideal op-amps at our disposal. What this means is that we need to choose op-amps with a gain–bandwidth product that is much higher than our frequency range of interest, or >1 MHz. The low-frequency delay of a first-order all-pass is $2RC$, so in this case we have chosen a DC group delay of 660 ns.

The delay-equalized filter meets the group delay specification. The peak-to-peak delay variation in the passband is approximately 900 ns (Figure 14.42), and hence, we meet the group delay specification.

The step response of the original and delay-equalized circuits (Figure 14.43) shows that the equalized circuit has less overshoot in the step response, as expected. However, the delay through the filter is increased (due to the all-pass network).

Example 14.7: An alternate design using the Butterworth filter

Next is a seventh-order Butterworth design that meets the group delay specification in the previous example *without* any further all-pass filtering. In this case, the

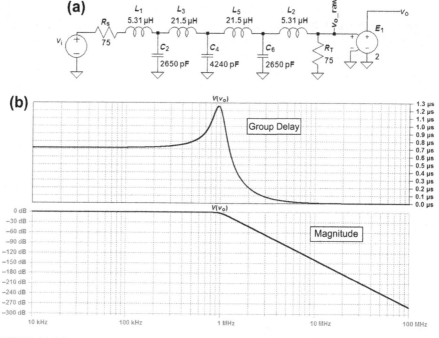

FIGURE 14.44

$N = 7$ Butterworth ladder filter response for Example 14.7. (a) Circuit. (b) Peak–peak group delay variation is <1000 ns ($\sim$535 ns). Note that a gain of +2 to compensate for the resistive divider is assumed following the filter. (For color version of this figure, the reader is referred to the online version of this book.)

Butterworth turns out to be the simpler design, even though it has a higher filter order, since a delay equalizer is not needed.

An $N=7$ Butterworth LPF has $\omega_c = 1$ rad/s component values of $R_s = 1$, $L_1 = 0.445$, $C_2 = 1.247$, $L_3 = 1.8019$, $C_4 = 2.0$, $L_5 = 1.8019$, $C_6 = 1.247$, $L_7 = 0.445$, and $R_T = 1$. The unnormalized values for a cutoff frequency of 1 MHz and source and termination resistors of 75 Ω are shown in the circuit of Figure 14.44(a). We note that we meet the magnitude specification and the group delay specification (Figure 14.44(b)) using this filter, without any additional group delay equalization.

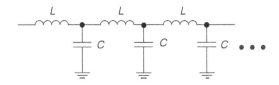

FIGURE 14.45

A finite length LC ladder implementing a time delay (Example 14.8).

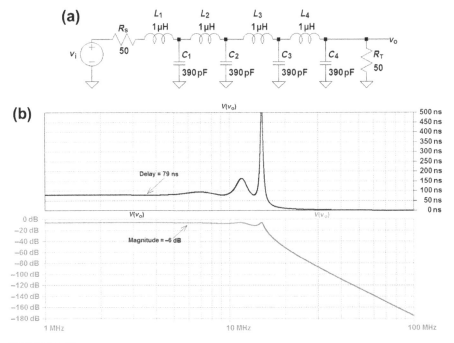

FIGURE 14.46

LC delay line designed for a 79-ns delay (Example 14.8). (a) Circuit with $Z_0 = 50.6\ \Omega$ and delay of each LC section $= 19.75$ ns. (b) Resultant magnitude and delay. (For color version of this figure, the reader is referred to the online version of this book.)

Example 14.8: Delay line with LC sections

A doubly terminated LC line can behave as a time delay. If we build a lumped equivalent of transmission line (Figure 14.45) and terminate the transmission line at both ends with its characteristic impedance the resultant circuit behaves like a finite-bandwidth delay line. Shown in Figure 14.46(a) is an LC delay line providing a 79-ns delay (in the frequency range less than about 10 MHz). Each LC section provides a time delay of $\sqrt{LC}$, and we need to be careful to have source and termination resistors equal to the characteristic impedance $Z_0 = \sqrt{L/C}$ of the LC line. If we add more LC sections, the resultant delay increases, and the delay line cutoff frequency increases as well.

You can also buy "tapped delay lines", which provide constant group delay over a limited bandwidth range.

Active LPF implementations

We can also directly implement the filter transfer functions using active filters with op-amps. One circuit for implementing a generic complex pole pair is the Sallen–Key circuit. The Sallen–Key (Figure 14.47), invented by Sallen and Key in 1955 at MIT's Lincoln Labs, generates the transfer function:

$$\frac{v_o}{v_i} = \frac{1}{R_1 R_2 C_1 C_2 s^2 + (R_1 + R_2)C_1 s + 1} \tag{14.16}$$

We can fit our standard second-order transfer function to this equation and recognize that the natural frequency and damping ratio are

$$\omega_n = \frac{1}{\sqrt{R_1 R_2 C_1 C_2}}$$

$$\zeta = \left(\frac{\omega_n}{2}\right)(R_1 + R_2)C_1 \tag{14.17}$$

To implement a fourth-order filter, you could cascade two Sallen–Key circuits, provided you know where all the filter poles are. You can implement a fifth-order filter by cascading two Sallen–Key circuits (to get the complex poles you need) with one more simple RC pole (on the real axis).

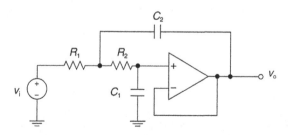

FIGURE 14.47

The Sallen–Key circuit (with DC gain = 1).

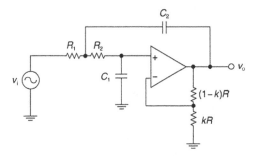

FIGURE 14.48

The Sallen–Key circuit (with adjustable Q). The adjustment can be made by implementing the $(1 - k)R$ and kR resistor with a potentiometer.

A variation on the theme is the Sallen–Key circuit with adjustable DC gain, which also adjusts the damping (and Q) of the filter, as shown in Figure 14.48. The transfer function of this filter is

$$\frac{v_{out}}{v_{in}} = \left(\frac{1}{k}\right) \frac{1}{R_1 R_2 C_1 C_2 s^2 + \left[R_2 C_2 + R_1 C_2 + R_1 C_1 \left(1 - \frac{1}{k}\right)\right] s + 1} \quad (14.18)$$

The Sallen–Key circuit has the advantage that it requires the least bandwidth performance of the op-amp as compared to that required by other filter types (multiple feedback, state variable filter, etc.) since the op-amp is used as a follower. It is somewhat difficult to tune. In the next few examples, we will consider several filters designed with Sallen–Key sections.

Example 14.9: 40-Hz Sallen–Key with adjustable Q

In Figure 14.49, we see the design of a 40-Hz LPF with adjustable Q. The Q and peaking are adjusted by tweaking the potentiometer.

Example 14.10: Active LPF

Let us use some filter software to design an active LPF with the following specifications:

- Passband DC: 1 kHz
- Stopband: -60 dB down at 2.5 kHz

Using TI's FilterPro[5] software, we find that a sixth-order Chebyshev filter with 1-dB passband ripple will meet the specifications. We will implement this with three op-amp Sallen–Key sections (Figure 14.50). In Figure 14.51, we see the filter

---

[5]FilterPro version 3.10 (2013), which is copyrighted and available from Texas Instruments, Inc.

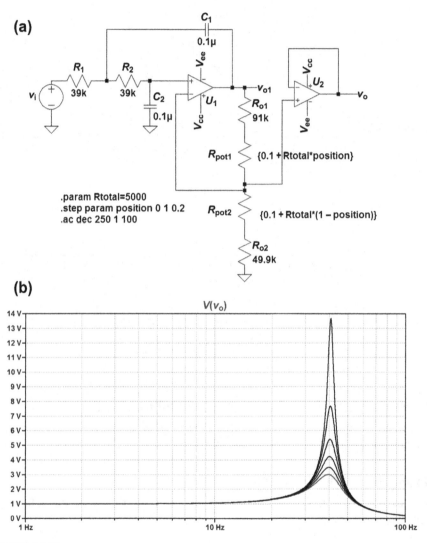

FIGURE 14.49

A 40-Hz Sallen–Key filter with adjustable Q (Example 14.9). (a) Circuit. (b) LTSPICE simulation showing Q adjustable from 2.9 to 14. (Note the LTSPICE implementation of a potentiometer.) (For color version of this figure, the reader is referred to the online version of this book.)

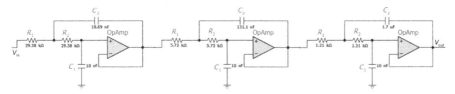

FIGURE 14.50

An $N=6$ Chebyshev filter, implemented with three Sallen−Key sections (Example 14.10). (For color version of this figure, the reader is referred to the online version of this book.)

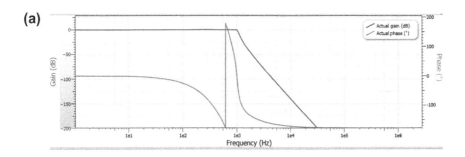

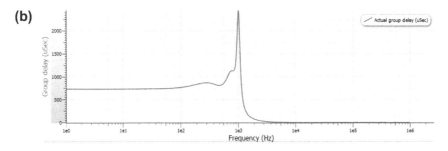

FIGURE 14.51

Chebyshev $N=6$ active low-pass filter response (Example 14.10). (a) Magnitude and phase. (b) Group delay. (For color version of this figure, the reader is referred to the online version of this book.)

response, assuming ideal op-amps. (You will need to select op-amps with enough gain−bandwidth product to do the job.)

Some comments on high-pass and band-pass filters

Of course, there are several filter types other than the LPF. In Figure 14.52, we see the high-pass filter, band-pass filter, and band-reject filter. The high-pass filter

(a)

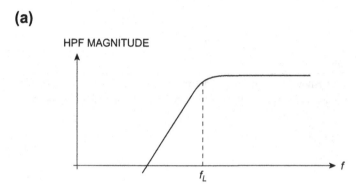

(b)

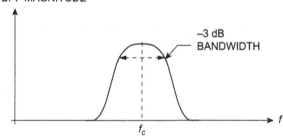

(c)

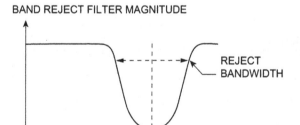

FIGURE 14.52

Other filter types. (a) High-pass filter. (b) Band-pass filter. (c) Band-reject filter.

attenuates low frequencies. The band-pass filter passes a finite-width band of frequencies, rejecting frequencies both higher and lower than the center frequency. Band-pass filters can be used to pick a narrow band signal out of a noisy signal. The band-reject filter is sometimes used to reject noise sources, such as 60-Hz noise.

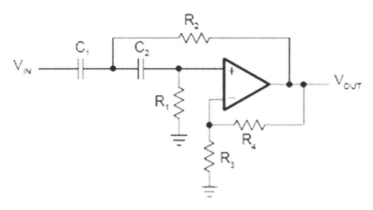

FIGURE 14.53

Sallen–Key high-pass filter section.

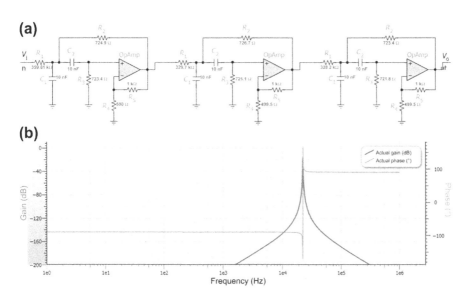

FIGURE 14.54

A 22-kHz band-pass filter (Example 14.11). (a) The implementation is a Bessel filter, built with Sallen–Key stages. (b) Magnitude and phase responses. (For color version of this figure, the reader is referred to the online version of this book.)

A high-pass Sallen–Key section is shown in Figure 14.53. You can also design band-pass and notch filters using the Sallen–Key, but they may be difficult to tune effectively.

Example 14.11: A 22-kHz band-pass filter

Shown in Figure 14.54(a) is a Sallen–Key implementation of an $N=6$ band-pass filter with a center frequency of 22 kHz and a passband bandwidth of 100 Hz. The filter response is Bessel.

Example 14.12: A 60-Hz band-stop filter

Shown in Figure 14.55(a) is a multiple-feedback implementation of an $N=4$ band-stop filter with a center frequency of 60 Hz.

Example 14.13: An $N=4$ active all-pass filter with a 175-µs time delay

Shown in Figure 14.56(a) is a multiple-feedback implementation of a fourth-order Bessel all-pass filter. The circuit has a gain of unity, and the group delay

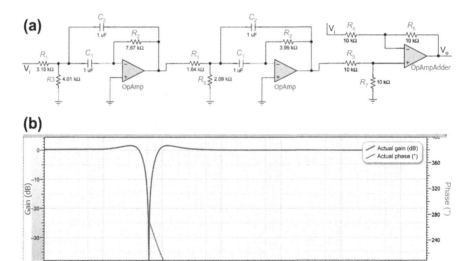

FIGURE 14.55

A 22-kHz band-pass filter (Example 14.12). (a) The implementation is a Bessel filter, built with Sallen–Key stages. (b) Magnitude and phase responses. (For color version of this figure, the reader is referred to the online version of this book.)

(a)

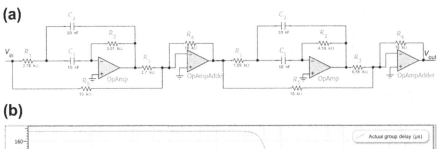

(b)

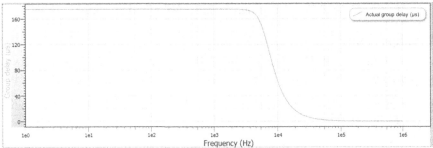

FIGURE 14.56

A fourth-order Bessel all-pass filter. (a) Implementation with multiple-feedback stages. (b) Group delay response. (For color version of this figure, the reader is referred to the online version of this book.)

response (Figure 14.56(b)) shows a 175-μs time delay for frequencies from DC to >1 kHz.

Chapter 14 problems

Problem 14.1

Design a seventh-order Butterworth LPF with a cutoff frequency of 4.5 MHz. Implement this using an LC ladder with a 50-Ω source impedance. Assume that you follow the filter with a gain of +2 to compensate for the −6-dB loss due to the source and termination resistances.

Problem 14.2

Calculate the attenuation (in decibels) of the 4.5-MHz filter in Problem 14.2 at 13.5 MHz.

Problem 14.3

Design a fifth-order Butterworth filter with a cutoff frequency of 50 kHz, using Sallen−Key sections.

Problem 14.4

Compare the step responses of a seventh-order 4.5-MHz Butterworth filter and a seventh-order Chebyshev filter with a cutoff frequency 4.5 MHz. Comment on the rise time and overshoot of each filter.

Problem 14.5

Design a sixth-order Chebyshev filter with a 0.5-dB ripple and cutoff frequency of 3.58 MHz. Find the peak–peak group delay variation in this filter.

Problem 14.6

An $N = 5$ Butterworth filter design with equal source and termination resistances is shown in Figure 14.57. The a -3-dB frequency of the filter shown is 1 rad/s (0.16 Hz).

a. Calculate the frequency (in radians per second) at which the output attenuation drops to -40 dB for the filter as built above.

b. Now, assume a source and termination resistance of 75 Ω. You now need a filter with a -3-dB cutoff at 10 MHz (not radians per second!). Find the unnormalized Butterworth values and sketch the circuit.

c. Sketch the group delay vs. frequency of the 10-MHz filter. In your sketch, include at least the following data points: DC, 5, 10, and 15 MHz. Indicate the frequency at which the group delay peaks.

Problem 14.7

An LPF prototype with -3-dB cutoff frequency has a group delay at DC of 1 s. The prototype filter is then transformed, and a filter with the same type and order is created, but now with a -3-dB cutoff frequency of 4.5 MHz. What is the group delay at DC of the new filter?

Problem 14.8

a. What is the minimum Butterworth filter order N to meet the following specification: Cutoff frequency: $f_c = 100$ kHz and -60-dB attenuation at $f = 300$ kHz

b. What is the low-frequency group delay of this filter?

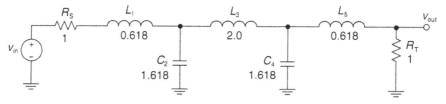

FIGURE 14.57

$N = 5$, 1 rad/s Butterworth low-pass filter.

Further reading

[1] Balch B. A simple technique boosts performance of active filters. *EDN* 1988:277−86.

[2] Blinchikoff H, Zverev A. *Filtering in the time and frequency domains*. John Wiley; 1976.

[3] Burton LT, Treleaven D. Active filter design using generalized impedance converters. *EDN* February 5, 1973:68−75.

[4] Chambers W. Know your options and requirements when designing filters. *EDN* 1991:129−38.

[5] Corral C. Designing elliptic filters with maximum selectivity. *EDN* 2000:101−9.

[6] Corrington MS. Transient response of filters. *RCA Rev* 1949;**10**(3):397−429.

[7] Downs R. Vintage filter scheme yields low distortion in new audio designs. *EDN* 1991:267−72.

[8] Sallen RP, Key EL. A practical method of designing RC active filters. *IRE Trans Circuit Theory* 1955:74−85.

[9] Steer Jr R. Antialiasing filters reduce errors in A/D converters. *EDN* 1989:171−86.

[10] Tow J. A step-by-step active-filter design. *IEEE Spectrum* 1969:64−8.

[11] Williams A, Taylor F. *Electronic filter design handbook*. McGraw-Hill; 1988.

[12] Yager C, Laber C. Create a high-frequency complex filter. *Electron Des* 1989:123−7.

[13] Zverev A. *Handbook of filter synthesis*. John Wiley; 1967.

Passive Components, Prototyping Issues, and a Case Study in PC Board Layout

> *"Real resistors and real capacitors have several reactive or lossy components. Is this part of our circuit design work? Yeah. So we have to engineer to get the system working despite component imperfections. Sometimes wise component choice is a very important part of our job."*
>
> — **Bob Pease, "Analog Circuits (World Class Designs)"**

IN THIS CHAPTER

▶ Some of the subtleties of passive components including construction techniques of these devices and parasitic effects are considered. We will cover some details about resistors, capacitors, and inductors. Then, we will use what we have learned in an illustrative discussion of printed circuit (PC) board layout issues.

Resistors

At first blush, a resistor is a resistor is a resistor. However, we will now delve into some of the subtleties of these devices. The impedance of an ideal resistor does not depend on operating frequency[1] and is:

$$Z_{\text{resistor,ideal}} = R \tag{15.1}$$

A real-world resistor (Figure 15.1) includes a parasitic inductance due to the geometry (size and shape) of the leads and a parasitic capacitance across the resistor. The impedance of the resistor, including these parasitic elements, is:[2]

$$Z_{\text{resistor,real}}(s) = \frac{Ls + R}{LCs^2 + RCs + 1} = \frac{R\left(1 + \frac{L}{R}s\right)}{LCs^2 + RCs + 1} \tag{15.2}$$

---

[1]Another way to look at this is that the current in an ideal resistor and a voltage across this ideal resistor are in phase.

[2]One way to sanity check this result is to consider the limit as $L \to 0$ and $C \to 0$; we want the impedance to be exactly R and this is indeed the case.

Intuitive Analog Circuit Design. http://dx.doi.org/10.1016/B978-0-12-405866-8.00015-2

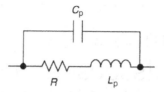

FIGURE 15.1

Resistor R showing parasitic elements including series inductance L_p and parallel capacitance C_p.

We can put this in "$j\omega$" form by making the substitution $s = j\omega$, resulting in:

$$Z_{resistor,real}(j\omega) = \frac{j\omega L + R}{(1 - \omega^2 LC) + j\omega RC} \tag{15.3}$$

For a large-valued resistor,[3] the RC time constant dominates. This is because a large resistor will swamp out the value of the parasitic inductance. For a low-valued resistor, the L/R time constant dominates because the resistor effectively shorts out the parasitic capacitance. The magnitude of the impedance of the real-world resistor is:

$$|Z_{resistor,real}| = \sqrt{\frac{(\omega L)^2 + R^2}{(1 - \omega^2 LC)^2 + (\omega RC)^2}} \tag{15.4}$$

In Figure 15.2, we see the impedance of a resistor with $R = 1$ MΩ, $C_p = 0.2$ pF, and $L_p = 10$ nH. Note that for this relatively large resistor ($R >> Z_o = 223$ Ω), parasitic capacitive effects dominate; the impedance rolls off at frequencies above approximately 1 MHz.

In Figure 15.3, we see the impedance of a low-valued resistor with $R = 10$ Ω, $C = 0.2$ pF, and $L = 10$ nH. Note that parasitic inductive effects dominate; the impedance increases at frequencies above approximately 10 Mrad/s.

It is difficult to quantify exactly the values of parasitic elements, but in standard through-hole resistors, you might expect fractions of a picofarad of parasitic capacitance and a few nanohenries of parasitic inductance.[4] You need to consider such parasitics as series inductance and parallel resistance in high-frequency circuits.

---

[3] We can see that "large valued" in this case is large enough so that $RC >> L/R$, or equivalently $R >> \sqrt{\frac{L}{C}}$. The term $\sqrt{\frac{L}{C}}$ comes up over and over again in RLC circuits and transmission lines and is called the *characteristic impedance* Z_o of this circuit.

[4] A very rough rule of thumb for the inductance of component leads above a ground plane is 10 nH/cm of lead length. So, it behooves you to keep lead lengths short if you want to minimize parasitic inductance. Of course, you can test your resistor using an impedance analyzer (such as the Hewlett Packard HP4192) and extract the parameters for your device.

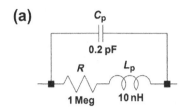

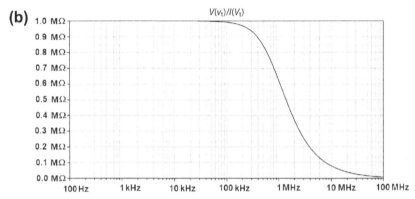

FIGURE 15.2

Large resistor with parasitics. (a) Model with $R = 1\ \text{M}\Omega$, $C_p = 0.2\ \text{pF}$, and $L_p = 10\ \text{nH}$.
(b) Magnitude of the impedance showing capacitive behavior above about 1 MHz.
(For color version of this figure, the reader is referred to the online version of this book.)

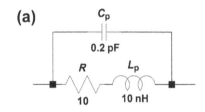

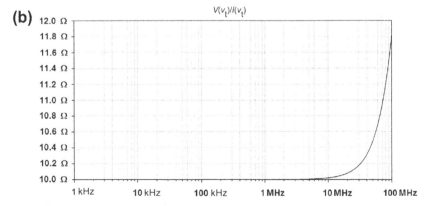

FIGURE 15.3

Small resistor with parasitics. (a) Model with $R = 10\ \Omega$, $C = 0.2\ \text{pF}$, and $L = 10\ \text{nH}$.
(b) Magnitude of the impedance showing inductive behavior above about 10 MHz.
(For color version of this figure, the reader is referred to the online version of this book.)

Table 15.1 Comparison of Surface-Mount Resistor Ratings

Resistor	Size (inches)	Typical Wattage Rating (mW)[5]	Typical Working Voltage Rating (V)[6]
0201	0.02 × 0.01	50	15
0402	0.04 × 0.02	50–62.5	50
0603	0.06 × 0.03	62.5–100	75
0805	0.08 × 0.05	100–250	100
1206	0.12 × 0.06	125–250	200
1210	0.12 × 0.10	250–333	200
1812	0.18 × 0.12	500	200
2010	0.20 × 0.10	500	200
2512	0.25 × 0.12	1000	250

Comments on surface-mount resistors

You can miniaturize your circuit and reduce somewhat the effects of parasitic inductance by using surface-mount resistors. Surface-mount resistors come in a variety of sizes, ranging from "0201" size, through "0402" size, and all the way up to "2512" size. The resistor size numbers indicate the length and width of the resistor. For instance, a 0805 resistor is 0.08″ in length and 0.05″ in width.

There is a tradeoff to be made with regard to resistor size, wattage, and working voltage rating, as shown in Table 15.1. For instance, a longer resistor will have a higher working voltage rating, due to a higher voltage breakdown.

Comments on resistor types

As the designer, you also have decisions to make regarding the type of resistor you put in your circuit. For instance, do you choose carbon composition, carbon film, metal film, wire-wound, or some other type of resistor?

Carbon composition resistors, sometimes called "carbon comp", are old-style resistors that have been used for years in electronics. The main advantage of carbon composition resistors is their ability to withstand high-current transient surges. They do, however, have the disadvantage of a relatively high temperature coefficient of resistivity. Remember that the resistance of a resistor varies with temperature and that the resistance can be expressed as:

$$R(T) = R_0(1 + \alpha(T - T_0)) \tag{15.5}$$

---

[5] There is some variation in wattage rating from manufacturer to manufacturer; so be sure to check the particular device datasheet.
[6] Again, use these numbers for comparison purposes only and check the specific manufacturer.

- $R(T)$ is the resistance at your operating temperature
- R_o is the reference resistance at temperature T_o
- α is the temperature coefficient of resistivity

Carbon composition resistors also have a tendency to drift in value with time, especially if they are overstressed with high power. Carbon composition resistors have largely been replaced in modern electronics by metal film and carbon film resistors. Film resistors have the advantage of a lower temperature coefficient of resistivity. Film resistors are, however, more susceptible to damage by electrical overloads.

Wire-wound resistors, as the name implies, have a length of wire with a known resistance wound on a cylindrical core. Wire-wound resistors are largely used where high wattage capability is needed. They do suffer from a relatively large series inductance due to the way they are manufactured with wound wires.

Foil resistors are precision resistors with low noise and temperature coefficients.

A comparison of resistor types is given in Table 15.2. (*Note: Resistor noise is covered extensively in Chapter 16.*)

Photographs of different types of resistors are shown in Figure 15.4.

Example 15.1: Resistance vs. temperature

A precision wire-wound resistor has a resistance of 10 Ω at 25 °C and a temperature coefficient of +10 ppm/°C. What is the resistance of this resistor at 75 °C?

Table 15.2 Comparison of Resistor Types

Resistor	Typical Power Rating	Temperature Coefficient	Comments
Carbon composition	0.25 W ~ 2 W	>1000 ppm/°C	Old-style resistors. Typically replaced by carbon film or metal film in new designs. Poor long-term stability, noise, and temperature coefficient
Carbon film	0.125 W ~ 2 W	Typically −50 to −1000 ppm/°C	Better noise performance than carbon composition
Metal film	0.125 W ~ 2 W	Typically +50 to +300 ppm/°C	Low noise; low power rating
Wire-wound resistor	Typically >5 W, up to hundreds of watts	Typically +100 ppm/°C	Typically used for high-wattage resistors. Be careful of high parasitic inductance. Lowest noise, high cost
Foil resistors	Typically low power, precision applications	Lowest temperature coefficient, as low as <1 ppm/°C	Typically used for precision circuits; most expensive; lowest noise

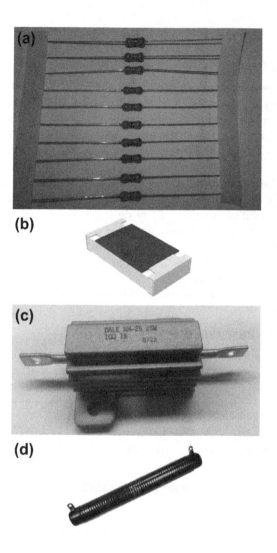

FIGURE 15.4

Photos of some resistors. (a) General-purpose, ¼-W metal film. (b) Surface-mount, 0805 size. (c) 25-W wire-wound power resistor. (d) 225-W wire-wound power resistor. (For color version of this figure, the reader is referred to the online version of this book.)

Solution: The approximate resistance is given by:

$$R(T = 75\,°\text{C}) = (10\,\Omega)(1 + 10^{-5}(75 - 25)) = 10.005\,\Omega \qquad (15.6)$$

You need to check that the temperature coefficient (10^{-5}/°C) is constant over this entire temperature range for this calculation to be valid.

Comments on the parasitic inductance of wire loops

Parasitic inductance exists in a lot of places, which can cause trouble in analog circuits. For instance the wiring inductance of component leads such as transistors and capacitors can result in undesired resonances and bandwidth limitations. The inductance of power supply leads can result in unwanted voltage ripple and spiking. The list goes on—so next we will investigate how to calculate, or at least estimate, inductance.

The calculation of inductance of wire loops is surprisingly difficult. For instance, in the case of circular loops, elliptic integrals must be evaluated as shown by

Table 15.3 Inductance of Various Wire Shapes

Wire Shape	Inductance Calculation	References
Loop of round wire with $a >> R$	$L \approx \mu_o a \left[\ln\left(\frac{8a}{R}\right) - 2 \right]$	Maxwell, A Treatise on Electricity and Magnetism C. Paul, Inductance Loop and Partial
Parallel wire line	$L \approx \frac{\mu_o l}{\pi} \ln\left[\frac{d}{R} + \frac{1}{4} - \frac{d}{l} \right]$	Grover, Inductance Calculations, pp. 39
Wire loop with side length D and copper width $2w$, $D >> w$, negligible thickness	$L \approx \frac{2\mu_o D}{\pi} \left[\sinh^{-1}\left(\frac{D}{w}\right) - 1 \right]$	M. Zahn, Electromagnetic Field Theory: A Problem Solving Approach, pp. 343
Generic wire loop with cross-sectional area A	Rough approximation: $L \approx \mu_o \sqrt{\pi A}$	T. Lee, The Design of CMOS Radio-Frequency Integrated Circuits, 2nd edition

Maxwell. Fortunately, there are some excellent references describing the calculation (either in closed form, or estimations) of the inductance of wire loops. The "Bible" of inductance calculations is Frederick Grover's *Inductance Calculations* from 1946. A sampling of some of the calculations of inductance found in various references is shown in Table 15.3.

Example 15.2: Inductance of a wire loop

What is the approximate inductance of a square loop of PC board trace, with the length of each square side being 1 cm and the trace width 1 mm?

Solution: The inductance is roughly found using the Lee approximation:

$$L \approx \mu_o \sqrt{\pi A} \approx \left(4\pi \times 10^{-7} \text{ H/m}\right) \sqrt{\pi (0.01 \text{ m})^2} = 22 \times 10^{-9} \text{ H} = 22 \text{ nH} \quad (15.7)$$

If we use the Zahn equation, we get:

$$L \approx \frac{2\mu_o D}{\pi} \left[\sinh^{-1}\left(\frac{D}{w}\right) - 1\right] = 21.5 \text{ nH} \quad (15.8)$$

The Lee approximation is quite often a good way to quickly ballpark-estimate the inductance of a wire loop.

Capacitors

Just as resistors suffer from parasitic components, so do capacitors. A model of a real-world capacitor constructed as a parallel plate filled with a dielectric is shown[7] in Figure 15.5(a). The resistance R_s is the series resistance of the leads. The parallel plate is filled with a dielectric that has a finite electrical conductivity.[8] This results in a dielectric resistance R_d that is in parallel with the desired capacitance C, as shown in Figure 15.5(b).

The input impedance to the real-world capacitor is:

$$Z_i(s) = R_s + \frac{R_d}{R_d Cs + 1} \quad (15.9)$$

---

[7]In this example, we have ignored the series inductance for simplicity. We are also ignoring the effects of dielectric relaxation, also called "soakage" when referring to capacitors. This effect is described in detail in Bob Pease's article "Understand Capacitor Soakage to Optimize Analog Systems". Dielectric absorption is an issue in high-precision analog circuits such as sample-and-holds. Polystyrene, polypropylene, and teflon capacitors have low dielectric absorption.
[8]For further information on the lossy capacitor see, e.g., Markus Zahn, *Electromagnetic Field Theory: A Problem Solving Approach*, Krieger reprint 1987, pp. 184–194. Note that in this initial model we do not include the effects of series inductance. We will consider this in more detail later.

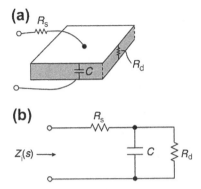

FIGURE 15.5

Capacitor showing parasitic elements. (a) Parallel-plate capacitor filled with dielectric with finite electrical conductivity. This capacitor has series resistance R_s and dielectric resistance R_d that is in parallel with the lumped capacitance C. (b) Electrical model.

We can expand this result to find the real and imaginary parts of the input impedance as follows:

$$
\begin{aligned}
Z_i(j\omega) &= R_s + \frac{R_d}{j\omega R_d C + 1} \\
&= R_s + \frac{R_d(1 - j\omega R_d C)}{1 + \omega^2 R_d^2 C^2} \\
&= \left[R_s + \frac{R_d}{1 + \omega^2 R_d^2 C^2} \right] - j\left[\frac{\omega R_d^2 C}{1 + \omega^2 R_d^2 C^2} \right]
\end{aligned}
\tag{15.10}
$$

The first term (the real part) is sometimes called the *equivalent series resistance* (ESR) of the capacitor, or:

$$
R_{ESR} = \left[R_s + \frac{R_d}{1 + \omega^2 R_d^2 C^2} \right]
\tag{15.11}
$$

Note that the ESR decreases as frequency increases. The equivalent capacitance is:

$$
C_{eq} = C\left[1 + \frac{1}{\omega^2 R_d^2 C^2} \right]
\tag{15.12}
$$

A simplified model of the capacitor showing the ESR is shown in Figure 15.6(a). To this model we have also added an equivalent series inductance (L_{ESL}). The value of the series inductance depends on the geometry of the internal construction of the capacitor as well as the lead length of the device as it is

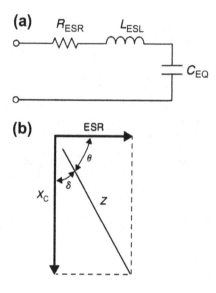

FIGURE 15.6

(a) Capacitor with equivalent series resistance (R_{ESR}) and equivalent series inductance (L_{ESL}). (b) Plot illustrating the DF. The DF is the tangent of the angle between the impedance of the capacitor X_C and the overall impedance Z.

connected in the circuit. A ballpark rule of thumb for ESL of a capacitance is 10 nH/cm of lead length. The ESL of some electrolytic capacitors is somewhat higher due to the foil winding inside the can.

The dissipation factor (DF) is another figure of merit often found on capacitor datasheets. The DF is given by:

$$DF = \omega C R_{ESR} \tag{15.13}$$

We note that the DF is the inverse of the Q of the capacitor. If we plot ESR, capacitive reactance (X_C), and total capacitor impedance (Z) as in Figure 15.6(b), we see that there is a phase angle between the capacitive reactance and the impedance of the capacitor. The DF is the tangent of this angle, or:

$$DF = \tan(\delta) = \frac{R_{ESR}}{X_C} \tag{15.14}$$

The impedance of an ideal capacitor is:

$$Z_{cap,ideal} = \frac{1}{j\omega C} \tag{15.15}$$

For a real-world capacitor (ignoring dielectric loss), the impedance is:

$$Z_{cap,real} = \frac{1}{j\omega C} + R + j\omega L = \frac{(1 - \omega^2 LC) + j\omega RC}{j\omega C} \tag{15.16}$$

The magnitude of the impedance is:

$$\left|Z_{cap,real}\right| = \frac{\sqrt{(1 - \omega^2 LC)^2 + (\omega RC)^2}}{\omega C} \tag{15.17}$$

The impedance plot of an electrolytic capacitor, comparing the ideal with the actual impedance, is shown in Figure 15.7 for $C = 100\ \mu F$, $L = 25\ nH$, and $R = 0.01\ \Omega$. We note that at frequencies above 100 kHz, the impedance of the capacitor is inductive.

Photos of various capacitor types are shown in Figure 15.8.

A comparison of different capacitor types is shown in Table 15.4.

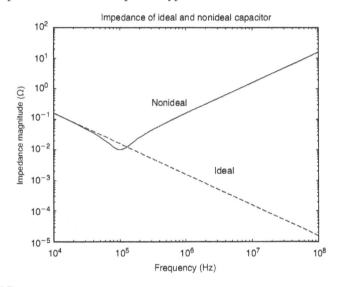

FIGURE 15.7

Impedance plot of electrolytic capacitor with $C = 100\ \mu F$, $L = 25\ nH$, and $R = 0.01\ \Omega$. Dotted line is the impedance of an ideal 100-μF capacitor.

Inductors

The impedance of an ideal inductor is:

$$Z_{inductor,ideal} = j\omega L \tag{15.18}$$

For a real-world inductor, the impedance is modified by the resistance of the copper wire[9] and the interwinding capacitance (Figure 15.9(a)). This impedance is:

$$Z_{cap,real} = \frac{j\omega L + R}{(1 - \omega^2 LC) + j\omega RC} \tag{15.19}$$

---

[9]This analysis ignores core losses (if any) and other high-frequency effects such as current crowding due to skin effect.

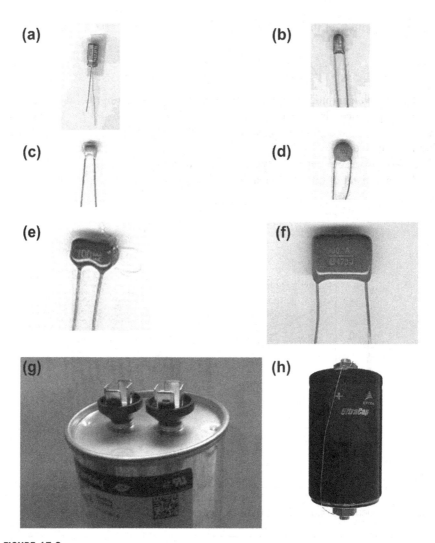

FIGURE 15.8

Photos of different types of capacitors: (a) 10-μF 25-V electrolytic. (b) 4.7-μF 16-V tantalum. (c) 0.1-μF 50-V ceramic. (d) Old-style ceramic disk capacitor. (e) 10-pF mica capacitor. (f) 0.047-μF 100-V film capacitor. (g) 120-μF 240-V AC motor starting film capacitor. (h) 5000-F 2.5-V "supercapacitor". (For color version of this figure, the reader is referred to the online version of this book.)

Table 15.4 Comparison of Capacitor Types

Capacitor	Typical Application	Advantages	Disadvantages
Mica	High-frequency, low-valued capacitors for filters and signal processing	Low loss; low inductance; relatively temperature stable	Low values of capacitance per unit volume
Ceramic	General-purpose filtering; medium-frequency signal processing	Low cost; wide range of values	High temperature coefficient of capacitance
Polystyrene, polypropylene film	Typically used in high-current applications	High-power handling capability; low dielectric loss	Large physical size
Tantalum	Power supply decoupling	Low ESR	High leakage current; polarized; poor stability and accuracy; limited to about 100 V rating; typically more expensive than electrolytics
Aluminum electrolytic	Power supply decoupling, motor starting	Large values, high-current handling capability, high voltage	High inductance; high leakage current; polarized; poor stability and accuracy

The magnitude of this impedance is:

$$\left| Z_{\text{inductor,real}} \right| = \sqrt{\frac{(\omega L)^2 + R^2}{(1 - \omega^2 LC)^2 + (\omega RC)^2}} \qquad (15.20)$$

The impedance plot for an inductor, comparing the ideal with the actual impedance, is shown in Figure 15.9(b) for $L = 100$ μH, $C = 25$ pF, and $R = 0.1$ Ω. Note that the self-resonant frequency of the inductor (at approximately 3.2 MHz) is clearly shown. Above the self-resonant frequency, the impedance of this inductor is capacitive.

A photograph of some off-the-shelf inductors is shown in Figure 15.10(a).

Discussion of some PC board layout issues

PC board layout and routing is a task that is sometimes left to the last minute in a design cycle. Doing a good PC board layout requires attention to many details, including:

- Knowledge of where you want the high frequency, high current, or sensitive circuitry to be.

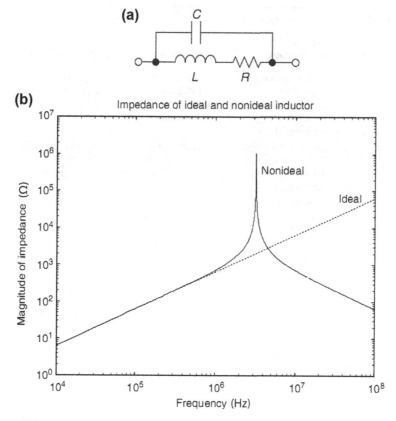

FIGURE 15.9

(a) Inductor showing parasitic elements of series resistance R and interwinding capacitance C. (b) Impedance plot of real-world inductor with $L = 100$ μH, $C = 25$ pF, and $R = 0.1$ Ω. The dotted line is the impedance of an ideal 100-μH inductor.

- Some information on component limitations.
- Information on noise sources.
- Real-world constraints, such as PC board form factor and location of connectors and mounting holes.
- Other constraints such as PC board design rules mandating minimum trace widths, trace-to-trace spacing, and the like.

Following is a nonexhaustive discussion of some of these design issues.

Power supply bypassing

The need for power supply bypassing from integrated circuits (IC) arises from the fact that there is no such thing as a perfect, zero-impedance ground. Consider the

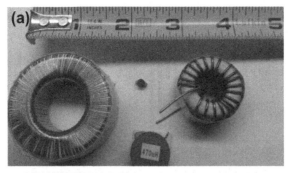

FIGURE 15.10

Inductors. (a) Some off-the-shelf inductors. (b) Custom inductors designed by the author for his Ph.D. thesis.

model of IC_1 in Figure 15.11, the details of which are unimportant for purposes of this discussion. The IC draws DC power from the supply (V_{supply}) through wires or ground and power planes. The series inductance and resistance of the interconnection to the supply is shown.

IC_1 draws a fast switching current with a high di/dt (modeled as current source $i(t)$). The hope in bypassing the IC is that proper selection and placement of the bypass capacitor C_B will force transient currents to circulate locally near the IC and hence voltage transients on the supply lines will be limited. Of course, we want the DC component of the IC_1 current to travel back to the power supply. However, if we send fast current pulses back to the power supply, we will induce voltages on the power supply lines to IC_2 and IC_3 due to distributed resistance and inductance.

The key to selection and placement of bypass capacitor C_B is to choose a capacitor that is sufficiently sized to do the job and to place it in close proximity to the power and

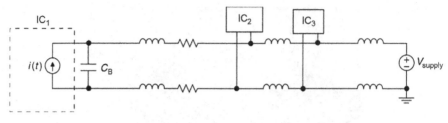

FIGURE 15.11

Model showing integrated circuit drawing current and its bypass path. C_B is the bypass capacitor for IC_1.

ground pins of IC_1. This will minimize the inductance of the bypass path and ensure that the high-frequency switching currents circulate locally near the IC.

Another reason to be careful about proper bypassing is that some ICs are susceptible to significant errors if the power supply voltage is glitching. For instance, remember that op-amps have finite power supply rejection ratio (PSRR); some of the stray signal on the power supply lines makes its way to the output of the op-amp, and PSRR gets worse with increasing frequency. The typical PSRR for an op-amp is greater than 80 dB at DC but drops quickly as you go to higher and higher frequency.

One way to reduce the impedance of current return paths is to use a ground plane, discussed in the following section.

Ground planes

A ground is a return path for current. It is desirable that this return path should have as low an impedance as possible, to reduce transient induced voltage drops and electromagnetic emissions. In the world of two-layer PC boards, it is difficult to have a dedicated ground plane since you generally want a couple of PC board layers available for routing signals. In multilayer boards, it is easy to dedicate an unbroken ground plane on an internal PC board layer as ground.

The use of a ground plane helps to reduce the inductance of signal-carrying traces on the PC board. One technique is to have high current and high di/dt traces directly above an unbroken ground plane. You can also make the traces wide if you want to reduce the inductance.

PC board trace widths

PC board traces must be sized appropriately (both in width and thickness, or copper weight[10]) to carry the current that you need without excessive temperature

---

[10]"Copper weight" tells you how thick the PC board trace is. Typical low-power analog boards use ½-ounce or 1-ounce copper. High-power boards may use 2-ounce copper or higher.
[11]From Douglas Brooks (1988).

Table 15.5 PC Board Copper Weight vs. Thickness	
Copper Weight (oz.)	**Copper Thickness**
0.5	0.0007″ (0.7 mils)
1.0	0.0014″ (1.4 mils)
2.0	0.0028″ (2.8 mils)

rise. A rule of thumb is that a 10-mil-wide, 1-ounce PC board trace can carry in excess of 500 mA with a 20 °C temperature rise above ambient. PC board copper weight vs. trace thickness is shown in Table 15.5. An estimate of the current-carrying capability for 20 °C temperature rise of PC board traces is shown in Figure 15.12. The fusing current (Figure 15.13) for PC board traces is significantly higher.

Approximate inductance of a PC board trace above a ground plane

The inductance of a PC board trace above a ground plane can be very roughly calculated by assuming a microstrip configuration. For a microstrip line of length l, width

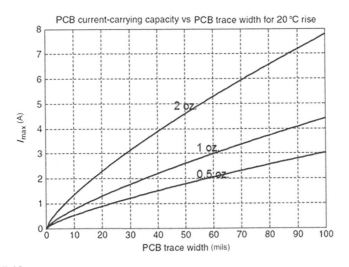

FIGURE 15.12

Approximate current-carrying capability of 0.5-oz., 1.0-oz., and 2.0-oz. PCB traces with 20 °C temperature rise.[11] PCB, PC board.

[12]From Douglas Brooks (1988).

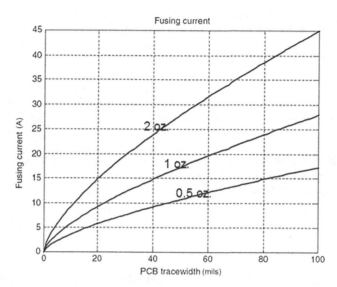

FIGURE 15.13

Approximate fusing current of 0.5-oz., 1.0-oz., and 2.0-oz. PCB[12] traces.

w, and strip-to-strip spacing $d << w$ (Figure 15.14(a)), the inductance is (very roughly[13]):

$$L \approx \mu_o \frac{lh}{w} \tag{15.21}$$

Using this approximation for a line with $w = 0.01''$ (0.0254 cm) and $h = 0.005''$ (0.0127 cm), we estimate an inductance of 6.3 nH/cm of length. Note that this approximation becomes less and less accurate as the trace height h increases above the ground plane. The inductance is also very frequency dependent as at higher frequency the ground plane crowds under the PC board trace and the inductance is lower than at DC. So, use this for ballpark estimates only.

A two-dimensional (2D) finite element analysis[14] (Figure 15.14(b)) estimates this inductance to be somewhat lower at approximately 3.9 nH/cm of length.

Example 15.3: PCB trace inductance at DC and 1 MHz

In Table 15.6, we see inductance calculated for a hypothetical PC board trace at height $h = 15$ mils above a ground plane. The DC inductance is quite a bit higher than the 1 MHz inductance, at all track widths w.

---

[13]For more detailed calculations for inductances of all kinds of geometries, see Frederick Grover's (1946) excellent reference *Inductance Calculations*. The author gratefully thanks Prof. Dave Perreault from MIT for recommending this book when we were both graduate students.

[14]Plots and analysis done with Finite Element Method Magnetics, a finite element package created by Dr David Meeker at Foster-Miller.

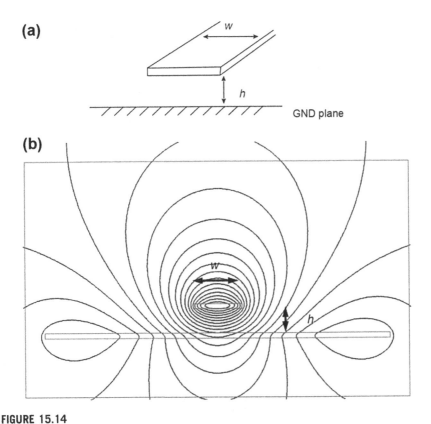

FIGURE 15.14

2D finite-element analysis (FEA) model of $w = 0.01''$ PC board trace $0.005''$ above a ground (GND) plane. (a) Geometry. (b) 2D FEA model. (For color version of this figure, the reader is referred to the online version of this book.)

Some personal thoughts on prototyping tools

Before you go to a full PC board layout, you will probably want to prototype critical parts of your circuit. Through-hole plug-in protoboard (Figure 15.15(a)) is OK for low-frequency circuits, but you need to be careful to properly bypass supply leads. For high frequencies, you also have to be aware that there is about 2 pF of parasitic capacitance between adjacent pins on the board, which can limit bandwidth significantly in high-speed circuits. (Think about what would happen to your high-frequency circuit if there was an extra 2 pF across the $C\mu$ of the gain transistor.)

Copper-clad prototyping board (usually called "vectorboard") is a great board to build a high-frequency circuit on (Figure 15.15(b)). There is a built-in ground plane that you can "sky-wire" components above. A pad cutter is used to ream out the

Table 15.6 Inductance Calculated from 2D FEA for Example 15.3 for a PC Board Trace; Trace Height Above Ground Plane $h = 15$ mils; Trace Width w as Shown; Trace Length 1″

w (mils)	L (nH) from FEA @ DC	L (nH) from FEA @ 1 MHz
25	16.6	8.5
50	13.5	5.86
75	11.7	4.5
100	10.3	3.7

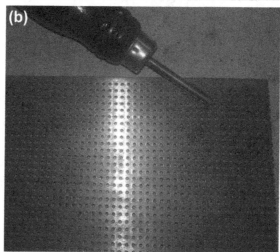

FIGURE 15.15

Prototyping boards. (a) Through-hole plug-in protoboard. (b) Copper-clad prototyping board and a pad cutter. (For color version of this figure, the reader is referred to the online version of this book.)

copper rings surrounding the holes, so that you can put through-hole components in the board without shorting. A prototype circuit built on copper-clad vectorboard is shown in Figure 15.16.

FIGURE 15.16

Prototype circuit built on copper-clad vectorboard at the author's laboratory, using the "sky-wire" technique. (For color version of this figure, the reader is referred to the online version of this book.)

If you are interested in prototyping a circuit using surface-mounted integrated circuits, there are sockets available that you can plug into a protoboard, or sky-wire onto a copper-clad vectorboard. Beware, however, of the parasitic inductance and capacitance of the sockets, which can affect operation of your circuit. The parasitics especially affect the operation of high-speed (>50–100 MHz) circuits.

Other prototyping tools (in no particular order) are:

A good benchtop soldering iron with various tips: adjustable power is best. Use small tips for precision work and large tips for soldering big devices or sections of ground plane.

Copper tape and copper foil: you can buy copper tape with peel-apart sticky side to make large traces on a protoboard. You can build electrostatic shields with copper foil.

Solder sucker and copper braid: for mopping up solder.

A good benchtop power supply with multiple outputs: I like −15 V and +30 V adjustable.

Universal serial bus (USB) oscilloscope: you can get a USB scope for very little money that plugs into your laptop. It may not have the bandwidth and voltage range of a good-quality Tek scope, but for quick and dirty work (and for putting waveforms in presentations), it is invaluable.

A heat gun: used for melting shrink wrap or thermal testing your circuit.

Battery pack: a battery pack with +, −, and GND outputs can be used to test low-noise circuits.

Laser thermometer: finds the hot spots in your circuit.

A good multimeter: for measuring voltage and current (dual ranges, 300 mA and 10 A).

Variac (a.k.a. variable autotransformer): for testing 60-Hz AC circuits at various input voltages.

Capacitor and resistor kits: available for very little money. When you need a 22-W resistor, you need a 22-W resistor…

Example 15.4: Design case study—high-speed semiconductor laser diode driver

This section considers the design, analysis, and PC board layout of a high-speed switching semiconductor laser diode system, which may be used as a modulated infrared light source. Direct modulation is a method by which the laser light power output of a semiconductor laser diode is changed by varying the diode current. To use a diode as a high-speed modulated light source, the laser is biased with a small DC current near the *lasing threshold* and a modulation current is superimposed. The light power output of the semiconductor diode is proportional to the laser current in excess of the threshold current. The direct modulation method is used for laser communication, for fiber-optic links, for industrial applications such as material cutting, and in such commercial products as compact disk players and medical laser printers.

Electrically, a semiconductor laser behaves like a diode, with a *V/I* curve shown in Figure 15.17(a). Since the semiconductor diode is made from gallium arsenide (GaAs) rather than silicon, the voltage "knee" when the diode turns on is approximately 1.5 V.

Under normal operation, the diode is driven by a current source so that the diode current remains constant even if the diode voltage drifts with time and temperature. The optical power output vs. diode current is shown in Figure 15.17(b) for a high-power laser diode. For very low currents, the diode does not produce laser light and there is very little optical power. (In fact, for current I_{th}, the laser behaves like an light-emitting diode (LED) and there is some very small amount of optical power emitted.) Once the diode current is increased to a value known as the threshold current (I_{th}) the diode begins lasing and the optical power output is proportional to the current in excess of the laser threshold current. For a 2-W laser diode, the operating laser current is approximately 2.5 A, as shown in Figure 15.17(b). If the laser current is increased further, the laser may be damaged by a process known as catastrophic optical damage where excess heating destroys[15] the laser-emitting area.

Semiconductor laser diodes are inherently fast devices. The intrinsic lasing processes may be modulated at very high rates by variation of the injected current. For representative diodes, the laser power transfer function (optical power output due to

---

[15]This curve is representative of one particular high-power laser diode used by the author; there are other lasers with different power levels and operating currents.

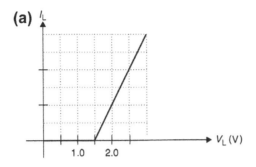

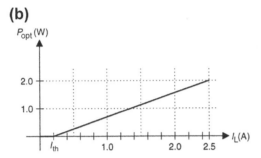

FIGURE 15.17

Laser diode curves for Example 15.4. (a) Representative laser diode $V-I$ curve showing laser voltage V_L and laser current I_L. (b) Representative high-power laser diode power–current curve for a 2-W laser diode. The horizontal axis (I_L) is laser diode current. The vertical axis is optical power output from the laser. Below the lasing threshold I_{th} the diode behaves as an LED.

current excitation) is flat out to several hundreds of megahertz, or even higher (Figure 15.18), depending on the details of the diode construction and the current bias level. The resonance near 10^{10} Hz is due to quantum relaxation processes. Therefore, the high light modulation speed indicated may be achieved in practice if the laser current is changed sufficiently fast. Next we will discuss how to switch laser current with high current and fast risetimes.

Driver implementation

One possible circuit topology suitable for driving a laser diode is shown in Figure 15.19. The laser is fed by two DC current sources I_{BIAS} and I_{th}, corresponding to a laser PEAK current and THRESHOLD currents. When V_{B1} is LOW and V_{B2} is HIGH, Q_1 is OFF and Q_2 is ON, the total current in the laser diode is $I_{PK} + I_{th}$. The resistor in the collector of Q_1 dissipates power, so Q_1 will not be damaged.

The author was responsible for the design of a semiconductor diode laser modulator capable of delivering 2.5-A pulses to a low-impedance load with risetime and

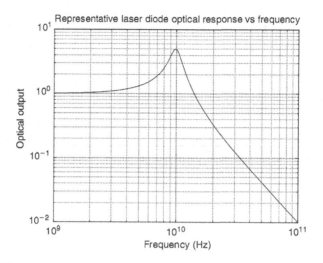

FIGURE 15.18

Representative laser diode intrinsic light output frequency response with laser resonance at 1 GHz. Horizontal axis—frequency (Hz).

falltime of less than 20 ns. The purpose of the circuit board was to drive semiconductor diode lasers for high-speed printing.[16]

There were several design challenges inherent in this design. First, the laser signal is a high-current, fast-risetime set of current pulses with any repetition rate from DC up to 10 MHz, with any duty cycle. This means that extreme care must

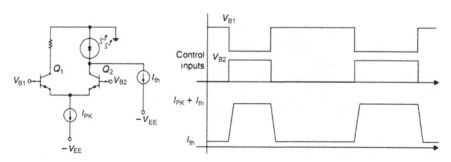

FIGURE 15.19

Laser driver circuit implemented as an emitter-coupled switch, and waveforms. The top traces are the control inputs V_{B1} and V_{B2}. The bottom trace is the laser current.

[16]For more details on the design, see Marc Thompson and Martin Schlecht, "High Power Laser Diode Driver Based on Power Converter Technology", *IEEE Transactions on Power Electronics,* vol. 12, no. 1, January 1997, pp. 46–52 and U.S. Patent 5,444,728 (issued 8/22/95).

be taken to ensure low-inductance path from the switching elements on the PC board to the laser. The design of the switch on the PC board presents interesting thermal problems as well.

A simplified schematic of the switching transistor array is shown in Figure 15.20. In order to provide low-inductance paths as well as good thermal management, the fast switch was broken up into a dozen smaller emitter-coupled switches, each pair implemented with a pair of 2N2222 transistors. Note that in emitter-coupled pair Q_{1A} and Q_{1B} only one transistor is on at a time; when DRIVE is HIGH and $\overline{\text{DRIVE}}$ is LOW, Q_{1A} is ON and Q_{1B} is OFF (and hence the laser is off, and idling at the threshold current I_{th}). When DRIVE is LOW and $\overline{\text{DRIVE}}$ is HIGH, Q_{1B} is ON and Q_{1A} is OFF and the total laser current is $I_{th} + I_{PK}$.

The critical high-speed and high-current switching paths are highlighted[17] in bold in Figure 15.20. Each of the transistor arrays switches up to a maximum of more than 200 mA. It is mandatory to keep the interconnection inductance between transistors and to the laser diode low in order that the transistor arrays can switch as fast as they are capable. Remember from previous chapters that emitter-coupled switches are inherently fast, provided that you provide sufficient base drive capability.[18]

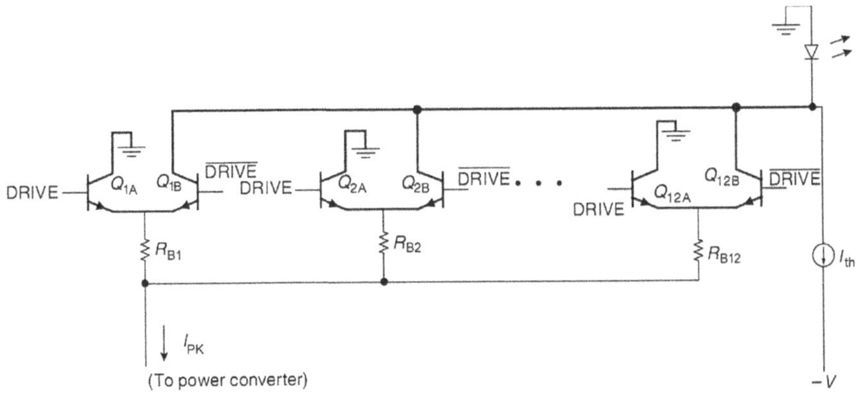

FIGURE 15.20

Switching transistor array capable of fast-risetime switching of up to 2.5 A. The array is composed of 12 pairs of high-speed switching transistors. The critical high-speed and high-current switching paths are shown in bold.

---

[17]When doing a PC board layout, the traces in bold are good candidates to be implemented as wide traces over an unbroken ground plane, to reduce parasitic inductance.

[18]See, e.g. Chapter 10 where we showed that the switching speed of the emitter-coupled pair for signal transistors is a few nanoseconds. Of course, this assumes that we have a good PC board layout so that parasitic inductances do not slow things down significantly.

Resistors R_{B1}, R_{B2}, up to R_{B12} are low-valued *ballast* resistors and ensure that the transistor pairs share the current among them equally.[19]

Another design challenge was that the current rating of the -12-V power supply was only 1 A, and we want to deliver 2.5 A to the lasers. Therefore, a DC/DC converter was needed to step down the voltage and step up the current. A simplified implementation of this is shown Figure 15.21. A detailed discussion of the DC/DC converter is beyond the scope of this book,[20] but this circuit steps DOWN the voltage and steps UP the current. Hence we draw less current from the -12-V power supplies than delivered to the lasers. We note that this is a high-speed switching circuit and hence we need to take special care in the layout of the MOSFET and diode. The power MOSFET is switched on and off at a high frequency with a variable duty cycle to regulate the current, as shown.

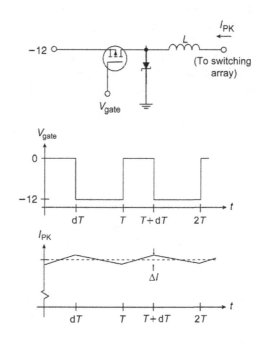

FIGURE 15.21

Simplified schematic of the DC/DC converter.

In this design, the form factor of the PC board was mandated to be 3.5″ × 4.5″, as this design was replacing a preexisting design and the PC boards had to be backward-compatible. The form factor and mounting holes are shown in Figure 15.22.

---

[19]The transistors in this design are not matched, and we want each transistor pair to shoulder an equal value of the load. The ballast resistors, on the order of 1 Ω, forces sharing between transistor pairs.
[20]Described in the author's Master's thesis from MIT.

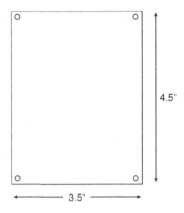

FIGURE 15.22

Form factor of PC board, 3.5″ × 4.5″, showing mounting holes.

The connector locations (Figure 15.23) were also set prior to the layout. The connectors are as follows (clockwise from bottom left):

- *Power*: +12 V @ 200 mA, −12 V @ 1 A, and two ground pins.
- *Laser diode connection*: A microstrip cable was soldered directly to the PC board to provide a low-impedance path to the lasers.

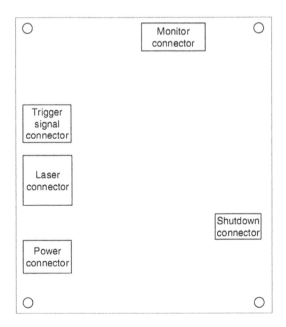

FIGURE 15.23

PC board connector locations.

- *Trigger signal*: This is a TTL-level signal that turns the laser diode ON and OFF. When the trigger signal is high, the laser is ON. Repetition rates for the TTL signal are from DC up to 10 MHz.
- *Monitor*: This connector is used to buffer and amplify a photodiode signal used to monitor the optical power output of the laser.
- *Shutdown*: Another TTL-level signal that is used to completely shut down the laser.

Next, the real estate for the various PC board traces was allocated as in Figure 15.24. We note that the high-current and high-speed circuitry is segregated from the low-level analog instrumentation circuitry. Furthermore, the PC board was multilayer, ensuring that an unbroken ground plane could be used under the high-speed circuitry. A breakdown of the PC board layers is as follows:

- Top layer: analog signals
- Internal layer #1: GND
- Internal layer #2: −12 V
- Bottom layer: analog signals, +12 V

An internal layer was dedicated to −12 V since there were significant switching currents drawn from the DC/DC converter.

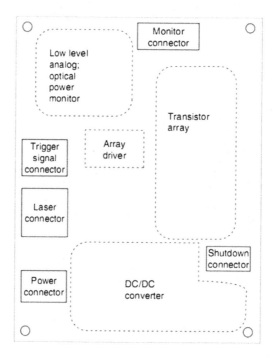

FIGURE 15.24

PC board allocation of real estate.

A photograph of the resultant PC board is shown in Figure 15.25. An oscilloscope photograph of the laser light output[21] is shown in Figure 15.26. We note that the laser is switching 2 W peak–peak, corresponding to a switched current of 2.5 A peak–peak. The risetime and falltime are less than 20 ns.

FIGURE 15.25

PC board showing top side (component side) final layout. Connection to laser diode is not shown.

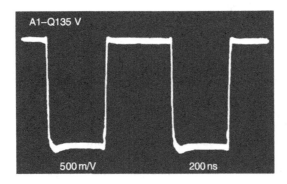

FIGURE 15.26

Scope photograph showing risetime and falltime of laser optical power. Horizontal, 200 ns per division; vertical, 2 W full scale. The resultant laser light 10–90% risetime and falltime are less than 20 ns.

---

[21]The light output was measured using an extremely fast photodetector. Since the laser is an inherently fast device, the light output shape is representative of the shape of the current pulses to the laser.

Chapter 15 problems

Problem 15.1

This problem concerns real-world capacitors. Every capacitor has some parasitic elements, that is, there is no such thing as "ideal capacitor". For instance, the leads that connect a capacitor to your circuit have a parasitic inductance that is in series with the capacitor. This parasitic series inductance affects how the capacitor operates at high frequency. Also, the capacitor has some internal resistance, also in series with the capacitance, known as the ESR.

a. Draw the lumped circuit model for the real-world capacitor.

b. The *very approximate* inductance of a pair of leads is approximately 1 µH/m of lead length. For a total lead length of 1″, calculate the self-resonant frequency of the capacitor for $C = 1$ µF and for $C = 1000$ µF.

c. Sketch a Bode (log–log) plot of a capacitor impedance for $C = 1000$ µF and 1″ leads, assuming an ESR $= 10$ mΩ.

d. Why should you keep lead lengths short?

Problem 15.2

A PC board trace is 0.1″ wide, 6″ long, and 0.06″ above a ground plane. Estimate the trace inductance and trace resistance at room temperature. Assume that the trace is composed of 1-ounce copper.

Problem 15.3

A microstrip line is built with two copper strips each 1 cm wide separated by a 0.005″-thick kapton insulator. Estimate the inductance of a 6″ long piece of microstrip.

Problem 15.4

A PC board-mounted inductor has an ideal value of $L = 1$ µH. However, the measured series resistance of this device is 0.5 Ω, and a parallel-resonant frequency of 15 MHz is measured using an impedance analyzer. Generate an appropriate lumped circuit model for this inductor.

Problem 15.5

A film capacitor has an ideal value of 2.2 µF. You initially charge the capacitor to 250 V, then disconnect the charging power supply, and, using an oscilloscope, discover that the capacitor voltage discharges to 100 V in 60 s. Find a lumped circuit model of this capacitor, utilizing these findings.

Problem 15.6

A MOSFET gate driver IC has a very low impedance and drives a MOSFET gate where the input impedance may be modeled as a capacitance of 1000 pF. The gate driver is 2″ away on a PC board, resulting in an approximate series inductance driving the gate from the driver of 50 nH. Draw the circuit model, and find the value of series resistance that you place at the output of the gate driver to achieve critical damping of the gate circuit. Simulate your circuit and find the 10—90% risetime of the gate signal.

Problem 15.7

How much of a voltage spike is created in a MOSFET circuit, switching 10 A in 50 ns, with a parasitic drain inductance of 10 nH?

Further reading

[1]   Bartoli M, Reatti A, Kazimierczuk M. *High-frequency models of ferrite core inductors.* In *International conference on industrial electronics, control and instrumentation, (IECON '94),* vol. 3; September 5—9, 1994. 1670—1675.

[2]   Brooks, D. Fusing currents—when traces melt without a trace. Available at: http://www.ultracad.com, printed in Printed circuit design, December 1998;15(12):53.

[3]   Cao Y, Groves R, Huang X, Zamdmer N, Plouchart J, Wachnik R, et al. Frequency-independent equivalent-circuit model for on-chip spiral inductors. *IEEE J Solid-State Circuits* 2003;**38**(3):419—26.

[4]   Demurie SN, DeMey G. Parasitic capacitance effects of planar resistors. *IEEE Trans Compon Packag Manuf Technol Part A* 1989;**12**(3):348—51.

[5]   Dolan JE, Bolton HR. Capacitor ESR measurement technique. In: *Eighth IEEE pulsed power conference 1991* June 16—19, 1991. p. 228—31.

[6]   Franco S. Polypropylene capacitors for snubber applications. In: *Proceedings of the thirty-first IAS annual meeting (IAS '96)* October 6—10, 1996. p. 1337—42.

[7]   Galbraith J. Reliable precision wirewound resistor design. *IRE Trans Compon Parts* 1956;**3**(3):116—9.

[8]   Grover FW. *Inductance calculations: working formulas and tables.* New York: Dover Publications, Inc.; 1946.

[9]   Jutty MK, Swaminathan V, Kazimierczuk MK. Frequency characteristics of ferrite core inductors. In: *Proceedings of the electrical electronics insulation conference and electrical manufacturing & coil winding conference, 1993* October 4—7, 1993. p. 369—72.

[10] Lee TH. *The design of CMOS radio-frequency integrated circuits.* 2nd ed. Cambridge University Press; 2003.

[11] Madou A, Martens L. Electrical behavior of decoupling capacitors embedded in multilayered PCBs. *IEEE Trans Electromagn Compat* 2001;**43**(4):549—66.

[12] Manka W. Alternative methods for determining chip inductor parameters. *IEEE Trans Parts, Hybrids, Packag* 1977;**13**(4):378—85.

[13] Massarini A, Kazimierczuk MK. Self-capacitance of inductors. *IEEE Trans Power Electron* 1997;**12**(4):671—6.

[14] Maxwell JC. A treatise on electricity and magnetism. Originally published 1873, reprinted by Dover Publishing; 1954.

[15] Naishadharn K. Experimental equivalent-circuit modeling of SMD inductors for printed circuit applications. *IEEE Trans Electromagn Compat* 2001;**43**(4):557—65.

[16] Neugebauer TC, Phinney JW, Perreault DJ. Filters and components with inductance cancellation. *IEEE Trans Ind Appl* 2004;**40**(2):483—91.

[17] Paul C. *Inductance: loop and partial.* John Wiley; 2009.

[18] Philips, Inc. Package lead inductance considerations in high-speed applications, Philips Application Note AN212.

[19] Reed EK. Tantalum chip capacitor reliability in high surge and ripple current applications. In: *1994 electronic components and technology conference* May 1—4, 1994. p. 861—8.

[20] Sakabe Y, Hayashi M, Ozaki T, Canner JP. High frequency measurement of multilayer ceramic capacitors. *IEEE Trans Compon Packag Manuf Technol, Part B: Adv Packag* 1996;**19**(1):7—14.

[21] Sarjeant WJ, Zirnheld J, MacDougall FW. Capacitors. *IEEE Trans Plasma Sci* 1998;**26**(5):1368—92.

[22] Sinclair I. *Passive components for circuit design.* Newnes; 2001.

[23] Smith LD, Hockanson D. Distributed SPICE circuit model for ceramic capacitors. In: *Proceedings of the 2001 electronic components and technology conference* May 29—June 1, 2001. p. 523—8.

[24] Stroud J. Equivalent series resistance-the fourth parameter for tantalum capacitors. In: *Proceedings of the 1990 electronic components and technology conference* May 20—23, 1990. p. 1009—12.

[25] Thompson M, Schlecht M. High power laser diode driver based on power converter technology. *IEEE Trans Power Electron* 1997;**12**(1):46—52.

[26] Wedlock BD, Roberge JK. *Electronic components and measurements.* Prentice-Hall; 1969.

[27] Ulrich RK, Brown WD, Ang SS, Barlow FD, Elshabini A, Lenihan TG, et al. Getting aggressive with passive devices. *IEEE Circuits Devices Mag* 2000;**16**(5):16—25.

[28] Venkataramanan G. *Characterization of capacitors for power circuit decoupling applications*In *Industry applications conference, 1998*, vol. 2; October 12—15, 1998. 1142—1148.

[29] Wadell BC. Modeling circuit parasitics 1. *IEEE Instrum Meas Mag* 1998;**1**(1):31—3.

[30] Wadell BC. Modeling circuit parasitics 2. *IEEE Instrum Meas Mag* 1998;**1**(2):6—8.

[31] Wadell BC. Modeling circuit parasitics 3. *IEEE Instrum Meas Mag* 1998;**1**(3):28—31.

[32] Wadell BC. Modeling circuit parasitics 4. *IEEE Instrum Meas Mag* 1998;**1**(4):36—8.

[33] Yu Q, Holmes TW. A study on stray capacitance modeling of inductors by using the finite element method. *IEEE Trans Electromagn Compat* 2001;**43**(1):88—93.

[34] Zahn M. *Electromagnetic field theory: a problem solving approach.* Krieger; 2003.

Noise

*"Statistical fluctuation of electric charge exists in all conductors, producing
random variation of potential between the ends of the conductor."*
—J. B. Johnson, **"Thermal Agitation of Electricity in Conductors", Physical Review,
vol. 32, pp. 97–109 (1928)**

IN THIS CHAPTER

▶ The basics of noise, including thermal (white noise), shot noise, and $1/f$ noise are described.
Several examples of operational amplifier (op-amp) noise calculations are given, focusing on
practical engineering approximations. We also draw the distinction between noise and other
electrical disturbances such as magnetic and electric field coupling, hum, and ringing.

Noise in resistors and integrated circuits is a fact of life. Noise in your circuit
limits the smallest signal that you can effectively detect and process. The following
is a very basic description of the various noise sources and characteristics in electrical circuits, and typical noise calculations.

Thermal (a.k.a. "Johnson" or "White") noise in resistors

Thermal[1] noise is caused by the thermal motion of charges in a conductor, akin to
Brownian motion. Thermal noise, which is produced by all resistors regardless of
type, is a noise signal that has zero average value, is broadband with a flat spectral density versus frequency, and the noise power increases with temperature.
Since noise is a statistical, random phenomenon, we need to consider the
mean-square and root-mean-square (RMS) quantities of noise voltage and noise
power.

The physics of thermal noise was first described[2] by J. B. Johnson at Bell Labs in
1928, and hence, we sometimes call thermal noise "Johnson" noise. A noisy resistor
can be modeled as shown in Figure 16.1(a) as an ideal resistor in series with a white

---

[1]The terms thermal noise, white noise, and Johnson noise are used interchangeably.
[2]Mr Nyquist helped out too.

Intuitive Analog Circuit Design. http://dx.doi.org/10.1016/B978-0-12-405866-8.00016-4

(a)   **(b)**

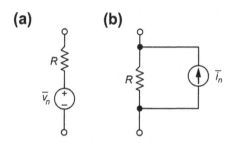

FIGURE 16.1

A resistor with two equivalent noise models. (a) A voltage source in series with a resistor. (b) A current source in parallel with a resistor.

noise-generating voltage source. A noisy resistor has a thermal noise "power spectral density" of

$$4kTR \tag{16.1}$$

independent[3] of frequency, with the interesting units of mean-squared volts per hertz. The thermal voltage noise spectral density is

$$\sqrt{4kTR} \tag{16.2}$$

with units of volts per root hertz. To find the total mean-square noise voltage created by the noisy resistor, we need to integrate the power density spectrum over the circuit[4] bandwidth. The mean-square value of the noise voltage generator (in units of volts squared) is found by integrating the power spectral density over all frequencies, which is easy in this case because the white noise spectral density is flat versus frequency. Let us assume that we are interested in the noise over the frequency band from direct current (DC) to Δf; the mean-square noise voltage is then (in units of volts squared):

$$\overline{v_n^2} = \int_0^{\Delta f} 4kTR\,df = 4kTR\Delta f \tag{16.3}$$

where k is Boltzmann's constant, T is the absolute temperature (in Kelvins), R is the resistance value, and Δf is the bandwidth of interest. The noise voltage RMS amplitude is given by (now in units of volts):

$$\overline{v_n} = \sqrt{4kTR\Delta f} \tag{16.4}$$

---

[3]Well, up to very, very high frequencies much higher than gigahertz.
[4]More in a little bit on what "bandwidth" is.

(a)

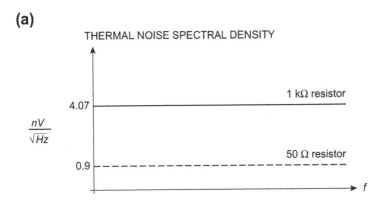

(b)

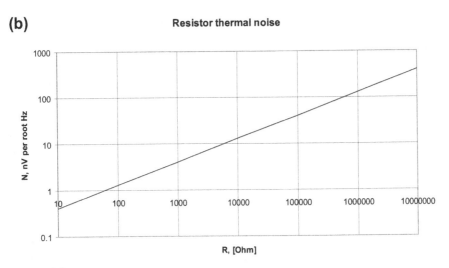

FIGURE 16.2

Thermal noise voltage spectral density for (a) 1-kΩ and 50-Ω resistors and (b) resistors in the range of 10 Ω–10 MΩ. (For color version of this figure, the reader is referred to the online version of this book.)

As an example, a 1-kΩ resistor at room temperature (300 K) over a 1-MHz bandwidth has an RMS thermal noise of

$$\overline{v_n} = \sqrt{4(1.38 \times 10^{-23})(300)(1000)(10^6)} = 4.07 \ \mu V \tag{16.5}$$

A handy thing to remember for on-the-fly noise calculations is that the room-temperature thermal noise voltage spectral density ($\sqrt{4kTR}$) of a 1-kΩ resistor is 4.07 nV per root hertz, independent of frequency (Figure 16.2). The thermal noise voltage of a 50-Ω resistor is $\sqrt{20}$ times smaller, or about 0.9 nV per root hertz.

We can, of course, represent the noisy resistor by its Norton equivalent, resulting in the circuit of Figure 16.1(b).

$$\overline{i_n} = \sqrt{\frac{4kT\Delta f}{R}} \tag{16.6}$$

The choice of which circuit representation to use for the noisy resistor is up to the designer, primarily on the basis of computation simplicity. (In the case of a 1-kΩ resistor, the noise current source $\overline{i_n}$ has a value of 4.07 nA over a 1-MHz bandwidth.) In Figure 16.2(a), we see the voltage spectral density (RMS volts per root hertz) for 1-kΩ and 50-Ω resistors. In Figure 16.2(b), we see the voltage spectral density for Johnson noise in resistors over the range 10 Ω–10 MΩ. The important take-away here is that larger-valued resistors are noisier than are smaller resistors.

If you connect the noisy resistor at the input of an amplifier, the resistor noise gets amplified. Let us say we connect our 1-kΩ resistor at the input of an amplifier with a gain of 1000 and a bandwidth of 1 MHz. The resultant total output noise (in RMS) is 4.07 mV, assuming that the amplifier itself is noiseless.

Ideally, passive components (inductors and capacitors) do not produce any thermal noise. However, if there is a series resistance, there is thermal noise created by the parasitic resistance.

An electrical waveform of thermal noise is shown in Figure 16.3; if you put this waveform through a speaker, you will hear the familiar white noise "hiss" that you may have heard if you have ever used a white noise generator (or had an old-style analog TV on late at night after the station goes off the air).

Example 16.1: Thermal noise at the terminals of a real-world resistor

Let us assume that we have a resistor, which is shunted by its real-world parasitic capacitance (Figure 16.4). We will calculate the RMS value of the noise generated by this combination over all frequencies. (The capacitor does not generate any thermal noise of its own.) To find the mean-square noise voltage at the terminals of the resistor, we integrate the resistor noise spectral density over all frequencies to get

$$\overline{v_{nt}^2} = \int_0^\infty \frac{4kTR}{1 + (2\pi fRC)^2} df = \frac{kTR}{RC} = \frac{kT}{C} \tag{16.7}$$

We note that the RC combination low-pass filters the noise. As a numerical example, consider a 1-kΩ resistor shunted by 0.1 pF of parasitic capacitance. The RMS noise voltage at the terminals of the resistor (at $T = 300$ K) is

$$\overline{v_{nt}} = \sqrt{\frac{(1.38 \times 10^{-23})(300)}{(10^{-13})}} = 203 \ \mu V \tag{16.8}$$

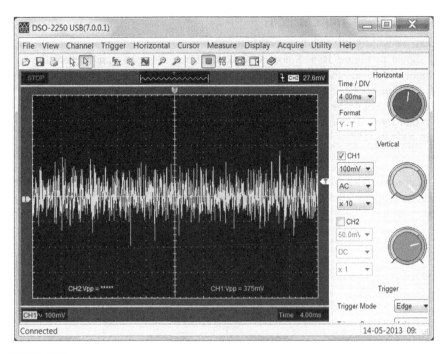

FIGURE 16.3

White noise generated by an Android phone white noise generator. (For color version of this figure, the reader is referred to the online version of this book.)

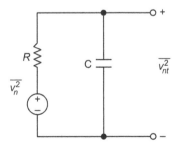

FIGURE 16.4

Resistor R shunted by its parasitic capacitance C for Example 16.1. We use this model to find the noise voltage $\overline{v_{nt}^2}$ at the physical output terminals of the resistor.

How do noise sources add?

Next, we will figure out how to combine two series-connected voltage noise sources, as shown in Figure 16.5. Since noise is measured in mean-square quantities, we cannot simply add the voltage sources together. We recall that the total RMS value of any number of combined waveforms (if the waveforms are "uncorrelated") is the square root of the sum of the mean-squared values of the individual waveforms. We find for two series-connected noise sources that

$$\overline{v_{n,total}} = \sqrt{\overline{v_{n1}^2} + \overline{v_{n2}^2}} \qquad (16.9)$$

Let us assume that we have a 4.07-nV (RMS) noise source connected in series with an uncorrelated 0.9-nV noise source. The net RMS noise from this series connection is

$$\overline{v_{n,total}} = \sqrt{(4.07\ \text{nV})^2 + (0.9\ \text{nV})^2} = 4.17\ \text{nV} \qquad (16.10)$$

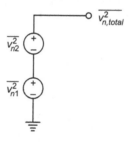

FIGURE 16.5

Two voltage noise sources in series.

Noisy resistors in series or in parallel

To find the total noise created by noisy resistors in series (Figure 16.6(a)), we recall the expression for the noise created by series-connected sources:

$$\overline{v_{n,total}} = \sqrt{\overline{v_{n1}^2} + \overline{v_{n2}^2}} = \sqrt{4kTR_1\Delta f + 4kTR_2\Delta f} = \sqrt{4kT(R_1 + R_2)\Delta f} \quad (16.11)$$

This means that we can treat series-connected resistors (R_1 and R_2) as a single resistor of value ($R_1 + R_2$) for noise calculations. Similarly (left as an exercise for the reader (!)), for noise calculations, we can combine parallel resistors (Figure 16.6(b)) into a single resistor being the parallel combination of R_1 and R_2. If you have a bunch of resistors in your circuit, combining series and parallel resistors can simplify noise computation considerably.

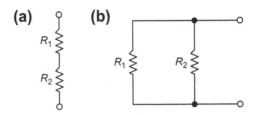

FIGURE 16.6

Noisy resistors in (a) series and (b) in parallel.

What bandwidth do you use when calculating total noise?

In previous discussions, we glossed over the details of what bandwidth to use when calculating noise by assuming a "bandwidth of interest" Δf. Let us consider an example of a circuit with a -3-dB bandwidth f_h as shown in Figure 16.7. If we use f_h as our bandwidth for noise calculations, the noise will be underestimated, since some noise will pass through the circuit at frequencies higher than f_h due to the finite roll-off of the filter. The "equivalent noise bandwidth" f_B is the bandwidth of a brick wall filter that will give you the same noise through your circuit.

If you calculate the noise through a first-order system, you will find that the equivalent noise bandwidth exceeds the -3-dB bandwidth by a factor of $\pi/2$, or by about 1.57. For a second-order system, the equivalent noise bandwidth is perhaps 20% higher than the -3-dB bandwidth (depending on the damping of the poles). As your circuit filtering order gets higher, the noise equivalent bandwidth f_B gets closer and closer to the -3-dB bandwidth.

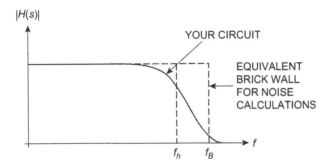

FIGURE 16.7

Circuit illustrating equivalent noise bandwidth f_B in relation to the -3-dB cutoff frequency f_h of your circuit.

Example 16.2: Estimating white noise RMS amplitude from oscilloscope measurements

The amplitude of a white noise source has a Gaussian amplitude distribution. This means that there is a better than 99% chance that the voltage amplitude at any given time is within six times the RMS value. A handy rule of thumb for a ballpark estimation of the RMS value of a white noise signal is to take the peak–peak value seen on an oscilloscope and divide by a factor of 6. As an example, consider the white noise voltage shown in Figure 16.8. The peak-to-peak amplitude is about 200 μV, meaning that the RMS noise voltage is about 33.3 μV.

Schottky ("shot") noise

Shot noise is associated with a DC current through a junction and the random times that individual charges cross a junction. Shot noise is a broadband white noise, with a power spectral density of

$$2qI_{DC} \tag{16.12}$$

in units of amperes squared per hertz. The total mean-square shot noise current is found by integrating the spectral density over the bandwidth Δf:

$$\overline{i_{ns}^2} = \int_0^{\Delta f} 2qI_{DC}df = 2qI_{DC}\Delta f \tag{16.13}$$

where I_{DC} is the diode DC current and q is the electronic charge. A 10-mA DC current through a diode junction results in about 20 pA/$\sqrt{Hz}$ of shot noise. The total current through a noisy diode is shown in Figure 16.9.

If we look at a typical diode, there are a couple of noise sources: the shot noise associated with the junction, and the thermal noise associated with any resistive parts of the diode. The circuit model for the noisy diode is shown in Figure 16.10.

1/f ("pink" or "flicker") noise

This type of noise is thought to be caused by impurities on semiconductor surfaces and by other phenomena. Johnson discovered 1/f noise in vacuum tubes around 1925 and noted that at low frequency the spectrum of this noise was not flat; rather the noise increased at low frequencies with roughly[5] a 1/f dependence. In Figure 16.11, we see the comparison of 1/f noise with white noise.

---

[5]Since the 1/f noise is not well understood, it is usually found by measurement.

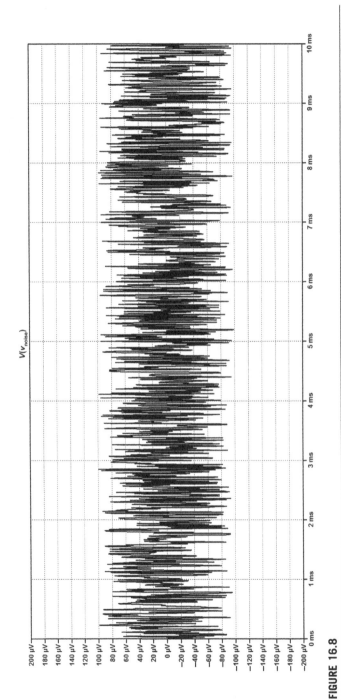

FIGURE 16.8

Oscilloscope output showing white noise for Example 16.2. This waveform has a peak–peak value of about 200 µV, and hence an RMS value of about 33 µV.

DIODE CURRENT

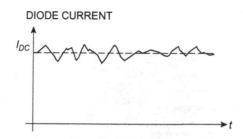

FIGURE 16.9

Illustration of the shot noise in a diode. The diode is biased with DC current I_{DC}. Current vs. time, with noise exaggerated.

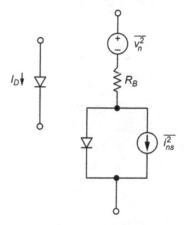

FIGURE 16.10

Model of a noisy diode including a shot noise current source, and Johnson voltage noise created by the resistive part (R_B) of the junction.

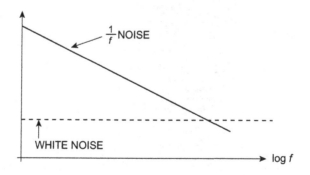

FIGURE 16.11

Comparison of thermal ("white") noise (constant vs. frequency) and 1/f noise.

The $1/f$ noise shows up in many different processes: semiconductors, electrolytes, metal films, and in mechanical and biological systems. Consequently, the $1/f$ noise is not well understood, and is usually expressed empirically. Mathematically, the $1/f$ noise power spectral density can be characterized by the following equation:

$$\overline{N^2} = \frac{\beta \Delta f}{f^n} \tag{16.14}$$

with β being a parameter that depends on your type of device, and n an exponent close to 1. (Note that the $1/f$ noise exists for both voltage and current.) To calculate noise power, we integrate the power spectral density function over a band of frequencies. Let us do this calculation assuming a $1/f$ dependence of the noise:

$$P = \int_{f_1}^{f_2} \beta \frac{df}{f} = \beta \ln\left(\frac{f_2}{f_1}\right) \tag{16.15}$$

The result is that the total noise power depends on the natural log of the ratio of the high and low ends of the frequency band of interest.

Excess noise in resistors

Resistors exhibit $1/f$ noise, often called "excess noise" because the $1/f$ component is in addition to the thermal noise. Excess noise in resistors occurs when there is a DC bias, and increases with a DC current flow. Excess noise is thought to be due to current flowing unevenly through the granular structure of the resistor. Excess noise is the worst in the old-style carbon composition resistors, and is better in carbon film, still better in metal films, and the best in wire-wound and foil resistors.

"Popcorn" noise (a.k.a. "burst" noise)

Popcorn noise is thought to be associated with contaminants in semiconductor junctions, and is seen in diodes and resistors. The amplitude of a typical popcorn noise signal is shown in Figure 16.12, where little square waves of noise appear on the top of the normal signal. The noise is two level, and generally occurs at audio frequencies—hence the name "popcorn noise", for the sound the noise signal makes when run through a speaker.

Bipolar transistor noise

In a transistor, there are noise sources due to physical resistors and shot noise in junctions. In Figure 16.13, we see the transistor low-frequency incremental

"POPCORN" NOISE AMPLITUDE

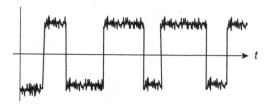

FIGURE 16.12

"Popcorn" noise vs. time.

model. Resistors r_x, r_c, and r_e (base, collector, and emitter) are physical resistors and create Johnson noise. There is shot noise associated with the base–emitter and collector junctions. Resistors r_π, r_o, and r_μ are derived incremental parameters (not physical resistors) and hence they do not create any noise.

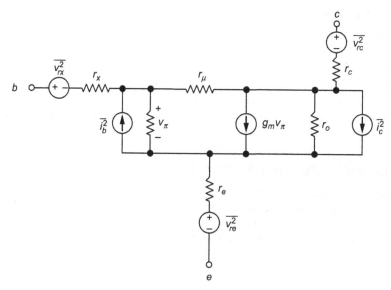

FIGURE 16.13

A transistor incremental model showing transistor noise sources. There is thermal (Johnson) noise due to the three physical transistor resistances. There is shot noise due to the current flow across junctions. There are 3 Johnson noise voltage sources and 2 shot noise current sources in this model.

The Johnson noise source values are

$$\overline{v_{rx}^2} = 4kTr_x\Delta f$$
$$\overline{v_{rc}^2} = 4kTr_c\Delta f \qquad (16.16)$$
$$\overline{v_{re}^2} = 4kTr_e\Delta f$$

Normally, the thermal noise of the base resistance r_x is much higher than the thermal noise of the emitter and collector resistances. The shot noise source values are

$$\overline{i_b^2} = 2qI_B\Delta f$$
$$\overline{i_c^2} = 2qI_C\Delta f \qquad (16.17)$$

The transistor also has a $1/f$ noise at low frequencies, not accounted for in this model.

Field-effect transistor noise

In summary, the drain current of a metal-oxide semiconductor field-effect transistor (MOSFET) exhibits thermal noise and $1/f$ noise and the gate exhibits broadband noise. For further information on MOSFET noise, the curious reader is invited to read Tom Lee's book on complementary metal-oxide semiconductor (CMOS) for further details. Regarding junction gate field-effect transistor (JFET) noise, the Siliconix book on JFETs does a good job.

Op-amp noise model

Various noise sources already described exist in op-amps. Typically, manufacturers will give you information about the noise sources, with the sources modeled as being

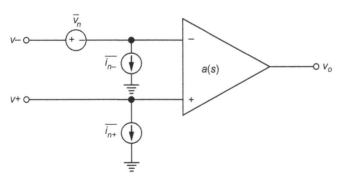

FIGURE 16.14

Ideal op-amp $a(s)$ with voltage and current noise sources referred to the input.

at the input of the op-amp, as shown in Figure 16.14. Typically, the noise voltage is in units of nanovolts per root hertz $(nV/\sqrt{Hz})$ and the input current noise is in units of femtoamperes $(10^{-15} A)$ per root hertz to picoamperes per root hertz $(fA/\sqrt{Hz}$ to $pA/\sqrt{Hz})$.

Some characteristics of the (wildly variable depending on the particular op-amp) noise are as follows:

- Input noise voltage spectral density at the input of op-amps is typically in the range $1-20 \, nV/\sqrt{Hz}$.
- Input noise current spectral density at the input of op-amps is typically in the range $< 1 \, fA/\sqrt{Hz}$ (for field-effect transistors (FETs) or CMOS op-amps) to more than $pA/\sqrt{Hz}$ for high-speed bipolar op-amps.
- Typically, bipolar op-amps have a lower input voltage noise than do JFET and CMOS-input op-amps. JFET and CMOS-input op-amps typically have a lower input current noise.
- Bipolar op-amps with a low voltage noise are sometimes run with high input-stage collector current (to reduce voltage noise at the input). This does, however, increase the current noise.
- At low frequencies, op-amp noise is dominated by $1/f$ noise; at high frequencies, the noise is white. The noise spectral density for a generic op-amp is as given in Figure 16.15. The $1/f$ corner frequency is found on a datasheet and is typically $1 \, Hz-1 \, kHz$. The noise voltage referred to as the input of the TLO84 FET input op-amp is shown in Figure 16.16, where we see a $1/f$ "knee" at about $1 \, kHz$. Above $1 \, kHz$, the white noise voltage noise is about $18 \, nV/\sqrt{Hz}$.

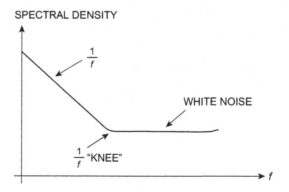

FIGURE 16.15

Op-amp input-referred noise. Both input current and input voltage noise exhibit the $1/f$ behavior; the $1/f$ "knee" generally occurs at different frequencies for voltage and current.

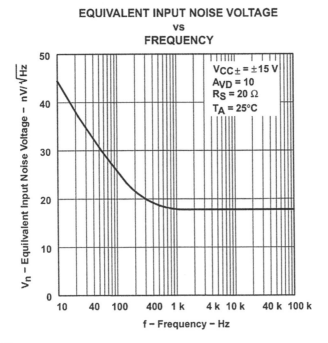

EQUIVALENT INPUT NOISE VOLTAGE
vs
FREQUENCY

$V_{CC\pm} = \pm 15$ V
$A_{VD} = 10$
$R_S = 20\ \Omega$
$T_A = 25°C$

V_n – Equivalent Input Noise Voltage – nV/$\sqrt{Hz}$

f – Frequency – Hz

FIGURE 16.16

Datasheet excerpt for the JFET-input TL084 op-amp, illustrating the $1/f$ voltage noise below about 1 kHz. Above 1 kHz, the voltage noise is white noise with an amplitude of about 18 nV/$\sqrt{Hz}$.

Reprinted with permission from the National Semiconductor.

Example 16.3. Noise calculation in an op-amp circuit

Let us find the noise at the output of an inverting amplifier (Figure 16.17(a)). In Figure 16.17(b), we see all the various noise sources. There is Johnson noise due to resistors R_1, R_2, and R_3; there is voltage noise referred to the input for the op-amp; and there is current noise at both op-amp inputs. The output noise due to each noise source is shown in Table 16.1 where we solve for each noise source individually. The total mean-square output noise *power* is the sum of the individual terms. The output noise *voltage* is the square root of the sum of the individual output powers.

Selecting a noise-optimized op-amp

We see from the previous examples that the noise voltages and currents add up (in RMS fashion) to give us the total output noise of the op-amp circuit. To minimize output op-amp noise, we need to keep resistors to a minimum value, if possible. If you have a circuit that runs from a finite source impedance as in the op-amp follower of Figure 16.18, you need to select an op-amp with optimized values of input voltage and current noise.

(a)

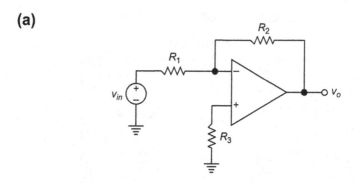

(b)

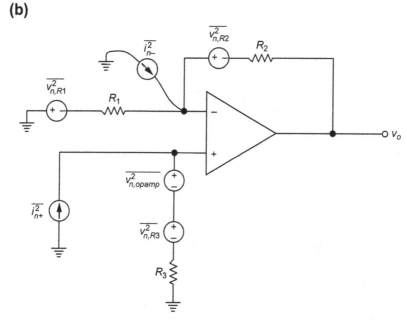

FIGURE 16.17

Op-amp circuit for calculating output noise. (a) Op-amp inverter. (b) Noise model including Johnson noise sources due to three resistors, and input-referred voltage and current noise sources for the op-amp.

The total output voltage noise is from three separate sources.

- The intrinsic voltage noise of the op-amp
- The noise current of the op-amp, multiplied by the source resistance R_s (which appears as a noise voltage at the input of the op-amp).
- The Johnson noise of the source resistor.

Table 16.1 Tabulation of Various Noise Sources for Example 16.3

Noise Source	Noise Power at the Op-Amp Output
Johnson noise from R_1, RMS value is $\overline{v^2_{n,R1}} = 4kTR_1\Delta f$	$\left(\frac{R_2}{R_1}\right)^2 \overline{v^2_{n,R1}}$
Johnson noise from R_2, $\overline{v^2_{n,R2}} = 4kTR_2\Delta f$	$\overline{v^2_{n,R2}}$
Johnson noise from R_3, $\overline{v^2_{n,R3}} = 4kTR_3\Delta f$	$\left(\frac{R_1+R_2}{R_1}\right)^2 \overline{v^2_{n,R3}}$
Op-amp input-referred voltage noise (RMS voltage) $\overline{v^2_{n,op-amp}}$ which is found by integrating the voltage noise spectral density (including any 1/f component) over the bandwidth	$\left(\frac{R_1+R_2}{R_1}\right)^2 \overline{v^2_{n,op-amp}}$
Op-amp current RMS noise at the +input, $\overline{i^2_{n+}}$ found by integrating the current noise spectral density over the bandwidth	$\left(\frac{R_1+R_2}{R_1}\right)^2 \overline{i^2_{n+}} R_3^2$
Op-amp current noise at the −input, $\overline{i^2_{n-}}$	$\overline{i^2_{n-}} R_2^2$

Rules of thumb for the selection of a noise-optimized op-amp are as follows:

- If your source resistance R_s is relatively "small" (whatever that means), select an op-amp with a low-noise voltage, since the specification on input noise current is not so important. (This may steer you toward a bipolar-input op-amp.) A 50-Ω source resistance (typical of many radiofrequency and video circuits) is a "relatively small" source resistance.

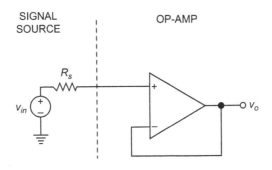

SIGNAL SOURCE   OP-AMP

FIGURE 16.18

A generic op-amp circuit driven by a signal source. The signal source has an output resistance R_s.

Table 16.2 Hypothetical Bipolar and CMOS Op-Amp Noise Specifications, with All Noise Sources Referred to the Input

Noise Source	Bipolar Op-Amp	CMOS Op-Amp
Input-referred voltage noise	$1\,\text{nV}/\sqrt{\text{Hz}}$	$20\,\text{nV}/\sqrt{\text{Hz}}$
Input-referred current noise	$1\,\text{pA}/\sqrt{\text{Hz}}$	$0.01\,\text{pA}/\sqrt{\text{Hz}}$

- If your source resistance is "large", you may want to look at an op-amp with a very low input current noise. (These design criteria may steer you toward JFET-input or CMOS op-amps.)

Example 16.4. Op-amp selection example

Let us assume that you have a bipolar and CMOS op-amp with the ballpark noise parameters given in Table 16.2. We see that the bipolar op-amp has a better voltage noise specification; the CMOS op-amp has a better current noise specification. We will also assume in this example that $1/f$ noise is negligible.

Let us select the better op-amp for a circuit with (a) a source resistance of $50\,\Omega$ and (b) a source resistance of $100\,\text{k}\Omega$.

For part (a), with a source resistance of $50\,\Omega$, we will use the bipolar op-amp since we are not so worried about the input noise current with the small source resistance. The op-amp intrinsic input voltage noise is $1\,\text{nV}/\sqrt{\text{Hz}}$, and there is an additional voltage noise due to the input current of $1\,\text{pA}/\sqrt{\text{Hz}} \times 50\,\Omega = 0.05\,\text{nV}/\sqrt{\text{Hz}}$. Including the Johnson noise from the resistor ($0.9\,\text{nV}/\sqrt{\text{Hz}}$), the total input-referred voltage noise using the bipolar op-amp is about $1.3\,\text{nV}/\sqrt{\text{Hz}}$, dominated by the intrinsic op-amp input voltage noise. (If we chose the CMOS option, we would have a total input-referred noise of about $20\,\text{nV}/\sqrt{\text{Hz}}$, again dominated by the input voltage noise of the op-amp.)

For part (b), with a 100-$k\Omega$ source resistance, we will use the CMOS op-amp. With $20\,\text{nV}/\sqrt{\text{Hz}}$ for the op-amp intrinsic voltage noise, and current noise $0.01\,\text{pA}/\sqrt{\text{Hz}} \times 100\,\text{k}\Omega = 1\,\text{nV}/\sqrt{\text{Hz}}$, the total input-referred noise (again ignoring the Johnson noise of the 100-$k\Omega$ resistor) is about $20\,\text{nV}/\sqrt{\text{Hz}}$. In this case, we have to include the Johnson noise of the resistor since it is relatively large ($40\,\text{nV}/\sqrt{\text{Hz}}$), bringing up the total input-referred noise to about $44.7\,\text{nV}/\sqrt{\text{Hz}}$.

If we chose the bipolar op-amp instead, the input-referred noise would be dominated by the current noise, with $1\,\text{pA}/\sqrt{\text{Hz}} \times 100\,\text{k}\Omega = 100\,\text{nV}/\sqrt{\text{Hz}}$. Including the Johnson noise of the 100-$k\Omega$ source resistor ($40\,\text{nV}/\sqrt{\text{Hz}}$), the total input voltage noise is about $108\,\text{nV}/\sqrt{\text{Hz}}$. A summary of the results of this thought-experiment is shown in Table 16.3.

Table **16.3** Total Input-Referred Voltage Noise for Example 16.4		
Source Resistance	**Bipolar Op-Amp**	**CMOS Op-Amp**
$R_s = 50\ \Omega$	$\sim 1.3\ \text{nV}/\sqrt{\text{Hz}}$	$\sim 20\ \text{nV}/\sqrt{\text{Hz}}$
$R_s = 100\ \text{k}\Omega$	$\sim 108\ \text{nV}/\sqrt{\text{Hz}}$	$\sim 44.7\ \text{nV}/\sqrt{\text{Hz}}$

Signal-to-noise ratio

The signal-to-noise ratio (sometimes abbreviated "SNR" or "S/N") is a measure of the relative amplitude of a signal power $\overline{v_s^2}$ to the noise power $\overline{v_n^2}$ that is on the signal. The definition is in terms of power, so the classic equation for SNR uses mean-square signal and noise, or

$$\text{SNR} = \frac{\overline{v_s^2}}{\overline{v_n^2}} \tag{16.18}$$

SNR is often expressed in decibels. In Figure 16.19, we see a 1-kHz signal with an SNR of 0 dB. The signal is just barely visible in the noise.

Noise figure

Noise Figure is the ratio of the SNR at the input of an amplifier to the SNR at the output. It is then a measure of how much your amplifier degrades the noise performance of your system. An ideal, noiseless amplifier has a noise figure of 1, or 0 dB.

Example 16.5: Noise in cascaded amplifiers

If we cascade gain stages (Figure 16.20), the noise from the first amplifier stage gets amplified by both the first and second gain stages.

At the output, the total noise is

$$\overline{v_{o,noise}^2} = A_1^2 A_2^2 \overline{v_{n1}^2} + A_2^2 \overline{v_{n2}^2} \tag{16.19}$$

which is generally dominated by the noise of the first gain stage. As a result, noise in cascaded amplifiers is sometimes analyzed by considering only the noise of the first stage.

Example 16.6: High gain amplifier using TL084

Let us figure out the total output noise for an amplifier (gain of +101) built with a TL084 op-amp (Figure 16.21). First, let us find the total equivalent input voltage noise for the op-amp, using the marked-up chart in Figure 16.22. The gain-bandwidth product of the TL084 is about 3 MHz, so we expect a closed-loop

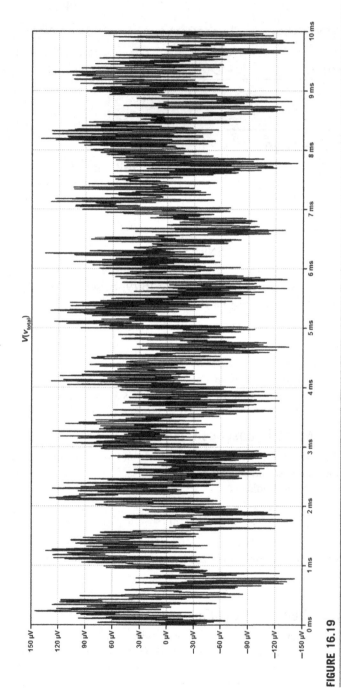

FIGURE 16.19

Signal with an signal-to-noise ratio of 1 (or 0 dB). The peak-to-peak noise voltage is 200 μV, corresponding to an RMS noise voltage amplitude of about 33 μV. The 1-kHz signal amplitude is 94.3 μV pp (with an RMS voltage of 33 μV).

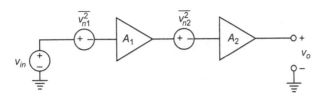

FIGURE 16.20

Cascaded noisy amplifiers for Example 16.5.

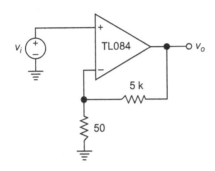

FIGURE 16.21

Gain of +101 amplifier built with a TL084 JFET-input op-amp (Example 16.6).

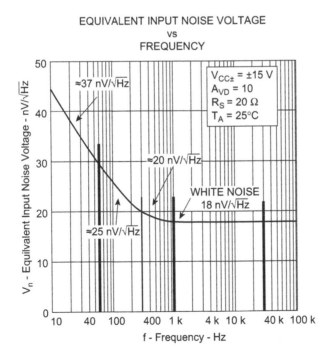

FIGURE 16.22

Equivalent input noise voltage for the TL084, with markups by the author (Example 16.6).

bandwidth of about 30 kHz, and this is out "bandwidth of interest", modified by a factor of 1.57 to account for the notion of the equivalent brick wall filter. Let us assume that the low-frequency band we are interested in starts at 10 Hz. (This design could be a first-cut for an audio amplifier, but (*spoiler alert*) we will find out that the noise may be too high for an audio application.)

Next, we will use the noise voltage chart (with the author's markups) to figure out the total voltage noise referred to the input. From 1 to 30 kHz, the calculation is easy because the white noise voltage spectrum is constant at 18 nV/$\sqrt{\text{Hz}}$. In the range of 10 Hz–1 kHz, we have broken up the 1/f noise spectrum into three individual "chunks", approximating an integral of the 1/f noise. (Note: if you look at the spectrum from 10 to 1 kHz, the noise voltage does not exactly follow a 1/f profile; rather it is 1/f^n where we could figure out n if we wanted to.) The calculations of the four chunks of the noise voltage integral are shown in Table 16.4. The total op-amp input noise voltage is 5017 nV, or about 5 mV.

Next, let us figure out if we can ignore the Johnson noise from the resistors and the input noise current from the op-amp. From the datasheet, we read that the equivalent input noise current for this amplifier is a fairly low 0.01 pA/$\sqrt{\text{Hz}}$, for frequencies >1 kHz. Multiplied by the 5-kΩ feedback resistor, we get a noise (at the output) of about 0.05 nV/$\sqrt{\text{Hz}}$, negligible compared to the op-amp voltage noise. The thermal noise from the 5-kΩ resistor is about 9 nV/$\sqrt{\text{Hz}}$, again negligible compared to the op-amp voltage noise. The noise from the 50-Ω resistor is negligible, even though it is multiplied by the op-amp gain from the inverting input (−100).

To find the total output noise, we simply multiply the voltage noise referred at the input (∼5 mV) by the closed-loop gain of the circuit (+101), for a total output noise of 505 mV. (This would be a pretty noisy circuit.) If we wanted to reduce the total output noise, we could do the following:

- Limit the bandwidth on the low end to reduce 1/f noise.
- Limit the bandwidth on the high end to reduce white noise.
- Use an op-amp with a lower input-referred white noise. (Maybe a bipolar op-amp, since the resistors surrounding the op-amp are relatively small. But if we do this, we would need to see if we can still ignore the op-amp input noise current.)

Table 16.4 Total Input-Referred Noise Voltage for Example 16.6

Frequency Range	Noise Calculation
1060 Hz	37 nV/$\sqrt{\text{Hz}}$ × $\sqrt{50\ \text{Hz}}$ = 261 nV
60–300 Hz	25 nV/$\sqrt{\text{Hz}}$ × $\sqrt{240\ \text{Hz}}$ = 387 nV
300 Hz–1 kHz	20 nV/$\sqrt{\text{Hz}}$ × $\sqrt{700\ \text{Hz}}$ = 529 nV
1–30 kHz	18 nV/$\sqrt{\text{Hz}}$ × $\sqrt{29,000 \times 1.57\ \text{Hz}}$ = 3840 nV
Total integral	5017 nV

If we kept the TL084, and increased the low-frequency breakpoint to 60 Hz (by high-pass filtering, and getting rid of about 261 nV of the $1/f$ noise) and reduced the high-frequency bandwidth from 30 to 20 kHz (by low-pass filtering, getting rid of about 731 nV of the input-referred noise) we can reduce the total output noise from 500 to about 400 mV. If we instead used a low-noise op-amp (with white noise of 2 nV/$\sqrt{\text{Hz}}$ and a better $1/f$ noise), we can reduce the total output RMS noise by maybe a factor of 10, to around 40 mV.

Things that are not noise

Many phenomena that really are not noise are often called "noise".

- *Sixty hertz hum.* If your circuit picks up 60-Hz magnetic fields or exhibits a ripple at 60 Hz due to a ground loop, that is not noise, it is a "hum". (It is called a "hum" because of the familiar 60-Hz noise that some audio power amplifiers make.)
- *Ringing.* If you have an underdamped signal that rings a lot, that is not noise, it is "ringing". See Figure 16.23, a ringing output of a capacitively loaded op-amp. Unterminated transmission lines (or transmission lines with unmatched terminations) can ring as well. See Figure 16.24, where we see the signal at the end of 24 ft of a 50-Ω coax cable, with no termination at the end.
- *Parasitic oscillation.* If you have an emitter follower that unintentionally oscillates at a high frequency, that is not noise, it is an "oscillation" (Figure 16.25).

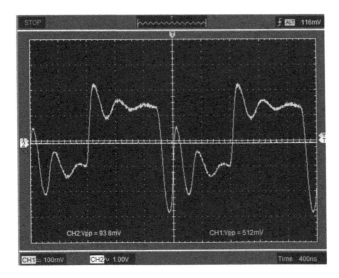

FIGURE 16.23

Ringing of a capacitively loaded op-amp circuit (from Chapter 11). (For color version of this figure, the reader is referred to the online version of this book.)

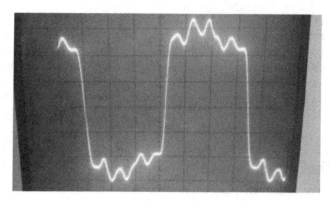

FIGURE 16.24

Ringing of an unterminated transmission line (from Chapter 17). (For color version of this figure, the reader is referred to the online version of this book.)

- *Electric field (capacitive) pickup.* If your circuit has high-frequency, high-voltage signals coupling into it, that is not noise, it is a capacitive pickup (Figure 16.26).
- *Magnetic field (magnetic) pickup.* If you have a wire loop near a time-varying magnetic field, and you pick up a voltage, that is not noise, it is magnetic field pickup (see Faraday's Law of Induction, and Figure 16.27).

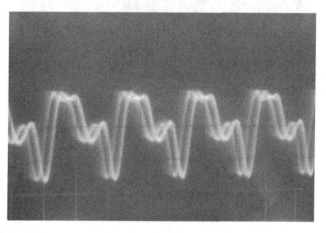

FIGURE 16.25

A 2N3904 emitter follower with a 100-MHz oscillation (from Chapter 7). (For color version of this figure, the reader is referred to the online version of this book.)

(a)

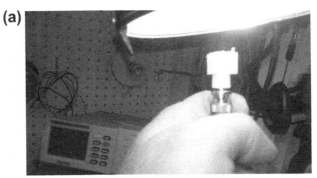

(b)

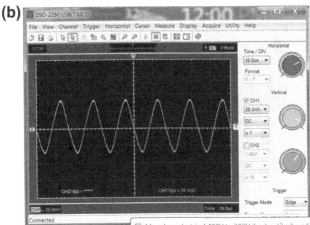

FIGURE 16.26

Electric field pickup demo. (a) An electric field probe near a compact-fluorescent light bulb. (b) Picked-up voltage at 80 kHz. (For color version of this figure, the reader is referred to the online version of this book.)

Chapter 16 problems

Problem 16.1

Calculate the thermal noise from a 1-MΩ resistor with a brick wall filter bandwidth of 1 MHz.

Problem 16.2

A microphone with an output resistance of 10 kΩ is connected to the input of an audio amplifier with a gain of 1000 and an input-referred white noise voltage density

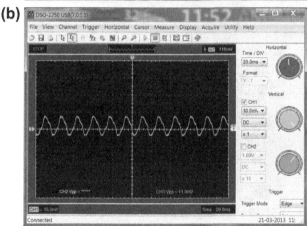

FIGURE 16.27

Magnetic field pickup demo. (a) Big drive coil driven with 60-Hz current, and a little pickup coil inside the drive coil. (b) Generated voltage at 60 Hz due to Faraday's Law of Induction. (For color version of this figure, the reader is referred to the online version of this book.)

of $10\,\text{nV}/\sqrt{\text{Hz}}$. Assume that the amplifier noise current is negligible, and that the -3-dB point of the amplifier is 20 kHz. Estimate the RMS value of the noise at the output of the amplifier. What is the SNR for an input signal of 1 mV (RMS)?

Further reading

[1] Adler RB Semiconductor noise, Lincoln Laboratory Technical Report, 1955.

[2] Analog devices, Inc. Op amp noise, Tutorial MT-047, 2008.

[3] Brisebois G. Op amp selection guide for optimum noise performance, Linear Technology Design Note 355.

[4] Erdi G. Amplifier techniques for combining low noise, precision, and high-speed performance. *IEEE J Solid-State Circuits* 1981;**SC-16**(6):653–61.

[5] Erdi G. Operational amplifier selection guide for optimum noise performance, Linear Technology DN-6, January 1988.

[6] Johnson JB. Thermal agitation of electricity in conductors. *Phys Rev* 1928;**32**.

[7] Leach Jr WM. Fundamentals of low-noise analog circuit design. *Proc IEEE* 1994;**82**(10):1515–38.

[8] Lee TH. *The design of CMOS radio-frequency integrated circuits*. 2nd ed. Cambridge University Press; 2003. Chapter 11 covers noise sources, and noise in MOSFETs in some detail.

[9] National Semiconductor. Noise specs confusing? Application Note AN-104.

[10] Pease RA. *Troubleshooting analog circuits*. Butterworth-Heinemann; 1991.

[11] Rich A. Noise calculations in op amp circuits, Linear Technology Design Note 15, September 1988.

[12] Siliconix, Inc.. *Designing with field-effect transistors*. McGraw-Hill; 1981.

[13] Vergers C. *Handbook of electrical noise measurement and technology*. 2nd ed. TAB Books; 1987.

Other Useful Design Techniques and Loose Ends

17

"… for Nature cannot be fooled."
—**Richard Feynman, "Personal Observations on the Reliability of the Shuttle"**

IN THIS CHAPTER

▶ Here at the end is a potpourri of (hopefully) useful design techniques.

Thermal circuits

All electrical engineers know Ohm's law, $V = IR$: that a current "flows" through a resistor, and that the current is forced by an electrical "pressure", or voltage. The constant of proportionality between the voltage and the current is the circuit resistance R, in ohms. Given that we all accept Ohm's law, analogies may be made with other physical systems in an attempt to generate simple models for complicated physical processes.

With a little thought, one can compare the flow of current in a resistor to the flow of water through a pipe, where the current is analogous to the water flow rate, and the voltage is analogous to the pressure differential across the pipe that forces the water to flow. The resistance to flow is dependent on the diameter of the pipe, its length, the viscosity of the fluid, and other parameters.

With a little more thought, one can model heat flow with simple circuit analogies. A warm body may be heated or cooled by three mechanisms: conduction, convection, and radiation. Conduction represents the transfer of energy when heat is transferred by a solid; for instance, heat (in watts) is transferred down a water pipe when heat is transferred down the pipe body from a warm indoors to a cold outdoors. Convection is heat transfer due to a moving fluid or air (e.g. blowing on a hot bowl of soup to cool it off). Radiation is an electromagnetic effect, and is nonlinear with temperature,[1] and therefore is not simply modeled by lumped models.

---

[1]Radiation from a warm body is proportional to T^4, where T is the temperature in Kelvins. In many instances, for example, at room temperature, radiation can be neglected compared to conductive and convective heat transfer. But, for forced-air convection (e.g. rapid air flow over the fins of a heat sink), the *convective* heat transfer may be the dominant effect.

Intuitive Analog Circuit Design. http://dx.doi.org/10.1016/B978-0-12-405866-8.00017-6

Using relatively simple models of static and time-varying heat transfer, one can model temperature effects. These techniques are useful for heat-sink design, PC thermal design, and for gaining a basic understanding of heat transfer processes.

Steady-state model of conductive heat transfer

In many cases of interest, where there is no air flow and at near room temperatures, conduction will be the dominant heat transfer effect. For instance, inside an IC package, there is no air flow, yet heat is transferred from the electrical junctions where the heat is being generated, out to the external world (e.g. to a heat sink or the PC board). To relate this to a simple circuit model, consider a simple resistor circuit, and its thermal circuit analogy (Figure 17.1).

If we examine the electrical case, current (I, in amperes) flows, and is forced by a voltage differential $V_2 - V_1$; the current flows from a higher voltage to the lower voltage, and is given by

$$I = \frac{V_2 - V_1}{R_{elec}} \tag{17.1}$$

We know that if voltage V_2 is greater than V_1, then the current will flow from V_2 to V_1. The proportionality constant between current and voltage is the electrical resistance in ohms (R_{elec}). The electrical resistance is given by

$$R_{elec} = \frac{l}{\sigma A} \tag{17.2}$$

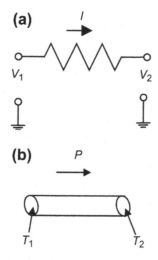

FIGURE 17.1

Analogy between current flow in a resistor (a) and one-dimensional heat flow through a long, skinny solid (b).

where l is the length of the resistor, A is the cross-sectional area through which the current flows, and σ is the electrical conductivity of the material that makes up the resistor, in $\Omega^{-1}m^{-1}$. Current flows more easily if the electrical resistor is shorter and thicker.

Next, let us consider the thermal case. Consider a long, skinny rod of a material, which is made of a good heat conductor such as a metal. It is in good thermal contact with two heat sinks, one at either end of the rod. We assume that there is a temperature difference between the two ends of the rod. For this simple case, it may be approximated that heat only flows in one direction. Heat (P, in watts) flows, and is forced by a temperature differential $T_2 - T_1$; a constant power flows from the higher temperature to the lower temperature, and is given by

$$P = \frac{T_2 - T_1}{R_{TH}} \tag{17.3}$$

The constant of proportionality between heat flow and temperature is the thermal resistance R_{TH}, which has units of degrees per watt. This is the proportionality constant that tells you how many degrees you need to put across an element to cause 1 W of heat to flow. Most metals, such as copper and aluminum, have a high thermal conductivity (i.e. they transfer heat very well).

Analogous to electrical resistance, thermal resistance is given by

$$R_{TH} = \frac{l}{kA} \tag{17.4}$$

where l is the length of the material through which heat flows, A is the cross-sectional area through which the heat flows, and k is the thermal conductivity of the material that makes up the resistor, in watts per meter per degree. As in the electrical case, thermal resistance is lower if the conducting path is shorter and has more area.

Still air, which is a good insulator, has a thermal conductivity $k \approx 0.03$ W/m °C. Copper, which is a good conductor of heat, has $k \approx 400$ W/m °C. These numbers illustrate why styrofoam, which is full of trapped air, is a good insulator, whereas copper and aluminum are not. This is one reason why metals are not often used as thermal insulators!

Thermal energy storage

The effects of energy storage in thermal systems are also demonstrated by analogy. Just as energy is stored in a capacitor as charge on the capacitor plates, energy is stored in the mechanical structure of a mass, in the vibration of the atoms. A body at a higher temperature has a higher amount of stored energy than the same body at a lower temperature.

In the electrical case, when a linear capacitor is charged from an initial voltage (V_i) to a final voltage (V_f), charge (q, in coulombs) is stored on the plates of the capacitor, as

$$q = C(V_f - V_i) \tag{17.5}$$

where C is the electrical capacitance, in coulombs per volt. During charging, a current flows into the capacitor, as

$$i = \frac{dq}{dt} = C\frac{dv(t)}{dt} \tag{17.6}$$

The charge that flows during the charging interval is stored on the plates of the capacitor.

In the thermal case, the thermal capacitance of a mass determines how well the mass stores energy; the unit of heat capacity is joules per degree. The thermal capacitance is given by

$$C_{TH} = Mc_v = V\rho c_v \tag{17.7}$$

where ρ is the density of the material (kilograms per cubic meter), M is the total mass of the material that is being heated or cooled (kilograms), V is the volume of the material (cubic meters) that is being heated or cooled, and c_v is the specific heat, which is a material property (joules per kilogram per degree).

When a mass is heated from an initial temperature (T_i) to a final temperature (T_f), energy (E, in joules, or watt seconds) must be added to the mass to heat it up, as

$$E = C_{TH}(T_f - T_i) \tag{17.8}$$

where C_{TH} is the heat capacity, in joules per degree. This is analogous to the storage of the charge in an electrical capacitor. During the heating process, a power (watts) flows into the mass, as

$$P(t) = \frac{dE}{dt} = C_{TH}\frac{dT}{dt} \tag{17.9}$$

where dT is the temperature change in the time interval dt. The energy that flows during the heating interval is stored in the mass of the body being heated.

To use the simple transient model, consider the arrangement shown in Figure 17.2. We see a block of material that is being heated by a source of energy (perhaps a blow torch) delivers P_o watts of power to the mass being heated.

The big "wall" that the block is attached to is a heat sink at an ambient temperature T_A. It is assumed that the heat sink is large and sufficiently cooled so that its temperature does not change during the duration of the test. What is the temperature profile of the block of material after the heat source is applied?

Using the circuit analogies, the heater is a source of power $P(t)$; this is modeled as a step in applied heat (analogous to a circuit current) with amplitude P_o (watts). Circuit "ground" is the ambient temperature T_A. Heat is conducted away from the block into the heat sink through a thermal resistance R_{TH}. Simultaneously, energy is stored in the thermal capacitance of the block C_{TH}.

The heat source is modeled as a "step" of power (Figure 17.3). When the heat is first turned on (at time $t = 0$), the block is at an ambient temperature, or at the same temperature as of the heat sink, T_A. The temperature of the block $T_B(t)$ increases as

$$T_B(t) = T_o R_{TH}\left(1 - e^{-\frac{t}{\tau_{TH}}}\right) \tag{17.10}$$

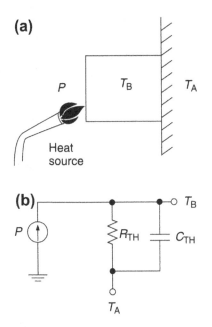

FIGURE 17.2

Use of a simple transient model to find a crude estimate of block temperature vs. time. (a) Physical arrangement showing a blow torch delivering power $P(t)$ to a block that is heated up. The temperature at the center of the block is T_B and the block is mounted to a heat sink that operates at a temperature T_A. (b) A circuit model showing thermal resistance R_{TH} and thermal capacitance C_{TH}.

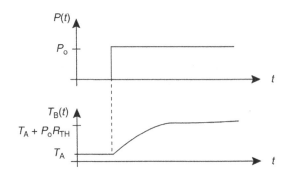

FIGURE 17.3

Use of a simple transient model. Input "step" of power $P(t)$ and temperature of block $T_B(t)$.

The temperature of the block reaches the final temperature with the RC thermal time constant (τ_{TH}), just as in the step response of a current source driven RC electrical circuit. The circuit will continue "charging" (i.e. the block's temperature will increase) until the input power is exactly balanced by the heat conducted away through R_{TH} to the heat sink. After a few time constants, the block approaches the final temperature:

$$T_F = P_o R_{TH} \tag{17.11}$$

Now, to be sure, some of the details of heat transfer are glossed over using this simple circuit; for instance, there is some heat transferred between the block and the surrounding air. However, if the heat sink is well designed, there will be much more efficient transfer of heat to the heat sink than to the air (after all, that is what a heat sink is meant for). Second, the lumped circuit model is valid only for "low frequencies", or for temperatures that change on a time scale much longer than the thermal time constant. This means that the temperature throughout the body changes the same everywhere. The specific heat and thermal conductivity of materials vary with temperature, but over a limited temperature range, they can be approximated as being constant.

For engineering calculations, and if you check your approximations, these calculations give you very useful results. You can always perform a finite element analysis to get more exact results.

To summarize, using relatively simple models, static and dynamic heat transfer can be modeled with simple circuit analogies. Important questions may be answered, such as the following:

- When you apply a source of heat to a material, what is the total temperature rise in the material?
- What do those specifications in manufacturers' data sheets mean for junction-to-pin and junction-to-ambient thermal resistance?
- On what time scale does the temperature change when a material is heated and how fast does the temperature reach its final value?

The electrical-to-thermal analogies are summarized in Table 17.1.

Using thermal circuit analogies to determine the static semiconductor junction temperature

The thermal model of a semiconductor mounted to a heat sink is shown in Figure 17.4. The transistor dissipates power, and this is indicated by the current source with a value P. The ambient temperature is T_A, and the heat-sink surface temperature and case temperature of the semiconductor are T_s and T_c, respectively. The transistor junction temperature is T_j, and our goal in designing a heat-sink system is to guarantee that the junction temperature does not exceed a safe level.

Table 17.1 Summary of Electrical and Thermal Analogies

	Electrical System	Thermal System
Stored quantity	Charge q	Energy E
"Flow" quantity	Current I	Power P
Pressure which forces flow	Voltage V	Temperature T
Resistance to flow	$R = \frac{1}{\sigma A}$	$R_{TH} = \frac{l}{kA}$
Capacitance	$C = \varepsilon \frac{A}{d}$	$C_{TH} = \rho c_v V$
Time constant	$\tau_{elec} = RC$	$\tau_{TH} = R_{TH} C_{TH}$

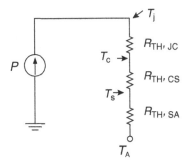

FIGURE 17.4

Thermal model of a semiconductor mounted to a heat sink.

The thermal resistances for heat conduction through the transistor to the heat-sink surface are $R_{TH,JC}$ and $R_{TH,CS}$. $R_{TH,JC}$ is the thermal resistance from the junction of the transistor (where the power is dissipated) to the case of the transistor. The thermal resistance $R_{TH,CS}$ is the thermal resistance from the case of the transistor to the heat sink. This value of thermal resistance depends on the contact area, how well you torque the transistor down to the heat sink, and on the type of thermal interface[2] material that you use.

The thermal resistance from the heat sink to ambient air ($R_{TH,SA}$) is a function of the heat-sink area and whether you cool the heat sink with forced air or not. Heat-sink manufacturers will specify this number for a given heat-sink area and air flow.

Mechanical circuit analogies

In many instances, solutions for natural frequency and mode shapes of electrical circuits can be found by considering the behavior of analogous mechanical systems.

---

[2]For example, do you use thermal grease, or a mica pad, etc.

Operation of the mechanical system, consisting of masses (m), dampers (c), and springs (k) may be easy to visualize by using simple physical reasoning. The solution can then be mapped to an analogous system consisting of inductors (L), resistors (R), and capacitors (C).

In the vibrating mechanical system, energy is transferred back and forth between kinetic energy in the mass and potential energy stored in the springs. Equivalently, in the electromagnetic system, energy is sloshed back and forth between magnetic energy stored in inductors and electrical energy stored in the capacitors. A simple example explains this duality.

Mechanical system

The mechanical system for this example consists of two masses on frictionless rollers, connected by a massless spring (Figure 17.5). The force exerted by the spring is kx, where k is the mechanical spring constant in newtons per meter. The two state variables for this system are the horizontal positions of the two masses, x_1 and x_2, defined with respect to a fixed position on the Earth.

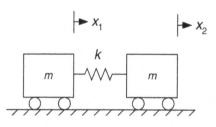

FIGURE 17.5

Mechanical two degree-of-freedom circuit.

Considering the first mass on the left, Newton's law gives[3]

$$m\ddot{x}_1 = k(x_2 - x_1)$$

Electrical system

The electrical dual for this system consists of an L-C-L circuit (Figure 17.8). This is the equivalent circuit of a section of a lossless transmission line. The two state variables for this system are the loop currents i_1 and i_2, defined as shown.

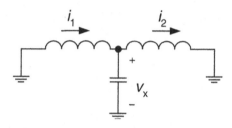

FIGURE 17.8

An electrical two degree-of-freedom circuit.

Considering the first inductor on the left, we get

$$L\dot{i}_1 = -v_x$$

Taking the derivative results in

$$L\ddot{i}_1 = -\frac{dv_x}{dt} = \frac{-(i_1 - i_2)}{C}$$

---

[3]Remember the notation: $\ddot{x}$ is the second derivative with respect to time of the variable x. $\dot{x}$ is the first derivative.

Applying the same reasoning to the right-hand mass results in the coupled equations of motion for the two masses:

$$m\ddot{x}_1 + k(x_1 - x_2) = 0$$
$$m\ddot{x}_2 + k(x_2 - x_1) = 0$$

Assuming sinusoidal variations in position, with $x = X_o e^{j\omega t}$, we get

$$-\omega^2 \begin{bmatrix} m & 0 \\ 0 & m \end{bmatrix} \begin{Bmatrix} x_1 \\ x_2 \end{Bmatrix} + \begin{bmatrix} k & -k \\ -k & k \end{bmatrix}$$
$$\times \begin{Bmatrix} x_1 \\ x_2 \end{Bmatrix} = 0$$

or

$$\begin{bmatrix} -\omega^2 m + k & -k \\ -k & -\omega^2 m + k \end{bmatrix} \begin{Bmatrix} x_1 \\ x_2 \end{Bmatrix} = 0$$

The natural frequencies are found by solving the determinant of the above matrix. Solutions for this set of simultaneous equation result in two natural frequencies (ω) and mode shapes for the position of the masses (the x_1, x_2 vector). Solving for allowable natural frequencies results in

Natural Frequency	Mode Shape
$\omega_a = 0$	$\{x\} = \begin{Bmatrix} 1 \\ 1 \end{Bmatrix}$
$\omega_b = \sqrt{\frac{2k}{m}}$	$\{x\} = \begin{Bmatrix} 1 \\ -1 \end{Bmatrix}$

Now, the above set of natural frequencies and mode shapes can be easily found by inspection. For instance, the first mode shape corresponds to the case where both masses move to the right (or left, depending on the initial conditions) in unison. In this case, there

Applying the same reasoning to the right-hand inductor results in the following coupled equations:

$$L\ddot{i}_1 + \frac{1}{C}(i_1 - i_2) = 0$$

$$L\ddot{i}_2 + \frac{1}{C}(i_2 - i_1) = 0$$

Assuming sinusoidal variations in current, with $i = I_o e^{j\omega t}$, we get

$$-\omega^2 \begin{bmatrix} L & 0 \\ 0 & L \end{bmatrix} \begin{Bmatrix} i_i \\ i_2 \end{Bmatrix} + \begin{bmatrix} \frac{1}{C} & -\frac{1}{C} \\ -\frac{1}{C} & \frac{1}{C} \end{bmatrix}$$
$$\times \begin{Bmatrix} i_1 \\ i_2 \end{Bmatrix} = 0$$

or

$$\begin{bmatrix} -\omega^2 L + \frac{1}{C} & -\frac{1}{C} \\ -\frac{1}{C} & -\omega^2 L + \frac{1}{C} \end{bmatrix} \begin{Bmatrix} i_1 \\ i_2 \end{Bmatrix} = 0$$

Solutions for this set of simultaneous equation result in two natural frequencies (ω) and mode shapes for inductor current (the i_1, i_2 vector). Solving for allowable natural frequencies results in

Natural Frequency	Mode Shape
$\omega_a = 0$	$\{i\} = \begin{Bmatrix} 1 \\ 1 \end{Bmatrix}$
$\omega_b = \sqrt{\frac{2}{LC}}$	$\{i\} = \begin{Bmatrix} 1 \\ -1 \end{Bmatrix}$

Now, the above set of natural frequencies and mode shapes can be easily found by inspection. For instance, the first mode shape corresponds to the case where both loop currents flow clockwise (or counterclockwise, depending on the

is no stretching of the spring, and no vibration. The spring could be replaced by a massless rigid rod (Figure 17.6) without disturbing the movement of the masses, which are both in simple translation. Therefore, the natural frequency $= 0$.

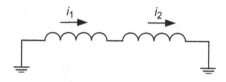

FIGURE 17.9

First (lower frequency mode).

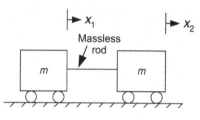

FIGURE 17.6

First (lower frequency mode).

For the second mode, both masses vibrate, but 180° out of phase. Therefore, the middle portion of the spring does not move (i.e. this position is a node). By symmetry, we could put a brick wall there and break (Figure 17.7) up the single spring into two springs of a spring constant $2k$, and solve for the vibration of each half-circuit independently.

initial conditions). In this case, there is no current flow in the capacitor, and no vibration. In fact, for this mode of operation, the capacitor can be removed from the circuit (Figure 17.9) since there is no current flow in it. Therefore, the natural frequency $= 0$.

For the second mode, there is oscillation in i_1 and i_2, but 180° out of phase. We can model the two half-circuits by breaking up the capacitor into two capacitors (Figure 17.10), each of a value $C/2$.

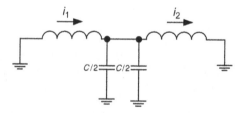

FIGURE 17.10

Second (higher frequency mode).

By symmetry, there is no current flow across the boundary. Therefore, the two half-circuits each vibrate independently.

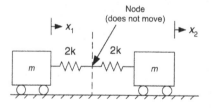

FIGURE 17.7

Second (higher frequency mode).

These same techniques can be used for systems with different boundary conditions, and for systems with more degrees of freedom.

Example 17.1: Using mechanical circuit analogies

As an example, consider the circuit in Figure 17.11(a). The expected natural frequency for the first mode of oscillation is zero, corresponding to the case when inductor currents i_1 and i_2 are of equal value and are in the same direction. For the first mode, the capacitor current is forever zero, and hence, the capacitor voltage remains constant. Therefore, there is no sinusoidal oscillation for the first mode.

In the second mode of oscillation i_1 and i_2 are in opposite directions and of equal value, and we refer to the circuits of Figure 17.11(b) and (c). The second mode of oscillation has an expected natural frequency of

$$\omega_{o,mode2} = \sqrt{\frac{2}{LC}} = 1.41 \text{ rad/s} \tag{17.12}$$

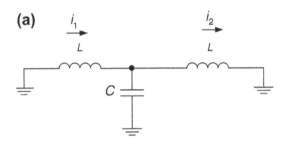

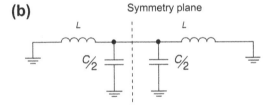

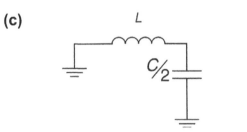

FIGURE 17.11

Circuit for example 17.1. (a) Original circuit. (b) Original circuit, redrawn showing the symmetry plane. For the second mode of oscillation, there is no current across the symmetry plane. (c) Simplified circuit for the second mode of oscillation, exploiting symmetry.

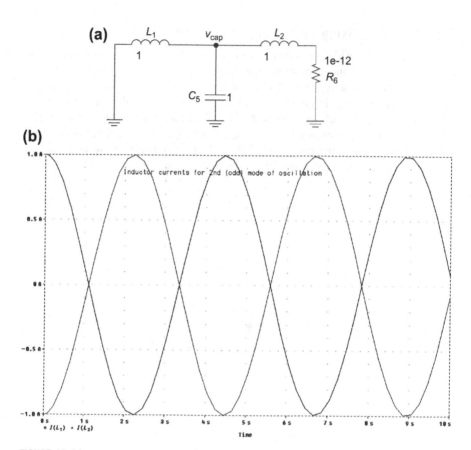

FIGURE 17.12

Second mode of oscillation. (a) PSPICE circuit. (b) PSPICE result for the second mode, showing inductor currents of an equal magnitude and 180° out of phase.

equivalent to a natural frequency of 0.225 Hz. The PSPICE output (Figure 17.12[4]) shows the inductor currents with initial conditions set $i_1 = 1$ and $i_2 = -1$. As expected, the circuit oscillates at 0.225 Hz.

For the second mode of operation, there are two independent circuits, each with an inductance L and a capacitance $C/2$. Therefore, the characteristic impedance of each circuit is

$$Z_0 = \sqrt{\frac{L}{C/2}} = \sqrt{\frac{2L}{C}} = 1.41 \ \Omega \tag{17.13}$$

---

[4]The small resistor is added because SPICE has convergence problems if there are voltage loops containing inductors.

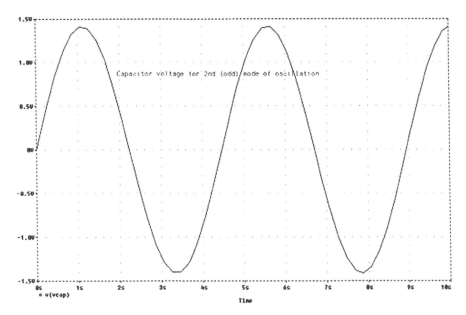

FIGURE 17.13

Capacitor voltage for the second mode of oscillation.

Therefore, we expect the peak-to-peak amplitude of oscillation of the capacitor voltage to be $2\sqrt{2}$, as shown in Figure 17.13.

The translinear principle

Translinear[5] circuits exploit the well-known exponential relationship between transistor base–emitter voltage and collector current, a relationship that holds over many orders of magnitude of the collector current. Consider the translinear circuit of Figure 17.14. We have a loop of V_{BE}s as indicated by the arrows. If we apply Kirchhoff's voltage law around the V_{BE} loop we get

$$-V_{BE1} - V_{BE2} + V_{BE3} + V_{BE4} = 0 \tag{17.14}$$

Let us recall the logarithmic expression for V_{BE} of a transistor operated in the forward-active region:

$$V_{BE} \approx \frac{kT}{q} \ln\left(\frac{I_C}{I_S}\right) \tag{17.15}$$

[5]The term "translinear" was coined by Barrie Gilbert in or around 1968.

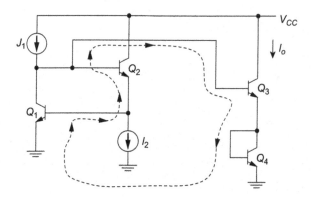

FIGURE 17.14

A translinear circuit, with the dotted line showing the "Gilbert loop". The output current I_o is equal to the square root of the product of the two inputs I_1 and I_2.

where I_S is the reverse saturation current of the transistor. In the translinear circuit above, if all transistors are identical, we can write

$$-\frac{kT}{q}\ln\left(\frac{I_{C1}}{I_S}\right) - \frac{kT}{q}\ln\left(\frac{I_{C2}}{I_S}\right) + \frac{kT}{q}\ln\left(\frac{I_{C3}}{I_S}\right) + \frac{kT}{q}\ln\left(\frac{I_{C4}}{I_S}\right) = 0 \tag{17.16}$$

$$\Rightarrow \ln(I_{C1}I_{C2}) = \ln(I_o^2)$$

The solution for this is

$$I_o = \sqrt{I_{C1}I_{C2}} \tag{17.17}$$

For a circuit of this type with a loop of V_{BES} and identical transistors, we can state the translinear principle:

The product of the clockwise currents equals the product of the counterclockwise currents, or

$$\prod \frac{I_C}{I_S}\bigg|_{cw} = \prod \frac{I_C}{I_S}\bigg|_{ccw} \tag{17.18}$$

Let us apply the translinear principle to another translinear circuit (Figure 17.15). This circuit has a loop of V_{BES} including Q_1, Q_2, Q_3, and Q_4. Going around the loop, we see that I_{C1} and I_{C2} are counterclockwise currents, and I_{C3} and I_{C4} are clockwise currents. Therefore, for this circuit,

$$I_{C1}I_{C2} = I_{C3}I_{C4} \tag{17.19}$$

This means that we can express the output current I_o as

$$I_o = \sqrt{\frac{I_{IN}^2}{I_{BIAS}}} \tag{17.20}$$

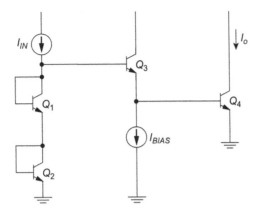

FIGURE 17.15

Another translinear circuit that performs a squaring function.

Input impedance of an infinitely long resistive ladder

Let us find the input impedance of an infinitely long ladder with series element of impedance Z and shunt element with admittance Y (Figure 17.16). Since this ladder is infinitely long, we can say that the impedance at position 1 is the input impedance at position 2. Mathematically, we can express this as

$$Z_{in} = Z + \frac{1}{Y} \| Z_{in} = Z + \frac{Z_{in}}{1 + YZ_{in}} \qquad (17.21)$$

We can rearrange this to get a quadratic equation for Z_{in}:

$$Z_{in}^2 - ZZ_{in} - Z/Y = 0 \qquad (17.22)$$

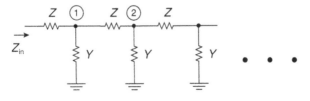

FIGURE 17.16

An infinitely long resistive ladder.

Solving for the input impedance shows[6]

$$Z_{in} = \frac{Z}{2}\left(1 + \sqrt{1 + \frac{4}{ZY}}\right)$$

(17.23)

Transmission lines 101

The interesting result from the infinitely long resistive ladder can be used to find the input impedance of the infinitely long transmission line (Figure 17.17). The elements L and C are inductance and capacitance per unit length of the line, respectively. We recognize that using the formulation above, we get

$$Z = j\omega L$$
$$Y = j\omega C$$

(17.24)

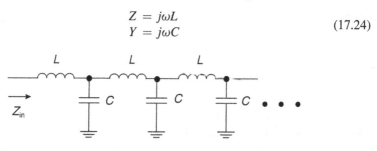

FIGURE 17.17

Ideal transmission line.

This means that the solution for the input impedance Z_{in} of the ideal transmission line is

$$Z_{in} = \frac{j\omega L}{2}\left(1 + \sqrt{1 + \frac{4}{(j\omega L)(j\omega C)}}\right)$$

(17.25)

For an ideal transmission line, there are lots of L and C lumps, so each lump is very small. We can express this as L and $C \to 0$. This means that we can find the input impedance in this limit as

$$Z_{in} = \frac{j\omega L}{2}\left(\sqrt{\frac{4}{(j\omega L)(j\omega C)}}\right) = \sqrt{\frac{L}{C}} \equiv Z_0$$

(17.26)

This characteristic impedance Z_0 gives us the ratio of the voltage to the current along the line. Also, if you check the units, you will find that the delay of the transmission line (in units of seconds per unit length) is

$$t_{do} = \sqrt{LC}$$

(17.27)

---

[6]We have thrown away one of the solutions to this quadratic that has no physical meaning.

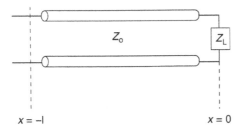

FIGURE 17.18

Terminated transmission line. The line of length l has a characteristic impedance Z_o and is terminated with an impedance Z_L.

Finding the input impedance of a finite-length transmission line

A transmission line can be modeled as a lumped circuit (i.e. with inductors and capacitors) if the length of the line is much smaller than a wavelength. Let us consider a transmission line of a characteristic impedance Z_o, length l, and terminated by an impedance Z_L at $x = 0$, as shown in Figure 17.18. The basic transmission line equations are

$$V(x, \omega) = V_+ e^{\frac{-j\omega x}{c}} = V_- e^{\frac{-j\omega x}{c}}$$

$$I(x, \omega) = \frac{V_+}{Z_o} e^{\frac{-j\omega x}{c}} - \frac{V_-}{Z_o} e^{\frac{-j\omega x}{c}}$$

(17.28)

where V_+ is the forward-traveling wave on the line, V_- is the reflected wave, and c is the speed of wave propagation down the line.

The reflection coefficient is the ratio of the reflected wave amplitude to the forward wave amplitude, or

$$\Gamma = \frac{V_-}{V_+}$$

(17.29)

We can use this in the basic transmission line equation to find

$$V(x, \omega) = V_+ \left(e^{\frac{-j\omega x}{c}} + \Gamma e^{\frac{j\omega x}{c}} \right)$$

$$I(x, \omega) = \frac{V_+}{Z_o} \left(e^{\frac{-j\omega x}{c}} - \Gamma e^{\frac{j\omega x}{c}} \right)$$

(17.30)

At $x = 0$, we have the boundary condition that $V/I = Z_L$, or

$$Z_o \left(\frac{1 + \Gamma}{1 - \Gamma} \right) = Z_L$$

(17.31)

Using this boundary condition enables us to solve for the reflection coefficient as

$$\Gamma = \frac{Z_L - Z_0}{Z_L + Z_0} \tag{17.32}$$

This result makes sense, because we know that on a matched transmission line (with $Z_L = Z_0$) there are no reflections from the end of the line. With an unmatched line ($Z_L \neq Z_0$), there are reflections. We can now use the reflection coefficient result to help solve the impedance at every point along the line:

$$Z(x, \omega) = \frac{V(x, \omega)}{I(x, \omega)} = \frac{V_+ \left(e^{\frac{-j\omega x}{c}} + \Gamma e^{\frac{j\omega x}{c}} \right)}{\frac{V_+}{Z_0} \left(e^{\frac{-j\omega x}{c}} - \Gamma e^{\frac{j\omega x}{c}} \right)} = Z_0 \frac{\left(e^{\frac{-j\omega x}{c}} + \left(\frac{Z_L - Z_0}{Z_L + Z_0} \right) e^{\frac{j\omega x}{c}} \right)}{\left(e^{\frac{-j\omega x}{c}} - \left(\frac{Z_L - Z_0}{Z_L + Z_0} \right) e^{\frac{j\omega x}{c}} \right)} \tag{17.33}$$

By multiplying through by $Z_L + Z_0$ we get

$$Z(x, \omega) = Z_0 \frac{\left((Z_L + Z_0) e^{\frac{-j\omega x}{c}} + (Z_L - Z_0) e^{\frac{j\omega x}{c}} \right)}{\left((Z_L + Z_0) e^{\frac{-j\omega x}{c}} - (Z_L - Z_0) e^{\frac{j\omega x}{c}} \right)} \tag{17.34}$$

Next, we make use of the formulas:

$$\cos(a) = \frac{1}{2} \left(e^{ja} + e^{-ja} \right)$$

$$\sin(a) = \frac{1}{2j} \left(e^{ja} - e^{-ja} \right) \tag{17.35}$$

to reduce the above equations to

$$Z(x, \omega) = Z_0 \frac{Z_L \cos\left(\frac{\omega x}{c}\right) - jZ_0 \sin\left(\frac{\omega x}{c}\right)}{Z_0 \cos\left(\frac{\omega x}{c}\right) - jZ_L \sin\left(\frac{\omega x}{c}\right)} \tag{17.36}$$

Next, let's find the input impedance looking into the input of the finite-length transmission line (at $x = -l$). This is found by[7]

$$Z(-l, \omega) = Z_0 \frac{Z_L \cos\left(\frac{\omega l}{c}\right) + jZ_0 \sin\left(\frac{\omega l}{c}\right)}{Z_0 \cos\left(\frac{\omega l}{c}\right) + jZ_L \sin\left(\frac{\omega l}{c}\right)} = Z_0 \frac{\frac{Z_L}{Z_0} + j \tan\left(\frac{\omega l}{c}\right)}{1 + j\frac{Z_L}{Z_0} \tan\left(\frac{\omega l}{c}\right)} \tag{17.37}$$

Next, let us assume that we are operating at a frequency low enough[8] so that $\omega l \ll c$. This also means that we have a "short" transmission line. We can then approximate the tangent in the numerator and the denominator by the Taylor series expansion as follows:

$$\tan(x) = x + \frac{x^3}{3} + \frac{2x^5}{15} + \ldots \approx x \quad \text{if} \quad x \ll 1 \tag{17.38}$$

---

[7] Remember that $\cos(-l) = \cos(l)$ and that $\sin(-l) = -\sin(l)$ since cosine is an even function and sine is an odd function. Also, remember that $\sin(x)/\cos(x) = \tan(x)$.

[8] Or, the line is short enough.

Also, remember that for small x the following holds:

$$\frac{1}{1+x} \approx 1 - x \quad \text{if} \quad x \ll 1 \tag{17.39}$$

So, we can further estimate the impedance looking into the line as

$$Z(-l,\omega) \approx Z_0 \frac{\frac{Z_L}{Z_0}+j\left(\frac{\omega l}{c}\right)}{1+j\frac{Z_L}{Z_0}\left(\frac{\omega l}{c}\right)} \approx \frac{Z_L + jZ_0\left(\frac{\omega l}{c}\right)}{1+j\frac{Z_L}{Z_0}\left(\frac{\omega l}{c}\right)} \tag{17.40}$$

Let us see what we get if we assume that the line is terminated with a low impedance, with $Z_L \ll Z_0$. In this limiting case of a short transmission line terminated with a low impedance,

$$Z(-l,\omega) \approx Z_L + j\omega\left(\frac{Z_0 l}{c}\right) \approx Z_L + j\omega L_{eq} \tag{17.41}$$

This equivalent circuit is shown in Figure 17.19(a) where we see that the approximate lumped inductance is $L_{eq} \approx Z_0 l/c$.

Now, in the case where the short transmission line is terminated with a high impedance $Z_L \gg Z_0$, the approximate impedance for the short line is

$$Z(-l,\omega) \approx \frac{Z_L + jZ_0\left(\frac{\omega l}{c}\right)}{j\frac{Z_L}{Z_0}\left(\frac{\omega l}{c}\right)} \approx \frac{1}{Z_L} + \frac{1}{j\omega\left(\frac{1}{Z_0 c}\right)} \approx \frac{1}{Z_L} + \frac{1}{j\omega C_{eq}} \tag{17.42}$$

The equivalent circuit for this case is shown in Figure 17.19(b) where we see an equivalent capacitance $C_{eq} \approx l/(Z_0 c)$ in parallel with the load impedance.

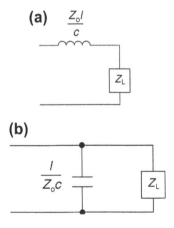

(a) $\frac{Z_0 l}{c}$

Z_L

(b) $\frac{l}{Z_0 c}$

Z_L

FIGURE 17.19

Equivalent circuit of short, unmatched transmission line. (a) With $Z_L \ll Z_0$. (b) With $Z_L \gg Z_0$.

Example 17.2: Terminated and unterminated transmission lines

Coax cables do a good job of transmitting pulse signals, provided that the cables are terminated properly at the end. Shown in Figure 17.20(a) is a 3-MHz square wave at the end of 24 ft of an RG58 BNC cable, with a 50-Ω termination resistor at the end of the transmission line, at the scope input. The signal at the scope input has a risetime of about 10 ns, limited by the risetime of the function generator used.

RG58 has a characteristic impedance of 50 Ω, so you may want to drive it with a voltage source (in this case, a function generator) that has a 50-Ω source impedance, and terminate the cable with 50 Ω as well. We can figure out the time delay (per foot) of the RG58 cable if we know the characteristic impedance ($Z_o = 50\ \Omega$) and the capacitance per foot ($C_o \approx 26$ pF/ft). The equations relating L_o, C_o, Z_o, and time delay t_{do} are

$$Z_o = \sqrt{\frac{L_o}{C_o}}$$

$$L_o = Z_o^2 C_o = (50^2)(26 \times 10^{-12}/\text{ft}) = 65\ \text{nH/ft} \tag{17.43}$$

$$t_{do} = \sqrt{L_o C_o} \approx 1.3\ \text{ns/ft}$$

The time delay of 24 ft of RG58 cable is about 31 ns. Since the risetime of the generator is short compared to the delay, we expect signal degradation to occur if the termination resistor is removed. This is indeed the case, as shown in Figure 17.20(b) where we remove the 50-Ω termination resistor. The signal integrity of the unmatched line is poor. The signal integrity would get even worse if the unterminated line was longer.

Example 17.3: Transmission line calculation

Find the approximate equivalent circuit, valid at low frequencies for the following scenarios: Assume that the wave propagation speed $c = 3 \times 10^8$ m/s.

1. One-meter long transmission line, with a characteristic impedance $Z = 75\ \Omega$, terminated with $Z_L = 5\ \Omega$.
2. One-meter long transmission line, with a characteristic impedance $Z = 75\ \Omega$, terminated with $Z_L = 1000\ \Omega$.

Solution:

1. For $Z_L = 50\ \Omega$, the approximate value of series inductance is

$$L_{eq} \approx \frac{Z_o l}{c} = \frac{(75)(1)}{3 \times 10^8} \approx 250\ \text{nH} \tag{17.44}$$

2. In the second scenario, the approximate value of the shunt capacitance is

$$C_{eq} \approx \frac{l}{Z_o c} = \frac{(1)}{(75)(3 \times 10^8)} \approx 44 \text{ pF} \qquad (17.45)$$

The equivalent circuits for the two cases are shown in Figure 17.21.

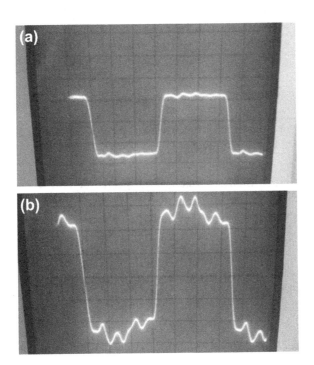

FIGURE 17.20

The 3-MHz signal passing through 18 ft of the RG58 (50-Ω) cable (Example 17.2). (a) Signal at the end of the line when the cable is properly terminated with 50 Ω, horizontal scale 50 ns/ div. (b) Signal when the cable is unterminated. (For color version of this figure, the reader is referred to the online version of this book.)

Node equations and Cramer's rule

Cramer's rule is a useful linear algebraic "recipe" that can help you solve any linear system of constant–coefficient equations, and hence to find the transfer function of a transistor amplifier. From a linear algebra point of view, the system of equations

$$[A]\{x\} = \{b\} \qquad (17.46)$$

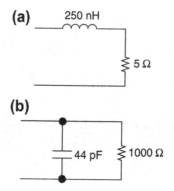

FIGURE 17.21

Equivalent circuit of a short, unmatched transmission line of Example 17.3. (a) With $Z_L = 5\,\Omega \ll Z_o$. (b) With $Z_L = 1000\,\Omega \gg Z_o$.

can be solved using Cramer's rule, where $[A]$ is an $m \times m$ matrix, $\{x\}$ is a column vector of m unknowns and $\{b\}$ is a column vector of inputs. Expanding this matrix notation results in a system with four equations and four unknowns:

$$[A]\{x\} = \begin{bmatrix} a_{11} & a_{12} & a_{13} & a_{14} \\ a_{21} & a_{22} & a_{23} & a_{24} \\ a_{31} & a_{32} & a_{33} & a_{34} \\ a_{41} & a_{42} & a_{43} & a_{44} \end{bmatrix} \begin{Bmatrix} x_1 \\ x_2 \\ x_3 \\ x_4 \end{Bmatrix} = \begin{Bmatrix} b_1 \\ b_2 \\ b_3 \\ b_4 \end{Bmatrix} \qquad (17.47)$$

Using Cramer's rule, if you want to find unknown x_3 you have to solve the following:

$$x_3 = \frac{\det \begin{bmatrix} a_{11} & a_{12} & b_1 & a_{14} \\ a_{21} & a_{22} & b_2 & a_{24} \\ a_{31} & a_{32} & b_3 & a_{34} \\ a_{41} & a_{42} & b_4 & a_{44} \end{bmatrix}}{\det \begin{bmatrix} a_{11} & a_{12} & a_{13} & a_{14} \\ a_{21} & a_{22} & a_{23} & a_{24} \\ a_{31} & a_{32} & a_{33} & a_{34} \\ a_{41} & a_{42} & a_{43} & a_{44} \end{bmatrix}} \qquad (17.48)$$

This looks like a mess, but can be easily manipulated using simple linear algebra. The term "det" denotes the "determinant" of a matrix. For a 2×2 matrix, the determinant is found as follows:

$$\det \begin{bmatrix} a_{11} & a_{12} \\ a_{21} & a_{22} \end{bmatrix} = a_{11}a_{22} - a_{12}a_{21} \qquad (17.49)$$

For a 3×3 matrix, the determinant is

$$\det \begin{bmatrix} a_{11} & a_{12} & a_{13} \\ a_{21} & a_{22} & a_{23} \\ a_{31} & a_{32} & a_{33} \end{bmatrix} = a_{11}(a_{22}a_{33} - a_{23}a_{32}) - a_{12}(a_{21}a_{33} - a_{23}a_{31})$$

$$+ a_{13}(a_{21}a_{32} - a_{22}a_{31}) \tag{17.50}$$

For an $N \times N$ matrix, it is more complicated but still doable. Consult a linear algebra textbook for the recipe.

Example 17.4: Using Cramer's rule to solve simultaneous linear equations

Let us use a simple linear algebra example to show the use of Cramer's rule. We will solve the following simultaneous linear equations for x, y, and z.

$$\begin{aligned} x + y + z &= 5 \\ x - y + 3z &= -3 \\ 2x + 2y + 3z &= 10 \end{aligned} \tag{17.51}$$

Putting this into matrix form results in

$$\begin{bmatrix} 1 & 1 & 1 \\ 1 & -1 & 3 \\ 2 & 2 & 3 \end{bmatrix} \begin{Bmatrix} x \\ y \\ z \end{Bmatrix} = \begin{bmatrix} 5 \\ -3 \\ 10 \end{bmatrix} \tag{17.52}$$

We solve for x, y, and z as follows:

$$x = \frac{\det \begin{bmatrix} 5 & 1 & 1 \\ -3 & -1 & 3 \\ 10 & 2 & 3 \end{bmatrix}}{\det \begin{bmatrix} 1 & 1 & 1 \\ 1 & -1 & 3 \\ 2 & 2 & 3 \end{bmatrix}} = \frac{(5)(-3-6) - 1(-9-30) + 1(-6+10)}{1(-3-6) - 1(3-6) + 1(2+2)} = \frac{-2}{-2} = 1$$

$$y = \frac{\det \begin{bmatrix} 1 & 5 & 1 \\ 1 & -3 & 3 \\ 2 & 10 & 3 \end{bmatrix}}{\det \begin{bmatrix} 1 & 1 & 1 \\ 1 & -1 & 3 \\ 2 & 2 & 3 \end{bmatrix}} = \frac{(1)(-9-30) - (5)(3-6) + (1)(10+6)}{-2} = \frac{-8}{-2} = 4$$

$$z = \frac{\det \begin{bmatrix} 1 & 1 & 5 \\ 1 & -1 & -3 \\ 2 & 2 & 10 \end{bmatrix}}{\det \begin{bmatrix} 1 & 1 & 1 \\ 1 & -1 & 3 \\ 2 & 2 & 3 \end{bmatrix}} = \frac{(1)(-10+6) - (1)(10+6) + (5)(2+2)}{-2} = \frac{0}{-2}$$

$$= 0$$

(17.53)

By direct substitution, we find that the values $x = 1$, $y = 4$, and $z = 0$ are indeed all correct.

Another simple example illustrates the use of Cramer's rule in the circuit realm. Shown in Figure 17.22 is the small-signal model for a common-emitter amplifier. Let us generate the node equations and solve for the transfer function using Cramer's rule.

We will lump R_s and r_x together in a single effective source resistance R'_s, with $G'_s = 1/R'_s$. At node v_π, we sum up currents at the junction resulting in

$$(v_i - v_\pi)G'_s - v_\pi C_\pi s - v_\pi r_\pi + (v_o - v_\pi)C_\mu s = 0 \qquad (17.54)$$

At node v_o, the node equation is

$$(v_\pi - v_o)C_\mu s - v_o G_L - g_m v_\pi = 0 \qquad (17.55)$$

Rewriting and collecting terms result in

$$-v_\pi \left[G'_s + (C_\pi + C_\mu)s + r_\pi \right] + v_o C_\mu s = -v_i G'_s$$
$$v_\pi \left[C_\mu s - g_m \right] - v_o \left[G_L + C_\mu s \right] = 0 \qquad (17.56)$$

This can be put into the standard matrix form:

$$\begin{bmatrix} -\left[G'_s + (C_\pi + C_\mu)s + r_\pi \right] & C_\mu s \\ C_\mu s - g_m & -\left[G_L + C_\mu s \right] \end{bmatrix} \begin{Bmatrix} v_\pi \\ v_o \end{Bmatrix} = \begin{Bmatrix} -v_i G'_s \\ 0 \end{Bmatrix} \qquad (17.57)$$

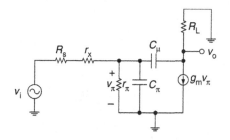

FIGURE 17.22

A small-signal model of a common-emitter amplifier (Example 17.4).

Using Cramer's rule, we find the output voltage to be

$$v_o = \frac{\det\left(\begin{bmatrix} -[G_s' + (C_\pi + C_\mu)s + r_\pi] & -v_i G_s' \\ C_\mu s - g_m & 0 \end{bmatrix}\right)}{\det\left(\begin{bmatrix} -[G_s' + (C_\pi + C_\mu)s + r_\pi] & C_\mu s \\ C_\mu s - g_m & -[G_L + C_\mu s] \end{bmatrix}\right)}$$ (17.58)

This result was summarized in earlier chapters where we found the transfer function for this amplifier.

Finding natural frequencies of LRC circuits

Consider the circuit in Figure 17.23. For convenience, conductances are used instead of resistance, where $g = 1/R$.

We can find the natural frequencies of this system by inspection, by considering the cases of "even" and "odd" initial conditions for v_1 and v_2. For initial conditions when $v_1 = v_2$, there is no current in g_2, and hence, we have two independent half circuits, each with a natural frequency:

$$\omega_{o,mode1} = \frac{g_1}{C}$$ (17.59)

For odd initial conditions with $v_1 = -v_2$, the center point of g_2 is at the ground potential, and hence, we can ground this point. The resulting natural frequency is then

$$\omega_{o,mode2} = \frac{g_1 + 2g_2}{C}$$ (17.60)

The standard state matrix form for a system of differential equations[9] is

$$\{\dot{x}\} = [A]\{x\}$$ (17.61)

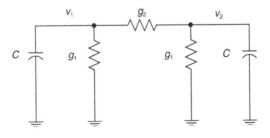

FIGURE 17.23

A symmetric RC circuit.

---

[9]Remember the notation $\{\dot{x}\}$ being the time derivative of the $\{x\}$ matrix.

where $\{x\}$ is the state vector and the *eigenvalues* of the $n \times n$ system matrix $[A]$ are the roots of the characteristic equation (hence, the poles of the system response). The mode shapes are the *eigenvectors* of $[A]$.

In the s-plane, the matrix equations for an RC circuit are

$$[sC + G]\{v\} = \{i\} \tag{17.62}$$

where $\{i\}$ is the current excitation vector. Natural frequencies (poles) are found by solving the homogenous case, where

$$[sC + G]\{v\} = \{0\} \tag{17.63}$$

This can be rewritten as

$$[sC]\{v\} = -[G]\{v\} \tag{17.64}$$

which we recognize as the same state-space formulation! Therefore, the natural frequencies are found by evaluating the determinant of the admittance matrix, or

$$\det[sC + G] = 0 \tag{17.65}$$

Equivalently,

$$\det[sI - A] = 0 \tag{17.66}$$

where I is the identity matrix and the system matrix $A = G/C$.

Example 17.5: Finding natural frequencies and mode shapes using MATLAB

Next, we will set up some matrices using the circuit of Figure 17.24 and solve them using MATLAB. By summing currents, we can derive the state equations for this circuit. For instance, at the v_1 node, we have:

$$-C\frac{dv_1}{dt} - g_1v_1 + (v_2 - v_1)g_2 = 0 \tag{17.67}$$

Similarly, at the v_2 node,

$$-C\frac{dv_2}{dt} - g_1v_2 + (v_1 - v_2)g_2 = 0 \tag{17.68}$$

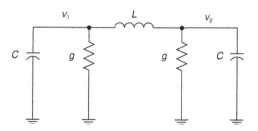

FIGURE 17.24

Symmetric LRC circuit for Example 17.5.

For this system, suitable state variables are v_1 and v_2, and the state vector is

$$\{x\} = \begin{Bmatrix} v_1 \\ v_2 \end{Bmatrix} \qquad (17.69)$$

In state space form, the equations for the above network are

$$\begin{Bmatrix} \dot{v}_1 \\ \dot{v}_2 \end{Bmatrix} = - \begin{bmatrix} \frac{g_1+g_2}{C} & \frac{-g_2}{C} \\ \frac{-g_2}{C} & \frac{g_1+g_2}{C} \end{bmatrix} \begin{Bmatrix} v_1 \\ v_2 \end{Bmatrix} \qquad (17.70)$$

The following is a MATLAB script for finding the natural frequencies of this system, assuming that $g_1 = 1/\Omega$, $g_2 = 10/\Omega$, and $C = 1$ F:

```
function RC
g1=1; g2=10;
% Conductance
C=1;
% Capacitors
% Form A matrix
A=-[(g1+g2)/C - g2/C;
    - g2/C (g1+g2)/C];
% Find natural frequencies and mode shapes
[Modeshapes, NatFreqs] = eig (A);
% Normalize modeshapes
N=size (Modeshapes,1);
for i=1:N
    Modeshapes (:,i)=Modeshapes (:,i)/max (abs (Modeshapes (:,i))); end
% Display modeshapes
Natural_Frequencies=NatFreqs
Mode_Shapes=Modeshapes
```

Here is the result of running this MATLAB script, where we find the natural frequencies and mode shapes. The natural frequencies are the diagonals of the natural frequency matrix. The mode shapes are the columns of the eigenvector matrix. For instance, for $\omega_0 = -1$ rad/s, the mode shape is [1 1], corresponding to the case when the initial conditions are set as $v_1 = v_2$. For $\omega_0 = -20$ rad/s, the mode shape is [1 −1], corresponding to the case when $v_1 = -v_2$.

```
» RC
Natural_Frequencies =
   -21        0
     0       -1
Mode_Shapes =
     1        1
    -1        1
```

Note that the natural frequencies are the same as those calculated by the "inspection" method.

We will repeat the MATLAB formulation using the symmetric LRC circuit of Figure 17.24. This circuit has one real and two imaginary poles, as we will see. Omitting the details of the node equations, we find the state matrix to be

$$\begin{Bmatrix} \dot{v}_1 \\ \dot{i} \\ \dot{v}_2 \end{Bmatrix} = - \begin{bmatrix} g/C & 1/C & 0 \\ -1/L & 0 & 1/L \\ 0 & -1/C & g/C \end{bmatrix} \begin{Bmatrix} v_1 \\ i \\ v_2 \end{Bmatrix} \tag{17.71}$$

A MATLAB script follows for solving these simultaneous equations.

```
function lrc
g=1;        %Conductance
C=1;        %Capacitors
L=1;        %inductance
A=-[ g/C 1/C 0
    -1/L 0 1/L
     0 -1/C g/C];   %Form A matrix
% Find natural frequencies and mode shapes
[Modeshapes, NatFreqs] = eig (A);
% Normalize modeshapes
N=size (Modeshapes,1);
for i=1:N
Modeshapes (:,i)=Mode shapes (:,i)/max (abs (Modeshapes (:,i)));
end
% Display modeshapes
Natural_Frequencies=NatFreqs
Mode_Shapes=Modeshapes

» lrc
Natural_Frequencies =
   -1.0000              0                    0
    0          -0.5000 + 1.3229i            0
    0                   0           -0.5000 - 1.3229i
Mode_Shapes =
    1.0000          0.7071               0.7071
    0.0000     -0.3536 - 0.9354i     -0.3536 + 0.9354i
    1.0000         -0.7071              -0.7071
```

Note the above natural frequencies and mode shapes. The first natural frequency ($\omega = -1$) has a mode shape of [1 0 1], meaning that $v_1 = v_2$ and that $i = 0$. The second and third natural frequencies are $\omega_{2,3} = -0.5 \pm 1.3229j$, corresponding to a pair of complex poles (and oscillatory behavior for this mode). The mode shapes are also complex, meaning that v_1, v_2, and i are out of phase for this mode.

Let us next try this method on an RC ladder (Figure 17.25).

To find the natural frequencies and mode shapes of the RC ladder, ground the input and form the G and C matrices. In state-space form this results in

$$\begin{Bmatrix} \dot{v}_1 \\ \dot{v}_2 \\ \dot{v}_3 \\ \dot{v}_4 \end{Bmatrix} = -\frac{G}{C} \begin{bmatrix} 2 & -1 & 0 & 0 \\ -1 & 2 & -1 & 0 \\ 0 & -1 & 2 & 1 \\ 0 & 0 & -1 & 2 \end{bmatrix} \begin{Bmatrix} v_1 \\ v_2 \\ v_3 \\ v_4 \end{Bmatrix} \tag{17.72}$$

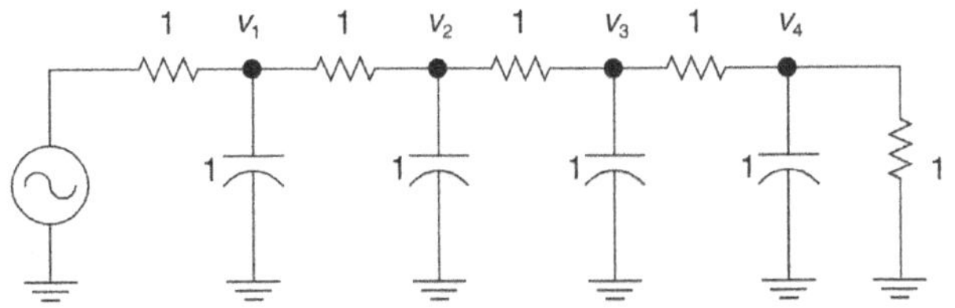

FIGURE 17.25

RC Ladder.

A MATLAB solution is as follows:

```
» rcladder
A =
     -2                1                0                0
      1               -2                1                0
      0                1               -2                1
      0                0                1               -2
Natural_Frequencies =
   -1.3820                0                0                0
         0          -1.3820                0                0
         0                0          -2.6180                0
         0                0                0          -3.6180
Mode_Shapes =
    1.0000          -0.6180          -1.0000          -0.6180
    0.6180          -1.0000           0.6180           1.0000
   -0.6180          -1.0000           0.6180          -1.0000
   -1.0000          -0.6180          -1.0000           0.6180
```

From this analysis, the lowest frequency pole is at $\omega = -0.38$ rad/s. This pole will dominate the input–output response.

Some comments on scaling laws in nature

Scaling laws and dimensional analysis are very valuable tools for determining how structures, circuits, and processes scale. The rationale for using scaling laws is as follows: building a full-scale prototype is often impractical and/or dangerous. An

Table 17.2 List of Symbols

B	Magnetic flux density (T)
C_d	Aerodynamic/hydrodynamic drag coefficient
E_k	Kinetic energy $= \frac{1}{2} Mv^2$ (J)
f_d	Drag force (N)
l	Length scale
P	Power (J, or W/s)
v	Velocity (m/s)
σ	Electrical conductivity ($1/\Omega$ m)
ε	Dielectric permittivity (F/m)
μ	Magnetic permeability (H/m)
μ_o	Magnetic permeability of free space ($4\pi \times 10^{-7}$ H/m)

alternate strategy is to build a smaller-scale model and use scaling laws to determine how the full-scale system will behave. This method of scale modeling is often used in

- Aerodynamics (wind tunnel tests)
- Hydrodynamics
- Magnetic (i.e. Maglev)
- Rocketry
- Power electronics

Many interesting scaling laws can be inferred by studying simple examples in engineering and nature. A list of commonly used symbols in engineering is shown in Table 17.2.

Geometric scaling

By elementary mathematics, in similar figures, the surface area scales as the square of the linear dimension and the mass scales as the cube of the linear dimension.[10] In a sphere, the surface area is $4\pi r^2$ while the volume is $\frac{4}{3}\pi r^3$, and hence, the ratio of the volume to the surface is $r/3$. In other words, the ratio of the volume to the surface area scales as the length scale l.

Fish/ship speed (Froude's law)

If every length scale l is increased by the same factor, how does the speed of a ship (or a fish for that matter) scale? Well, the propulsion power that an engine can provide is proportional to the mass of the engine, which scales as l^3, or

$$P_p = k_1 l^3 \tag{17.73}$$

---

[10]This example, as well as some of the examples to follow, are derived from Darcy W. Thompson's, *On Growth and Form*.

The drag force on the hull is proportional to the velocity. The proportionality constant is the drag constant C_d, which is proportional to the hull area.

$$f_D = C_d v = k_2 l^2 v \tag{17.74}$$

In a steady state, the propulsion power balances the power due to the drag force on the hull, with power equal to force multiplied by velocity. Hence, the drag power is

$$P_D = f_d v = k_2 l^2 v^2 \tag{17.75}$$

In the steady state, the propulsion power and the drag power are equal, resulting in

$$k_1 l^3 = k_2 l^2 v^2 \tag{17.76}$$

Manipulating this expression results in

$$v^2 = \frac{k_1}{k_2} l \Rightarrow v \propto \sqrt{l} \tag{17.77}$$

This result, known as Froude's law, shows that if everything else is equal, the maximum velocity of a ship scales as the square root of the length scale. For example, if a 100-ft long ferry can travel at 10 mph, a scaled-up ship that is 1000 ft long will have a maximum speed of approximately 32 mph.

Fruit

How do the sizes of plants scale? For instance, the stalk of a piece of fruit hanging on a tree (Figure 17.26) can withstand a maximum stress (or force per unit area) in newtons per square meter. The force exerted on the stalk scales as the mass of the fruit, or

$$f \propto l^3 \tag{17.78}$$

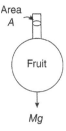

Area
A

Fruit

Mg

FIGURE 17.26

Cartoon sketch of a piece of fruit hanging from a tree, with the stalk having cross-sectional area A.

The area of the stalk scales as l^2. Therefore, the force/area scales as l. The results from this scaling law are

- Enormous structures, if scaled up directly, will collapse under their own weight.
- Old trees have huge bases.
- The Eiffel tower is tapered.
- Large fruit lie on the ground.

Bending moments

By elementary mechanics, the bending of a beam under its own weight scales as the length squared, if the cross-sectional area remains constant. Said another way, a 12-ft 2 × 4 fixed at one end will have four times the deflection of a 6-ft 2 × 4 fixed at one end. In nature, to counteract this effect, as the size of an animal increases the limbs tend to become thicker and shorter.

Size and heat in bodies (Bergman's law)

Heat production in a body (i.e. power) is proportional to mass, or l^3. Heat loss is proportional to the surface area, or l^2. The ratio

$$\frac{\text{Heat loss}}{\text{Heat production}} \propto \frac{1}{l} \qquad (17.79)$$

shows that small animals have a proportionally higher heat loss than do large animals. This is why birds and mice need to eat lots of food in proportion to their body weight. A man eats 1/50 of his body weight per day, whereas a mouse eats approximately half its body weight in food. Another result is that there are no tiny animals in the Arctic.

Size and jumping (Borelli's law)

When an animal jumps, the leg muscles impart a force impulse to the animal; this impulse is proportional to muscle mass, or l^3. Therefore, the velocity is

$$v \propto \frac{\text{Muscular force impulse}}{\text{Mass moved}} = \text{constant} \qquad (17.80)$$

Borelli's law shows why all animals can jump approximately the same height.

Walking speed (Froude's law)

Assuming that everyone's legs travel through the same angle α, a model for determining the scaling law for walking speed is found in Figure 17.27. The distance per oscillation for one step is proportional to A/B. Assuming that the legs behave as pendulums, the time per oscillation scales as

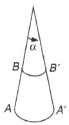

FIGURE 17.27

Model for determining walking speed by using Froude's law.

$$t \propto \sqrt{\frac{A}{B}}$$ (17.81)

The walking speed v is found by

$$v \propto \frac{\text{distance}}{\text{time}} \propto \sqrt{\frac{A}{B}}$$ (17.82)

This again is Froude's law.

Capacitors

How does the RC time constant of a capacitor scale with size? A model for a parallel-plate capacitor is shown in Figure 17.28(a). Parallel plates are spaced d apart, with a material inside with dielectric permittivity ε and electrical conductivity σ. An electrical model for this capacitor is shown in Figure 17.28(b). The capacitance is given by

$$C = \frac{\varepsilon A}{d}$$ (17.83)

where A is the area of the plates. The resistance is

$$R = \frac{d}{\sigma A}$$ (17.84)

Therefore, the RC time constant is

$$RC = \frac{\varepsilon}{\sigma}$$ (17.85)

independent of size.

Inductors

A single-loop inductor is shown in Figure 17.29.

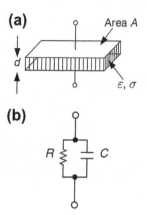

FIGURE 17.28

Capacitor model. (a) Parallel-plate geometry filled with a lossy dielectric. (b) Electrical model.

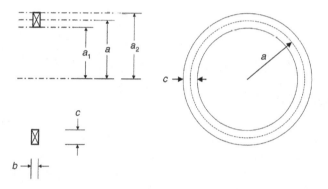

FIGURE 17.29

Inductor model. The inductor has a mean radius a, axial thickness b, and radial thickness c.

The inductor has a mean radius a, axial thickness b, and radial width c. For a thin hoop with $c \ll a$, the resistance of the loop is

$$R \approx \frac{2\pi a}{bcgs} \tag{17.86}$$

The inductance of the loop is found by using the inductance reference by Frederick Grover, *Inductance Calculations*:

$$L \approx kaPF \tag{17.87}$$

with

$k = $ constant
$P = $ function of $c/2a$
$F = $ function of b/a

The time constant of an inductor is L/R, which is found by

$$\frac{L}{R} \approx \frac{kPFbc\sigma}{2\pi} \propto l^2 \tag{17.88}$$

An interesting result pops out here: magnetic scaling laws show that large magnetic elements are more efficient in energy conversion than are smaller ones. For this disk coil geometry, this effect can be quantified by considering the ratio of inductance to resistance. The inductance L is approximately proportional to a as shown above. The resistance of the coil is proportional to a/bc, the ratio of the current path length is proportional to the coil cross-sectional area. Therefore, the ratio of inductance to resistance is proportional to bc, or the cross-sectional area of the coil. If all coil lengths are scaled up by the same factor l, then this ratio increases by the factor l^2, or the length squared. This scaling law shows that the efficiency of inductors (and hence, electric motors) improves with increasing size as l^2.

Lift force of electromagnet

The lift force f_l of an electromagnet (Figure 17.30) is found by analyzing the magnetic pressure, and is approximately

$$f_l \approx \frac{B^2 A}{2\mu_o} \tag{17.89}$$

where B is the magnetic flux density (Tesla), A is the total area of the two poles of the electromagnet, and μ_o is the magnetic permeability of free space.

How does this lift force scale with the magnet length scale? The B field is found approximately by

$$B \approx \frac{\mu_o NI}{2g} \propto \frac{N}{l} \tag{17.90}$$

where g is the air gap. The number of turns scales with the area, or

$$N \propto l^2 \tag{17.91}$$

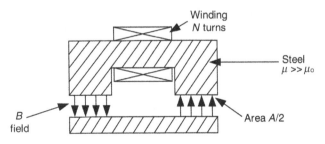

FIGURE 17.30

Electromagnet.

The lift force scales as B^2A, or as l^4, while the weight of the magnet scales as l^3. The lift/weight ratio therefore scales as l. We might infer from this scaling law that a tiny electromagnet cannot lift its own weight.

Some personal comments on the use and abuse of SPICE modeling

I am not a big fan of SPICE modeling. It can not only be useful (that is why I use it to model lots of circuits in this book), but it can also be a misleading tool. I tell my students that the only people that should be allowed to use SPICE are those who already know the ballpark answer to their problem. SPICE is useful for visualizing answers and for figuring out sensitivities of circuits to component variations, but in the author's opinion, you should have a pretty good idea as to how your circuit works before you model it with SPICE. Do not arbitrarily trust the answer that the computer spits out—double-check it with pencil and paper and prototype it if you are serious about the circuit.

For instance, if you model a high-speed op-amp circuit, SPICE may say that you will get a closed loop bandwidth of 200 MHz. Then, you build it on a protoboard and your bandwidth is 150 MHz (or it oscillates). What is the culprit? Maybe the SPICE model you have of the component is wrong. Maybe you have not put in all the parasitic capacitances and inductances. Maybe you have not properly bypassed the power supply leads.

You need to be very, very careful about the SPICE device models you use. I have seen SPICE models for the 2N3904 transistor (Figure 17.31) with 10 Ω of base resistance (Rb = 10 is way too low) and a BETA of 416.4 (too high). Also, in this model, T_f and C_{jc} are way too low. For some kinds of transistor circuits, as discussed in Chapter 7, the incorrect value of Rb (a.k.a. r_x in your small-signal model) will cause your gain and bandwidth to be grossly in error.

Transistor circuits may oscillate on the breadboard, but not in an SPICE simulation. Precision circuits, with all of their mysterious noise and thermal effects, may not be modeled well with SPICE. (Does your SPICE op-amp model properly model

```
.model 2N3904 NPN(Is=6.734f Xti=3 Eg=1.11
+               Vaf=74.03 Bf=416.4 Ne=1.259
+               Ise=6.734f Ikf=66.78m Xtb=1.5
+               Br=.7371 Nc=2 Isc=0 Ikr=0 Rc=1
+               Cjc=3.638p Mjc=.3085 Vjc=.75
+               Fc=.5 Cje=4.493p Mje=.2593
+               Vje=.75 Tr=239.5n Tf=301.2p
+               Itf=.4 Vtf=4 Xtf=2 Rb=10)
```

FIGURE 17.31

Transistor SPICE model, use with caution.

$1/f$ noise? What about popcorn noise in your transistor circuits?) SPICE device models of transistors and op-amps should always be viewed with a critical eye. Make sure you put parasitic elements (resistance, inductance) in your capacitor models and (resistance, capacitance) in your inductor models. Transistor lead length translates to a parasitic inductance (more lead length, more inductance). If you need to know the lead length inductance, estimate it with some of the techniques given earlier in this book, or simulate it with magnetic field modelers such as FEMM[11] (2D) or FastHenry[12] (3D).

For further reading on the topic of SPICE use and abuse, please view the wonderful screeds of the late Bob Pease. Do not let SPICE be a crutch—figure out how your circuit works with pencil and paper, then simulate it if necessary and definitely prototype it. Compare your simulation results with the prototype, and if there is a good match, maybe, just maybe your SPICE model has some useful level of accuracy. Trust, but verify. Do not let novice engineers do lots and lots of SPICE simulations without them understanding exactly what they are doing. Treat SPICE as a tool that you need training to use—after all, you would not let an untrained person use a chainsaw, would you?

Chapter 17 problems
Problem 17.1

For the LC circuit in Figure 17.32:

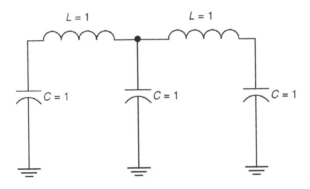

FIGURE 17.32

Circuit for Problem 17.1.

---

[11]I have used FEMM (Finite Element Method Magnetics) with excellent results for over 10 years for magnetic stuff. It now has the capability to do electrostatic and thermal modeling and supports AC and nonlinear modeling.

[12]I have used FastHenry to model the inductance of PC board inductors. This is very fast and useful, but no high magnetic permeability material is supported. Designed at MIT to model the inductance of integrated circuit interconnections, this freeware is now available on the web.

a. By analogy, find the natural frequencies of the LC circuit.

b. Simulate your results using LTSPICE. Excite each mode independently by the proper setting of inductor current initial conditions.

Problem 17.2

Using Cramer's rule, solve the following simultaneous linear equations:

$$x + y = -1$$
$$2x - 3y = 13$$

Problem 17.3

Marc is 6 ft tall and Lisa is 5 ft tall.[13] Marc walks a mile in 20 min at his most comfortable walking speed. At her most comfortable walking speed, approximately how long does it take Lisa to walk a mile? Roughly how long does it take their 4-ft-tall daughter Sophie to walk a mile?

Problem 17.4

Find the thermal resistance through an aluminum plate with dimensions T (in the direction of heat flow) $= 1$ mm, and cross-sectional area $A = 100$ mm^2. Assume that the thermal conductivity of aluminum $k = 240$ W/(m °C).

Problem 17.5

A 2N3906 transistor is in a circuit in which the transistor dissipates 100 mW of energy. The ambient temperature is 40 °C. What is the junction temperature of the device? Assume that the thermal resistance, junction to ambient, is 200 °C/W. Draw the thermal circuit.

Problem 17.6

a. Draw an appropriate *static* thermal model for the transistor mounted to a heat sink (Figure 17.33). Include the junction temperature, case temperature, and heat-sink temperature.

b. Assume that the total loss in the transistor is 10 W. Assuming the following parameters: $T_A = 50$ °C; $R_{TH,JC} = 1$ °C/W; $R_{TH,CS} = 1$ °C/W; Calculate the maximum thermal resistance from the heat sink to ambient, $R_{TH,SA}$, to keep the junction temperature of the transistor below 150 °C.

---

[13]Names changed to protect the innocent.

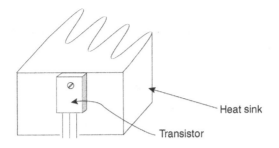

FIGURE 17.33

Transistor mounted to a heat sink for Problem 17.6.

Problem 17.7

Using the symmetric RC circuit of Figure 17.34 with node voltages of interest (v_1 and v_2) shown.

a. Assume that at time $t = 0$ the circuit has initial conditions $v_1 = v_2 = 1$ V. Sketch the responses of $v_1(t)$ and $v_2(t)$ for $t > 0$.

b. Assume that at $t = 0$ that $v_1 = 1$ and $v_2 = -1$ V. Sketch the responses of $v_1(t)$ and $v_2(t)$ for $t > 0$.

Problem 17.8

The engine block in an automobile has the following parameters:

- Mass: 100 kg of aluminum

- Power dissipated when the engine is idling: 5000 W

- Aluminum heat capacity $c_p = 900$ J/kg °C

a. When you first turn the engine on, the engine is at ambient temperature ($T_A = 25$ °C) and there is no water circulating through the engine (since the engine is cold). With no water circulation, the thermal resistance from

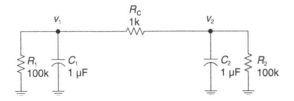

FIGURE 17.34

Circuit for Problem 17.7.

engine-to-ambient is 0.1 °C/W. Generate a lumped thermal model of the system, assuming that the temperature of the motor is uniform throughout. Note that thermal capacitance is Mc_p (joules per degree centigrade).

b. The engine is "warmed up" when the temperature of the engine reaches 100 °C. How long does the engine take to warm up?

c. During running, assume that the temperature of the engine reaches a final value of 125 °C. Plot the temperature of the engine after it is shut off.

Problem 17.9

This problem concerns a power resistor mounted in a cross-shaped block of copper, as shown in Figure 17.35. The area of each member is 1×1 cm as shown, with the length of each member being 5 cm long. The cross-sectional area of the resistor is small compared to the area of the copper (hence you can consider the resistor to be a point source of power). The four members mount to an outside wall that has a constant temperature of 25 °C. A layer of thermal grease with an average thickness of 0.02 mm is used to thermally couple the edge of the copper bars to the wall. Since this layer is thin, you may neglect the thermal capacitance of the thermal grease layer.

a. Generate a lumped thermal model of the system.

b. If you energize the resistor at $t = 0$ with 100 W of power, what is the final resistor temperature as $t \rightarrow \infty$? (Assume that the thermal joint between the resistor and the copper members is perfect.)

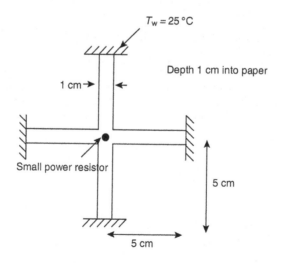

FIGURE 17.35

Thermal system for Problem 17.9.

c. In a separate experiment, before $t = 0$, the wall, the copper, and the resistor are at a uniform temperature, and there is no current flow in the power resistor. Assume that the power resistor is energized and dissipates 100 W, beginning at $t = 0$. Sketch the resistor temperature for all times, showing breakpoints, time constants, etc. given your thermal model and the simplifying assumptions made. (Hint: parts (2) and (3) can be made significantly simpler by exploiting symmetry, breaking up the large problem into smaller chunks, using intuitive reasoning as to the power flow paths, etc.).

Further reading

[1] Brown WL, Szeto A. Verifying SPICE results with hand calculations: handling common discrepancies. *IEEE Trans Educ* 1994;**37**(4):358−68.

[2] Dowell RI. Hijacked by SPICE. *IEEE Solid-State Circuits Mag* Spring 2011:14−6.

[3] FastHenry circuit solver. Available from: http://www.fastfieldsolvers.com/download.htm.

[4] Fogel M, editor. *The automatic control systems/robotics problem solver*. Piscataway NJ: Research and Education Association; 1990.

[5] General Electric Corp. SCR handbook. 6th ed. Good general-purpose manual with lots of practical advice on heat sink design, and mounting and cooling techniques for power devices.

[6] Gilbert B. A new wide-band amplifier technique. *IEEE J Solid-State Circuits* 1968;**SC-3**(4):353−65.

[7] Gilbert B. A precise four-quadrant multiplier with subnanosecond response. *IEEE J Solid-State Circuits* 1968;**SC-3**(4):365−73.

[8] Gilbert B. A DC-500 MHz amplifier multiplier principle, 1968 ISSCC digest of technical papers, p. 114−5.

[9] Gilbert B. Translinear circuits: a proposed classification. *Electron Lett* 1975;**1**:14−6.

[10] Gilbert B. Current-mode circuits from a translinear viewpoint: a tutorial. In: Toumazou C, Lidgey FJ, Haigh DG, editors. *Analogue IC design: the current-mode approach*. London: Peter Peregrinus; 1990. p. 11−91.

[11] Gilbert B. Translinear circuits—25 years on, part I: the foundations. *Electron Eng* 1993;**65**(800):21−4.

[12] Gilbert B. Translinear circuits: an historical review. *Analog Integr Circuits Signal Process* 1996;**9**(2):95−118.

[13] Grover F. Inductance calculations, originally published by D. Van Nostrand, 1946, reprinted by Dover, 2004. Classic text on inductance calculations.

[14] Jezierski E. On electrical analogues of mechatronic systems. In: *Proceedings of the 2001 second international workshop on motion and control* October 18−20, 2001. p. 181−8.

[15] Kassakian J, Schlecht M, Verghese G. *Principles of power electronics*. Prentice-Hall; 1991. A well-regarded reference for all aspects of power electronics design, including magnetics.

[16] Klein RE. Teaching linear systems theory using Cramer's rule. *IEEE Trans Educ* 1990;**33**(3):258−67.

[17] Langlois PJ. Graphical analysis of delay line waveforms: a tutorial. *IEEE Trans Educ* 1995;**38**(1):27–32.

[18] Mathworks, Inc., *MATLAB reference guide*; 1992.

[19] Meeker D. Finite element method magnetics (FEMM), 2D magnetic, electrostatic and thermal modeler. Available at the time of this writing at: http://www.femm.info/wiki/DavidMeeker.

[20] Middlebrook RD. Null double injection and the extra element theorem. *IEEE Trans Educ* 1989;**32**(3):167–80.

[21] Middlebrook RD. The two extra element theorem. In: *Proceedings of the IEEE frontiers in education twenty-first annual conference*. Purdue University; September 21–24, 1991. p. 702–8.

[22] Middlebrook RD. Low-entropy expressions: the key to design-oriented analysis. In: *Proceedings of the IEEE frontiers in education twenty-first annual conference*. Purdue University; September 21–24, 1991. p. 399–403.

[23] Pease RA. *Sugar and spice and nothing nice? and Appendix G: more on SPICE, found in Troubleshooting analog circuits*. Butterworth-Heinemann; 1993 and What's all this spicey stuff anyhow? parts I, II, 2.5, III and IV, found in Electronic Design Magazine.

[24] Rao SS. *Mechanical vibrations*. 3rd ed. Addison-Wesley; 1995.

[25] Rothkopf E. Teaching for understanding—analogies for learning in electrical engineering, 1995 IEEE Frontiers in Education Conference, session 2b4, p. 2b4.9–13.

[26] Seevinck E, Wiegerink RJ. Generalized translinear circuit principle. *IEEE J Solid-State Circuits* 1991;**26**(8):1098–102.

[27] Shen LC, Kong JA. *Applied electromagnetism*. Brooks/Cole; 1983.

[28] Teoh CS, Davis LE. A coupled pendula system as an analogy to coupled transmission lines. *IEEE Trans Educ* 1996;**39**(4):548–57.

[29] Thompson DW. *On growth and form*. Dover Publications (reprinted); 1992.

[30] Thornton RD. Electronic circuits modeling, analysis, simulation; intuition and design, MIT Course Notes, January 1993.

[31] Vorperian V. Improved circuit-analysis techniques require minimum algebra. *EDN* 1995:125–34.

[32] Zahn M. *Electromagnetic field theory: a problem solving approach*. Reprinted by Krieger; 1987.

Appendices

Appendix 1: Some useful approximations and identities

Item	Calculation	Comment
Power series expansion for an exponential	$e^x = 1 + x + \dfrac{x^2}{2!} + \dfrac{x^3}{3!} + \cdots$	
	$e^x \approx 1 + x \quad \text{for} \quad x \ll 1$	Used in finding small signal models and linearized equations at an operating point
Derivative of the arctangent	$\dfrac{d}{dx}(\tan^{-1} u) = \dfrac{1}{1 + u^2}\dfrac{du}{dx}$	Used in finding group delay from the angle of a transfer function
Power series expansion of the arctangent	$\tan^{-1}(x) = \pi/2 - 1/x + 1/(3x^3) - \cdots$ for $x > 1$	
Power series expansion for tangent	$\tan(x) = x + \dfrac{x^3}{3} + \dfrac{2x^5}{15} + \cdots \approx x$ if $x \ll 1$	
	$\dfrac{1}{1+x} \approx 1 - x \quad \text{for} \quad x \ll 1$	Try it out on a calculator: $\dfrac{1}{1 + 0.1} = 0.909090\ldots$
	$\dfrac{1}{\sqrt{1+x}} \approx 1 - \dfrac{x}{2} \quad \text{for } x \ll 1$	$\dfrac{1}{\sqrt{1 + 0.1}} = 0.95346\ldots$
Sine and cosine as a function of complex exponentials	$\cos(a) = \dfrac{1}{2}(e^{ja} + e^{-ja})$ $\sin(a) = \dfrac{1}{2j}(e^{ja} - e^{-ja})$	

Appendix 2: p, μ, m, k and M

Item	Name	Value
p	Pico	10^{-9}
μ	Micro	10^{-6}
m	Milli	10^{-3}
k	Kilo	10^{3}
M	Mega	10^{6}

Appendix 3: MATLAB scripts for control system examples
MATLAB script for gain of +1 and +10 amplifiers

```
function c1
% Control system example #1
% Calculates parameters for gain of +1 and gain of +10 amplifiers
% Marc Thompson, 10/22/99
% Open loop transfer function a (s)
ao=1e5;                             % DC gain
d=conv ([0.1 1],[1e-6 1]);          % Poles
a=tf (ao, d);                       % Create transfer function a (s)
bode (a)                            % Plot Bode plot of a (s)
title ('file: c1.m: Transfer function of a (s)')
figure;
pzmap (a);                          % Plot pole/zero map of a (s)
title ('file: c1.m: Pole map of a (s)');
damp (a)                            % Find natural frequency and damping
                                    ratio of a (s)
% gain of +1
f=1;                                % Feedback gain of +1
margin (a*f)                        % Find phase and gain margin
title ('Phase margin calculation for gain of +1 amplifier'); figure;
f=tf (f,1);                         % Create feedback f (s)
cloop=feedback (a, f)               % Close the loop, find transfer function H (s)
bode (cloop)
title ('Bode plot of closed-loop transfer function for gain of +1 amplifier'); figure
step (cloop)
title ('Step response for gain of +1 amplifier'); grid; figure
pzmap (cloop)
```

title ('Pole map of closed-loop gain of +1 amplifier'); grid; figure;

damp (cloop)   % Find natural frequency and damping
                                                    ratio of

H (s)

% gain of +10

f = 0.1;

margin (ao*f, d); title ('Phase margin calculation for gain of +10 amplifier'); figure;

f = tf (f,1);

cloop = feedback (a, f); cloop gain of 10 = cloop

bode (cloop); title ('Bode plot of closed-loop transfer function for gain of +10 amplifier');
figure

step (cloop); grid; title ('Step response for gain of +10 amplifier'); pzmap (cloop); title
('Pole map of closed-loop gain of +10 amplifier'); grid; damp (cloop)

MATLAB script for integral control example

```
function c3
% Control example 3
% driving reactive load
L=10e-6;
C=10e-6;
R=10;
Zo = sqrt(L/C);                            % Characteristic Impedance
Q=R/Zo
% Calculate PLANT num=1;
denom=[L*C L/R 1];
plant=tf(num, denom)
damp(plant)                                % Find poles and damping ratio
bode(plant); title('file: c3.m; REACTIVE LOAD EXAMPLE')
% Integral control, attempt                #1
Gain=4e3;                                  % Integrator gain
denom=[1 0];
Gc=tf(Gain, denom);                        % Form Gc(s)
Forw=series(plant, Gc);                    % Cascade with plant
margin(Forw);                              % Find gain and phase margin
F=tf(1,1);
Cloop=feedback(Forw, F,−1)
figure; step(Cloop); title('STEP RESPONSE, CONTROLLER #1'); grid
%Integral control, attempt                 #2
figure
d=[1/5e4 1];
LPF=tf(1, d);                              % Add lowpass filter to damp
                                           complex pole pair
```

```
Gc=series(LPF, Gc);
Forw=series(plant, Gc);
margin(Forw); title('PHASE MARGIN CALC, CONTROLLER #2');
F=tf(1,1);
Cloop=feedback(Forw, F,−1)
figure; step(Cloop); title('STEP RESPONSE, CONTROLLER #2'),-grid
```

MATLAB script for MOSFET current source example

```
function moscursource
% Analysis of MOSFET current source
% Marc Thompson, 3/28/00
% LOAD
RL=1;
% MOSFET model
gm = 8.6;                              % transconductance
Cgs = 1040e-12;
Cgd = 160e-12;
% OPAMP model
rout = 100;                            % output resistance of opamp
ao=2*pi*4*1e6;                         % GBP = 4 MHz
denom=[1/ao 0];
highpole=[1/ao 1];
d=conv(denom, highpole);
opamp=tf(1, d)
% MOSFET follower model
Rsense = 0.08;                         % current sense resistor
Ao=gm*Rsense/(1+gm*Rsense);            % gain of follower
% MOSFET OCTC calculation
Rgs=(rout+Rsense)/(1+gm*Rsense);
Tgs=Rgs*Cgs;
GM=gm/(1+gm*Rsense);
Rgd=rout+RL+(GM*rout*RL);
Tgd=Rgd*Cgd;
T=Tgs+Tgd                              % sum of OCTCs
mosfetpole=1/T
mosfet=tf(Ao,[T 1])
% Find loop transmission
LT=series(opamp, mosfet);
margin(LT);
title('moscursource. FREQUENCY RESPONSE OF UNCOMPENSATED MOSFET
CURRENT SOURCE') figure;
```

```
% close the loop
f=tf(1,1)
uncomp=feedback(LT, f,−1);
step(uncomp); grid;
title ('moscursource. STEP RESPONSE OF UNCOMPENSATED MOSFET CURRENT
SOURCE')
figure
% add lag compensation
Rf=470000; C=1e-9;
Ri=47000;
numlag=[Ri*C 1];
denomlag=[(Ri+Rf)*C 1];
f=tf(numlag, denomlag)
comp=feedback(LT, f,−1);
comp=series(f, comp);
step(comp); grid;
title('moscursource. STEP RESPONSE OF LAG COMPENSATED MOSFET CURRENT
SOURCE')
```

MATLAB script for Maglev example

```
function maglev
    % Maglev example
    % Marc Thompson 4/3/00
    % Maglev plant
    wn=sqrt(1e3);
    num=1.96e-6;
    denom =[1/wn^2 0 1];
    plant=tf(num, denom);
    step(plant);
    title('Step response of uncompensated Maglev system'); grid; figure
    % feedback compensation
    Kp=1e5;
    Kv=1e4;
    num=[Kv Kp];
    f=tf(num,1);
    % Closed-loop
    sys=feedback(plant, f,−1)
    damp(sys)
    step(sys);
    grid;
    title('Step response of compensated Maglev system')
```

Index

Note: Page numbers with "f" denote figures; "t" tables; "b" boxes.

Printed in the United States
By Bookmasters